MODERN PRACTICES IN RADIATION THERAPY

Edited by **Gopishankar Natanasabapathi**

Modern Practices in Radiation Therapy
Edited by Gopishankar Natanasabapathi

Published by InTech

Janeza Trdine 9, 51000 Rijeka, Croatia

Publishing Process Manager Iva Simcic

Technical Editor Teodora Smiljanic

Additional hard copies can be obtained from orders@intechopen.com

Modern Practices in Radiation Therapy, Edited by Gopishankar Natanasabapathi
p. cm.
ISBN 978-953-51-0427-8

For EU product safety concerns:
IN TECH d.o.o., Prolaz Marije Krucifikse Kozulić 3, 51000 Rijeka, Croatia,
info@intechopen.com or visit our website at intechopen.com.

Meet the editor

Dr Gopishankar has more than fourteen years of teaching and working experience as Medical Physicist-cum-Radiological Safety Officer for the Departments of Neurosurgery and Radiation Oncology at All India Institute of Medical Sciences, New Delhi, India. His research interests are Gel Dosimetry, Radiochromic Film Dosimetry and its application on advanced radiotherapy techniques like SRS/SRT, Intensity Modulated Radiotherapy (IMRT) and other advanced conformal treatments. He has several international presentations and publications. He is a life time member of UICC. He had done PhD in the field of Gel dosimetry for its application in Radiotherapy at his working institute. He also has done his Master's in Computer Application. He is at present member of several research projects and is involved in guiding PhD research scholars.

Contents

Preface

Cherish the help of men of skill,
Who ward and safe-guard you from ill.
Thiruvalluvar (An Indian Poet)

Cancer is a dreadful disease that confiscates million of people's life every year. It has created trepidation in the human minds for significant amount of time. General perception about cancer is it often leads to death. A large number of cancer patients today can expect to recover from this increasingly treatable illness. This achievement is due to significant advances over the last 50 years in the technology for treating cancer with radiation. While radiation therapy technology has progressed considerably in the last half-century, the basic goal of such treatment is unchanged: To target and kill cancer cells while exposing the surrounding healthy tissue to as little as possible. Radiation therapy kills cancer cells by damaging their DNA either directly or indirectly by creating free radicals within the cells that can in turn damage the DNA. Radiation may be delivered by a higher energy radiation generating equipments to shrink tumors and kill cancer cells. Does radiation therapy kill only cancer cells? The answer is no. It can also damage normal cells leading to side effects as well.

How far has radiation therapy technology progressed and how is the future of radiation therapy. Does this treatment modality for cancer have any role in treating tumors which usually prefer other treatments? All answers for these questions are found in this book entitled "**Modern Practices in Radiation Therapy**". This book contains 19 exceptional chapters contributed by renowned world-class radiotherapy professionals and researchers who have overwhelming knowledge in this field. To make this more interesting, all the chapters were further grouped into sections so that the readers could pursue their specific subjects of interest in radiation treatment.

Section I entitled "**External Beam RT and New Practices**" brings together chapters related to external beam radiotherapy which is defined as the methodology for treating tumors with radiation generation equipments like linear accelerators, cobalt units, etc. In recent times a remarkable advancement has happened in this treatment technique. This section groups chapters discussing relatively new type of external beam radiation therapy delivery system such as Stereotactic Body Radiotherapy

(SBRT), Involved-Field Radiation Therapy (IF-RT), a rapid clinical work flow STAT RAD using tomotherapy system and in addition it discusses about segmentation techniques of anatomical structures for planning in External beam RT which is also useful in Brachytherapy planning as well.

Section II entitled "**Particle Therapy**" has blended chapters pertinent to treatment modalities such as ion beam therapy. Main advantage of this technique is that it provides supreme dose conformity. Chapter 5 discusses about beam tracking system for moving targets treatment using ion beam therapy. Chapter 6 is about influence of neutron in charged particle therapy. Chapter 7 enumerates stopping power data which is determines the characteristics of ion beam therapy.

Section III entitled "**Brachytherapy and Intraoperative Radiation Treatments**" has unified chapters related to delivery of radiation locally to the tumor with rapid dose fall-off in the surrounding normal tissue. New technical developments in brachytherapy such as transperineal seed implantation and Intra-operative radiotherapy, is discussed in this section.

Section IV entitled "**Scope of Radiation Therapy for Specific Diseases**" contains two chapters; first one reveals the recent advances in the treatment of multiple myeloma (MM) such as targeted radiotherapy. Second chapter of this section mentions about underutilized radiation therapy modality for skin cancer which could be effective treatment for this disease if proper communication is established between the dermatologist's and radiation oncologist's.

Section V entitled "**Radiation Induced Effects and Overcoming Strategies**" congregates chapters discussing complications associated with radiation treatment and methods to protect normal tissue from radiation damage. There is one chapter in this section which reveals facts about anti-tumor effect at a non irradiated location in patients.

Section VI entitled "**Emerging Dosimeters and New QA Practices**" focuses on topics which are essential to determine and enhance the quality of the radiation equipment for patient treatment. With the introduction of new technology into the field of radiation oncology, a need arises to have a quality assurance program that is customized to these newer treatment modalities. The goal of a QA program for radiotherapy equipment is to assure that the machine characteristics do not deviate significantly from their baseline values acquired at the time of acceptance and commissioning. In early times radiation measurements were restricted to point measurements or two dimensional measurements. Advanced treatment techniques exhibit more complex radiation patterns which are characterized with steep dose gradients.

Section VII entitled "**Enhancing Patient Care in RT**" contains a single chapter about communication which is the key factor for providing better patient care. How it influences cancer patients is well discussed in this section.

In 2008, there were an estimated 12.7 million cases of cancer diagnosed and 7.6 million deaths from cancer around the world. Cancer survival tends to be poorer in developing countries, most likely because of combination of a late stage at diagnosis and limited access to timely and standard treatment. A considerable proportion of the worldwide burden of cancer could be prevented through the application of existing cancer control knowledge and by implementing methods for early detection and treatment. Emergence of advanced technologies is giving hope to more patients in recent times due to fewer side effects. It is expected that search for the origin and treatment of this disease will continue over the next quarter century in much the same manner as it has, by adding more complexity to scientific literature that is already complex almost beyond measure. Main goal of this book "**Modern Practices in Radiation Therapy**" is to provide contemporary knowledge and serve as a stepping stone for treating cancer patients efficiently in future.

Gopishankar Natanasabapathi
Gamma Knife Unit, Neurosciences Centre,
All India Institute of Medical Sciences, New Delhi,
India

Part 1

External Beam RT and New Practices

1

Stereotactic Body Radiotherapy for Pancreatic Adenocarcinoma: Set-Up Error Correction Using Internal Markers and Its Association with the Patient's Body Mass Index

Chi Lin, Shifeng Chen and Michael J. Baine
University of Nebraska Medical Center, USA

1. Introduction

Approximately 44,000 patients will develop new pancreatic cancers in the US in 2011 and 38,000 patients will die from the disease (ACS). Prognosis is directly related to the extent of tumor. The median survivals for these patients range from 11-18 months for those with localized disease, 10-12 months for those with locally advanced disease, and 5-7 months for those with metastatic disease, respectively (Evans DBAJ 2011). Although surgical resection is the only treatment associated with long-term survival, patients with resectable diseases usually account for only 20-25% of cases at diagnosis.

Despite resection, local regional recurrence and distant metastases occur in up to 50% of patients and two-year survival rates range from 20-40% with surgery alone. In 1974, the Gastrointestinal Tumor Study Group prospectively randomized patients after curative resection of pancreatic adenocarcinoma to adjuvant chemoradiation versus observation. The results of this study indicated a doubling of median and quadrupling of long-term survival with adjuvant chemoradiation (median, 20 vs. 11 months; 5-year survival, 19% vs. 5%). A US Intergroup study compared gemcitabine vs. infusional 5-FU chemotherapy for one month prior to and three months after chemoradiation, consisting of continuous infusional 5-FU, as adjuvant therapy after pancreatic cancer resection; outcome in those with tumor located in the pancreatic head was the primary study endpoint (Regine et al. 2008). The gemcitabine plus chemoradiation arm was superior to the 5-FU plus chemoradiation arm, with a median survival of 20.6 months vs. 16.9 months and survival at 3-years of 32% vs. 21%. This survival advantage came at a cost of appreciable toxicity, with grade 3-4 hematologic and non-hematologic toxicities occurring in 58% and 58% of subjects, respectively. Oettle et al compared gemcitabine given at 1000 mg/m^2 weekly for 3 of 4 weeks x 6 cycles to no additional therapy in 368 patients with resected pancreatic cancer (Oettle et al. 2007). Adjuvant gemcitabine was associated with a significant improvement in disease-free survival (13.4 vs 6.9 months), and a trend towards improvement in overall survival (median 22.1 vs 20.2 months); 34% of those receiving gemcitabine were alive at 3 yr vs. 20.5% with surgery alone. Grade 3-4 hematologic and non-hematologic toxicities occurred in fewer than 5% of subjects receiving gemcitabine.

While these studies indicate improvement with adjuvant therapy, there is still need to improve upon these results. A disadvantage of adjuvant therapy is that as many as 25% of patients have their treatment either delayed or forgone due to post-operative complications (Yeo CJ 1997; Spitz et al. 1997; Klinkenbijl et al. 1999). In an effort to increase the number of patients receiving adjuvant therapy, chemotherapy and radiation therapy can be administered pre-operatively (neoadjuvantly) to potential surgical candidates. Additional potential benefits of pre-operative therapy include the delivery of therapy to well-oxygenated tissues, the potential to downstage tumors (particularly when the lesion is borderline resectable or unresectable because of regional factors such as tumor involvement of the superior mesenteric vein or portal vein, or tumor abutment/encasement of the superior mesenteric artery or celiac trunk or gastroduodenal artery up to hepatic artery), and the opportunity to observe patients for the development of metastatic disease during therapy. After maximal tumor shrinkage and no interval development of metastatic disease, surgery can be considered.

The current standard neoadjuvant regimen includes several months of chemotherapy followed by 5 - 6 weeks of radiation therapy concurrent with radiation sensitizing chemotherapy, followed by a 4 - 6 week therapy break prior to surgery. This chemoradiation regimen is fairly debilitating. ECOG (Pisters et al. 2000) conducted a phase II trial of preoperative conventional (50.4 Gy, 1.8 Gy/fraction) chemoradiation, showing that 51% of patients had toxicity-related hospital admissions. Treatment-related toxicities were found to be proportional to the irradiated volume and radiation dose. At M.D. Anderson, an accelerated radiotherapy schedule using 30 Gy in 10 fractions appeared to be more tolerable and equally effective (Breslin et al. 2001; Pisters et al. 1998). A recent randomized trial (Bujko et al. 2006) has compared preoperative short-course radiotherapy with preoperative conventionally fractionated chemoradiation for rectal cancer. The results showed no difference in actuarial 4-year overall survival (67.2% in the short-course group vs. 66.2% in the chemoradiation group, P = 0.960), disease-free survival (58.4% vs. 55.6%, P = 0.820), and crude incidence of local recurrence (9.0% vs. 14.2%, P = 0.170). The study also reported similar late toxicity (10.1% vs. 7.1%, P = 0.360) and higher early radiation toxicity in the chemoradiation group (18.2% vs. 3.2%, P < 0.001). These data suggest the equivalence in efficacy between short course and long course neoadjuvant therapy. Koong et al. (Koong et al. 2004) has conducted a phase I study of stereotactic radiosurgery in patients with unresectable pancreatic cancer. Fifteen patients were treated at 3 dose levels (3 patients received 15 Gy in 1 fraction, 5 patients received 20 Gy in 1 fraction, and 7 patients received 25 Gy in 1 fraction). No Grade 3 or higher acute GI toxicity was observed. In the 6 evaluable patients who received 25 Gy, the median survival was 8 months. All patients in the study had local control until death or progressed systemically as the site of first progression. This study suggests the feasibility of stereotactic radiosurgery in pancreatic cancer.

Following the methodology of Koong et al, one can apply the linear-quadratic formulism for radiation cell killing to "equate" schemes that vary the dose/fraction and number of fractions. This concept of biologically equivalent dose says that the total effect is given by:

$$(nd)\left\{1+d\Big/\frac{\alpha}{\beta}\right\}$$

Where n is the # of fractions and d is the dose/fraction. The "alpha-beta ratio" characterizes the radiation response of a particular tissue; a higher value is indicative of a tissue that responds acutely to the effects of radiation. Due to their highly proliferative nature, most tumors fall into this category. Because prolonging the treatment time introduces a sparing (repair) effect in acutely responding tissues, there is significant motivation to deliver radiation in larger fractions over a shorter time.

As the duodenum is in closest proximity to the majority of the pancreatic head tumors, it is impossible to avoid treating this structure to a relatively high radiation dose. Koong et al's data suggests that it is possible to irradiate a small volume of duodenum to a dose of 22.5 Gy in one fraction with acceptable toxicity. While the dose-fractionation scheme employed by Koong et al resulted in no significant morbidity, we proposed a phase I study of hypofractionated stereotactic body radiotherapy as part of a neoadjuvant regimen in patients with locally advanced pancreatic cancer using a more conservative starting dose of 5 Gy x 5.

The types of geometric uncertainties that should be considered in stereotactic body radiotherapy include tumor motion and patient position (setup error). Discrepancies between the actual and planned positions of targets and organs-at-risk during stereotactic body radiotherapy can lead to reduced doses to the tumor and/or increased doses to normal tissues than planned, potentially reducing the local control probability and/or increasing toxicity. Therefore, accurate and precise target localization is critical for hypofractionated stereotactic body radiotherapy. Studies found that the bony anatomy is a poor surrogate for intraabdominal (Herfarth et al. 2000) and intrathoracic (Guckenberger et al. 2006; Sonke, Lebesque, and van Herk 2008) targets. Therefore, direct tumor localization is important. Unfortunately, soft tissues are not seen on Exac-Trac (BrainLAB, Heimstetten, Germany) X-ray images. Thus, fiducial markers for the pancreatic cancer are required. The purpose of the current study is to assess daily set-up error using the Exac-Trac system and implanted pancreatic fiducial markers during stereotactic body radiotherapy for patients with locally advanced pancreatic adenocarcinoma in the current ongoing institutional phase I study and to evaluate the effect of body mass index (BMI) on set-up error correction.

2. Methods

2.1 Patients

Included in this study are adult patients (≥ 19 years old) who had a Karnofsky performance status of ≥ 60 and underwent stereotactic body radiotherapy planning and treatment between October 2008 and February 2011 as part of an institutional research ethics board-approved study of neoadjuvant hypofractionated stereotactic body radiotherapy following chemotherapy in patients with borderline resectable or unresectable pancreatic adenocarcinoma. Daily isocenter positioning correction was investigated in 26 patients treated with 5 fractions of SBRT for locally advanced pancreatic cancer. Two fiducial markers were implanted into the pancreatic head approximately two centimeters apart. With daily Exac-Trac images, 3 dimensional couch shifts were made by matching corresponding fiducial markers to the digitally reconstructed radiograph from a simulation CT scan. BMI was calculated by Weight (kg)/Height^2 (m^2) and categorized into normal weight 18.5 -25 (kg/m^2) and overweight/obese >25 (kg/m^2).

2.2 Stereotactic body radiotherapy planning and treatment

2.2.1 Patient's positioning

The treatment position of the patient was supine, with their arms above their head. The immobilization device (Medical Intelligence blue bag) was molded into an immobilizing bed for the intended patient's entire body to make sure that the patients' position was the same during planning, simulation and treatment.

2.2.2 Patient data acquisition

A treatment planning free breathing CT scan with IV contrast was required to define tumor, clinical, and planning target volumes. A respiratory sorted treatment planning 4D CT scan was then acquired with the patient in the same position and immobilized using the same device as used for treatment. All tissues to be irradiated were included in the CT scan, with a slice thickness of 3 mm. Conventional MRI scans (T1 and T2) were included to assist in definition of target volumes. FDG PET-CT, if available, was also included in the treatment planning. The Gross Tumor Volume (GTV), Clinical Target Volume (CTV), Planning Target Volume (PTV), and organs-at-risk were outlined on all CT slices in which the structures exist.

2.2.3 Volumes

The GTV was defined as all known gross disease determined from CT, clinical information, endoscopic findings, FDG PET-CT and/or conventional MRI. The Integrated Tumor Volume based on CT/MRI/PET (GTV_{fusion}) was defined as gross disease on the free breathing CT scan, MRI scan and FDG-PET scan. These scans were correlated via image fusion technique. The volume was delineated by the treating physician on the above scans separately. The GTV_{CT}, GTV_{MRI} and GTV_{PET} (if done) were eventually fused together to generate GTV_{fusion}. Patients who had the maximal dimension of the GTV_{fusion} > 8 cm were not eligible for the study. The CTV was defined as the GTVs plus areas considered containing potential microscopic disease. In this study, we had no intension to treat the potential microscopic disease with stereotactic body radiotherapy, therefore the CTV was defined as GTVs (i.e. both the primary tumor and the lymph nodes containing clinical or radiographic evidence of metastases) plus areas between $GTV_{primary}$ and $GTV_{lymph\ nodes}$. The integrated CTV was created with 4D CT information to compensate for internal organ motion. The PTV provided a margin around integrated CTV to compensate for the variability of treatment set-up. Organs-at-Risk were defined as follows: the skin surface, the unspecified tissue (the tissue within the skin surface and outside all other critical normal structures and PTVs was designated as unspecified tissue), spinal cord (spinal cord contours were defined at least 5 mm larger in the radial dimension than the spinal cord itself, i.e. the cord diameter on any given slice was 10 mm larger than the cord itself), duodenum, stomach, liver, right kidney, left kidney, small bowels excluding duodenum, and spleen.

2.2.4 The treatment technique

The Novalis accelerator (BrainLAB, Heimstetten, Germany) was used to deliver stereotactic body radiotherapy. It incorporates stereotactic x-ray capabilities for verifying target position. This consists of two floor mounted x-ray tubes and two opposing amorphous silicon flat panel detectors mounted to the ceiling. Each x-ray tube/detector pair is configured to image through the linac isocenter with a coronal field of view of approximately 18 cm in both the superior-inferior and left-right directions at isocenter. For soft tissue targets, the system is designed to be used with radio-opaque platinum markers implanted near the target. Two markers, 2 cm away from each other and placed close enough to the target anatomy so that they could be observed within the field of view of the x-ray localization system at the time of treatment, were implanted prior to CT imaging and treatment planning,. Specific patient breathing characteristics were determined during 4D CT. If the breathing pattern was adequate, respiratory-gated delivery (turning the beam on only at a specified phase of respiration) was used. This method "freezes" target motion and allows reduction of beam margins, thereby reducing the amount of irradiated normal tissue. The Novalis system is well suited to gated delivery and has been evaluated extensively by Tenn et al (Tenn, Solberg, and Medin 2005). The following is a brief procedural summary from that work which is incorporated into this study: The patient is set up in the treatment room and infrared reflective markers with adhesive bases are attached to their anterior surface so that breathing motion can be monitored. A second set of infrared reflective markers is rigidly attached to the treatment couch and used as a reference against which the movement of patient markers is measured. These rigidly mounted reflectors are also used to track couch location during the patient positioning process. The 3D movement of the patient's anterior surface is tracked via the infrared markers and the anterior-posterior component of this trajectory is used to monitor breathing motion. The system plots breathing motion versus time and a reference level is specified on this breathing trace. This designates the point in the breathing trace at which the verification x-ray images will be triggered. The two images are obtained sequentially at the instant the breathing trace crosses this level during exhale phase. Because the patient is localized based on these images, the gating level is set at the same phase in the breathing cycle at which the planning CT data was obtained. Within each image the user locates the positions of the implanted markers. From these positions the system reconstructs the 3D geometry of the implanted markers and determines the shifts necessary to bring them into alignment with the planning CT. The patient is subsequently positioned according to the calculated shifts. Finally, a gating window (beam-on region) during which the linac beam will be delivered is selected about the reference level. The system can gate the beam in both inhale and exhale phases of the breathing cycle. Subsequent x-ray images verifying the location of the implanted markers are obtained at the gating level continuously during treatment. If marker positions remain within tolerance limits, the target position may also be assumed to be correctly positioned. If they are outside the limit, the newly obtained images can be used to reposition the patient and maintain treatment accuracy.

2.2.5 Dose computation

The treatment plan used for each patient was based on an analysis of the volumetric dose, including dose volume histogram (DVH) analyses of the PTV and critical normal structures. Treatment planning was accomplished with multiple coplanar conformal beams or arcs to allow for a high degree of dose conformality. The uniformity requirement is +10%/-5% of the total dose at the prescription point within the tumor volume. The IMRT was used if it was of benefit for decreasing tissue complications. Beam's Eye View techniques were used to select the beam isocenter and direction to fully encompass the target volume while minimizing the inclusion of the critical organs in order to select the plan that minimizes the dose to normal tissues.

2.2.6 Dose specification

A 5-fraction dose was prescribed. The prescription dose was the isodose which encompasses at least 95% of PTV. DVHs were generated for all critical organs-at-risk. The dose to the kidneys was carefully monitored and kidney volumes were defined on simulation fields. The percent of total kidney volume (defined as the sum of the left and right kidney volumes) receiving 15 Gy (3 Gy per fraction) was required to be less than 35% of the total kidney volume. The maximum dose to any point within the spinal cord was not allowed to exceed 15 Gy (3 Gy per fraction). At least 700 ml or 35% of normal liver was required to receive a total dose less than 15 Gy (3 Gy per fraction). The maximum point dose to the stomach or small bowel except duodenum could not exceed 60% of prescription dose. An isodose distribution of the treatment at the central axis view indicating the position of kidneys, liver and spinal cord was required. Dose homogeneity was defined as follows: No more than 20% of PTV receive >110% of its prescribed dose; No more than 1% of PTV receive <93% of its prescribed dose; No more than 1% or 1 cc of the tissue outside the PTV receive >110% of the dose prescribed to the PTV.

2.2.7 Daily target verification

The locations of the implanted markers were verified on daily Exac-Trac X-Rays prior to the delivery of stereotactic body radiation therapy.

2.3 Statistical analysis

For each patient, the mean and standard deviation of daily 3-dimensional position shifts (lateral, longitudinal and vertical) were measured. The systematic error (the mean of all patients' means) and the random error (the standard deviation around the systemic error) were calculated for daily patient position shifts. The amplitude changes and variability in amplitude changes were also measured. Multivariate logistic regression was used to analyze the effect of patients' BMI on patient position changes. All statistical calculations were performed using SAS 9.2 (SAS Institute Inc., Cary, North Carolina, USA).

3. Results

3.1 Systematic and random daily couch shifts

A total of 127 treatments from 26 patients were studied. Table 1 provides a summary of the systematic and random couch shifts using implanted internal markers. The entire group mean (systematic) and standard deviation (random) of the couch shifts from the body surface markers are -0.4 ± 5.6 mm, -1.3 ± 6.6 mm and -0.3 ± 4.7 mm in lateral (left-right), longitudinal (superior-inferior) and vertical (anterior-posterior) directions, respectively. The mean systematic couch shifts > 0 occur in (13/26) 50%, (12/26) 46% and (10/26) 38% in the left-right, superior-inferior and anterior-posterior directions, respectively. The mean random couch shifts > 5mm occur in (7/26) 27%, (12/26) 46% and (5/26) 19% in the left-right, superior-inferior and anterior-posterior directions, respectively. The mean systematic couch shifts are significantly smaller than the mean random couch shifts in left-right (-0.3 ± 3.6 mm vs. 4.1 ± 2.8 mm, $p < 0.0001$), superior-inferior (-1.1 ± 4.1 mm vs. 5.5 ± 3.2 mm, $p < 0.0001$) and anterior-posterior (-0.1 ± 3.1 mm vs. 3.5 ± 2.0 mm, $p < 0.0001$) directions, respectively. The couch shifts for the majority of fractions are within ± 10 mm (Figure 1A-1C)

	Systematic Error Mean (mm) ± SD	Random Error Mean (mm) ± SD	P (X^2)
Lateral shift	-0.3 ± 3.6	4.1 ± 2.8	<0.0001
Longitudinal shift	-1.1 ± 4.1	5.5 ± 3.2	<0.0001
Vertical shift	-0.1 ± 3.1	3.5 ± 2.0	<0.0001

Table 1. The averages of systematic and random daily couch shifts three-dimensionally

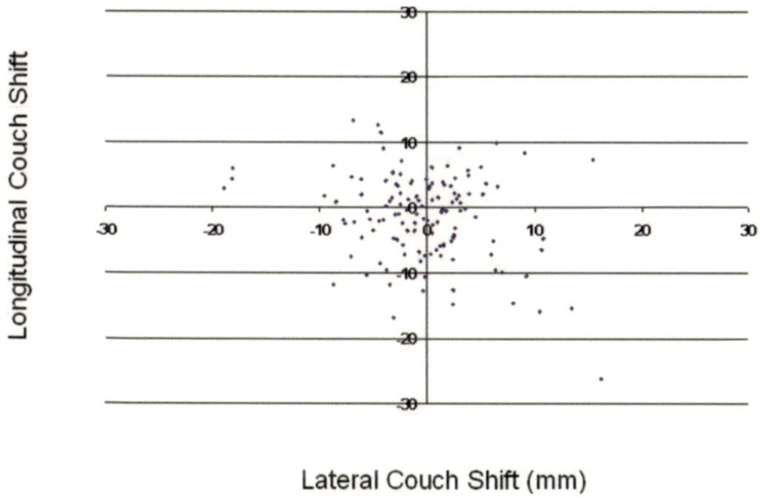

Fig. 1A. Longitudinal vs. Lateral couch shifts

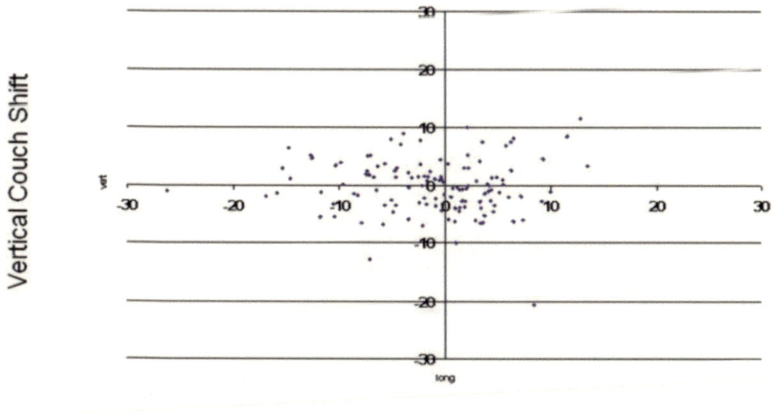

Fig. 1B. Vertical vs. Longitudinal couch shifts

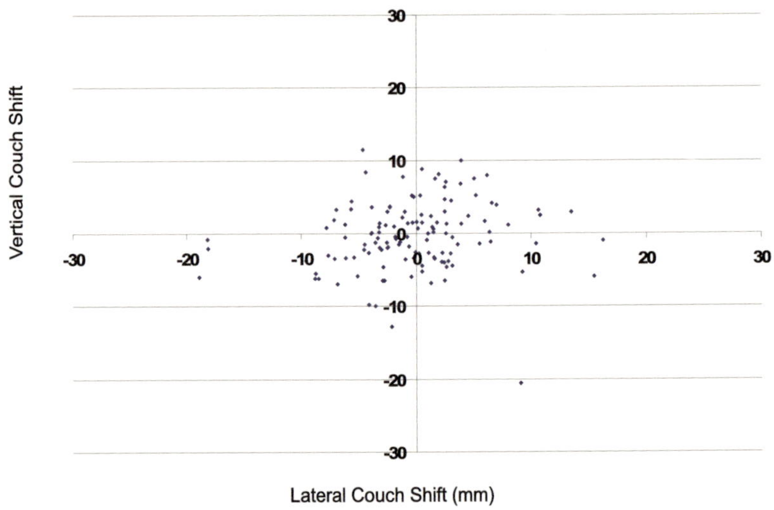

Fig. 1C. Vertical vs. Lateral couch shifts

3.2 Absolute systematic and random daily couch shifts

The amplitudes of the systemic and random daily couch shifts are summarized in table 2. The mean amplitudes of systematic couch shifts are significantly larger than the mean amplitude of random couch shift in left-right (4.1 ± 2.9 mm vs. 2.5 ± 1.3 mm, $p = 0.015$), superior-inferior (5.2 ± 3.1 mm vs. 3.2 ± 1.6 mm, $p = 0.007$) and anterior-posterior (3.6 ± 1.5 mm vs. 2.5 ± 1.6 mm, $p = 0.016$) directions, respectively. The amplitudes of couch shifts in the superior-inferior direction are significantly larger than those in the left-right ($p = 0.045$) or anterior-posterior directions ($p = 0.001$). The absolute couch shifts ≤ 3 mm, ≤ 5 mm and ≤ 10 mm occur in (51%, 71% and 93%), (37%, 60% and 87%) and (51%, 73% and 98%) in the left-right, superior-inferior and anterior-posterior directions, respectively (Figure 2A-2F). There is no correlation among 3 dimensional couch shifts (Figure 3A-3C).

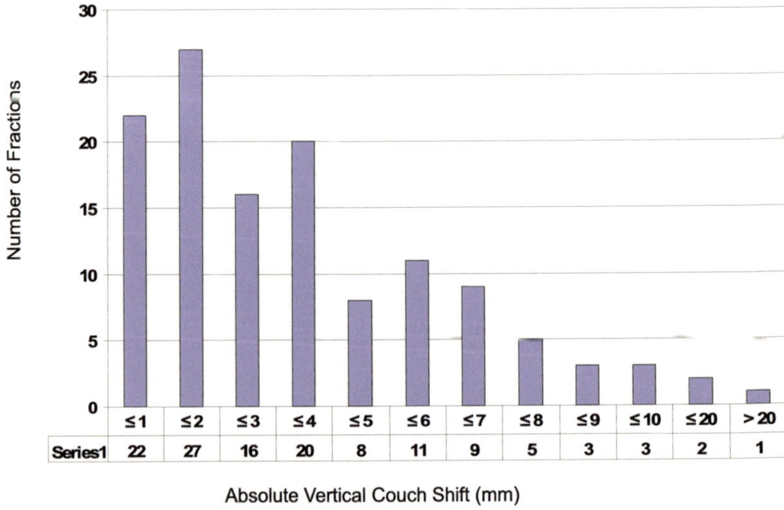

Fig. 2A. Distribution of absolute vertical couch shifts

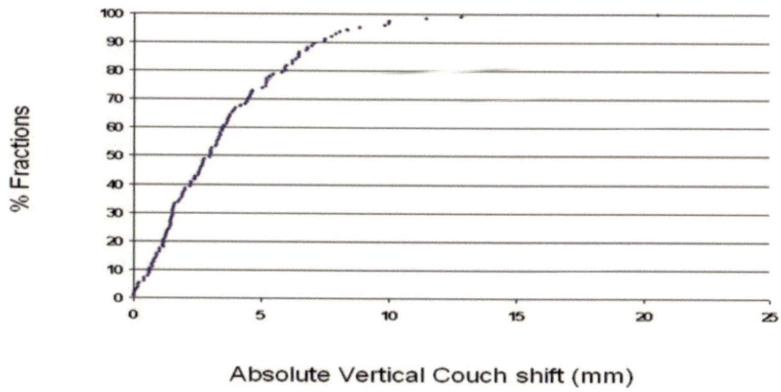

Fig. 2B. Cumulative distribution of absolute vertical couch shifts

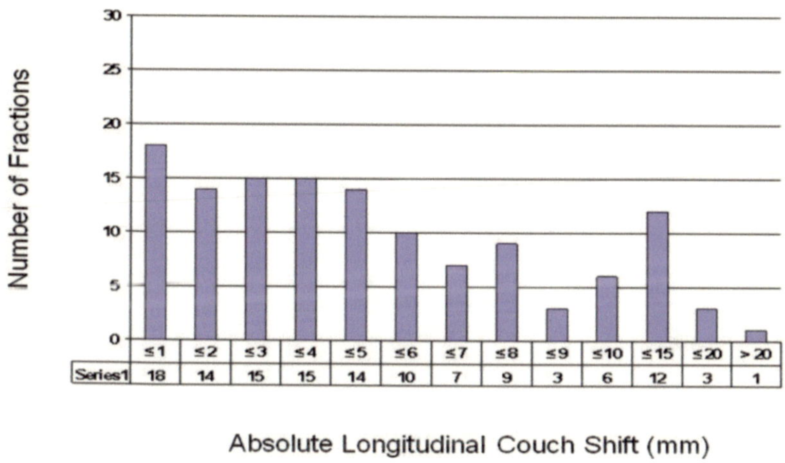

Fig. 2C. Distribution of absolute longitudinal couch shifts

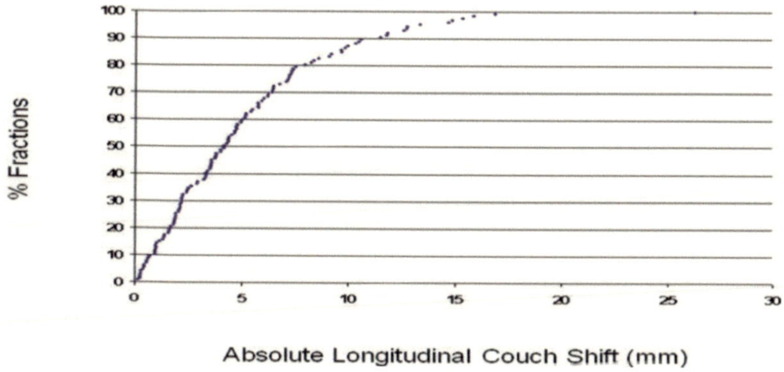

Fig. 2D. Cumulative distribution of absolute longitudinal couch shifts

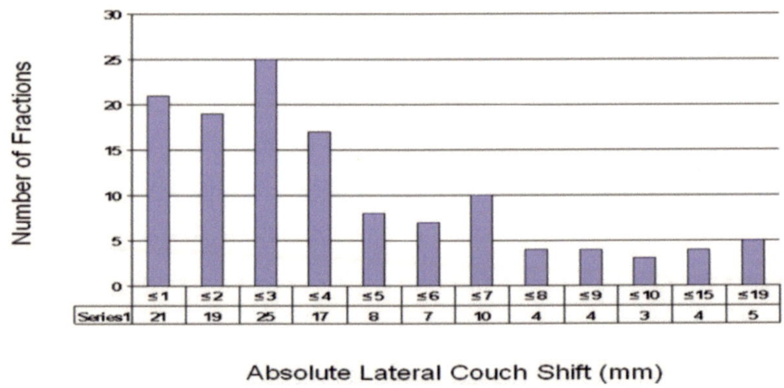

Fig. 2E. Distribution of absolute lateral couch shifts

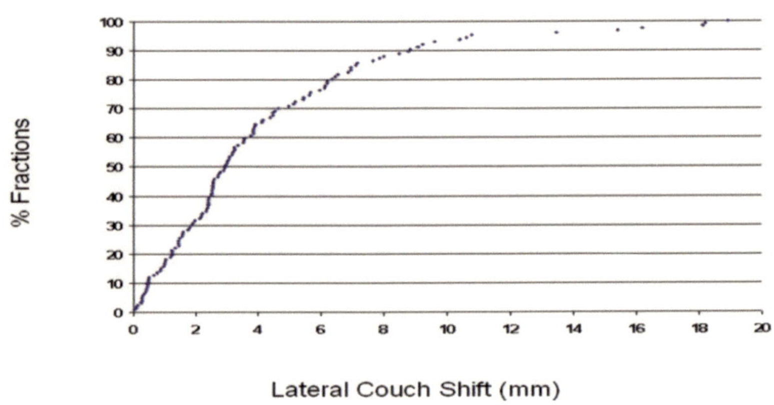

Fig. 2F. Cumulative distribution of absolute lateral couch shifts

Absolute Value (Amplitude)	Systematic Error Mean (mm) ± SD	Random Error Mean (mm) ± SD	P (X^2)
Lateral shift	4.1 ± 2.9	2.5 ± 1.3	0.015
Longitudinal shift	5.2 ± 3.1	3.2 ± 1.6	0.007
Vertical shift	3.6 ± 1.6	2.5 ± 1.6	0.016

Table 2. The averages of absolute systematic and random daily couch shifts three-dimensionally

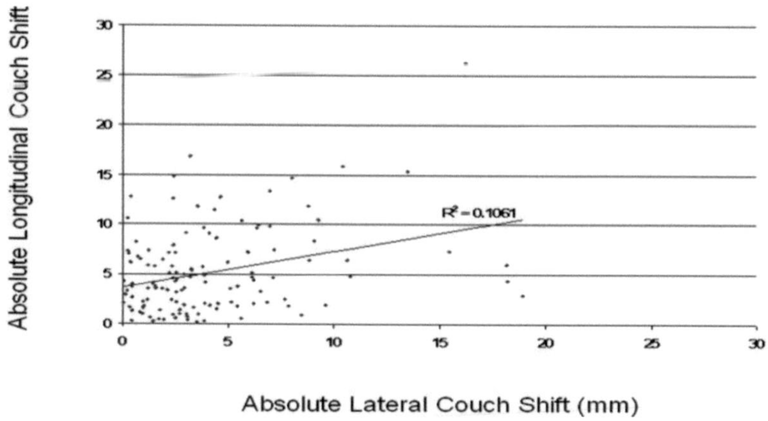

Fig. 3A. Absolute longitudinal vs. absolute lateral couch shifts

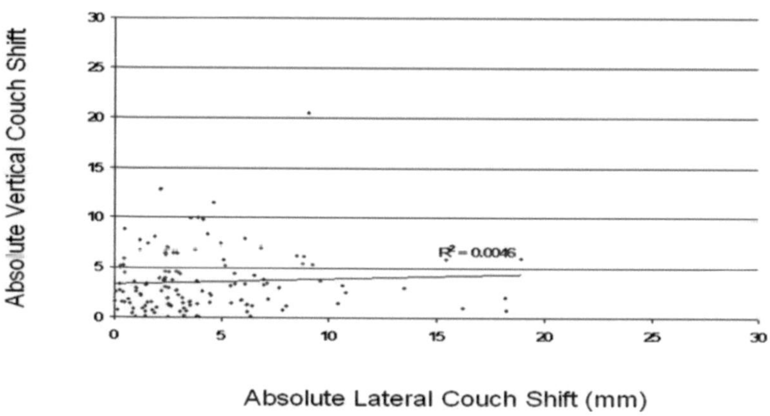

Fig. 3B. Absolute vertical vs. absolute lateral couch shifts

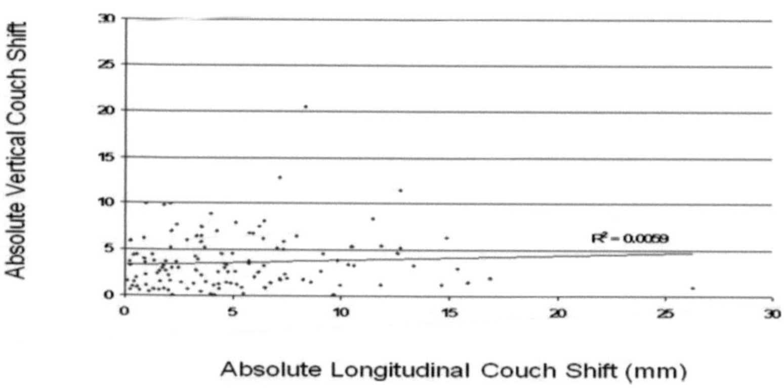

Fig. 3C. Absolute vertical vs. absolute longitudinal couch shifts

3.3 The magnitude of the pancreatic tumor motion vs. the amplitude of setup correction

Inter-fraction variability in the position of pancreatic tumors is generally considered to be resultant from pancreatic breathing motion and patient positioning. We examined the association of the magnitude of the changes in the pancreatic tumor breathing motion in 50% and 80% inhale and exhale with the amplitude of daily setup error correction and found no correlation between them ($R^2 \leq 0.012$). (Figure 4A-4C)

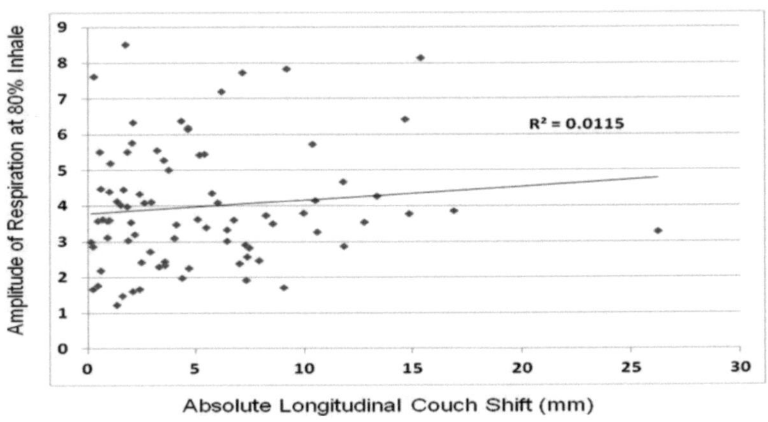

Fig. 4A. Relationship between amplitude of respiration and setup correction in the longitudinal direction

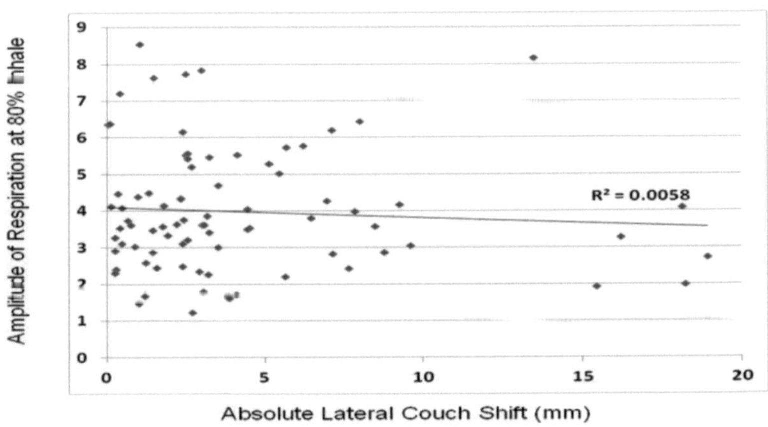

Fig. 4B. Relationship between amplitude of respiration and setup correction in the lateral direction

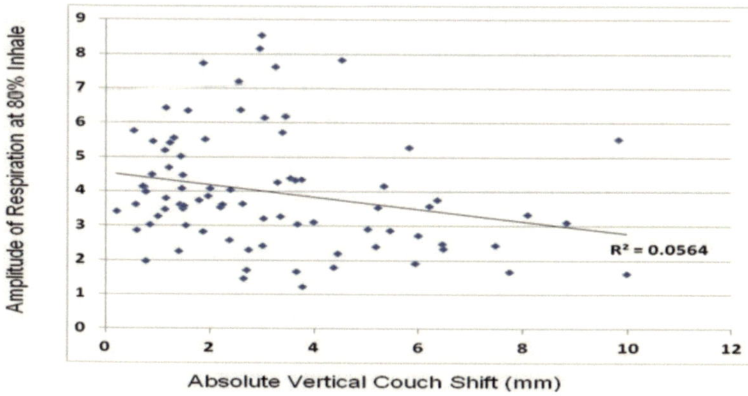

Fig. 4C. Relationship between amplitude of respiration and setup correction in the vertical direction

3.4 The effect of body mass index on daily couch shifts

The median age for this group of patients is 60 years old (range: 34 -79). Slightly more than half of the patients (14/26) are males. The BMIs for this group of patients range between 20 and 46 with a median value of 27 (kg/m^2). There are 8 patients with BMIs of 20-25 and 18 patients with the BMIs of 26-46. Table 3 shows that there is no difference in daily couch shift distribution between these two groups based on BMIs of > 25 or ≤ 25.

	BMI 20-25 N (%)	BMI 26-46 N (%)	P (X^2)
Lateral systematic shift			0.395
≤ 0	5 (62.5)	8 (44.4)	
> 0	3 (37.5)	10 (55.6)	
Longitudinal systematic shift			0.793
≤ 0	4 (50)	10 (55.6)	
> 0	4 (50)	8 (44.4)	
Vertical systematic shift			0.070
≤ 0	7 (87.5)	9 (50)	
> 0	1 (12.5)	9 (50)	
Lateral random shift			0.418
≤ 5 mm	5 (62.5)	14 (77.8)	
> 5 mm	3 (37.5)	4 (22.2)	
Longitudinal random shift			0.555
≤ 5 mm	5 (62.5)	9 (50)	
> 5 mm	3 (37.5)	9 (50)	
Vertical random shift			0.562
≤ 5 mm	7 (87.5)	14 (77.8)	
> 5 mm	1 (12.5)	4 (22.2)	

Table 3. Distributions of daily couch shifts in patients with body mass indexes of > 25 and ≤ 25

Table 4 shows the results of multivariate regression analysis, revealing that patients with a BMI ≤ 25 are less likely to have an anterior vertical couch shift from the initial positioning (OR: 0.35, 95% CI: 0.16-0.77, p = 0.009) than those with a BMI > 25 after adjusting for age and gender, suggesting less correction is needed due to less body relaxation and skin movement in patients with a BMI ≤ 25 (kg/m^2) than those with a BMI > 25. BMI has no effect on the left-right or superior-inferior couch shifts.

BMI 20-25 vs. BMI 26-46	Odd Ratio (95% CI)	P (X^2)
Lateral systematic shift	0.787 (0.371-1.671)	0.533
Longitudinal systematic shift	1.384 (0.657-2.914)	0.393
Vertical systematic shift	0.351 (0.160-1.773)	0.009
Lateral random shift	2.087 (0.333-13.077)	0.432
Longitudinal random shift	0.606 (0.105-3.513)	0.577
Vertical random shift	0.501 (0.043-5.838)	0.581

Table 4. The effect of body mass index on daily couch shifts by multivariate logistic regression analysis

Factors included in the regression models are age, gender and BMI.

4. Discussion

Accurate and precise patient positioning at the time of delivering stereotactic body radiotherapy is crucial. Image-guided respiratory-gated radiation therapy has been a major advancement in minimizing inter- and intra- fractional target variations. To minimize and correct for setup uncertainties and inter-fractional motion of extracranial tumors, various immobilization and localization techniques have been clinically implemented, including transabdominal ultrasonography (Lattanzi et al. 1999; Chandra et al. 2003; Langen et al. 2003), megavoltage imaging (Schiffner et al. 2007; Serago et al. 2006), kilovoltage imaging (Jaffray et al. 1999; Kupelian et al. 2008), use of an in-room computed tomography (CT)-linear accelerator system (Court et al. 2003; Wong et al. 2005), cone-beam CT (Jaffray et al. 1999; Kupelian et al. 2008), and placement of internal fiducials (Kupelian et al. 2008; Chen et al. 2007; Chung et al. 2004). Pancreatic tumor targets usually exhibit inter-fractional motion relative to the bony anatomy because of daily variation in stomach and duodenal filling and respiratory patterns. The bony anatomy can be imaged and aligned using in-room kilovoltage X-rays; however, with this approach, the position of the pancreatic tumor with respect to the bony anatomy is uncertain. Jayachandran et al. has compared the inter-fractional variation in pancreatic tumor position using bony anatomy to implanted fiducial markers and observed substantial residual uncertainty after alignment to bony anatomy when irradiating pancreatic tumors using respiratory gating (Jayachandran et al. 2010). They reported that bony anatomy matched tumor position in only 20% of the radiation treatments. This study evaluates the use of implanted platinum markers in pancreatic cancer patients for daily setup correction. We acquired treatment planning CT scans at least 1 week after two fiducial markers were implanted to allow time for inflammation or edema to subside. The positions of these markers were then used to guide the daily patient setup correction.

4.1 The magnitude of the inter-fractional setup correction for pancreatic cancer

In this study, inter-fractional shifts of > 5 mm are observed in 29%, 27% and 40% of fractions in left-right, anterior-posterior and superior-inferior directions. When we examined the percentage of fractions with the inter-fractional shifts of > 10 mm, we observed only 7% and 2% in the directions of left-right and anterior-posterior but 13% in the direction of superior-inferior. The median and maximum couch shifts are 4.2 and 26.2 mm, 3.0 and 20.5 mm and 3.0 and 18.9 mm in superior-inferior, anterior-posterior and left-right directions, respectively. The couch shift in the superior-inferior direction is significantly larger than that in the anterior-posterior ($p = 0.001$) and left-right ($p = 0.045$) directions. There is no difference in couch shifts between anterior-posterior and left-right directions ($p = 0.22$). (Table 5)

Shifts	Median (mm)	Maximum (mm)	≤ 3 mm N (%)	≤ 5 mm N (%)	≤ 10 mm N (%)	T-test
Left-Right	3.0	18.9	65/127 (51)	90/127 (71)	118/127 (93)	SI vs. LR p=0.001
Superior-Inferior	4.2	26.2	47/127 (37)	76/127 (60)	111/127 (87)	SI vs. AP p=0.045
Anterior-Posterior	3.0	20.5	65/127 (51)	93/127 (73)	124/127 (98)	LR vs. AP p=0.216

Table 5. Comparison of the amplitudes of three-dimensional shifts in median, maximum, 3 mm, 5 mm and 10 mm.

There is no correlation in either the direction or the amplitude of couch shifts among all three directions. In contrast to our findings, Jayachandran et al. reported that the maximum shifts needed in the anterior-posterior, left-right, and superior-inferior directions were 9 mm, 13 mm, and 19 mm, respectively when fiducial markers were used which are smaller than what we found in the current study (Jayachandran et al. 2010). On the other hand, some earlier studies have shown much larger inter-fractional pancreatic motions of up to 40 mm (Booth and Zavgorodni 1999; Horst et al. 2002). Allen et al. reported a maximum of 17.7 mm inter-fractional setup error in pancreatic cancer using daily on line CT scan images (Li et al. 2007). Feng et al. characterized pancreatic tumor motion using CINE MRI and found that tumor borders moved much more than expected. They indicated that to provide 99% geometric coverage, margins of 20mm inferiorly, 10 mm anteriorly, 7 mm superiorly, and 4 mm posteriorly are required if respiratory gating is not used (Feng et al. 2009).

4.2 Relationship between setup correction and amplitude of pancreatic tumor breathing motion

To our knowledge, this is the first study to evaluate the correlation of the amplitude of respiration to patient position displacement in pancreatic cancer. We did not find any association between the magnitude of the changes in the pancreatic tumor breathing motion and the amplitude of daily setup error correction. This is consistent with the results reported by Case et al. for liver tumors (Case et al. 2009).

4.3 Effect of body mass index on setup correction

This is the first study to examine the influence of BMI on setup correction for pancreatic cancer. We observed that patients with a BMI of > 25 have a greater possibility of needing vertical setup correction than those with a BMI of ≤ 25. On the other hand, BMI has no effect on the setup corrections in superior-inferior and left-right directions. Worm et al reported that intrafractional errors for liver and lung cancer were independent of patient's BMI (Worm et al.). A study on generic planning target margin for rectal cancer treatment setup variation did show that BMI was significantly associated with systemic superior-inferior ($p<0.05$) and anterior-posterior ($p<0.01$) variation and random left-right variation ($p<0.05$) (Robertson, Campbell, and Yan 2009).

5. Conclusion

Daily alignment using fiducial markers is an effective method of localizing pancreas displacement. It provides the option of reducing margins, thus limiting normal tissue toxicity and allowing the possibility of dose escalation for better long-term control. For those patients without daily image guided set-up correction, margins of |mean| + 2|standard deviation| (11.6 mm, 14.5 mm, and 9.7 mm in left-right, superior-inferior, and anterior-posterior directions, respectively) should be added to the planning target volume. Patients with BMI >25 (kg/m^2) may need a larger anterior-posterior margin for planning target volume than those with BMI ≤ 25 (kg/m^2).

6. Abbreviations

US: United States
ACS: American Cancer Society
5-FU: 5-Fluorouracil
ECOG: The Eastern Cooperative Oncology Group
CT: computed tomography
4D CT: 4 dimensional computed tomography
MRI: Magnetic resonance imaging
FDG PET: fluorodeoxyglucose Positron emission tomography
GTV: Gross tumor volume
CTV: Clinical target volume
PTV: Planning target volume
DVH: Dose volume histogram
IMRT: Intensity modulated radiation therapy
SD: Standard deviation
BMI: Body mass index

7. References

ACS. 2011, Cancer Facts& Figures, http://www.cancer.org/acs/groups/content/@epidemiologysurveilance/documents/document/acspc-029771.pdf.

Booth, J. T., and S. F. Zavgorodni. 1999. "Set-up error & organ motion uncertainty: a review." Australas Phys Eng Sci Med no. 22 (2):29-47.

Breslin, T. M., K. R. Hess, D. B. Harbison, M. E. Jean, K. R. Cleary, A. P. Dackiw, R. A. Wolff, J. L. Abbruzzese, N. A. Janjan, C. H. Crane, J. N. Vauthey, J. E. Lee, P. W. Pisters, and D. B. Evans. 2001. "Neoadjuvant chemoradiotherapy for adenocarcinoma of the pancreas: treatment variables and survival duration." Ann Surg Oncol no. 8 (2):123-32.

Bujko, K., M. P. Nowacki, A. Nasierowska-Guttmejer, W. Michalski, M. Bebenek, and M. Kryj. 2006. "Long-term results of a randomized trial comparing preoperative short-course radiotherapy with preoperative conventionally fractionated chemoradiation for rectal cancer." Br J Surg no. 93 (10):1215-23.

Case, R. B., J. J. Sonke, D. J. Moseley, J. Kim, K. K. Brock, and L. A. Dawson. 2009. "Inter- and intrafraction variability in liver position in non-breath-hold stereotactic body radiotherapy." Int J Radiat Oncol Biol Phys no. 75 (1):302-8.

Chandra, A., L. Dong, E. Huang, D. A. Kuban, L. O'Neill, I. Rosen, and A. Pollack. 2003. "Experience of ultrasound-based daily prostate localization." Int J Radiat Oncol Biol Phys no. 56 (2):436-47.

Chen, J., R. J. Lee, D. Handrahan, and W. T. Sause. 2007. "Intensity-modulated radiotherapy using implanted fiducial markers with daily portal imaging: assessment of prostate organ motion." Int J Radiat Oncol Biol Phys no. 68 (3):912-9.

Chung, P. W., T. Haycocks, T. Brown, Z. Cambridge, V. Kelly, H. Alasti, D. A. Jaffray, and C. N. Catton. 2004. "On-line aSi portal imaging of implanted fiducial markers for the reduction of interfraction error during conformal radiotherapy of prostate carcinoma." Int J Radiat Oncol Biol Phys no. 60 (1):329-34.

Court, L., I. Rosen, R. Mohan, and L. Dong. 2003. "Evaluation of mechanical precision and alignment uncertainties for an integrated CT/LINAC system." Med Phys no. 30 (6):1198-210.

Evans DBAJ, Willett CG. . 2011. Cancer of the pancreas. Edited by Rosenberg SA DeVita VT Jr HS. 1 vols. Vol. 1, Cancer principles and practice of oncology. New York: Lippincott Williams & Wilkins.

Feng, M., J. M. Balter, D. Normolle, S. Adusumilli, Y. Cao, T. L. Chenevert, and E. Ben-Josef. 2009. "Characterization of pancreatic tumor motion using cine MRI: surrogates for tumor position should be used with caution." Int J Radiat Oncol Biol Phys no. 74 (3):884-91.

Guckenberger, M., J. Meyer, J. Wilbert, K. Baier, G. Mueller, J. Wulf, and M. Flentje. 2006. "Cone-beam CT based image-guidance for extracranial stereotactic radiotherapy of intrapulmonary tumors." Acta Oncol no. 45 (7):897-906.

Herfarth, K. K., J. Debus, F. Lohr, M. L. Bahner, P. Fritz, A. Hoss, W. Schlegel, and M. F. Wannenmacher. 2000. "Extracranial stereotactic radiation therapy: set-up accuracy of patients treated for liver metastases." Int J Radiat Oncol Biol Phys no. 46 (2):329-35.

Horst, E., O. Micke, C. Moustakis, A. Schuck, U. Schafer, and N. A. Willich. 2002. "Conformal therapy for pancreatic cancer: variation of organ position due to gastrointestinal distention--implications for treatment planning." Radiology no. 222 (3):681-6.

Jaffray, D. A., D. G. Drake, M. Moreau, A. A. Martinez, and J. W. Wong. 1999. "A radiographic and tomographic imaging system integrated into a medical linear accelerator for localization of bone and soft-tissue targets." Int J Radiat Oncol Biol Phys no. 45 (3):773-89.

Jayachandran, P., A. Y. Minn, J. Van Dam, J. A. Norton, A. C. Koong, and D. T. Chang. 2010. "Interfractional uncertainty in the treatment of pancreatic cancer with radiation." Int J Radiat Oncol Biol Phys no. 76 (2):603-7.

Klinkenbijl, J. H., J. Jeekel, T. Sahmoud, R. van Pel, M. L. Couvreur, C. H. Veenhof, J. P. Arnaud, D. G. Gonzalez, L. T. de Wit, A. Hennipman, and J. Wils. 1999. "Adjuvant radiotherapy and 5-fluorouracil after curative resection of cancer of the pancreas and periampullary region: phase III trial of the EORTC gastrointestinal tract cancer cooperative group." Ann Surg no. 230 (6):776-82; discussion 782-4.

Koong, A. C., Q. T. Le, A. Ho, B. Fong, G. Fisher, C. Cho, J. Ford, J. Poen, I. C. Gibbs, V. K. Mehta, S. Kee, W. Trueblood, G. Yang, and J. A. Bastidas. 2004. "Phase I study of stereotactic radiosurgery in patients with locally advanced pancreatic cancer." Int J Radiat Oncol Biol Phys no. 58 (4):1017-21.

Kupelian, P. A., C. Lee, K. M. Langen, O. A. Zeidan, R. R. Manon, T. R. Willoughby, and S. L. Meeks. 2008. "Evaluation of image-guidance strategies in the treatment of localized prostate cancer." Int J Radiat Oncol Biol Phys no. 70 (4):1151-7.

Langen, K. M., J. Pouliot, C. Anezinos, M. Aubin, A. R. Gottschalk, I. C. Hsu, D. Lowther, Y. M. Liu, K. Shinohara, L. J. Verhey, V. Weinberg, and M. Roach, 3rd. 2003. "Evaluation of ultrasound-based prostate localization for image-guided radiotherapy." Int J Radiat Oncol Biol Phys no. 57 (3):635-44.

Lattanzi, J., S. McNeeley, W. Pinover, E. Horwitz, I. Das, T. E. Schultheiss, and G. E. Hanks. 1999. "A comparison of daily CT localization to a daily ultrasound-based system in prostate cancer." Int J Radiat Oncol Biol Phys no. 43 (4):719-25.

Li, X. A., X. S. Qi, M. Pitterle, K. Kalakota, K. Mueller, B. A. Erickson, D. Wang, C. J. Schultz, S. Y. Firat, and J. F. Wilson. 2007. "Interfractional variations in patient setup and anatomic change assessed by daily computed tomography." Int J Radiat Oncol Biol Phys no. 68 (2):581-91.

Oettle, H., S. Post, P. Neuhaus, K. Gellert, J. Langrehr, K. Ridwelski, H. Schramm, J. Fahlke, C. Zuelke, C. Burkart, K. Gutberlet, E. Kettner, H. Schmalenberg, K. Weigang-Koehler, W. O. Bechstein, M. Niedergethmann, I. Schmidt-Wolf, L. Roll, B. Doerken, and H. Riess. 2007. "Adjuvant chemotherapy with gemcitabine vs observation in patients undergoing curative-intent resection of pancreatic cancer: a randomized controlled trial." Jama no. 297 (3):267-77.

Pisters, P. W., J. L. Abbruzzese, N. A. Janjan, K. R. Cleary, C. Charnsangavej, M. S. Goswitz, T. A. Rich, I. Raijman, R. A. Wolff, R. Lenzi, J. E. Lee, and D. B. Evans. 1998. "Rapid-fractionation preoperative chemoradiation, pancreaticoduodenectomy, and intraoperative radiation therapy for resectable pancreatic adenocarcinoma." J Clin Oncol no. 16 (12):3843-50.

Pisters, P. W., W. A. Hudec, J. E. Lee, I. Raijman, S. Lahoti, N. A. Janjan, T. A. Rich, C. H. Crane, R. Lenzi, R. A. Wolff, J. L. Abbruzzese, and D. B. Evans. 2000. "Preoperative chemoradiation for patients with pancreatic cancer: toxicity of endobiliary stents." J Clin Oncol no. 18 (4):860-7.

Regine, W. F., K. A. Winter, R. A. Abrams, H. Safran, J. P. Hoffman, A. Konski, A. B. Benson, J. S. Macdonald, M. R. Kudrimoti, M. L. Fromm, M. G. Haddock, P. Schaefer, C. G. Willett, and T. A. Rich. 2008. "Fluorouracil vs gemcitabine chemotherapy before and after fluorouracil-based chemoradiation following resection of pancreatic adenocarcinoma: a randomized controlled trial." Jama no. 299 (9):1019-26.

Robertson, J. M., J. P. Campbell, and D. Yan. 2009. "Generic planning target margin for rectal cancer treatment setup variation." Int J Radiat Oncol Biol Phys no. 74 (5):1470-5.

Schiffner, D. C., A. R. Gottschalk, M. Lometti, M. Aubin, J. Pouliot, J. Speight, I. C. Hsu, K. Shinohara, and M. Roach, 3rd. 2007. "Daily electronic portal imaging of implanted gold seed fiducials in patients undergoing radiotherapy after radical prostatectomy." Int J Radiat Oncol Biol Phys no. 67 (2):610-9.

Serago, C. F., S. J. Buskirk, T. C. Igel, A. A. Gale, N. E. Serago, and J. D. Earle. 2006. "Comparison of daily megavoltage electronic portal imaging or kilovoltage imaging with marker seeds to ultrasound imaging or skin marks for prostate localization and treatment positioning in patients with prostate cancer." Int J Radiat Oncol Biol Phys no. 65 (5):1585-92.

Sonke, J. J., J. Lebesque, and M. van Herk. 2008. "Variability of four-dimensional computed tomography patient models." Int J Radiat Oncol Biol Phys no. 70 (2):590-8.

Spitz, F. R., J. L. Abbruzzese, J. E. Lee, P. W. Pisters, A. M. Lowy, C. J. Fenoglio, K. R. Cleary, N. A. Janjan, M. S. Goswitz, T. A. Rich, and D. B. Evans. 1997. "Preoperative and postoperative chemoradiation strategies in patients treated with pancreaticoduodenectomy for adenocarcinoma of the pancreas." J Clin Oncol no. 15 (3):928-37.

Tenn, S. E., T. D. Solberg, and P. M. Medin. 2005. "Targeting accuracy of an image guided gating system for stereotactic body radiotherapy." Phys Med Biol no. 50 (23):5443-62.

Wong, J. R., L. Grimm, M. Uematsu, R. Oren, C. W. Cheng, S. Merrick, and P. Schiff. 2005. "Image-guided radiotherapy for prostate cancer by CT-linear accelerator combination: prostate movements and dosimetric considerations." Int J Radiat Oncol Biol Phys no. 61 (2):561-9.

Worm, E. S., A. T. Hansen, J. B. Petersen, L. P. Muren, L. H. Praestegaard, and M. Hoyer. "Inter- and intrafractional localisation errors in cone-beam CT guided stereotactic radiation therapy of tumours in the liver and lung." Acta Oncol no. 49 (7):1177-83.

Yeo CJ, Cameron JL, Sohn TA, Lillemoe KD, Pitt HA, Talamini MA, Hruban RH, Ord SE, Sauter PK, Coleman J, Zahurak ML, Grochow LB, Abrams RA. 1997. "Six hundred fifty consecutive pancreaticoduodenectomies in the 1990s: pathology, complications, and outcomes." Ann Surg no. 226:248-257.

2

STAT RAD: A Potential Real-Time Radiation Therapy Workflow

David Wilson, Ke Sheng,
Wensha Yang, Ryan Jones,
Neal Dunlap and Paul Read
*University of Virginia,
USA*

1. Introduction

1.1 Epidemiology and cost of metastatic disease

The American Cancer Society estimates that approximately 1.5 million people in the United States will be diagnosed with cancer, and 560,000 will die of cancer in 2010 (Jemal et al., 2010). These numbers are projected to increase rapidly in the near future due to national demographics with a large number of Americans reaching retirement age over the next 15-20 years, resulting in a doubling of projected new cancer diagnoses in 2050 to 3 million (Hayat et al., 2007). Most cancer deaths involve extensive locoregional tumors or metastatic disease to brain, lung, liver, or bone causing pain, disability, and decreased quality of life. As treatments for cancer improve, patients are living longer with advanced cancer than ever before, and the management of metastatic disease is becoming increasingly more multi-disciplinary and complex with patients treated simultaneously with systemic therapy, surgery, and radiation. It is well documented that cancer-related pain is often inadequately controlled in the palliative care setting, and both the pain and opioid medication interfere with patient function and quality of life (Bruera & Kim, 2003; Cleeland et al., 1994; McGuire, 2004). Radiotherapy is an important treatment for the alleviation of pain and suffering for cancer patients. It prevents pathologic bone fractures, and palliates tumor-induced obstruction, bleeding, and pain that is not well palliated with pharmacologic treatment (Halperin et al., 2008).

The skeleton is one of the most common sites of metastatic disease and is often the first site affected by metastases and the most common origin of cancer-related pain (Schulman & Kohles, 2007; Coleman, 2006). It was estimated that in 2004, 250,000 cancer patients were afflicted with metastatic bone disease (Schulman & Kohles, 2007). Bone metastases are most common in patients with multiple myeloma, of whom 90% develop bone metastases (Lipton, 2010). Approximately 70% of patients dying of breast and prostate cancer have evidence of metastatic bone disease, and bone metastases are also common in thyroid, kidney, and lung cancers, occurring in 30-40% of these cancers (Coleman, 2006). Metastatic bone disease causes considerable morbidity in patients with cancer, resulting in pain, hypercalcemia, pathologic fractures, compression of the spinal cord or cauda equina, and spinal instability (Coleman, 2006).

The treatment of metastatic bone disease is financially costly. Schulman and Kohles estimated that the mean per patient direct cost for cancer patients after diagnoses with metastatic bone disease was $75,329 compared to $31,455 for cancer-matched controls without metastatic bone disease (Schulman & Kohles, 2007). Using this data, the authors estimated that the national cost burden for patients with metastatic bone disease was $12.6 billion in 2004, which was 17% of the NIH-reported $74 billion direct medical costs for cancer (Schulman & Kohles, 2007). These costs will clearly increase with our aging population and associated increase in cancer prevalence (Hayat et al., 2007). From a societal standpoint, looming Medicare financial constraints will likely result in reduced reimbursement for palliative services, driving the economic incentive to develop the next generation of more clinically efficient palliative radiotherapy workflows.

2. Standard palliative radiotherapy techniques

2.1 Lack of dose conformality

For 30-40 years, standard palliative radiotherapy treatment techniques have utilized simple opposed beam arrangements such as treating a patient with parallel opposed anterior and posterior beams. Although simple to plan and deliver, such techniques provide poor conformality, and large volumes of organs at risk (OARs) may receive the full prescribed dose depending on the area treated. See Figure 1. Radiation to these OARs (skin, lung, esophagus, trachea, stomach, small bowel, rectum, bladder, or genitals) may result in cough, dysphagia, odynophagia, nausea, vomiting, weight loss, fatigue, diarrhea, dysuria, erythema, and pruritus of the skin and genitals (Gaze et al., 1997; Langendijk et al., 2000). Despite being planned and delivered on sophisticated systems, these treatments are frequently only moderately effective, and cause significant toxicity to an already ill patient population with a limited life expectancy (Gaze et al., 1997).

2.2 Slow treatment planning and quality assurance workflow

Conventional simulation and treatment planning is performed over a several day process prior to the first delivered treatment. The patient is first seen in consultation and scheduled for a CT simulation on a subsequent day. During the CT simulation the patient is placed in the position in which they will ultimately be treated on a treatment unit, and immobilization and support devices are fabricated, after which they undergo a CT scan in the treatment position. He or she must then wait, sometimes several days, for the contouring of the CT simulation images, a process by which the radiation oncologist specifies the planning target volume (PTV) of the tumor to be treated and the regional OARs or adjacent tissues that may receive radiation resulting in toxicity. Following the contouring of the CT images, radiation treatment planning is performed, during which time medical dosimetrists and physicians determine the beam angles and treatment techniques to deliver the prescribed dose to the PTV while attempting to minimize dose to OARs if possible. Following treatment planning, quality assurance calculations and/or measurements are performed by medical physicists before delivery of the first treatment to ensure accuracy of delivering the planned dose and ensure patient safety. Finally, the first treatment is then delivered 3-7 working days after the initial consultation.

2.3 Inconvenient, modestly effective treatments

Although fractionation schedules in Europe are trending toward hypofractionation (fewer treatments), the most common palliative dose fractionation schedules in the United States vary between 20 and 30 gray (Gy) in 5 -10 fractions delivered over 1 -2 weeks (Fairchild et al., 2009). Adding the one week pre-treatment work process to the 1-2 weeks of treatment delivery results in an overall duration of 2-3 weeks for completion of palliative treatment. Conventional radiotherapy, regardless of fractionation schedule, has been found to be modestly effective in treatment of bone metastases, resulting in an improvement in pain in only about 60% of patients (Wu et al., 2003; Chow et al., 2007). In a retrospective study of end stage cancer patients receiving palliative radiotherapy, Gripp et al found that half of the patients received treatment for >60% of their final days of life (Gripp et al., 2010). Thus, these often modestly effective treatments subject the patients to repeated visits to the treatment center and consume precious time and energy for ill patients and their families. Clearly it is important that we design more effective palliative treatments that are more efficient to plan and deliver, minimize acute toxicity, and require fewer total treatments and time.

2.4 Mark-and-start radiation therapy workflows

Traditional emergent radiation therapy workflows referred to as "mark and start" protocols were developed to rapidly treat patients with severe pain, spinal cord compression, superior vena cava syndrome, and life-threatening obstruction of major organs. They generally rely on a good understanding of surface anatomy to direct placement of square or rectangular treatment fields on the patient with the patient on the treatment couch. A port film is obtained to confirm that the target is being treated and to document the treatment volume. The treatment field is then marked on the patient and documentation photos are obtained. Following anatomic volume determination and verification, the prescription dose is converted to treatment unit monitor units which are calculated using the field size, treatment distance, treatment depth, and machine-specific output factors for a given photon energy. The best quality assurance practices are to have two people calculate the monitor units independently and to have at least one person perform the calculation by hand if a computer calculation program is used. Once the monitor units are calculated, the patient can be treated. Emergent treatments generally use one or two parallel opposed beams to deliver non-conformal dose with large volumes of non-target tissue being irradiated to the prescribed dose.

Since most patients treated with radiation therapy on an emergent basis are symptomatic with pain, bleeding, or obstruction, it can be difficult for them to lie still on a flat treatment table for prolonged periods of time. Therefore, the faster the clinical workflow, the better the patient will tolerate the process. Most new linear accelerators (LINACs) are equipped with kilovoltage imaging capabilities on the treatment unit which can make the initial field placement easier by functioning similar to a CT or fluoroscopic simulator. This can increase the efficiency of the process since accurate field placement is the most time consuming part of the "mark and start" workflow. Once the field is accurately marked, the monitor unit calculations take only a few minutes, and the patient can rapidly be treated.

Clearly, for emergency situations, a simple treatment option is highly desirable for any treatment system, especially for a system in a one-unit radiation oncology clinic. Some complex treatment systems have no easy methodology or workflow to treat patients emergently with simple fields if the patient has not undergone a separate CT simulation. This is due to the fact that they have no way to calculate a treatment plan without a contoured CT image dataset. In addition, some intensity modulated radiation therapy (IMRT) dedicated systems with their own CT treatment planning algorithms do not have an easy way to perform an independent calculation to verify the accuracy of the planned dose calculation. Due to these limitations, the treatment of the emergency patient on these systems generally requires performance of the standard workflow of CT simulation, CT contouring, dose calculation, dose verification with unit measurements, and then image guided treatment delivery to the patient. The development of novel and greatly expedited workflows for these systems that utilize conformal dose delivery would provide an improved method to treat emergency patients that could also be used to treat non-emergent palliative patients more rapidly. In this chapter, we propose the development of one potential rapid clinic workflow utilizing the TomoTherapy system called STAT RAD.

3. Stereotactic Body Radiotherapy (SBRT): A more effective, highly conformal hypofractionated palliative radiation technique

In the search for more effective and less toxic radiotherapy techniques, much attention has been focused on stereotactic body radiotherapy (SBRT). SBRT utilizes hypofractionated, highly conformal, high dose radiation delivery that has been modeled after intracranial stereotactic radiosurgery (SRS). Like SRS, SBRT uses multiple beams that converge on the target volume. This minimizes the volume of tissue receiving high dose to where the beams intersect, reducing dose to normal tissue. This allows for the delivery of ablative doses of radiation in a few fractions with acceptable toxicity (Read, 2007; Timmerman et al., 2010). SBRT is a proven method for treating lung cancer, yielding excellent rates of local control for non-small-cell lung cancer and resulting in 5-year survival rates potentially comparable to that of surgery (Timmerman et al., 2010; Onishi et al., 2010). In addition, the treatment of liver metastases with SBRT has yielded promising results, achieving local control rates at 2 years of approximately 70–90% (Dawood, Mahadevan, & Goodman 2009; Rusthoven et al., 2009).

SBRT has also been used in the palliative treatment of bone metastases to the spine with remarkable success. Multiple studies have used SBRT to safely deliver high doses of radiation to spinal metastases while significantly limiting dose to the spinal cord and achieving local control rates of >80% at one year (Gerszten et al., 2007; Nelson et al., 2009; Gibbs et al., 2007). Fractionations in these studies have ranged from 1 to 5 fractions delivering 4 – 24 Gy per individual fraction, with total doses between 10 to 30 Gy (Gerszten et al., 2007; Nelson et al., 2009; Gibbs et al., 2007). In the largest prospective study of spine SBRT by Gerszten, 336 cases were treated primarily to relieve pain, and they achieved significant pain improvement in 290 patients (86%). Nelson, Gibbs, and Ryu, have also reported pain reduction in greater than 80% of patients in their studies (Gerszten et al. 2007; Nelson et al., 2009; Gibbs et al., 2007; Ryu et al., 2008), much improved over the 60% in conventional radiotherapy (Wu et al., 2003; Chow et al., 2007). Not only do more people experience pain relief with SBRT, but the pain relief is reported to be more durable. Gagnon demonstrated statistically significant improvement in pain scores lasting throughout 4 years of follow-up (Gagnon et al., 2009). Ryu found the median duration of pain relief to be 13.6 months with SBRT (Ryu et al., 2008), which is a dramatic improvement compared to the average 3 to 6 months of palliation with conventional therapy (Gaze et al., 1997; Foro Arnalot et al., 2008). Additionally, spinal SBRT treatments have been effective in achieving local control in tumors typically resistant to radiotherapy, such as renal cell carcinoma and melanoma, reportedly due in part to radiation injury to the tumor vasculature (Gerszten et al., 2007; Gibbs et al., 2007; Ryu et al., 2008; Gagnon et al., 2009).

3.1 Adverse events with SBRT: Minimal toxicity

Though great success is seen in high dose, hypofractionated therapy, care must be taken to avoid incorrectly delivering the high dose radiation to normal tissue. Prevention of damage to normal tissue is ensured through careful patient immobilization, co-registration of multiple diagnostic imaging modalities (MRI, PET CT, contrast enhanced CT) to the kVCT simulation to accurately define the target and OARs, inverse treatment planning with the use of intensity modulated radiation therapy, patient-specific quality assurance, and CT image guidance at the time of treatment delivery. Nevertheless, common side effects of radiotherapy can occur with SBRT. However, the advantage of conformal radiation is that it spares high radiation dose to normal tissue with the relatively small target volumes employed in this technique compared to parallel opposed techniques in which prescription doses are delivered to all tissues, target and OARs, in the beam path through the patient. This advantage of SBRT has been demonstrated in many trials by reports of little to no toxicity (Gerszten et al., 2007; Gagnon et al., 2009), and is reinforced by the findings of McIntosh et al, who compared conformal TomoTherapy to conventional 3D conformal treatment techniques on an anthropomorphic phantom and showed that TomoTherapy plans significantly improved conformality and reduced dose to regional critical structures (McIntosh et al., 2010).

Most significant adverse events in spinal SBRT have occurred with treatments that used extremely high-doses (>20 Gy) in a single fraction. Gomez et al reported odynophagia and dysphagia in 1 patient who had received 22 Gy to the esophagus in a single dose, and another patient developed an esophageal ulcer and necrosis after receiving 24 Gy to his esophagus in one fraction (Gomez et al., 2009). Another patient developed bronchial stenosis after receiving 11 Gy to a bronchus in a single fraction. In another study with similarly high dose fractionation schedules, 39% of patients treated with 18 to 24 Gy in a single dose developed new or progressive vertebral fractures (Rose et al., 2009). However, their patient selection did not utilize a scoring system to identify patients at high risk for pathologic fracture, such as the Mirels scoring system (Cumming et al., 2009). In contrast, Gagnon et al, using mean doses of 26 Gy in 3 fractions in 200 patients, only had 2 patients (1%) develop vertebral fractures (Gagnon et al., 2009). Sahgal et al reported 5 cases of radiation myelopathy and concluded that for single fraction SBRT, up to 10 Gy to a maximum point to the thecal sac is safe (Sahgal et al., 2010).

3.2 Extrapolation of spinal SBRT-like dose distributions to non-spine metastases

Given the advances in radiation delivery with SBRT and its success in palliation of spine metastases, it is logical to apply these advancements in technology to extra-axial bone metastases; however, no trials have been published to date. This is due to the fact that SBRT is only reimbursed for limited indications such as spinal metastases. It is fair to hypothesize that the extrapolation of SBRT-like dose distributions to extra-axial bone metastases will improve pain control and that rapid institution of radiation will minimize the time patients are in pain and on high dose opioids that place them at risk for iatrogenic medical complications. By applying the concepts of spinal SBRT, highly conformal hypofractionated radiation therapy plans could be used to treat non-spinal metastases. This allows for increased dose per fraction and fewer total fractions with less toxicity compared to standard non-conformal palliative regimens. See Figures 1-2.

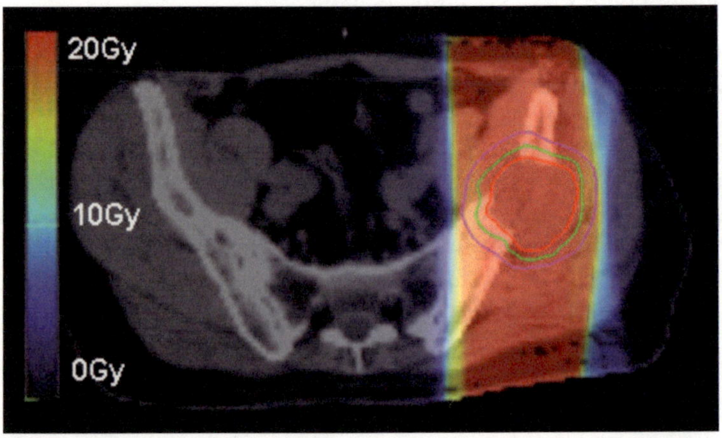

Fig. 1. Nonconformal Technique

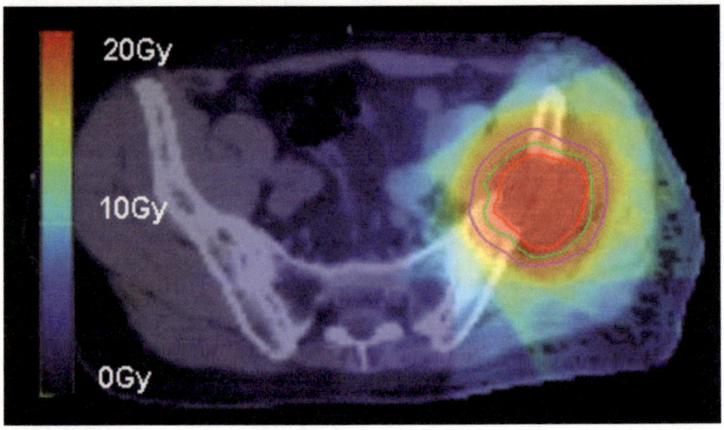

Fig. 2. Conformal Technique

3.3 Relative Biologic Effective Dose: A method to compare different dose fractionation schedules

Based on the linear-quadratic equation, one can calculate the biologic effective dose (BED) to compare dose delivery of different fractionation schedules using the equation:

$$BED = nd\ [1+d(alpha/beta)] \qquad (1)$$

n = number of fractions, d = dose per fraction, alpha/beta = the ratio of intrinsic radiosensitivity to repair capacity

As seen in Table 1, when compared with conventional fractionation schedules for palliative osseous metastases, such as 30 Gy in 10 fractions or 20 Gy in 5 fractions, high dose per fraction regimens deliver very similar BED to early responding tissues and slightly higher BED to late responding tissues. We believe that for reasonable rates of symptom relief and duration of palliation that palliative regimens should deliver a minimum BED of 25 Gy to most treatment targets. Twenty-four Gy in 3 fractions (8 Gy x 3 fractions) is an attractive palliative regimen that balances a high radiobiologic dose with the convenience of a highly hypofractioned regimen. This daily dose can generally be delivered in 10 minutes or less depending on target size and modulation allowing patients in pain to tolerate the treatment without moving. Finally, doses of 8 Gy or higher may be more effective against tumor histologies thought to be more radioresistant such as melanoma or renal cell carcinoma due to cytotoxic effects to the tumor vasculature. However, the dose per treatment, number of treatments, and total dose will depend on the patient's overall condition and tumor-specific factors including histology, location, proximity to critical OARs, and size. Relative BED provides a method to compare different dose fractionation schedules that can be used to correlate the treatment with patient outcomes.

	Total dose	# of fractions	Dose per fraction (Gy)	Alpha/beta	BED
Early Responding Tissues	30	10	3	10	39
	20	5	4	10	28
	24	3	8	10	43
Late Responding Tissues	30	10	3	3	60
	20	5	4	3	47
	24	3	8	3	88

Table 1. Comparison of BED in Different Palliative Fractionation Schedules

4. STAT RAD: A rapid palliative radiotherapy workflow in clinic development

Clearly a faster, more efficient workflow to treat metastatic disease is needed. Patients with widespread metastases frequently have short life expectancies and need treatments that minimize their time in clinic while providing rapid and durable pain relief for the remainder of their lives. Additionally, this need for efficiency will further rise with the increasing cancer burden and health care costs due to the aging baby-boom population.

Thus, at the University of Virginia we are piloting a new workflow called "STAT RAD" to rapidly deliver advanced radiotherapy to patients with metastatic disease on an internal review board approved clinical trial. This STAT RAD workflow offers same-day palliation in an approximately 6-hour time frame similar to a standard GammaKnife ® (Elekta, Stockholm, Sweden) workflow. STAT RAD is a highly coordinated conventional workflow that includes kVCT simulation, treatment planning, treatment plan quality assurance, and delivery of conformal hypofractionated radiotherapy in a single day. All treatments are planned and delivered on FDA-approved systems including the TomoTherapy treatment machine. This workflow allows patients to receive an entire course of palliative treatment from start to finish in a few days, a process that conventionally takes 2-3 weeks. Since patients are billed for each individual treatment, requiring fewer treatments reduces health care costs in addition to being more convenient. With the STAT RAD program we are now able to offer a unique workflow that delivers rapid, effective, and efficient palliative radiotherapy that is cost effective, less toxic, and more convenient for cancer patients and their families.

We have treated approximately 50 cancer patients with the conformal hypofractionated STAT RAD treatment regimen for a variety of palliative indications. We have treated patients with IMRT and 3D delivery using TomoHelical and TomoDirect planning modes. Retrospective review of these patients shows that the majority of these patients experienced rapid and durable palliation of symptoms with minimal toxicity (unpublished data). In general, patients are extremely satisfied with the speed at which their treatment is initiated and the convenience of the hypofractionated regimens.

In our current trial we are quantifying patient outcomes following treatment with the current STAT RAD workflow in an effort to determine its benefits and risks to patients. In addition, we are systematically evaluating and optimizing software and hardware necessary to make the STAT RAD workflow even more efficient.

4.1 Technologic rationale for the choice of the TomoTherapy platform

The TomoTherapy platform has been chosen for the STAT RAD workflow for a variety of reasons. TomoTherapy delivers highly conformal and homogenous dose distributions through modulation of dose from a bank of 64 binary 6.25-mm-wide collimator leaves capable of pneumatic opening or closing 51 times per revolution as the gantry revolves around the patient. The system can also treat patients with discreet beam angles (i.e. the radiation beam not rotating) in a mode called TomoDirect. Although all TomoTherapy treatment delivery is technically IMRT, the treatment planning can be done in either a 3D or IMRT mode allowing highly conformal treatments to be billed as 3D and thus used in the treatment of all patients with bone metastases. While the 3D planning mode limits the planning options for dose constraints on OARs, partial and complete blocking can be assigned to non-target structures. Partial blocking allows beams to exit through the structure after treating the PTV but not to enter through the structure prior to treating the PTV, and complete blocking restricts beams from entering or exiting through a structure. In addition, good preliminary data exists to support the use of the fan beam MVCT as a CT simulation image set for treatment planning and the use of CT detector-based exit dose methodology for quality assurance, making this system an excellent platform to pilot and optimize this workflow.

5. Scan-plan-verify-treat STAT RAD workflow: A novel and more efficient STAT RAD workflow

With recent advances in software and technology, we plan to further condense the STAT RAD workflow into the Scan-Plan-Verify-Treat workflow, a 30-minute process in which all steps (MVCT simulation, diagnostic image co-registration, treatment planning, and treatment delivery with real time quality assurance) are performed on the TomoTherapy unit. This advanced workflow will eliminate the need for the patients to undergo a kVCT simulation on a separate unit as well as make it unnecessary for the patient to leave the treatment table between the simulation and treatment.

Requirements for the clinical implementation of the Scan-Plan-Verify-Treat STAT RAD workflow are envisioned as follows:

1. **Scan:** MVCT simulation image acquisition (10 minutes) then rigid or deformable image co-registration of existing diagnostic image sets with pre-contoured target and OAR volumes to the MVCT simulation scan for contour transfer (3-5 minutes).
2. **Plan:** Rapid inverse treatment planning (3-5 minutes).
3. **Verify:** Real-time patient-specific quality assurance using CT detectors prior to or during treatment delivery (10 minutes).
4. **Treat:** Simple real-time patient motion tracking to monitor patient position real-time during the Scan-Plan-Verify-Treat process to ensure accurate delivery.

5.1 New image co-registration workflow

In the conventional workflow, target volumes and OARs are contoured on recent diagnostic images (MRI, PET-CT, or diagnostic CT that are already available in the patient's electronic radiology chart). After the patient undergoes a kVCT simulation, the contoured diagnostic images are rigidly or deformably co-registered to the kVCT simulation images, and the contours are transferred. This allows for high resolution diagnostic images to be used for tumor and normal tissue identification, which are not always possible to differentiate on CT simulation scans due to the resolution of standard wide bore CT simulation scanners and images that are frequently obtained without intravenous contrast. Multiple commercial image processing systems are available for this image processing, and we are currently using Velocity® (Atlanta, GA) image processing software. Following treatment planning, the patient then undergoes image guided treatment delivery, a process in which a daily MVCT scan is obtained on the TomoTherapy unit and co-registered to the planning kVCT scan. Patient setup shifts can then be made to ensure accurate patient setup, and the patient is treated. Therefore, this is a two image co-registration workflow. See Figure 3.

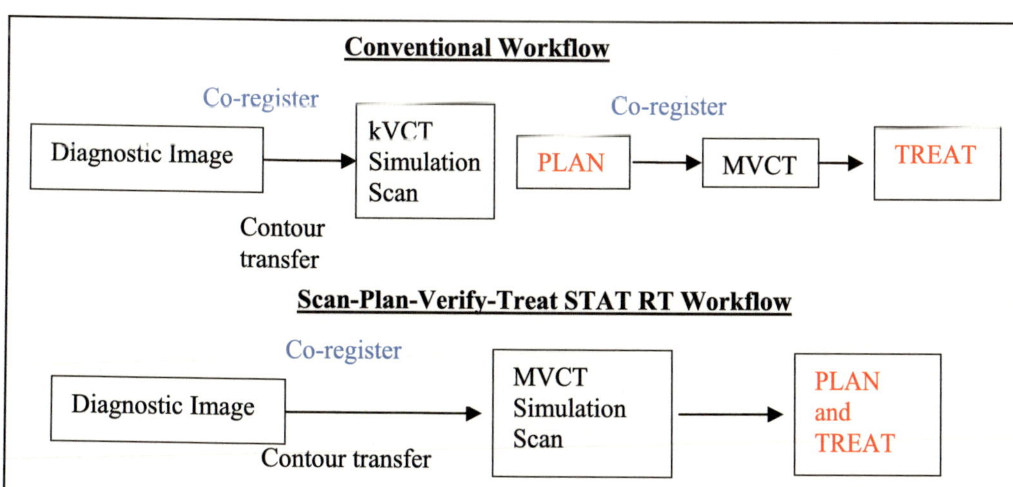

Fig. 3. Comparison of Image Co-Registration Workflows

A kVCT simulation scan has historically been used for simulation in the conventional workflow for both palliative and curative radiation planning. Compared to MVCT scans, it has higher resolution and allows the possibility for administration of iodinated IV and/or GI contrast, which makes it easier to identify soft tissues and bony anatomy for treatment planning. However, contrast agents are not generally given for kVCT simulations of patients for palliative treatment of metastases since the soft tissue and bone windows are adequate. Soft tissue and bone windows of MVCT scans have quite reasonable resolution and can easily be co-registered to higher resolution diagnostic studies for contour transfer. Software has been used clinically since 2004 to automatically co-register kVCT simulation images and daily MVCT scans for image guidance on a daily basis.

Our preliminary unpublished data confirms that the MVCT scan has sufficient resolution, particularly of bone anatomy, for accurate co-registration to contoured diagnostic images and that this one step co-registration process yields comparable agreement to the conventional two step image co-registration workflow with +/- 2-3 mm differences. See Figure 3. This level of agreement is consistent with results reported from image co-registration studies performed on a multi-institutional pediatric clinical trial with co-registration data of 51 patients from 45 institutions using 11 different image software systems. They reported an inherent uncertainty of 2 mm for MRI to CT co-registration (Ulin, Urie, & Cherlow 2010). Thus, preliminary data suggests that the optimization of this one step image co-registration workflow of diagnostic image sets to a MVCT simulation scan will be clinically similar to the conventional two image co-registration workflow. MVCT image guidance scans and kVCT simulation co-registration occurs routinely in the clinic and only takes a few seconds, therefore, we do not believe that this will be a rate-limiting step in the clinical implementation of the Scan-Plan-Verify-Treat STAT RAD workflow. Further optimization of the image co-registration workflow will make the 30-minute Scan-Plan-Verify-Treat STAT RAD workflow feasible.

5.2 Rapid inverse treatment planning on MVCT scans

CT image sets are used for radiation treatment planning because the electron density of tissues, which is required for calculating dose, is easily determined based on the Hounsfield units. The tissue electron density determination is essentially the same for MVCT and kVCT scans. It has previously been reported that as far as the dose calculations are concerned, treatment planning on either a kVCT simulation image set or a MVCT simulation image set yields treatment plans that are within 1% of each other (Langen et al., 2005).

We have recently published that the TomoTherapy STAT RT treatment planning module can calculate SBRT plans in just a few minutes (Dunlap et al., 2010). The computing speed of radiation treatment planning systems is about to take a quantum leap forward with the incorporation of new algorithms that will take advantage of the processing power of graphics processing units (GPU) whose more rapid and parallel calculating potential can improve treatment planning speed by 10-20 times (Hissoiny, Ozell, & Despres, 2010; Hissoiny, Ozell, and Despres, 2009). Same-day inverse treatment planning of IMRT or 3D TomoTherapy plans has not been a problem for patients treated with STAT RAD to date. We are comparing planning times for current FDA-approved treatment planning systems to those of newer, in-development GPU-based algorithms. In general, highly conformal 3D or IMRT plans can be generated in 2-3 minutes with GPU-based algorithms.

We have shown that accelerated treatment planning software for Helical TomoTherapy provides clinically acceptable dosimetry, with conformality and homogeneity that is superior to standard LINAC-based 3D conformal planning and is only slightly inferior to standard Helical TomoTherapy dosimetry (McIntosh et al., 2010). We have also shown that, with planning times of 2-5 minutes, this accelerated treatment planning software provides levels of dosimetric conformality, heterogeneity, and avoidance of organs at risk for simple SBRT treatments that are clinically equivalent to those generated with conventional Helical TomoTherapy treatment planning (Dunlap et al., 2010). This preliminary data supports that treatment planning speed is not likely to be rate limiting in the ultimate clinical implementation of the Scan-Plan-Verify-Treat STAT RAD workflow.

5.3 Novel CT-detector-based quality assurance methodology

Current standard of care TomoTherapy quality assurance methodology requires that each patient-specific treatment plan be delivered to a cylindrical plastic phantom with ion chamber and film measurement to ensure geometric and planar dose distribution accuracy using gamma criteria of 3%/3mm. However, this method does not measure the dose that the patient is receiving during treatment or provide full 3D dose verification. It causes another delay in delivering the first treatment to the patient as it requires approximately 30 minutes to complete, and is generally done by a medical physicist after daily clinic patient treatment is finished. A methodology to monitor the patient exit dose in real time would increase patient safety through verification of daily treatment accuracy as well as expedite the treatment workflow. Clearly, a real-time quality assurance methodology that does not require moving the patient off the TomoTherapy treatment couch for phantom measurements is essential for the development of a 30-minute Scan-Plan-Verify-Treat workflow. Current dose verification methodologies measuring dose at the time of patient treatment are limited to point measurements on the patient surface (Essers & Mijnheer, 1999), which is rarely in the target volume or a critical OAR, or through expensive implanted dosimeters (Beyer et al., 2007; Scarantino et al., 2004), which are not practical for most palliative patients. Since there is no method to directly measure the three dimensional dose in the patient, alternative approaches are being developed and tested in academic clinical settings. These alternative approaches reconstruct the delivered three dimensional dose distribution based on the measurement of either entrance or exit dose and back-projecting the measurements onto simulation or image guidance CT image sets.

The opportunity to reconstruct dose from information collected during treatment became available with the incorporation of radiation imaging detectors, such as electronic portal imaging devices (EPID) on linear-accelerators and CT detector arrays on TomoTherapy. Dose reconstruction using in-line EPID was first described by McNutt et al (McNutt, Mackie, & Paliwal 1997; McNutt et al., 1996). The EPID, when deployed during treatment, collects exit fluence from the patient and then back-projects this to X-ray fluence before entering the patient; then, the dose in the patient is re-computed using this entrance fluence and the planning CT images. However, there are many limitations to EPID-based dose verification. For example, the EPID was originally designed for semi-quantitative portal imaging; and for the purpose of dose reconstruction, it suffers from a narrow dynamic range, short life span, non-linearity in the dose response, ghost artifacts from low temporal resolution, and cross-plane scatter photon contribution to the measured fluence (Mijnheer, 2008). Investigators are currently working on methods to overcome these challenges.

The TomoTherapy unit has an in-line source-patient-detector geometry with CT ion chamber detectors that are used for daily MVCT scan image guidance for accurate patient positioning that remain in place during both imaging and treatment. These CT detectors can also be used to measure the patient exit dose fluence and back-project this onto a planning CT scan for volumetric or 3D dose reconstruction. Dose verification on TomoTherapy was first studied by Kapatoes et al., who calculated the entrance fluence from the exit dose using a transfer matrix, which is calculated based on the radiological path length from the source to the detector (Kapatoes et al., 2001; Kapatoes et al., 2001). The use of a CT ion chamber array has multiple advantages over EPID for exit fluence measurement. It is more durable, and has a much longer life span. It has a wider dynamic range and doesn't limit treatment positions. Finally, it is less sensitive to the noise from cross-plane scatter photons that complicate EPID-based dose reconstruction (Siewerdsen & Jaffray 2001).

Our pre-clinical evaluation of the CT detector-based exit radiation dose verification algorithm has been retrospectively studied by Sheng et al. using in-development software (Sheng et al., 2011). We compared planned and delivered doses with the conventional phantom quality assurance measurements for 24 patients and 347 treatment fractions. The concordance of planned to delivered dose calculated by the in-development software was shown to be +/- 5% (Sheng et al., 2011). This tolerance is within the standard of care of other current clinically available quality assurance methods. Further refinements are expected to improve dose monitoring accuracy for this or other algorithms.

5.4 Optical tracking methods for patient intra-fractional motion monitoring

Consistent patient positioning during CT image acquisition and treatment is critical to ensure accurate dose delivery. Physical immobilization devices such as external body frames, aquaplast masks and other body molds, and vac-lock vacuum bags are commonly used to ensure patient positioning reproducibility. X-ray or CT image guidance prior to radiation delivery on the treatment unit is routinely employed in the clinic. Methods for optical tracking of markers on the patient surface or tracking of the patient's skin surface itself are available to ensure consistent patient positioning after image guidance and during treatment, known as intra-fractional motion (Wagner et al., 2007; Wiersma et al., 2010). This provides a method without ionizing radiation for confirming patient position that can be used real-time during treatment delivery. With this information, if the patient's position moves outside of acceptable limits in any direction, treatment could be paused. A mechanism to ensure that the patient's position doesn't change between MVCT simulation and treatment delivery would obviate the need for a repeat image guidance MVCT scan just prior to delivery in the Scan-Plan-Verify-Treat workflow.

We have recently developed an in house optical tracking system using multiple OptiTrack FLEX:V100 cameras (Natural Point, Corvallis, OR). The camera utilizes 26 infrared light-emitting diodes and a charge coupled device to capture the reflective light from markers with special coating. By using multiple cameras, the 3D position of each reflective marker can be determined precisely. Multiple markers can be placed on a patient and monitored simultaneously. In the lab, localization precision of 0.1 mm was achieved (unpublished data). Through strategic positioning of the markers, movements of the head, neck, and extracranial locations can be closely monitored.

6. Clinical benefits and future directions of STAT RAD implementation

6.1 Additional benefits to patients with metastatic disease

Several advantages of this streamlined workflow are envisioned that will improve the care of patients with metastatic disease. The most obvious is that patients who live far from treatment centers can be offered palliative radiation therapy as an option. Take for example the case of a patient who lives 50 miles from a radiation oncology center. If they are seen in consultation, undergo a CT simulation on a second visit, and then are treated with 30 Gy in 10 fractions, they will have to drive 1200 miles for this treatment course. Clearly this is not practical for many ill patients in the last few months of life. If they can receive a conformal high dose palliative treatment in one day, it is much more likely that they will receive this treatment. We have been coordinating STAT RAD treatments on days that patients have appointments with other oncologists or specialists. Patients come to the radiation oncology clinic and undergo a consultation and CT simulation and then go to their other appointments while planning and quality assurance measurements are performed, and then they return to the radiation oncology clinic later that same day for treatment. Once the Scan-Plan-Verify-Treat STAT RAD workflow is available, we envision treating patients at the end of the scheduled workday on a same-day physician request basis. This service holds high utilization potential because many times physicians do not know if a patient is in significant need of palliation until they examine the patient at the time of a scheduled appointment.

Frequently patients are admitted to the hospital for management of cancer-related symptoms such as intractable pain, spinal cord compression, profuse tumor bleeding, or tumor related acute obstruction. These patients are frequently treated with palliative radiation therapy. The STAT RAD workflow enables patients to receive high dose and conformal treatments that start faster than conventional kVCT simulation workflows and can shorten the length of hospitalization to complete treatment.

Finally, this workflow makes the treatment of patients with oligometastatic disease more streamlined and practical because it enables SBRT-like dose distributions to be delivered to multiple lesions that currently cannot be treated with SBRT, such as nodal disease or non-spinal bone metastases.

6.2 Incorporation of translational technology development into routine clinic care for all patients

Several aspects of the Scan-Plan-Verify-Treat STAT RAD clinical development can be incorporated into the routine care of patients undergoing curative radiation therapy. Specifically, CT detector-based quality assurance of all treatments could be automated and performed daily. Such quality assurance could provide a warning if the delivered dose is greater than a threshold such as +/- 5% for a patient and trigger an investigation into the cause of this deviation. Quality assurance of each fraction of treatment would be a major advancement compared to current quality assurance methods of checking each plan prior to treatment. Using daily quality assurance to monitor changes in patient status such as significant weight loss in a head and neck cancer patient could trigger re-planning that could be done on an adaptive basis.

A simple system to monitor patient motion following image guidance could reduce the risk of geometric misses due to intra-fractional patient motion. If it were determined that a specific patient had more or less intra-fractional motion than accounted for in their PTV expansion, then their treatment plan could be adaptively re-planned to mirror their specific expansion requirements.

6.3 Future directions for spinal SBRT

The Scan-Plan-Verify-Treat STAT RAD workflow could easily be incorporated into the treatment of spinal SBRT patients. We have previously reported that treatment planning algorithms currently exist that can create highly conformal spinal SBRT plans in just a few minutes (Dunlap et al., 2010) and that the CT detector-based quality assurance algorithms can measure exit dose to within +/- 5% (Sheng et al., 2011). Real-time spinal SBRT simulation, planning, and delivery would eliminate the need for patients to be accurately re-set up and positioned between simulation and treatment to within two millimeters of accuracy which is the current accuracy of most co-registration-based image guided systems. In addition, differences in pitch, yaw, or roll of the patient between simulation and treatment delivery setups cannot be routinely corrected by most CT-based image guidance systems, requiring the patient to be re-positioned and re-imaged prior to treatment if they are out of tolerance. With this proposed workflow, the patient could be treated in the planning position, which could potentially eliminate these re-positioning error issues.

7. Conclusion

As the cancer burden in the population increases and heath care costs continue to rise, a faster, more efficient workflow is needed to treat patients with metastatic disease. Conformal hypofractionated treatment has demonstrated promising results for the palliation of bone metastases, and its incorporation into a workflow such as the STAT RAD workflow also improves patient convenience. In the near future, we believe the optimization of new software and hardware will enable a 30-minute Scan-Plan-Verify-Treat STAT RAD workflow to further maximize patient convenience and efficiency. This more efficient and cost effective workflow may result in more widespread incorporation of palliative radiation for cancer patients failing systemic therapy earlier in their disease process, reducing pain and functional loss and improving quality of life.

8. Abbreviations

3D - 3-dimensional
BED - biologic effective dose
CT – computed tomography
CTV – clinical target volume
EPID - electronic portal imaging device
FDA – Food and Drug Administration
GI - gastrointestinal
GPU - graphics processing unit
Gy - gray
IMRT- intensity modulated radiation therapy

IV - intravenous
kVCT – kilovoltage CT
LINAC - linear accelerator
MRI – magnetic resonance imaging
MVCT – megavoltage CT
OAR - organ at risk
PET CT – positron emission tomography CT
PTV - planning target volume
SBRT - stereotactic body radiotherapy
SRS - stereotactic radiosurgery
STAT RAD - urgent and rapid radiation treatment

9. References

Beyer, G. P., C. W. Scarantino, B. R. Prestidge, A. G. Sadeghi, M. S. Anscher, M. Miften, T. B. Carrea, M. Sims, & R. D. Black. (2007). Technical evaluation of radiation dose delivered in prostate cancer patients as measured by an implantable MOSFET dosimeter. *International Journal of Radiation Oncology, Biology, Physics*, Vol.69, No.3, pp. 925-35, ISSN 0360-3016

Bruera, E. & H. N. Kim. (2003). Cancer pain. *JAMA : The Journal of the American Medical Association*, Vol.290, No.18, pp. 2476-9, ISSN 0098-7484

Chow, E., K. Harris, G. Fan, M. Tsao, & W. M. Sze. (2007). Palliative radiotherapy trials for bone metastases: A systematic review. *Journal of Clinical Oncology : Official Journal of the American Society of Clinical Oncology*, Vol.25, No.11, pp. 1423-36, ISSN 1527-7755

Cleeland, C. S., R. Gonin, A. K. Hatfield, J. H. Edmonson, R. H. Blum, J. A. Stewart, & K. J. Pandya. (1994). Pain and its treatment in outpatients with metastatic cancer. *The New England Journal of Medicine*, Vol.330, No.9, pp. 592-6, ISSN 1533-4406

Coleman, R. E. (2006). Clinical features of metastatic bone disease and risk of skeletal morbidity. *Clinical Cancer Research : An Official Journal of the American Association for Cancer Research*, Vol.12, No.20, Pt.2, pp. 6243s-9s, ISSN 1557-3265

Cumming, D., J. Cumming, A. Vince, & R. Benson. (2009). Metastatic bone disease: The requirement for improvement in a multidisciplinary approach. *International Orthopaedics*, Vol.33, No.2, pp. 493-6, ISSN 1432-5195

Dawood, O., A. Mahadevan, & K. A. Goodman. (2009). Stereotactic body radiation therapy for liver metastases. *European Journal of Cancer*, Vol.45, No.17, pp. 2947-59, ISSN 0959-8049

Dunlap, N., A. McIntosh, K. Sheng, W. Yang, A. B. Turner, A. Shoushtari, J. Sheehan, et al. (2010). Helical tomotherapy-based STAT stereotactic body radiation therapy: Dosimetric evaluation for a real-time SBRT treatment planning and delivery program. *Medical Dosimetry*, Vol.35, No.4, pp. 312-9, ISSN 0958-3947

Essers, M. & B. J. Mijnheer. (1999). In vivo dosimetry during external photon beam radiotherapy. *International Journal of Radiation Oncology, Biology, Physics*, Vol.43, No.2, pp. 245-59, ISSN 0360-3016

Fairchild, A., E. Barnes, S. Ghosh, E. Ben-Josef, D. Roos, W. Hartsell, T. Holt, J. Wu, N. Janjan, & E. Chow. (2009). International patterns of practice in palliative radiotherapy for painful bone metastases: Evidence-based practice? *International Journal of Radiation Oncology, Biology, Physics*, Vol.75, No.5, pp. 1501-10, ISSN 0360-3016

Foro Arnalot, P., A. V. Fontanals, J. C. Galceran, F. Lynd, X. S. Latiesas, N. R. de Dios, A. R. Castillejo, et al. (2008). Randomized clinical trial with two palliative radiotherapy regimens in painful bone metastases: 30 gy in 10 fractions compared with 8 gy in single fraction. *Radiotherapy and Oncology : Journal of the European Society for Therapeutic Radiology and Oncology,* Vol.89, No.2, pp. 150-5, ISSN 0167-8140

Gagnon, G. J., N. M. Nasr, J. J. Liao, I. Molzahn, D. Marsh, D. McRae, & Sr Henderson FC. (2009). Treatment of spinal tumors using cyberknife fractionated stereotactic radiosurgery: Pain and quality-of-life assessment after treatment in 200 patients. *Neurosurgery,* Vol.64, No.2, pp. 306-7, ISSN 1524-4040

Gaze, M. N., C. G. Kelly, G. R. Kerr, A. Cull, V. J. Cowie, A. Gregor, G. C. Howard, & A. Rodger. (1997). Pain relief and quality of life following radiotherapy for bone metastases: A randomised trial of two fractionation schedules. *Radiotherapy and Oncology : Journal of the European Society for Therapeutic Radiology and Oncology,* Vol.45, No.2, pp. 109-16, ISSN 0167-8140

Gerszten, P. C., S. A. Burton, C. Ozhasoglu, & W. C. Welch. (2007). Radiosurgery for spinal metastases: Clinical experience in 500 cases from a single institution. *Spine,* Vol.32, No.2, pp. 193-9, ISSN 1528-1159

Gibbs, I. C., P. Kamnerdsupaphon, M. R. Ryu, R. Dodd, M. Kiernan, S. D. Chang, & J. R. Adler Jr. (2007). Image-guided robotic radiosurgery for spinal metastases. *Radiotherapy and Oncology : Journal of the European Society for Therapeutic Radiology and Oncology,* Vol.82, No.2, pp. 185-90, ISSN 0167-8140

Gomez, D. R., M. A. Hunt, A. Jackson, W. P. O'Meara, E. N. Bukanova, M. J. Zelefsky, Y. Yamada, & K. E. Rosenzweig. (2009). Low rate of thoracic toxicity in palliative paraspinal single-fraction stereotactic body radiation therapy. *Radiotherapy and Oncology : Journal of the European Society for Therapeutic Radiology and Oncology,* Vol.93, No.3, pp. 414-8, ISSN 0167-8140

Gripp, S., S. Mjartan, E. Boelke, & R. Willers. (2010). Palliative radiotherapy tailored to life expectancy in end-stage cancer patients: Reality or myth? *Cancer,* Vol.116, No.13, pp. 3251-6, ISSN 1097-0142

Halperin, Edward C., Carlos A. Perez, Luther W. Brady, & Ralph Erskine Conrad Memorial Fund. (Eds.). (2008). *Perez and brady's principles and practice of radiation oncology,* (5th ed.), Wolters Kluwer Health/Lippincott Williams & Wilkins, ISBN 978-0781763691, Philadelphia

Hayat, M. J., N. Howlader, M. E. Reichman, & B. K. Edwards. (2007). Cancer statistics, trends, and multiple primary cancer analyses from the surveillance, epidemiology, and end results (SEER) program. *The Oncologist,* Vol.12, No.1, pp. 20-37, ISSN 1083-7159

Hissoiny, S., B. Ozell, & P. Despres. (2010). A convolution-superposition dose calculation engine for GPUs. *Medical Physics,* Vol.37, No.3, pp. 1029-37, ISSN 0094-2405 (2009). Fast convolution-superposition dose calculation on graphics hardware. *Medical Physics,* Vol.36, No.6, pp. 1998-2005, ISSN 0094-2405

Jemal, A., R. Siegel, J. Xu, & E. Ward. (2010). Cancer statistics, 2010. *CA: A Cancer Journal for Clinicians,* ISSN 1542-4863

Kapatoes, J. M., G. H. Olivera, J. P. Balog, H. Keller, P. J. Reckwerdt, & T. R. Mackie. (2001). On the accuracy and effectiveness of dose reconstruction for tomotherapy. *Physics in Medicine and Biology,* Vol.46, No.4, pp. 943-66, ISSN 1361-6560

Kapatoes, J. M., G. H. Olivera, K. J. Ruchala, J. B. Smilowitz, P. J. Reckwerdt, & T. R. Mackie. (2001). A feasible method for clinical delivery verification and dose reconstruction in tomotherapy. *Medical Physics,* Vol.28, No.4, pp. 528-42, ISSN 0094-2405

Langen, K. M., S. L. Meeks, D. O. Poole, T. H. Wagner, T. R. Willoughby, P. A. Kupelian, K. J. Ruchala, J. Haimerl, & G. H. Olivera. (2005). The use of megavoltage CT (MVCT) images for dose recomputations. *Physics in Medicine and Biology,* Vol.50, No.18, pp. 4259-76, ISSN 1361-6560

Langendijk, J. A., G. P. ten Velde, N. K. Aaronson, J. M. de Jong, M. J. Muller, & E. F. Wouters. (2000). Quality of life after palliative radiotherapy in non-small cell lung cancer: A prospective study. *International Journal of Radiation Oncology, Biology, Physics,* Vol.47, No.1, pp. 149-55, ISSN 0360-3016

Lipton, A. (2010). Bone continuum of cancer. *American Journal of Clinical Oncology,* Vol.33, No.3 Suppl, pp. S1-7, ISSN 02773732

McGuire, D. B. (2004). Occurrence of cancer pain. *Journal of the National Cancer Institute.Monographs,* Vol.32, pp. 51-6, ISSN 1745-6614

McIntosh, A., N. Dunlap, K. Sheng, C. C. Geezey, C. B. Turner, L. Blackhall, G. Weiss, E. Lappinen, J. M. Larner, & P. W. Read. (2010). Helical tomotherapy-based STAT RT: Dosimetric evaluation for clinical implementation of a rapid radiation palliation program. *Medical Dosimetry,* Vol.35, No.4, pp. 280-6, ISSN 0958-3947

McNutt, T. R., T. R. Mackie, & B. R. Paliwal. (1997). Analysis and convergence of the iterative convolution/superposition dose reconstruction technique for multiple treatment beams and tomotherapy. *Medical Physics,* Vol.24, No.9, pp. 1465-76, ISSN 0094-2405

McNutt, T. R., T. R. Mackie, P. Reckwerdt, N. Papanikolaou, & B. R. Paliwal. (1996). Calculation of portal dose using the convolution/superposition method. *Medical Physics,* Vol.23, No.4, pp. 527-35, ISSN 0094-2405

Mijnheer, B. (2008). State of the art of in vivo dosimetry. *Radiation Protection Dosimetry,* Vol.131, No.1, pp. 117-22, ISSN 1742-3406

Nelson, J. W., D. S. Yoo, J. H. Sampson, R. E. Isaacs, N. A. Larrier, L. B. Marks, F. F. Yin, Q. J. Wu, Z. Wang, & J. P. Kirkpatrick. (2009). Stereotactic body radiotherapy for lesions of the spine and paraspinal regions. *International Journal of Radiation Oncology, Biology, Physics* Vol.73, No.5, pp. 1369-75, ISSN 0360-3016

Onishi, H., H. Shirato, Y. Nagata, M. Hiraoka, M. Fujino, K. Gomi, K. Karasawa, et al. (2010). Stereotactic body radiotherapy (SBRT) for operable stage I non-small-cell lung cancer: Can SBRT be comparable to surgery? *International Journal of Radiation Oncology, Biology, Physics,* (In press), ISSN 0360-3016

Read, P. W. (2007). Sterotactic body radiation therapy: 2007 update. *Community Oncology,* Vol.4, No.10, pp. 616-20, ISSN 1548-5315

Rose, P. S., I. Laufer, P. J. Boland, A. Hanover, M. H. Bilsky, J. Yamada, & E. Lis. (2009). Risk of fracture after single fraction image-guided intensity-modulated radiation therapy to spinal metastases. *Journal of Clinical Oncology : Official Journal of the American Society of Clinical Oncology,* Vol.27, No.30, pp. 5075-9, ISSN 1527-7755

Rusthoven, K. E., B. D. Kavanagh, H. Cardenes, V. W. Stieber, S. H. Burri, S. J. Feigenberg, M. A. Chidel, et al. (2009). Multi-institutional phase I/II trial of stereotactic body radiation therapy for liver metastases. *Journal of Clinical Oncology : Official Journal of the American Society of Clinical Oncology,* Vol.27, No.10, pp. 1572-8, ISSN 1527-7755

Ryu, S., R. Jin, J. Y. Jin, Q. Chen, J. Rock, J. Anderson, & B. Movsas. (2008). Pain control by image-guided radiosurgery for solitary spinal metastasis. *Journal of Pain and Symptom Management,* Vol.35, No.3, pp. 292-8, ISSN 0885-3924

Sahgal, A., L. Ma, I. Gibbs, P. C. Gerszten, S. Ryu, S. Soltys, V. Weinberg, et al. (2010). Spinal cord tolerance for stereotactic body radiotherapy. *International Journal of Radiation Oncology, Biology, Physics,* Vol.77, No.2, pp. 548-53, ISSN 0360-3016

Scarantino, C. W., D. M. Ruslander, C. J. Rini, G. G. Mann, H. T. Nagle, & R. D. Black. (2004). An implantable radiation dosimeter for use in external beam radiation therapy. *Medical Physics,* Vol.31, No.9, pp. 2658-71, ISSN 0094-2405

Schulman, K. L., & J. Kohles. (2007). Economic burden of metastatic bone disease in the U.S. *Cancer,* Vol.109, No.11, pp. 2334-42, ISSN 1097-0142

Sheng, K., R. Jones, W. Yang, B. Schneider, Q. Chen, G. Sobering, G. H. Olivera, & P. W. Read. (2011). 3D dose verification using tomotherapy CT detector array. *International Journal of Radiation Oncology, Biology, Physics,* (In press), ISSN 0360-3016

Siewerdsen, J. H., & D. A. Jaffray. (2001). Cone-beam computed tomography with a flat-panel imager: Magnitude and effects of x-ray scatter. *Medical Physics,* Vol.28, No.2, pp. 220-31, ISSN 0094-2405

Timmerman, R., R. Paulus, J. Galvin, J. Michalski, W. Straube, J. Bradley, A. Fakiris, et al. (2010). Stereotactic body radiation therapy for inoperable early stage lung cancer. *JAMA : The Journal of the American Medical Association,* Vol.303, No.11, pp. 1070-6, ISSN 0098-7484

Ulin, K., M. M. Urie, & J. M. Cherlow. (2010). Results of a multi-institutional benchmark test for cranial CT/MR image registration. *International Journal of Radiation Oncology, Biology, Physics,* Vol.77, No.5, pp. 1584-9, ISSN 0360-3016

Wagner, T. H., S. L. Meeks, F. J. Bova, W. A. Friedman, T. R. Willoughby, P. A. Kupelian, & W. Tome. (2007). Optical tracking technology in stereotactic radiation therapy. *Medical Dosimetry,* Vol.32, No.2, pp. 111-20, ISSN 0958-3947

Wiersma, R. D., Z. Wen, M. Sadinski, K. Farrey, & K. M. Yenice. (2010). Development of a frameless stereotactic radiosurgery system based on real-time 6D position monitoring and adaptive head motion compensation. *Physics in Medicine and Biology,* Vol.55, No.2, pp. 389-401, ISSN 1361-6560

Wu, J. S., R. Wong, M. Johnston, A. Bezjak, T. Whelan, & Cancer Care Ontario Practice Guidelines Initiative Supportive Care Group. (2003). Meta-analysis of dose-fractionation radiotherapy trials for the palliation of painful bone metastases. *International Journal of Radiation Oncology, Biology, Physics* Vol.55, No.3, pp. 594-605, ISSN 0360-3016

3

Segmentation Techniques of Anatomical Structures with Application in Radiotherapy Treatment Planning

S. Zimeras
University of the Aegean,
Department of Statistics and Actuarial, Financial Mathematics,
Karlovassi Samos,
Greece

1. Introduction

The definition of structures and the extraction of organ's shape are essential parts of medical imaging applications. These might be applications like diagnostic imaging, image guided surgery or radiation therapy. The aim of the volume definition process is to delineate a specific shape of an organ on a digital image as accurate as possible especially for 3D rendering, radiation therapy, and surgery planning. This can be done, either by manual user interaction or applying imaging processing techniques for the automatic detection of specific structures in the image using segmentation techniques. Segmentation is the process that separates an image into its important features (primitives) so that each of them can be addressed separately. This converts the planar pixel of the image into a distinguishable number of individual organs or tumour that can be clearly identified and manipulated. The segmentation process might involve complicate structures and in this case usually only an expert can perform the task of the identification manually on a slice-by-slice base. Humans can perform this task using complex analysis of shape, intensity, position, texture, and proximity to surrounding structures. In this work we present a set of tools that are implemented on several computer based medical application. Central focus of this work, are techniques used to improve time and interaction needed for a user when defining one or more structures based on segmentation techniques. These techniques involve interpolation methods for the manual volume definition and methods for the semi-automatic organ shape extraction.

2. Radiotherapy Treatment Planning (RTP)

The goal of radiotherapy treatment planning is to justify an effective treatment that will deliver a precise irradiation dose to the target volume without causing damage to the surrounding normal tissues. Therefore patient positioning, target volume definition and irradiation field placement are very critical steps while planning the irradiation process. Clinically a radiotherapy treatment plan is verified by Virtual Simulators (VS). (Sherouse et. al., 1987) first proposed the concept, often defined as CT-Sim to distinguish it from Sim-CT where a simulator is modified for CT use and by the late 1990s several designs and clinical assessments of CT virtual simulators have been reported (Sherouse et. al., 1987; Nagata et. al., 1990; Nishidai et. al., 1990; Rosenman et. al., 1991; Sherouse and Chaney, 1991; Perez et. al., 1994; Butker et. al., 1996; Chen et. al., 1996; Michalski et. al., 1996; Purdy, 1996; Ragan et. al., 1996; Rosenman, 1996) . Using VS, the clincal routine is modefied accordingly (Figure 1) (Conway and Robinson, 1997; Valicenti et. al., 1997; Cai et. al., 1997; Karangelis, 2004):

1. Collect patient's CT data including attached aluminium markers.
2. Transfer CT data to VS. The physician defines the tumour volume and the organs at risk and she/he will place the necessary fields relative to the tumour volume.
3. The simulation plan and the CT data are transferred via DICOM (Digital image and Communication in Medicine) server to the TPS for dose calculation and final treatment plan optimization.
4. Verify patient position on LINAC before irradiation.
5. Perform treatment on the treatment machine (Linear Accelerator or LINAC).

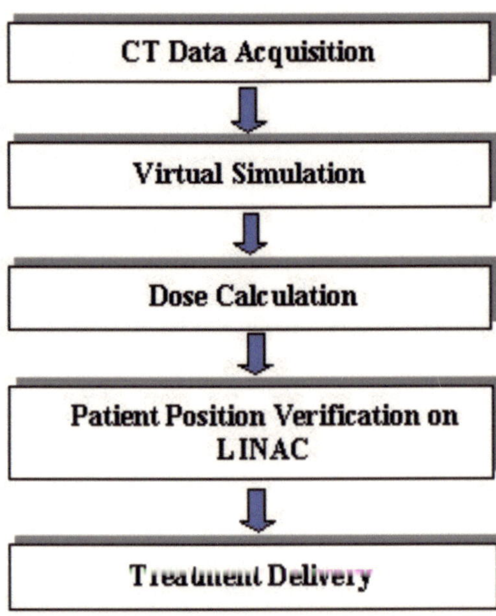

Fig. 1. Current clinical routine for external beam treatment delivery.

The important RTP imaging is the dynamic X-ray image on the Simulator monitor, which is generated from the viewpoint of the beam source, called Beam's Eye View (BEV) image, see the BEV view direction in Figure 2. In the system the corresponding window, called the BEV window, displays the interactive DRR images. Through the BEV window, the physicians can investigate the patient anatomy in the DRR image as they observe it on the Simulator monitor. Observer's Eye View (OEV) window visualizes the patient from the room's eye view, which is the same view as the physicians observes the patient in the Simulator room, see the OEV view direction in Figure 2. As we explained before, all of the images in both BEV and OEV windows are generated with the patient CT scan by the direct volume rendering (Figure 2).

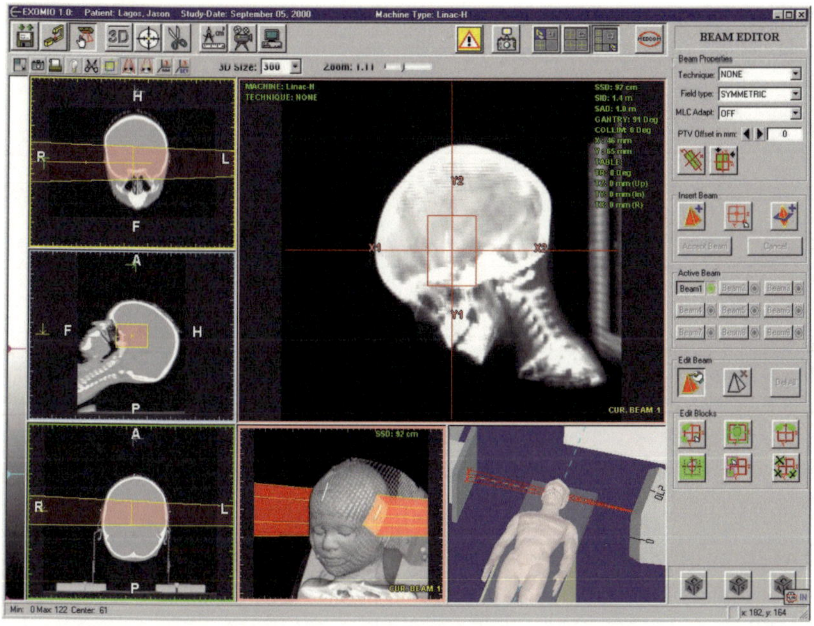

Fig. 2. The six-window layout of the system. On the left side the slices windows, the middle lower window is the OEV, the lower right the Room View and the upper right hosts the BEV.

3. Volume rendering

Volume rendering is the technique according to which a scalar field of data with discrete values, volume data, is selectively sampled in order to generate a useful image in relation to the sampled values. A volume data set is typically a set of discrete samples of one or more physical properties in a limited object space, $V(x)$, $x \in \boldsymbol{R}_n$, in which $\{x\}$ is a set of sampling points; n is the dimension of the sampling space, usually $n=3$, i.e. 3D volume data; V represents the sampling values, it can be a single-valued or multi-valued function. According to the distribution of sampling points (i.e. the structure of x), volume data sets are classified into structured and unstructured data sets. In medical imaging, volume data is usually a structured data set, typically organised as a stack of slices; V can be single valued (e.g. the CT Hounsfield value) or multi-valued (e.g. density, T1, T2 in MRI). The resulting data structure is an orthogonal 3D array of voxels, each representing a sampling value. A general pipeline of volume visualisation in medicine can include several steps (Sakas, 1993; Karangelis, 2004).

According to the distribution of sampling points, volume data sets are classified into structured and unstructured data sets. In medical imaging, volume data is usually a structured data set, typically organized as a stack of slices. The rendering methods which are proposed in this work are based on the work of (Sakas, 1993): (1) Transparent mode: Digitally Reconstructed Radiographies (DRR) (Sakas, 1993); (2) Surface reconstruction mode: iso-value, gradient (Sakas, 1995) and semi-transparent. DRR images (Cai, 1999) generated from CT digital volumes are often called using the term digitally reconstructed radiographs (DRRs) (Figure 3). The term DRR is used when we refer to those X-ray images that are

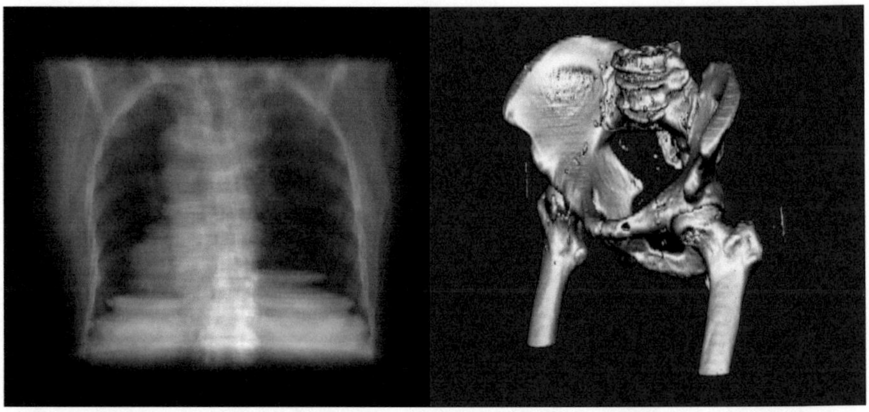

Fig. 3. left image. Reconstruction of CT volume using DRR; right image. Reconstruction using direct volume surface using direct gradient volume estimation (Karangelis, 2004; Zimeras and Karangelis, 2008)

generated with an unrealistic way using direct volume rendering techniques or to those images that are generated from volume data using a better approximation of the physical model. In both cases we try to simulate the attenuation of the X-ray through a medium, in our case through the digital patient's body. This can be achieved by using different transfer functions simulating the classical way of X-ray image reconstruction.

Assuming that a ray with initial energy I_o enters a volume, which has a thickness of Δ_s. and it is composed from different materials. Using discrete values the final energy of the ray is a product of the following equation (Max, 1995):

$$I(s) = I_0 \exp\{-(p_1 s_1 + p_2 s_2 + \ldots + p_x s_x)\} \tag{1}$$

where I(s) is the attenuated energy after leaving the volume, I_o the original energy of the ray, p_i the linear attenuation coefficient for the corresponding voxel material, s_i the corresponding distance from the ray entrance point to the ray exiting position of the voxel. In medical data the material type corresponds of course to the different tissue type. For detailed description of the model of the X-ray refer to (Karangelis, 2004). The contrast intensity of the final DRR image on the screen level can be calculated using the equation:

$$I_p = I_0 * (1 - L) + I_{background} * L \tag{2}$$

where $L = \exp\{-(p_1 s_1 + p_2 s_2 + \ldots + p_x s_x)\}$. In two dimensions the interpolation scheme will involve four data points' p_{00}, p_{01}, p_{10} and p_{11}. In this case L becomes:

$$L = \exp\left[F(p_0 p_1)\right] \tag{3}$$

where F given by the form

$$F(p_0 p_1) = \int_{p_0}^{p_1} K_l(t) dt \tag{4}$$

where K_l is the attenuation coefficient and l is the wavelength. This interpolation method is known as *bi-linear* interpolation and is commonly applied on image interpolation e.g. when magnifying raster images. In volumetric data linear interpolation is known as *tri-linear* interpolation and estimates the value of the result voxel considering the neighbourhood of eight (8) voxels (p_{000}, p_{001}, p_{011}, p_{010}, p_{100}, p_{110}, p_{101}, and p_{111}).

To reconstruct a DRR the digital CT data of the patient are used. Each voxel acquired using CT has an value that is called the Hounsfield unit which was named after the CT inventor. The HU can be estimated using the following formula:

$$HU = \frac{\left(\rho_{\mu} - \rho_{water}\right)}{\rho_{water}} * 1000 \tag{5}$$

where p_μ , p_{water} are the attenuation coefficient of the material for the specific voxel and the water respectively. When on the above equation one replaces the $p_\mu = p_{water}$, then will receive a $HU_{water} = 0$. In addition the air as material corresponds to -1000HU, since $p_{air} = 0$. The Hounsfield units have no upper limit but usually for medical scanners a range between -1024 to +3071 is provided. Apparently 4096 different HU values are provided and therefore to illustrate the complete range of the HU on a volume 12bit voxels are required.

Probably the most common methods for reconstructing surfaces directly from voxels are the gradient surface and iso-surface model. This rendering model is widely used in almost all medical data sets to render the surface detected by the gradient operator. (Levoy, 1988) presented this concept, which is effectively used in most medical imaging applications for the last decade. Depending on the data set size and on the size of the resulting 3D image, which influence the number of rays, a 3D view image can be calculated almost in real time (0.8sec to 2.3sec) (Figure 4)

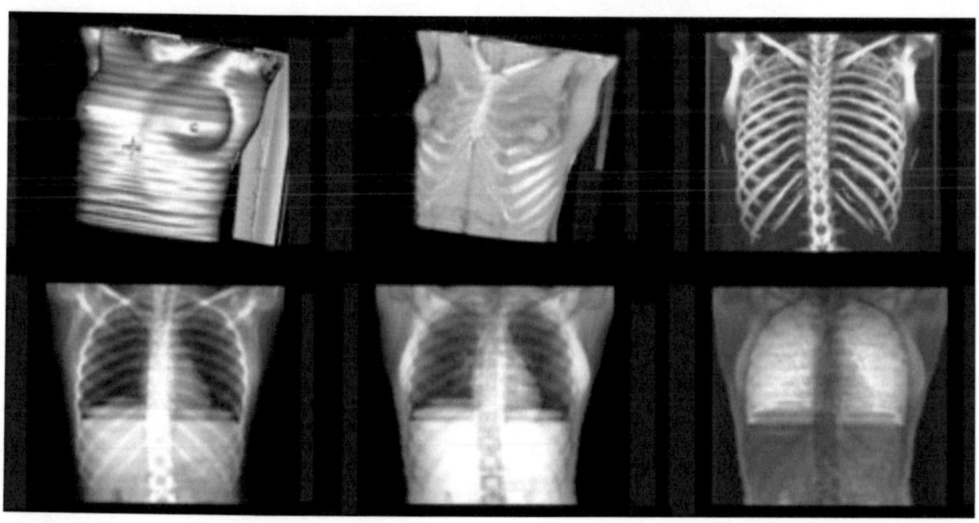

Fig. 4. Volume rendering modes. On the top row from left to right: isovalue mode, semitransparent mode and maximum intensity projection. On the lower row X-ray images reconstructed using different tissue ranges. From left to right: full tissue range, muscle tissues and lung tissues (Karangelis, 2004)

A significant part for the radiation field is the virtual light field projection on the patient skin [LuH99]. In physical simulators, a light source is located near the irradiation source. The orientation of the light intensity is diverted through the gantry head using a mirror aperture. The outcome of this process is the exact projection of the radiation field on the patient's skin. The two main axis of the field are indicated as line shadows. In case the radiation field is delineated using shielding blocks, then the light field area is also modified accordingly. This process described above should be performed in a similar manner in the virtual simulation process. In order to realise this principle we take advantage of the convexity of the tetrahedral objects and the Z-depth information derived during polygon scan conversion. Using the conventional Simulator the block shape was drawn manually on the patient's X-ray film, acquired from the BEV direction. Then this shape was digitized and its digital points were saved on the block-cutting machine. Both beam and block geometry is defined using combinations of 3D planes. Each beam is represented as a pyramid. The height of the pyramid is calculated according to the current machine specification; the base of the pyramid represents the irradiation field size projected to the image detector level and each side of the pyramid is assigned to a plane (Figure 5)

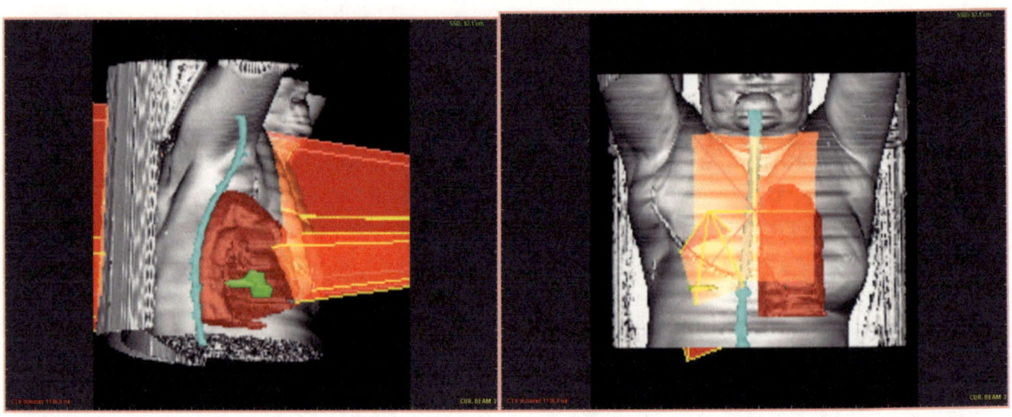

Fig. 5. 3D beam object reconstructed with patient's CT data. Green object: tumour

4. Segmentation techniques

Depending upon the application field, individual-processing steps may be neglected, combined, or reversed. Important to notice is that the final 3D rendering image can be obtained in two ways: either through the intermediate surface representation or through the volume representation (i.e. direct volume rendering segmentation is the process that separates an image into its important features (primitives) so that each of them can be addressed separately. This converts the planar pixel of the image into a distinguishable number of individual organs or tumour that can be clearly identified and manipulated. The segmentation process might involve complicate structures and in this case usually only an expert can perform the task of the identification manually on a slice-by-slice base. Humans can perform this task using complex analysis of shape, intensity, position, texture, and proximity to surrounding structures. All these features are differently qualified depending on the experience of the user. To generate a "complete", segmentation application numerous tools and algorithms must be combined (Kuszy et. al., 1995).

The first aim of the segmentation process in RTP is to define as accurately as possible the target that will be irradiated. This process is a manual process and usually is performed from an expert oncologist. With the terms of manual process, we mean that the physicians will exam the digital volume, slice-by-slice and using the mouse cursor will illustrate the shape of the tumour. The coordinates of the shape of the tumour are stored from the system for further processing. The target definition process is probably the most time consuming process involving an expert during the RT planning (Ketting et. al., 1997). The following describes a number of reasons proving that the automatic target volume definition is still very difficult to be performed from an expert system. The obvious reasons found in the daily clinical routine are:

1. The irregular tumour properties, like shape, texture, volume, relation with surrounding regions.
2. Location within the human body. A tumour theoretically could grow anywhere within the patient's body.
3. Tumour spreading and variation. Depending from the region the tumour is grown, disease cells might spread to the surrounding region. The disease cell distribution cannot be predicted and detected on the digital patient's data.
4. Artifacts in the acquired digital data. When treating elderly patients it is often the case to have prosthesis, usually metallic (heart irritating) like hip prosthesis. These patients cannot be examined in a MRI modality and therefore CT imaging is the only alternative for their RTP. Nevertheless, it is well known that metallic components generate severe artifacts in CT imaging that reduce image quality and blur tumour borders. Thus, the accurate manual target volume definition is even more necessary nowadays.

Classically image segmentation denotes the technique of extraction of images structures (regions or objects) so that the outlines of these structures will coincide as accurately as possible with the physical 2D object outlines. Image segmentation approaches may be performed in one of these ways: Manual segmentation methods include pixel selection, geometrical boundary selection and tracing. Given normal image resolution, selection of individual pixels is clearly impractical and rarely used.

Linear Interpolation: Linear interpolation between contours is the first approach used to provide an acceleration tool for the manual contouring. The mechanism of the linear interpolation is applied when between the key contours at least one slice exists. To perform the linear interpolation we create triangles between the contour points of the key contours as described in (Strassman, 2000). For this operation both contour's points must be rotated towards the same rotation direction. The interpolated contour points are created after calculating the intersection of each triangle side with the intermediate slice (Figure 6)

Orthogonal Contour Interpolation: The orthogonal contour interpolation serves to create a volume combining and interpolating orthogonal drawn contours. Principally the algorithm needs at least 2 orthogonal contours to work. The perpendicular plane to these two contours creates intersection points that are the key points to create the new interpolated contour. In this approach we use the cubic Spine interpolation. In case the lines are completely equal in size and their centres match or have very small distance then the result of the interpolation will be a circle. In any other case that the two vertical lines are unequal the result will be an ellipse (Figure 7).

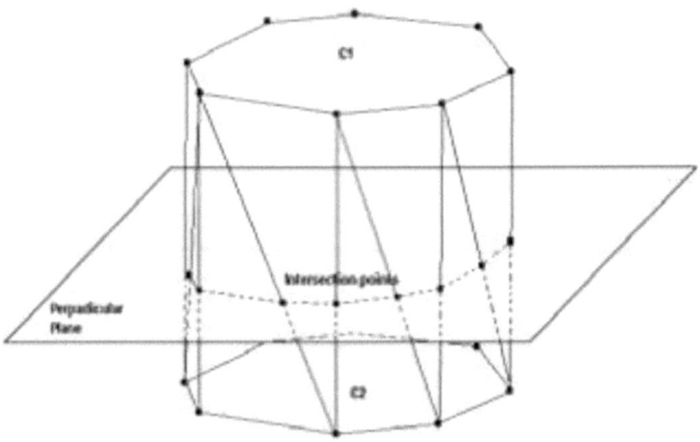

Fig. 6. On the left side a simple case where an interpolated contour is created from the plane intersection with the triangles (Zimeras and Karangelis, 2008)

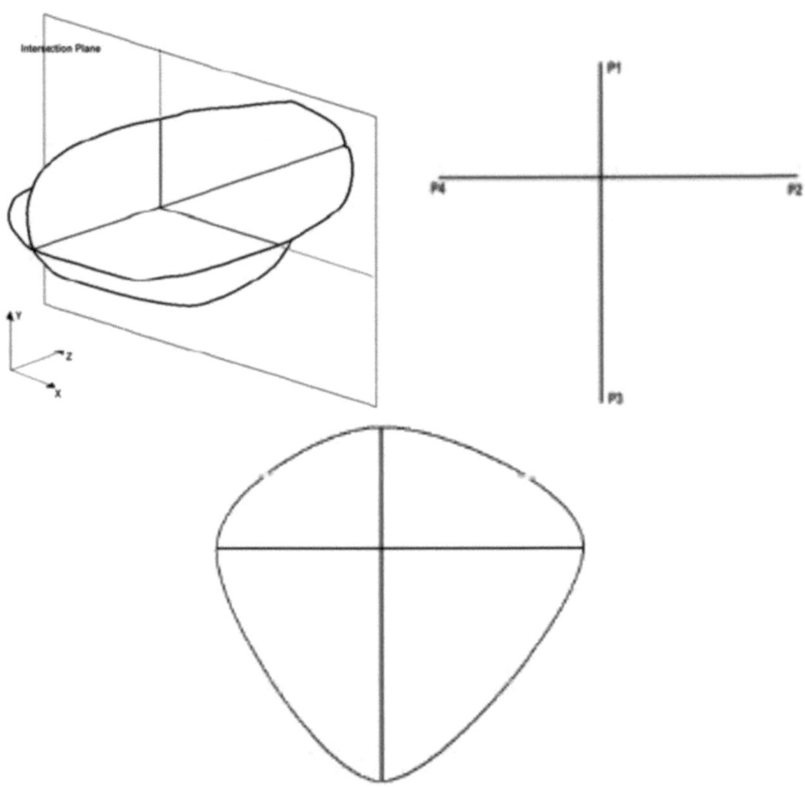

Fig. 7. The orthogonal contour interpolation. In (a), the two drawn contours and the intersection plane in Z direction. The result of the intersection is shown in (b). After applying the cubic Spline interpolation a new contour is created (c) (Zimeras and Karangelis, 2008).

Fully automatic segmentation methods are usually impractical due to image complexity and the variety of image types and interpretation. In addition, low contrast between structures generally causes many times robust automatic algorithms to fail.

Active Contour Model: Active Contour Models (ACMs) are adaptive contour representations, also known as snakes or deformable models (Terzopoulos and Fleischer, 1998). They are able to recover and represent physical contours of an image, and hence can be used as a model to determine object boundaries in static images as well as for tracking in image sequences. (Grosskopf et al., 1998; 2002) uses the Euler Time Integration to solve the optimisation problem. After initialisation by a user sketch, the contour is deformed to fit the actual object by simulating physical properties of an elastic material or fluid. This method is very reliable to overcome local minima and very fast due to its deterministic character.

So far only a very few techniques propose physicians one or more alternatives for volume definition (Ketting et. al., 1997; Belsh et. al., 1997). Recently (Pekar et. al., 2004) reported a method based on an adaptation of 3D deformable surface models to the boundaries of the anatomic structures of interest. The adaptation was based on a tradeoff between deformations of the model induced by its attraction to certain image features and the shape integrity of the model. Nevertheless, to make the concept clinically feasible, interactive tools were also introduced that allow quick correction in problematic areas in which the automated model adaptation may fail. Currently the tool used to perform the volume definition is the mouse cursor. The user defines a number of digital points on the image level, closing the first and the last point of the contour to generate this way a structure. The connectivity between the *key points* can be linear or higher order. The higher order connectivity can be achieved using interpolation models like Hermitte cubic, Spline curves, Bezier curves (Laurent, 1994; Spath, 1995; Cohen 2001), which are the most common and successfully used techniques for smoothing curves in the systems. Aim of the high order interpolation techniques is to reduce the amount of input points required to describe a smooth shape. Performing a simple comparison between linear and higher order interpolation, it can find out that the amount of points used to illustrate the shape of a structure using linear interpolation requires at least twice as many samples as by using the higher order interpolation algorithms. A common methodology used to combine high order interpolation and image edge properties is the use of active contour models. The active contour models or Snakes can be 2D image curves (Kass et. al., 1987; Blake and Isard, 1998) or 3D polygon meshes (Terzopoulos and Fleischer, 1998), which are adjusted from an initial approximation to the image or volume features by a movement caused by simulated forces. Image features provide the so-called external force. An internal tension of the curve resists against highly angled curvatures, which makes the Snakes movement robust against noise. After a starting position is given, the Snake adapts itself to shape by relaxation to the equilibrium of the external force and internal tension. Snakes have been proven efficient and fast for a number of applications in medicine involving different imaging modalities (McIne and Terzopoulos, 1996; Gross et. a., 1998; Behr et. al., 2000; Sakas et. al., 2001; Gross et. al., 2002).

The interpolation techniques described above, are usually applied only on a single slice level (2D). The use of high resolution CT data, allows the use of multiplanar reconstructions (MPR) for the sagittal and coronal direction, in relation with the patient anatomy. These two images are orthogonal to each other and perpendicular with the axial plane. The navigation though these images help in the observation of complex anatomy. The sagittal and coronal views often offer a better overview of organs 3D shape. Defining volumes in these directions could be of benefit since several organs are aligned along the longitudinal body axis. The problem we have to solve in our case is the generation of a surface and contours from structured closed parallel and non-parallel contours. The data points represent the contour points as they are generated from the user on the different levels of the axial slices or/and on the MPRs. The most common approaches used to reconstruct surfaces form parallel contours and are well established in medical imaging applications (Boissonat, 1988; Meyer et. al. 1992; Payne and Toga, 1994; Bajaj et. al. 1995; Weinstein, 2000). The limitation of these algorithms is that they cannot be applied on non-parallel contours. The problem of the non-parallel contours could be also formulated as generation of surfaces from scatter data, which are very common in industrial applications (Hoppe et. al. 1992; Ament et. al. 1998). For our application, we selected the approach presented from Turk (Turk and O'Brien 1999). Their method adapts earlier work on thin plate splines and radial basis interpolants to create a new technique in generating implicit surfaces. Their method allows direct specification of a complex surface from sparse, irregular scatter samples. The main restriction of the method is the relatively small number of sample data that can be handled. This drawback makes the above approach unsuitable for a number of applications that a large number of sampling points are needed. In this work, we demonstrate the use of implicit function interpolation to reconstruct 3D organ shapes from closed planar parallel and non-parallel contours that have been defined selectively by the user. The total number of contour points will be used as the input data to the implicit surface algorithm with arbitrary order. The number of these sampling points will not exceed the level of few hundred, and therefore the calculation times will be in acceptable ranges despite the complexity of the algorithm (Figure 8).

The output result of the reconstruction algorithm is provided in two forms: as a triangulated mesh or as multiple parallel contours extracted in arbitrary plane directions. For the RTP applications we focus mostly on the reconstruction of contours in the axial direction. The algorithm can be separated into different modules from the point editing until the reconstruction of the surface and contours as follows:

1. Collection and processing of the given contour points. This step involves the generation of the contour constraints and filtering of unwanted.
2. Calculation of the implicit functions in 3D. In this step the information produced in step one will be used to solve a linear equation system that will provide the coefficients representing our interpolation function.
3. Evaluating the implicit function over a 2D or 3D grid we extract 2D planar contours or 3D polygon meshes respectively.

Semi-automatic segmentations methods combine the benefit of bath manual and automatic segmentation techniques. By supplying initial information about the region of interest, the user may guide an otherwise automatic segmentation procedure. Any remaining errors introduced by automatic segmentation methods may be corrected by manual editing. In this work, a boundary tracking algorithm (Karangelis *et al.*, 2001, Zimeras and Karangelis, 2001, 2002) was implemented for the segmentation part.

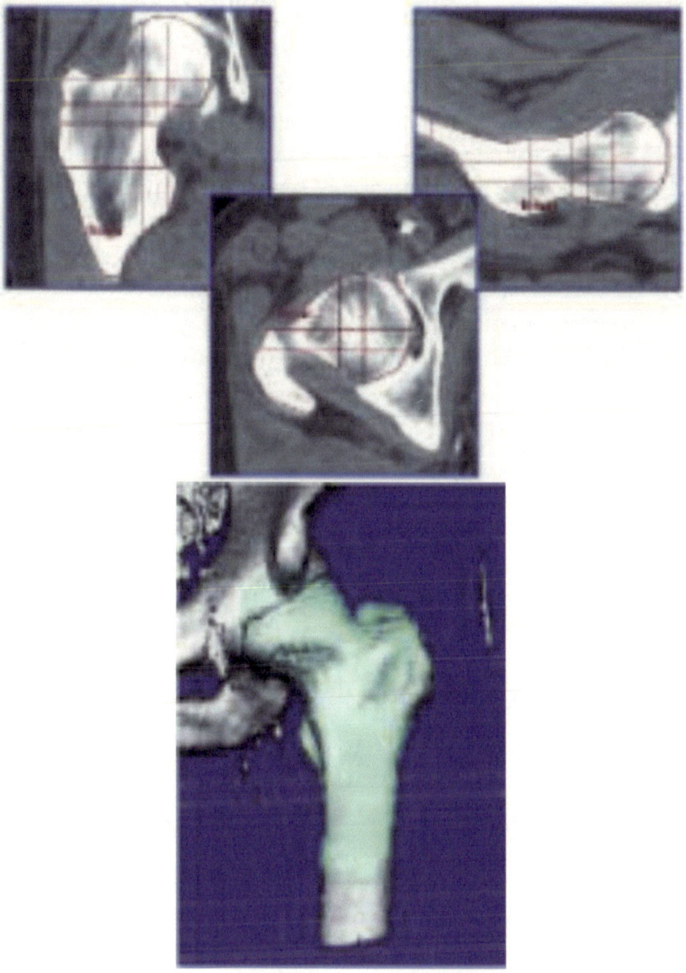

Fig. 8. Combination between Manual segmentation methods (between two slices and ACM (voxel reconstruction) on CT slices. In (a) from left to right: coronal, axial and sagittal contours. In (b) surface reconstruction (Zimeras and Karangelis, 2008).

The algorithm can automatically trace the organ through the complete volume of cross sections. False contours that are not corresponding to the spine shape and position can be rejected automatically from the system and can be replaced with linear interpolated contours considering as key contours those already found by the system. The boundary-tracking methods used, belong to the deterministic approaches and therefore there is the tendency to produce misleading results under some circumstances. To reduce that effect data pre-processing and the gradient volume of the original CT data can be used as input to the segmentation routine. Target volume and critical structure definition is a complex and time-consuming process in radiotherapy. The complexity varies for different anatomic sites. In plan evaluation, both the physicists and radiation oncologists interact closely to subjectively identify the plan most appropriate for the individual patient. In order to reduce the investment of time and effort by the radiation oncology staff, several image analysis tools are integrated. A function that significantly accelerates the contouring process is the linear interpolation between the original key-contours. The same principle can be applied for defining structures in both planar planes, sagittal and coronal. Organs with large differences in there intensities can be segmented semi-automatically. In terms of user effort the only action required from the user is the selection of an initial point from the algorithm on the original axial slices. The complete 3D geometry of the organ will be traced automatically. Some of the common organs with high sensitivity factor and vital importance are the lungs, the spinal cord and the trachea (Zimeras and Karangelis, 2001; Karangelis and Zimeras, 2002a, b; Zimeras *et al.* 2002). In addition to those organs, the external body contour can be extracted in a similar manner. The contours that are generated semi-automatic can be manipulated and modified at the same manner as those defined manually (Figure 9a).

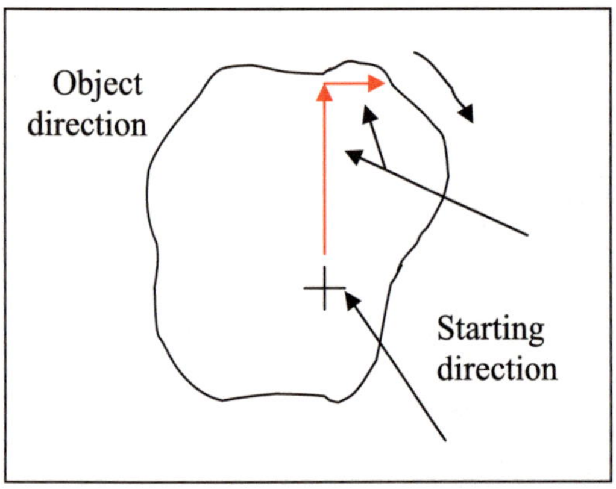

Fig. 9a. Boundary tracking method.

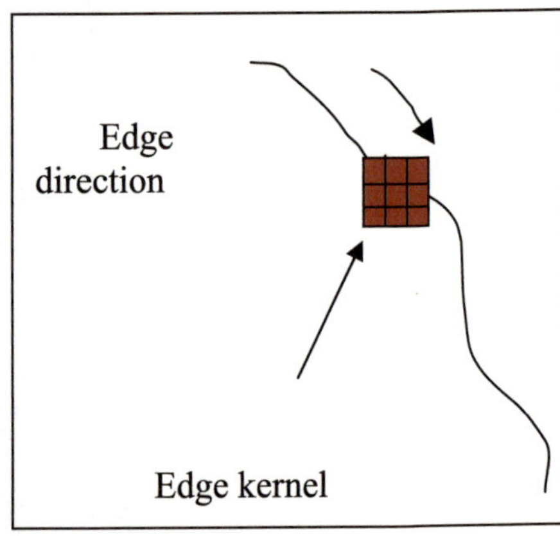

Fig. 9b. Contour filtering

After segmenting the regions of interest, second task to use a 3x3 mask edge detection operator (Figure 9b) for improving the appearance of the contours within these regions. Edge detection operators examine each pixel neighborhood and quantify the slope, and the direction of the grey-level transition. There are several ways to do this, most of which are based upon convolution with a set of directional derivative masks. For that purpose, a Sobel edge kernel was applied for finding sharp region boundaries, especially for these, which are changing greatly in intensity over short image distances. The two convolution kernels are given by:

$$S_x = \begin{bmatrix} -1 & -2 & -1 \\ 0 & 0 & 0 \\ 1 & 2 & 1 \end{bmatrix} \text{ and } S_y = \begin{bmatrix} -1 & 0 & 1 \\ -2 & 0 & 2 \\ -1 & 0 & 1 \end{bmatrix} \quad (6)$$

Each point in the image is convoluted with both kernels. One kernel (S_x) responds maximally to a generally vertical edge and the other (S_y) to a horizontal edge. The maximum value of the two convolutions is taken as the output value for that pixel.

In the analysis of the objects in images it is essential that we can distinguish between the objects of interest and "the rest." This latter group is also referred to as the background. The techniques that are used to find the objects of interest are usually referred to as segmentation techniques - segmenting the foreground from background. Image thresholding is a technique for converting a grayscale or color image to a binary image based upon a threshold value. If a pixel in the image has intensity less than the threshold value, the corresponding pixel in the resultant image is set to white. Otherwise, if the pixel is greater than or equal to the threshold intensity, the resulting pixel is set to black. For that purpose, a low pass filter was used based on the rectangular window or box function based on the rule (Figure 10):

$$I(i,j)=\begin{cases} 1,\ if\ \mathrm{I(i,j)} \geq T \Rightarrow object \\ 0,\ \text{otherwise} \Rightarrow \text{background} \end{cases}$$

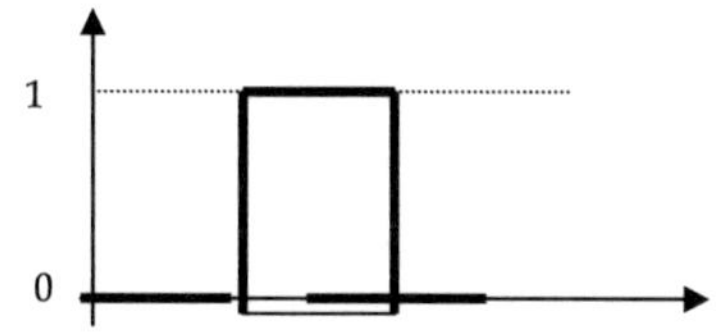

Fig. 10. Thresholding cases

where I(i,j) image matrix and T is the appropriate threshold. The output is the label "object" or "background" which, due to its dichotomous nature, can be represented as a Boolean variable "1" or "0".Summarizing the above procedures, the pseudo-code for the segmentation technique is presented below (Figure 11):

```
define the threshold labelling
          for i=1 to Whole_Voxel
          {
            find the starting point
            do
            {
step 1             if there is an object then
                   find  sharp edge using the binary image (0,1) then
                       define initial directions (UP and RIGHT)
                     calculate contour for the region of interest
                       apply contour filtering mask (Sobel filter)
                       reduce the number of points
                       calculate contour measures
                 else
                 change starting point
                 goto step 1
            }while region becomes small
```

Fig. 11. Pseudo-code for segmentation

The main drawback of the method is that is a binary approach and hence is very sensitive to gray value variations. If the threshold value is not selected properly then the system will fail to detect the appropriate organ shape. Most of the inaccuracies of the segmentation method require the user intervention to optimize the result. To overcome the limitation we calculate a secondary opacity volume from the original CT data that is very often used to visualize surfaces from scalar volume data in volume rendering.

In order to speed up the discrete contour drawing mode the user may draw a smaller number of key contours on distinct slices while the intermediate contours are interpolated linear. To perform this, a slope factor **A** is calculated between the key slices. The value of this is given from the equation:

$$A = \frac{(\text{Current\_slice\_pos - Previous\_contour\_pos})}{(\text{Previous\_contour\_pos - Next\_contour\_pos})} \tag{7}$$

Since each key contour does not have the same number of segment points the algorithm subdivides each contour into the same number of points (currently 100) in order to simplify the contour interpolation step. Then a starting point is selected -- usually this is the point of the 12o'clock position of the contour (Figure 3). To calculate the X and Y pixel (or voxel) position of the interpolated point we use the following formula:

$$\begin{aligned} X_{int} &= X_{previous\_cnt} + A^*(X_{next\_cnt} - X_{previous\_cnt}) \\ Y_{int} &= Y_{previous\_cnt} + A^*(Y_{next\_cnt} - Y_{previous\_cnt}) \end{aligned} \tag{8}$$

For visualizing volumetric medical data, sequences of 2D images are piled up to recreate the three-dimensional structure. Usually, in this step linear interpolation of adjacent slices is needed for the generation of new slices (Figure 12), since one of the problems resulting from image acquisition is the space between slices. This problem occurs because the sampling interval between slices is normally greater than the generated image resolution, and then the volume voxels are not cubic. After interpolation this size distortion is corrected, so that the visualization algorithm could generate correct proportion projections.

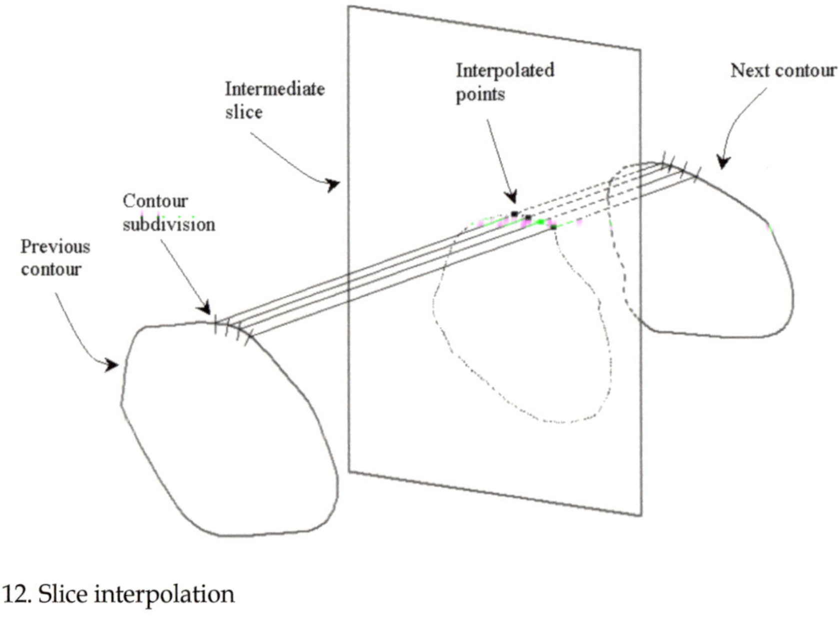

Fig. 12. Slice interpolation

Figure 13 illustrates different 2D and 3D reconstruction examples of segmented structures.

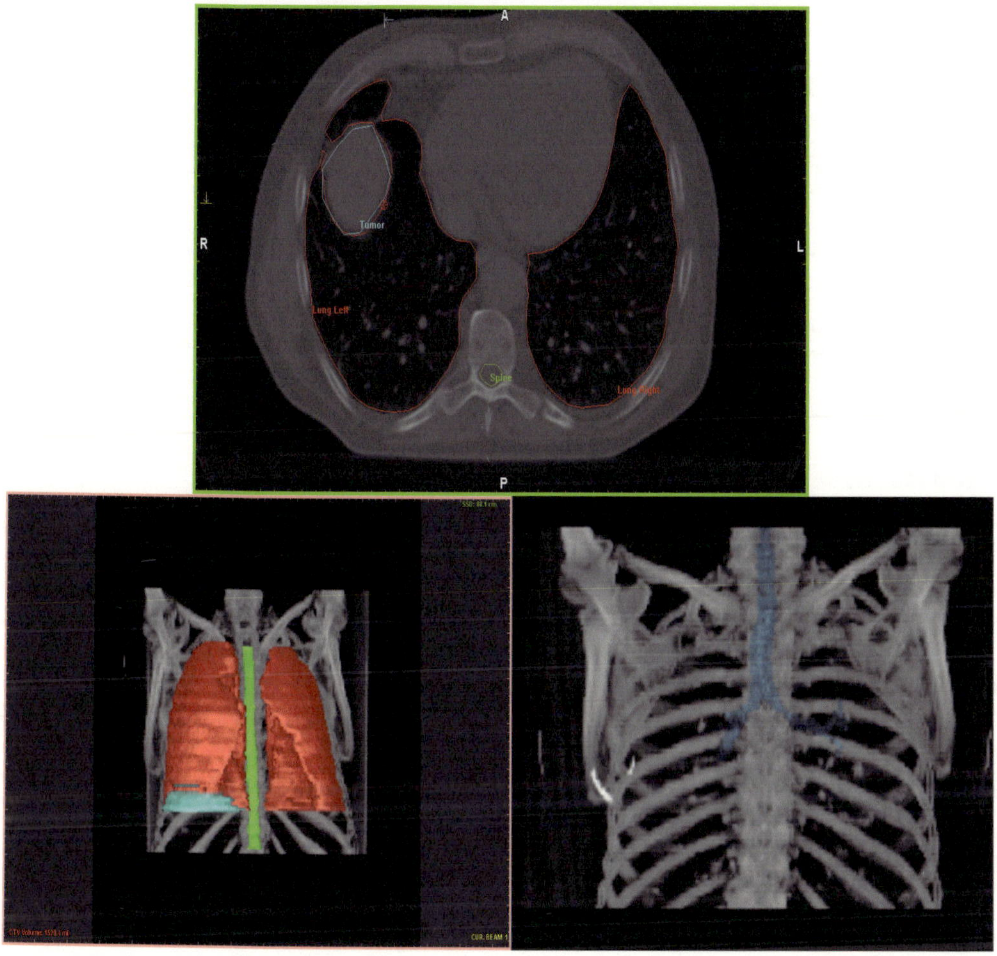

Fig. 13. Segmentation of different anatomical structures. 2D segmentation of the left and right lung; 3D segmentation of the left, right lung and spinal cord, 3D segmentation of the bronchus.

5. Conclusions

In this work, an effective semi-automatic method was presented, based on the *boundary tracking technique*, that improves the time when one or more structures are in use. The implemented algorithms can segment within a few seconds the complete volume of specific organs e.g. lungs, skin, spinal canal, bronchus and brain. The only interaction of the user is to select the starting point in the region of interest and the algorithm will track the object boundaries in 3 dimensions. For each particular organs of interests (lungs, skin, spine canal, bronchus and brain), a different segmentation techniques is proposed, but all of them are based on the *boundary tracking technique*. The 3D-shape is reconstructed out of the segmented regions, in order to illustrate the efficiency of the segmentation techniques. The benefit of the 3D illustration is the visual presentation of the organs. In many medical cases, illustration of the organ involves an anomaly, clinical problem or generally artifacts. Visual representation of the particular organ, in addition with the clinical examinations, could be a powerful tool to the doctors for diagnosis, medical treatment or surgery.

6. Acknowledgement

The author would like to thank Dr. Grigoris Karangelis and Prof Nikos Zamboglou for the useful scientific help and comments about the progress of this work. Also many thanks to Prof G. Sakas MedCom Company and Städtisches Klinikum Offenbach whom gave me equipments and medical data sets for the implementation of the above work. Finally the author would like to thank the EC and the Marie Curie Fellowship Association for this great opportunity to work with different people. This work had been supported by a Marie Curie Industry Host Fellowship Grant no: HPMI-CT-1999-00005.

7. References

A. Blake, and M. Isard; Active Contours: The Application of Techniques from Graphics, Vision, Control Theory and Statistics to Visual Tracking of Shapes in Motion. Springer-Verlag, London, 1998

B.A. Payne, and A.W. Toga; Surface Reconstruction By Multiaxial Triangulation, IEEE Computer Graphics and Applications, Vol: 14, pp:28-35, 1994

B.S. Kuszyk, D.R. Ney, and E.K. Fishman; The Current State of the Art in Three Dimensional Oncologic Imaging: An Overview, Int. J. Radiation Oncology Biol. Phys., Vol:33, pp:1029-1039, 1995

Butker, E.K.; Helton, D.J.; Keller, J.W.; Hughes, L.L.; Crenshaw, T.; Davis, L.W. A totally integrated simulation technique for three field breast treatment using a CT simulator. *Med Phys* 1996;23:1809-1814.

C. Bajaj, F. Bernardini, and G. Xu; Automatic Reconstruction of Surfaces and Scalar Field from 3D Scans, Computer Graphics, SIGGRAPH' 1995,pp: 109-118, 1995

C.H. Ketting, M.A. Seymour, I. Kalet et al.; Automated Planning Target Volume Generation: An Evaluation Pitting A Computer-Based Tool Against Human Experts, Int. J. Radiation Oncology Biol. Phys., Vol: 37, pp:697-704, 1997

Cai, W. Transfer functions in DRR volume rendering. CARS'99, Paris, France, June 1999, 23-26.

Cai, W.; Sakas, G.; Karangelis, G. Volume Interaction Techniques in the Virtual Simulation of Radiotherapy Treatment Planning. International Conference on Computer Graphics and Vision (Graphicon) Moscow, 1999.

Chen, G.T.Y.; Pelizzari, C.A.; Vijayakumar, S. Imaging: The Basis for Effective Therapy. *Front Radiat Ther Oncol* 1996;29:31-42.

Conway,. J.; Robinson, M.H. CT virtual simulation. *Brit J Radiol* 1997;70:106-118.

D. Terzopoulos, and K. Fleischer; Deformable Models, The Visual Computer, Vol: 4, pp: 306-331, 1988

D. Weinstein; Scanline Surfacing: Building Separating Surfaces from Planar Contours, In Proceeding of 11th IEEE Visualization'2000, 2000

E. Cohen, R.F. Riesenfeld, and G. Elber; Geometric Modeling with Splines: An Introduction, A.K. Peters Ltd, 2001

G. Karangelis and S. Zimeras (2002a). An accurate 3D segmentation method of the spinal canal applied on CT images *BVM 2002, Conference Proceedings, Springer-Verlag, Berlin, Germany, 366-369*

G. Karangelis and S. Zimeras (2002b). 3D segmentation method of the spinal cord applied on CT data, *Computer Graphics Topics, Vol 14, 28-29.*

G. Karangelis, S. Zimeras, E. Firle, Min Wang and G. Sakas, (2001). Volume definition tools for medical image applications, *4th MICCAI International Conference, Utrecht, Netherlands, M-A. Viergever, T. Dohi, M. Vannier (eds), Springer-Verlag, Lecture Notes in Computer Sciences 2208, 1295-1297.*

G. Sakas, G. Karangelis, and A. Pommert; Advanced Applications of Volume Visualization Methods in Medicine. In Advanced Signal Processing Handbook, Theory and Implementation for Radar, Sonar, and Medical Imaging Real-Time Systems, Edited by Stergios Stergiopoulos, CRC Press, ISBN 0-8493-3691-0, 2001

G. Sakas; Interactive Volume Rendering of Large Fields, The Visual Computer, Vol: 9, pp:425-438, 1993.

G. Turk, and J.F. O'Brien; Shape Transformation Using Variational Implicit Functions, Computer Graphics, SIGGRAPH'1999, pp. 335-342, 1999

H. Hoppe, T. DeRose, T. Duchamp, J. McDonald, and W. Stuetzle; Surface Reconstruction from Unorganised Points, Computer Graphics, SIGGRAPH' 1992, pp: 71-78, 1992

H. Spath; Two Dimensional Spline Interpolation Algorithms, A.K. Peters Ltd, 1995

J Strassman G., Kolotas C., Heyd R.: Navigation system for interstitial brachytherapy, Radiotherapy & Oncology, Vol 56, pp:49-57, 2000

J. Behr, S.M. Choi, S. Großkopf, H. Hong, S.A. Nam, Y. Peng, A. Hildebrand, M.H. Kim, and G. Sakas; Modeling, Visualization, and Interaction Techniques for Diagnosis and Treatment Planning in Cardiology, Computer & Graphics, Vol: 24, pp: 741-753, 2000

J. Boissonat; Shape Reconstruction from Planar Cross Sections, Computer Vision, Graphics and Image Processing, Vol: 44, pp:1-29, 1988

Karangelis G. 3D Simulation of external beam radiotherapy, Ph.D. Thesis, University of Darmstadt, 2004.

M. Kass, A. Witkin, and D. Terzopoulos; Snakes: Active Contour Models, IEEE First Int. Conf. Comput. Vision, pp: 259-268, 1987

M. Levoy et al.(1988). Display of surface from volume data. *IEEE CGA*, **8**.

Meyers, S. Skinner, and K. Sloan; Surfaces from contours, ACM Transactions On Graphics, Vol: 11, pp: 228-258, 1992

Michalski, J.M.; Purdy, J.A.; Harms, W.; Matthews, J.W. The CT-Simulation 3D Treatment Planning Process. *Front Radiat Ther Oncol* 1996;29:43-56.

N. Amenta, M. Bern, and M. Kamvysselis; A New Voronoi-Based Surface Reconstruction Algorithm, Computer Graphics, SIGGRAPH' 1998, pp: 415-421, 1998

N. Max; Optical Models For Direct Volume Rendering, IEEE Transactions On Visualization And Computer Graphics, Vol: 1, pp:99-108, 1995

Nagata, Y.; Nishidai, T.; Abe, M.; Takahashi, M.; Okajima, K.; Yamaoka, N.; Ishihara, H.; Kubo, Y.; Ohta, H.; Kazusa, C. CT Simulator: A new 3-D planning and simulating system for radiotherapy: Part 2. Clinical application. *Int J Rad Oncol Biol Phys* 1990;18:505-513.

Nishidai, T.; Nagata, Y.; Takahashi, M.; Abe, M.; Yamaoka, N.; Ishihara, H.; Kubo, Y.; Ohta, H.; Kazusa, C. CT Simulator: A new 3-D planning and simulating system for radiotherapy: Part 1. Description of system. *Int J Rad Oncol Biol Phys* 1990;18:499-504.

P.J. Laurent, A.Le Mehaute, and L.L. Schumaker; Curves and Surfaces in Geometric Design, A.K. Peters Ltd, 1994

Perez, C.; Purdy, J.A.; Harms, W.; Gerber, R.; Matthews, J.; Grigsby P.W.; Graham, M.L.; Emami, B.; Lee, HK.; Michalski, JF.; Baker, S. Design of a fully integrated three-dimensional computed tomography simulator and preliminary clinical evaluation. *Int J Rad Oncol Biol Phys* 1994;30:887-897.

Purdy, J.A. 3D Radiation Treatment Planning: A New Era. *Front Radiat Ther Oncol* 1996;29:1-16.

R. Belshi, D. Pontvert, J.C. Rosenwald, and G. Gaboriaud; Automatic Threedimensional Expansion of Structures Applied to Determination of the Clinical Target Volume in Conformal Radiotherapy, Int. J. Radiation Oncology Biol. Phys., Vol: 37, pp:689-696, 1997

Ragan, DP.; Forman, JD.; He, T.; Mesina, C. Clinical results of computerized tomography-based simulation with laser patient marking. *Int J Rad Oncol Biol Phys* 1996;34:691-695.

Rosenman, J. Where Will 3D Conformal Radiation Therapy Be at the End of the Decade? *Front Radiat Ther Oncol* 1996;29:264-271.

Rosenman, J.; Sailer, S.; Sherouse, G.; Chaney, EL.; Tepper, JE. Virtual simulation: Initial clinical results. *Int J Rad Oncol Biol Phys* 1991;20:843-851.

S. Grosskopf, J.L. Encarnação (Referent); G. Sakas (Referent); Realitätsnahe Modellierung und Visualisierung dynamischer medizinischer Bilddaten mittels aktiver Konturen, aktiver Regionen und deformierbarer Modelle, Darmstadt, Techn. Univ., Diss., 2002

S. Grosskopf, S.Y. Park, and M.H. Kim; Segmentation of Ultrasonic Images by Application of Active Contour Models, In Proceedings CARS 1998, Computer Assisted Radiology and Surgery, Tokyo, Japan, pp: 871-877, 1998

S. Zimeras and G. Karangelis, (2001). Semi-automatic segmentation techniques for CT medical data, *3d Caesarism Computer Aider Medicine, Bonn, Germany*

S. Zimeras, G. Karangelis (2008). Segmentation of anatomical structures using volume definition tools, *Lecture Notes in Computer Sciences 27, Computer Application in Modern Medicine, M., Springer-Verlag, 825-836*

S. Zimeras, G. Karangelis and E. Firle (2002). Object segmentation and shape reconstruction using computer-assisted segmentation tools, *Medical Imaging 2002, San-Diego, USA.* 9

Sakas, G., Grimm, M. and Savopoulos, A. Optimised maximum intensity projection (MIP), Rendering Techniques'95, Springer Verlag 1995:51-63.

Sherouse, G.; Chaney, E. The portable virtual simulator. *Int J Rad Oncol Biol Phys* 1991;21:475-482.

Sherouse, G.; Mosher, C.; Novins, K.; Rosemann, J.(1987) Chaney, E.L. Virtual simulation: concept and implementation. In: *Proceedings of 9th International Conference of the Use of Computers in Radiation Therapy (ICCR).* Scheveningen, The Netherlands: North Holland Publishing Co.; 1987:433-436.

T. McInerney, and D. Terzopoulos; Deformable Models in Medical Image Analysis: A Survey, Medical Image Analysis, Vol: 1, pp: 91-108, 1996

V. Pekar, T.R. McNutt, and M. R. Kaus; Automated Model-based Organ Delineation for Radiotherapy Planning in Prostatic Region, Int. J. Radiation Oncology Biol. Phys., Vol: 60, pp:973-980, 2004

Valicenti, R.K.; Waterman, F.M.; Corn, B.W.; Curran, W.J. A prospective randomized study addressing the need for physical simulation following virtual simulation. Int J Rad Oncol Biol Phys 1997;39:1131-1135.

4

Involved-Field Radiation Therapy (IF-RT) for Non-Small Cell Lung Cancer (NSCLC)

Tomoki Kimura
Department of Radiation Oncology, Hiroshima University, Graduated School of Biomedical Sciences, Japan

1. Introduction

1.1 Patterns of lymphatic spread of Non-Small Cell Lung Cancer (NSCLC)

The pattern and incidence of lymphatic spread of non-small cell lung cancer (NSCLC) is differentiated according to location, size and histologic type of primary tumors. As the primary tumor size increases, the incidence of lymphatic metastasis increases. Ogata et al reported the incidence of mediastinal lymphatic metastasis increased from 24% for tumors under 2 cm in size to more than 40% for tumors larger than 5 cm [1]. Hata et al analyzed 192 lymphoscintigraphies in 179 patients to determine the lymphatic drainage from each segmental bronchus into the mediastinum [2]. For the right lobes, most of the lymph flowed into the right supraclavicular nodes through the subcarina or right paratracheal nodes. There were few drainages to the left supraclavicular nodes through subcarinal nodes. In contrast, the lymphatic drainage from the left lung was more variable, and four routes were determined, as follows.

1. The route through the subaortic nodes.
2. The route runs along the left phrenic nerve through the para-aortic nodes to the left supraclavicular nodes.
3. The route runs along the left main bronchus to the left hilar or the left prevascular nodes. From the left hilar nodes, this route divides into two branches. One extends to the right supraclavicular nodes through the right upper paratracheal nodes. The other runs upwards along the left side paratracheal nodes
4. The route runs under the left main bronchus to the subcarinal nodes. After passing the subcarinal nodes, this route extends to the right supraclavicular nodes along the trachea through the pretracheal nodes and the left upper paratracheal nodes. Some branches extend upwards along the left side of the trachea to the upper paratracheal and supraclavicular nodes.

Fig. 1 shows the standard patterns of lymphatic drainage [2]. These results suggest that most of mediastinal lymph node metastasis were found ipsilaterally in the right primary lung cancer, and the mediastinal or supraclavicular lymph node metastasis were found bilaterally in the left primary lung cancer. Nohl-Oser examined the location of nodal involvement in 749 patients based on data obtained via mediastinoscopy, scalene lymph node biopsy and surgical specimen (Table 1) [3]. The right upper lobe tumors spread to the right upper and

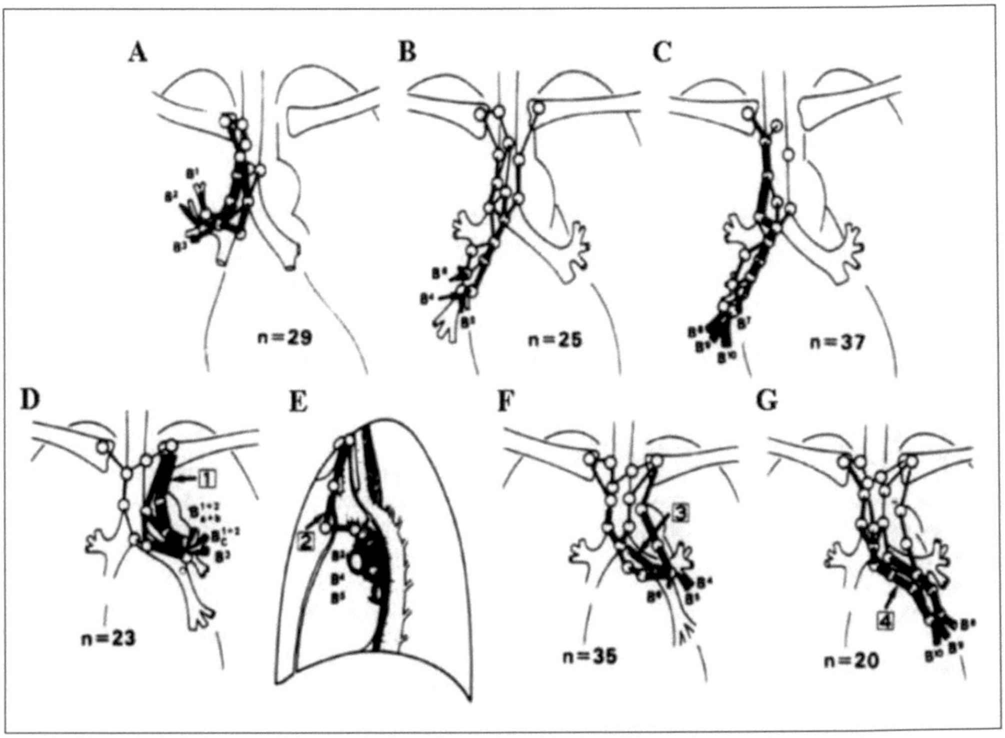

Fig. 1. A-G Standard patterns of lymphatic drainage. Each arrow shows the direction of the pathway of lymphatic drainage. The width of each the drain tubes indicates roughly the frequency of lymphatic drainage along each pathway. (From Hata E et al, References 2)

Lymph node location	Right upper (%)	Right lower (%)	Left upper (%)	Left lower (%)
Ipsilateral				
Scalene	27	10	13	5
Upper and lower tracheal	78	21	4	0
Tracheobronchial	36	9	46	15
Subcarinal	2	13	9	5
Contralateral				
Scalene	6	3	10	6
Upper and lower tracheal	1	1	10	6
Tracheobronchial	0	0	3	5
No. of patients	230	108	202	68

Table 1. Pattern of mediastinal lymph node metastases (From Nohl-Oser HC, References 3)

lower paratracheal and supraclavicular nodes and rarely to subcarinal nodes or to the contralateral nodes. In contrast, right lower lobe tumors spread to the right hilar and subcarinal nodes. Left lobe tumors might cross the midline, and the right mediastinal nodes might be invaded.

On the other hand, skip metastasis also occur more frequently in adenocarcinoma than in squamous cell carcinoma[4)]. According to several authors, among cases without hilar node dissemination, routine mediastinal dissection revealed 6% unexpected mediastinal lymph node involvement (pN2), and an average of 34% of pN2 cases have mediastinal dissemination without hilar lymph node involvement [5)].

2. The evidence of IF-RT – Is ENI needed?

In the standard radiation therapy for patients with unresectable advanced non-small-cell lung cancer (NSCLC), elective nodal irradiation (ENI) to the entire mediastinum, ipsilateral hilum and supraclavicular fossa has been deemed necessary due to anatomical lymphatic drainage and pathologic information regarding the high incidence of hilar and mediastinal node metastasis described Section 1. Recently, in order to improve the local control without increasing normal tissue toxicity, involved-field radiation therapy (IF-RT) using three or four dimensional conformal radiation therapy (3 or 4D-CRT) and intensity modulated radiation therapy (IMRT) technique for dose escalation is generally considered (Fig 2).

In IF-RT for advanced NSCLC, whether ENI is necessary or not has been controversial. The argument against the use of ENI may be summarized as follows [6, 7)]. 1) Failure is uncommon in nodal regions that are neither clinically involved nor specially targeted from many reports [8-12)]. 2) It appears that a dose greater than the conventional 60-70Gy is required to cure a larger fraction of NSCLC patients. 3) The use of ENI causes severe adverse effects, such as radiation pneumonitis and esophagitis. 4) Clear progress has been made in staging by using FDG-PET. On the other hand, the argument against the omission of ENI may be summarized as follows [13-15)]. 1) The incidence of pathologically proven nodal metastasis even in stage I NSCLC may be as high as 26% [16)], and the incidence of lymphatic invasion or metastasis rises with increasing tumor size [17)]. 2) Therefore many patients would die from distant metastasis or local failure, and ENF may not be often observed. 3) None of the studies on IF-RT provided pathologic confirmation of the status of nodal disease, nor data from autopsy findings. Thus, although there is a large discrepancy between IF-RT data and surgical data focused on ENF, actually many authors have reported retrospectively that ENF occurs in fewer than 10% of cases. In a phase I-II dose escalation study using IF-RT (RTOG 9311), the elective nodal failure rate was < 10% at last follow-up of 177 eligible patients [9)]. Senan et al reported 50 patients with unresectable stage IIIA or IIIB NSCLC were treated with sequential chemotherapy and IF-RT, and omitting elective mediastinal irradiation did not result in isolated nodal failure [11)]. Yu et al treated 80 patients 70 years or more with early stage (I / II) with IF-RT using intensity modulated radiation therapy (IMRT). Although 29 patients (36.7%) with ENF were identified, they concluded IF-RT using IMRT did not cause a significant amount of lymph node regions and improved outcomes in elderly patients [12)]. Matsuura et al reported 10 patients with locally advanced NSCLC (9 patients in stage IIIB) were treated with hypofractionated IF-RT (median dose 65 Gy / 26 fr), and no ENF was encountered with good feasibility [18)].

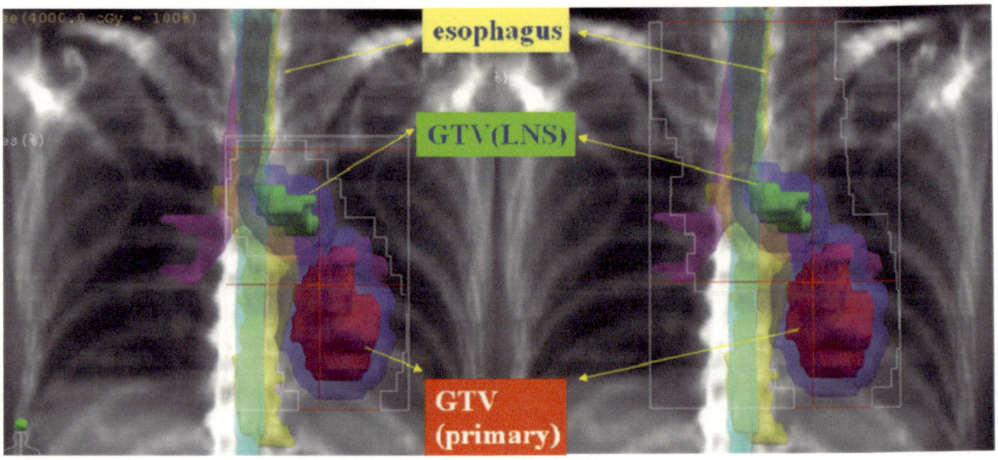

Fig. 2.

In the only prospective study comparing ENI with IF-RT, Yuan et al evaluated the effects of IF-RT in their prospective randomized trial in which 193 patients were randomly assigned to IF-RT to 68 to 74Gy or ENI to 60 to 64Gy using 3D-CRT, and reported that ENF was found in only 4% of patients in the ENI arm versus 7% in the IF-RT arm within 5 years [19]. Although the irradiation dose to elective nodal regions was higher in the ENI arm than in IF-RT arm, ENF was not significant in either arm (p=0.351). They also demonstrated an increase in local control with IF-RT of 8%, 8% and 15% at 1, 2 and 5 years, respectively. Fernandes et al reported a comparative analysis of ENI vs IF-RT. They concluded nodal failure rates in clinically uninvolved nodal stations were not increased with IF-RT when compared to ENI, and also resulted in significantly decreased esophageal toxicity, suggesting that IF-RT may allow for integration of concurrent systemic chemotherapy in greater proportion of patients with NSCLC [20]. Although we don't have the conclusions whether ENI is necessary or not, recent clinical trials of NSCLC have adopted IF-RT and IF-RT is going to be mainstream of radiation therapy for NSCLC.

3. Incidental irradiation of IF-RT

An interesting question is why the incidence of ENF is so low. It may be that incidental irradiation to clinically uninvolved nodal regions may help to explain the low incidence of ENF. Chen et al reported the results of IF-RT using 3D-CRT technique and examined incidental irradiation and ENF in thirty-five patients with inoperable early-stage NSCLC (T1-3N0M0) [21]. Although the incidental irradiation to regional nodal stations was low (fewer than 10% of all nodal regions received a dose of >40Gy), ENF was observed in only two patients who developed nodal relapse after local progression, and no patients failed initially at nodal sites. They concluded that the incidence of nodal failure was low and did not seem to be due to high-dose incidental irradiation. Rosenzweig et al reported the results of IF-RT in a large number (524) of patients with stage I-III (65% stage III) NSCLC [8]. Only 32 patients (6.2%), 42 nodal regions with ENF were identified, and among the 42 nodal regions, six regional failures (14%) were in nodal regions that had incidentally received >45Gy, which is a typical dose of ENI. Jeremic analyzing these data of Rosenzweig et al noted that only 14% of nodal failures occurred in regions receiving >45Gy, whereas nodal failures happened in 86% of patients when nodal regions received less than 45Gy ($p<0.01$) [13]. Chapet et al reported the results of IF-RT in 40 patients with stage III NSCLC, and analyzed incidental irradiation to non-involved nodal stations [22]. The doses of incidental irradiation at level 4R, 4L, 7 and 10I were relatively high. They concluded that significant incidental irradiation was observed, with this possibly helping to explain the low rate of regional recurrence observed when ENI is not applied with stage III NSCLC treated with 3D-CRT technique. Kimura et al also evaluated the incidental irradiation dose to elective nodal regions in 50 patients with locally advanced NSCLC who treated IF-RT and the pattern of ENF [23]. ENF was observed in 4 patients (8%) five nodal regions, and no mean dose to the nodal region exceeded 40 Gy. Although these reports were retrospective in nature, as the stage of NSCLC advanced, especially to stage III, we have the impression that high-dose incidental irradiation may contribute to the low incidence of ENF in the patients receiving IF-RT.

4. Treatment planning of IF-RT – Especially, impact of FDG-PET on radiation therapy volume delineation

Although there are some points about the practical treatment planning of IF-RT, the most important point is the judgment of metastatic lymph nodes on CT or FDG-PET. Therefore evidence already exists that PET-based patient selection can improve the apparent survival of patients treated with RT for NSCLC [24], and the routine omission of ENI without considering the accuracy of staging by using FDG-PET may not be advisable. Thus, FDG-PET should be recommended as a useful tool in enhancing staging accuracy and RT planning [25]. The use of FDG-PET also may contribute to the low incidence of ENF in IF-RT.

On the other hand, the SUV (standardized uptake value) cutoff value chosen have been controversial. Using SUV > 2.5 or regions of 40% maximum SUV, a lesion is usually considered malignant unless proved otherwise [24]. However, exclude use of SUV can be misleading [25]. We recommend to diagnose the positive lymph nodes by the following points consulting with the nuclear medicine physician [27]. 1) An increased uptake to a level greater than that in the mediastinal blood pool activity was considered to characterize malignancy. 2) FDG-PET image was performed at 1-hour (early) post-FDG injection and repeated 2-hours (delayed) after injection only in the thoracic area. Using dual-time point FDG-PET (combined early and delayed PET), we calculated the retention index (RI): (SUV delayed - SUV early)×100/ SUV early. RI values of more than 0% were taken to be the PET criterion for malignancy. Fig 3 shows an example.

Although there is no doubt the use of FDG-PET is effective on radiation treatment planning, we should pay attention on some pitfalls. Vanneste et al described one should be cautious to repeat the diagnostic FDG-PET scan for each patient if the time-interval between the staging FDG-PET-CT scan and the start of the irradiation is 4 weeks or more [28]. Additionally, we should reconsider CT diagnosis of metastatic lymph nodes define more than 1cm in the short axis, especially in regions where enlarged lymph nodes are rarely seen (para-aortic, retrocrural or pericardial fat).

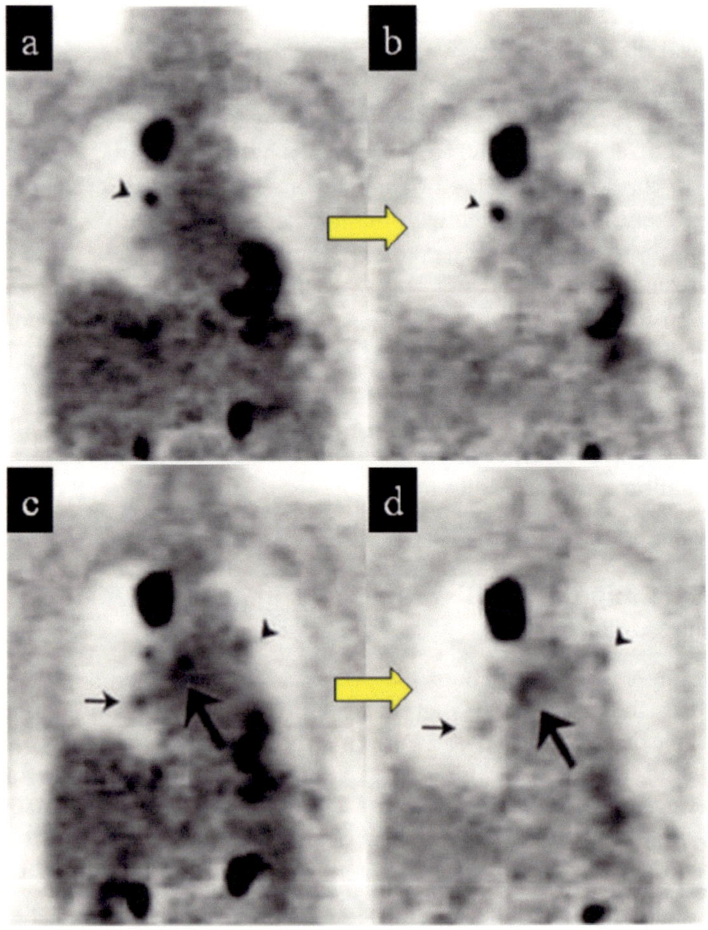

Fig. 3. The right primary lung cancer (squamous cell carcinoma) in a 79-years-old man. a) The early F-18- uorodeoxyglucose positron emission tomography (FDG-PET) image (coronal view) demonstrates focal accumulation in the right upper lobe and right tracheobronchial node (arrowhead). b) The delayed FDG-PET image (coronal view) shows intense FDG uptake at right tracheobronchial node (arrowhead). c) Another early FDG-PET image (coronal view) shows abnormal uptake at right peribronchial node (small arrow), subcarinal node (large arrow), and aortopulmonary node (arrow- head). Standardized uptake value for early images (SUVearly) was 5.86 at right tracheobronchial node, 3.37 at right peribronchial node, 3.92 at subcarinal node, and 3.41 at aortopulmonary node. d) Another delayed FDG-PET image (coronal view) shows slight uptake at right peribronchial node (small arrow) and aortopulmonary node (arrowhead) and mild uptake at subcarinal node (large arrow). Standardized uptake value for delayed images (SUVdelayed) was 7.36 at right tracheobronchial node and retention index (RI) was 25.6. SUVdelayed was 2.71 and RI -19.6 at right peribronchial node, SUVdelayed 3.56 and RI -9.2 at subcarinal node, and SUVdelayed 3.27 and RI -4.1 at aortopulmonary node. Nodal staging based on early and delayed FDG-PET demonstrated N3 stage. However, nodal staging based on combined delayed PET with RI value, demonstrated N2 stage. The surgical result also indicated N2 stage. (From Nishiyama Y et al, References 27)

5. Conclusions

However whether ENI is necessary or not has been still controversial for advanced NSCLC, "If one can't control gross disease, why enlarge the irradiated volumes to include areas that might harbor microscopic disease? [29)]" IF-RT is deemed an acceptable method for advanced NSCLC without increasing the risk of ENF or adverse effects, but further clinical trials are needed.

6. References

[1] Ogata T and Naruke T: Twenty years experience with lymph node dissection in patients with lung cancer. The mode of lymph node metastasis and the effect of dissection of nodes on the prognosis. In: Motta G (ed) Cancer of the lung and pleura: surgery considerations. Masson, Paris: 35-46, 1986.

[2] Hata E, Hayakawa K, Miyamoto H and Hayashida R. Rationale for extended lymphadenectomy for lung cancer. Theor Surg 5: 19-25, 1990.

[3] Nohl-Oser HC: Surgery of lung. In Nohl-Oser HC, Nissen R, Schreiber HW (eds). Surgery of the lung. Thieme, Stuttugart: 37-184, 1981.

[4] Libshiz HI, Mckenna RJ, and Mountain CF. Patterns of mediastinal metastasis in bronchogenic carcinoma. Chest 90: 229-232, 1986.

[5] Houtte PV, Mornex F, Rocmans P, et al: Lung cancer. In Gregoire V, Scalliet P, and Ang KK(ed): Clinical target volumes in conformal and intensity modulated radiation therapy. Springer: 91-103, 2004.

[6] Schild SE. Elective nodal irradiation (ENI) dosen't appear to provide a clear benefit for patients with unresectable non-small-cell lung cancer (NSCLC). Int J Radiat Oncol Biol Phys 2008; 72: 311-312.

[7] Williams TE, Thomas CR Jr and Turrisi III AT. Counterpoint: Better radiation treatment of non-small-cell lung cancer using new techniques without elective nodal irradiation. Seminars in Radiation Oncology 2000; 10: 315-323.

[8] Rosenzweig KE, Sura S, Jackson A, et al. Involved-Field Radiation Therapy for Inoperable Non-Small-Cell Lung Cancer. JCO 2007; 35: 5557-5561.

[9] Bradley J, Graham MV, Winter K, et al. Toxicity and outcome results of RTOG 9311: A phase I-II dose-escalation study using three-dimensional conformal radiotherapy in patients with inoperable non-small-cell lung carcinoma. Int J Radiat Oncol Biol Phys 2005; 61: 318-328.

[10] Bradley JD, Wahab S, Lockett MA, et al. Elective nodal failures are uncommon in medically inoperable patients with srage I non-small-cell lung carcinoma treated with limited radiotherapy fields. Int J Radiat Oncol Biol Phys 2003; 56: 342-347.

[11] Senan S, Burgers S, Samson MJ, et al. Can elective nodal irradiation be omitted in stage III non-small-cell lung cancer? Analysis of recurrences in a phase II study of induction chemotherapy and involved-field radiotherapy. Int J Radiat Oncol Biol Phys 2002; 54: 999-1006.

[12] Yu HM, Liu YF, Yu JM, et al. Involved-field radiotherapy is effective for patients 70 years old more with early stage non-small cell lung cancer. Radiother Oncol 2008; 87 : 29-34.

[13] Jeremic B. Low incidence of isolated nodal failures after involved-field radiation therapy for non-small-cell lung cancer: Blinded by the light? JCO 2007; 35: 5543-5545.

[14] Jeremic B. Incidental irradiation of nodal regions at risk during limited-field radiotherapy (RT) in dose-escalation studies in nonsmall cell lung cancer (NSCLC). Enough to convert no-elective into elective nodal irradiation (ENI)? Radiother Oncol 2004; 71: 123-125.

[15] Liengswangwong V and Bonner JA. Point: The potential importance of elective nodal irradiation in the treatment of non-small cell lung cancer. Seminars in Radiation Oncology 2000; 10: 308-314.

[16] Suzuki K, Nagai K, Yoshida J, et al. Clinical predictors of N2 disease in the setting of a negative computed tomographic scan in patients with lung cancer. J Thorac Cardiovasc Surg 1999; 177: 593-8.

[17] Ishida T, Yano T, Maeda K, et al. Strategy for lymphadenopathy in lung cancer 3cm or less in diameter. Ann Thorac Surg 1991; 50: 708-771.

[18] Matsuura K, Kimura T, Kashiwado K, et al. Results of a preliminary study using hypofractionated involved-field radiation therapy and concurrent carboplatin /paclitaxel in the treatment of locally advanced non-small-cell lung cancer. Int J Clin Oncol 2009; 14: 408-415.

[19] Yuan S, Sun X, Li M, et al. A randomized study of involved-field irradiation versus elective nodal irradiation in combination with concurrent chemotherapy for inoperable stage III nonsmall cell lung cancer. Am J Clin Oncol 2007; 30: 239- 244.

[20] Fernandes AT, Shen J, Finlay J, et al. Elective nodal irradiation (ENI) vs. involved field radiotherapy (IFRT) for locally advanced non-small cell lung cancer (NSVLC): A comparative analysis of toxicities and clinical outcomes. Radiother Oncol 2010; 95: 178-184.

[21] Chen M, Hayman JA, and Ten Haken RK, et al. Long-term results of high dose conformal radiotherapy for patients with medically inoperable T1-3N0 non-small-cell lung cancer. Is low incidence of regional failure due to incidental nodal irradiation? Int J Radiat Oncol Biol Phys 2006; 64: 120-126.

[22] Chapet O, Ten Haken RK, Quint L, et al. Incidental irradiation to non-involved nodal stations in patients with stage III non-small-cell lung cancer treated with 3-D conformal radiation therapy. Int J Radiat Oncol Biol Phys 2006; 66 (Suppl): S474-475.

[23] Kimura T, Togami T, Nishiyama Y, et al. The impact of Incidental Irradiation on Clinically Uninvolved Nodal Regions in Patients with Advanced Non-Small Cell Lung Cancer (NSCLC) Treated with Involved-Field Radiation Therapy (IF-RT) -Does Incidental Irradiation Contribute to the Low Incidence of Elective Nodal Failure (ENF)?- Int.J. Radiat. Oncol. Biol. Phys: 77； 337-343, 2010.

[24] Mac Manus M and Hicks RJ. The use of positron emission tomography (PET) in the staging/evaluation, treatment, and follow-up of patients with lung cancer: A critical review. Int J Radiat Oncol Biol Phys 2008; 72: 1289-1306.

[25] Hong R, Halama J, Bova D, et al. Correlation of PET standard uptake value and CT window-level thresholds for target delineation in CT-based radiation treatment planning. Int J Radiat Oncol Biol Phys 2007; 67: 720-726.

[26] Belderbos JSA, Kepka L, Kong FM, et al. Report from the international atomic agency (IAEA) consultants'meeting on elective nodal irradiation in lung cancer: Non-small-cell lung cancer (NSCLC). Int J Radiat Oncol Biol Phys 2008; 72: 335-342.

[27] Nishiyama Y, Yamamoto Y, Kimura N, et al. Dual-time-point FDG-PET for evaluation of lymph node metastasis in patients with non-small lung cancer. Ann Nucl Med 2008; 22: 245-250.

[28] Vanneste BGL, Haas RLM, Bard MPL, at al. Involved field radiotherapy for locally advanced non-small-cell lung cancer: isolated mediastinal node relapse. Lung Cancer 2010; 70: 218-220.

[29] Bradley J, Govindan R and Komaki R: Lung. In Principles and Practice of Radiation Oncology (4th edition), Perez CA, Halperin EC, Brady LW, Schmid-Ullrich RK (eds). Lippincott-Raven, Philadelphia-New York: 1201-1243, 1998.

Part 2

Particle Therapy

5

Scanned Ion Beam Therapy of Moving Targets with Beam Tracking

Nami Saito and Christoph Bert
GSI Helmholtz Centre for Heavy Ion Research, Germany

1. Introduction

Ion beam therapy has been offering beneficial dose conformity based on the fact that ions deposit large dose sharply at depth and significantly less dose at the entrance or behind the peak, known as the Bragg curve. In the decades, delivery systems of ion beams, especially proton and carbon ions, have been developed and used in number of clinics, and they have demonstrated excellent dose conformity on static tumours. The conventional ion beam delivery utilizes broad ion beams with a collimator for a beam shaping laterally and a patient specific compensator to define the beam depth. Therefore the conventional broad beam delivery is in principle not able to adapt beams to the target if the target is moving internally and changing its radiological depth. On the other hand, the scanned beam delivery system (Haberer et al. 1993;Pedroni et al. 1995) uses narrow ion beams to scan the target volume by controlling scanning magnets without needs of any beam shaping collimator. The scanned ion beam delivery system utilizes no patient specific compensator materials instead the system changes beam energy to control the beam depth, therefore it has an ability to modulate dose peak location flexibly on demand in three-dimensions (3D). Flexibility of such an active beam delivery system allows us to implement an advanced irradiation control system so called beam tracking that tracks moving targets continuously with ion beams not only in lateral directions but also longitudinally. The longitudinal adaptation is a particular objective for ion beam tracking different from the beam tracking in radiotherapy with photon where only lateral direction are adapted with multi-leaf collimators (Huang et al. 2008;Mao et al. 2008;Murphy 2004;Sawant et al. 2008).

In this chapter we describe an ion beam tracking system that is implemented at GSI Helmholtz centre for heavy ion research (GSI) in Germany based on the active scanned ion beam delivery system, so called the rasterscan (Haberer, Becher, Schardt, & Kraft 1993). At GSI, the rasterscan has been applied to treat static tumours in the head, neck and pelvis since 1997 (Kraft 2000;Schulz-Ertner et al. 2007). With the rasterscan technique, the target is actively scanned with precisely controlled narrow ion beams with horizontal- and vertical-scanning magnets. The target volume is split into virtual slices based on the beam path length calculated from the patient's computed tomography (CT) data. For each virtual slice, the beam with the corresponding energy to reach the slice is delivered from the synchrotron accelerator (typical pulse period ~5 s) with pre-selected intensity and beam size. The Bragg peak of the beam is widened to approximately 3 mm with a ripple filter (Weber and Kraft

1999), and the individual Bragg curves for all slices create a so-called extended Bragg peak with homogeneous depth dose distribution over the target. Lateral target coverage is achieved by deflecting the beam with dipole scanning magnets to several hundred raster positions per iso-energy slice. At each scan position, an optimized number of particles (Krämer et al. 2000;Krämer and Scholz 2000) is deposited in typically 10 ms, then the pencil beam is deflected to the next position. The beam position and the number of particles are measured continuously and recorded for each raster point. An interlock system interrupts treatments in the case of deviations between nominal and measured beam parameters or other unexpected events (Badura et al. 2000).

For intra-fractionally moving targets conventional treatments with scanned ion beams can cause unacceptable local under- or over-dosage due to interplay effects of sequential dose delivery and target motion (Bert et al. 2008;Phillips et al. 1992). Interplay effects require modifications of the conventional scanned beam delivery technique for moving targets. Various irradiation techniques, rescanning (Phillips, Pedroni, Blattmann, Boehringer, Coray, & Scheib 1992), beam gating (Bert et al. 2009;Lu et al. 2007;Minohara et al. 2000) and beam tracking (Bert et al. 2007;Grözinger et al. 2008;Keall et al. 2001) have been reported to reduce or even almost eliminate the detrimental dosimetric effects caused by interplay. Simulation studies (Bert 2006;Grözinger et al. 2006;Li et al. 2004) for the irradiation of moving targets with scanned ion beams showed that beam tracking has the potential to fully transfer the advantages of scanned ion beam particle therapy, i.e. 3D target conformation, to irradiation of moving targets such as thoracic tumours. In this chapter a treatment of moving targets is conceptually considered as a four-dimensional (4D) treatment that adds time information to 3D space information. Recent progress of time resolved CT (4DCT) plays important role in treatment planning of moving targets. Usage of the 4DCT is described in this chapter as 4D treatment planning (4DTP).

The principle of beam tracking is to adapt beams to a moving target continuously. Respiration induced anatomical changes cause not only displacement of the target position but also change the tissue configuration and thus the densities along the beam paths (Langen and Jones 2001;Mori et al. 2007).Therefore, for accurate beam tracking apart from lateral directions also particle ranges have to be adapted according to the actual anatomy. In the lateral direction, the amount of required adaptation is rather similar from raster point to raster point because scanned beam application is fast in comparison to respiratory motion. In the longitudinal direction, the amount of range adaptation depends on anatomical changes along the beam path and may differ even for adjacent raster points. Thus adaptation of particle ranges should, thus, ideally be applied within the time comparable to typical irradiation times of a raster point (approximately 10 ms).

A beam tracking system was designed and implemented on a horizontal fixed beam at GSI to compensate target motion by adapting ion beams laterally with the scanning magnets and longitudinally with a range shifter system for each beam port. In this chapter at the section 2 we describe the beam tracking system and the pre-process to perform tracking. Technical detail concerning the speed and accuracy of the beam tracking system can be found in the report (Saito et al. 2009b). In the section 2.1, as a pre-process of beam tracking, 4D treatment planning is described. Details of the 4D treatment planning method are described in (Bert and Rietzel 2007).

Beam tracking is a real-time system that responds to the target motion. Therefore a dedicated verification of the real-time process is required. With a simple motion phantom dosimetrical verification measurements have been performed. Beam tracking system delivered dose successfully to the moving detectors, radiographic films, ionization chambers (Bert, Saito, Schmidt, Chaudhri, Schardt, & Rietzel 2007), range telescopes (Sihver et al. 1998) and biological targets (Gemmel et al. 2008;Schmidt et al. 2008). Even though the dose delivery of beam tracking system has been confirmed with a motion phantom, dose delivery to target volume in a patient has to be ensured as a quality assurance (QA) process. In the section 3.1 we describe a proposed QA method dedicated for beam tracking. The QA process plays an important role to find possible errors on patient specific treatment plan under the beam tracking condition. The QA process supports to verify the beam tracking functionality and the created treatment plan for the patient. In the section 3.2 an online monitoring method to verify the delivered dose is presented. The online monitoring utilizes positron-emission-tomography (PET) as 4D in-beam PET. Initial tests of the 4D in-beam PET have been performed with retrospective analysis (Parodi et al. 2009).

The beam tracking system has been implemented so far only at GSI as an experimental phase. The technology of the beam tracking system is expected to be transferred to Heidelberg Ion Beam Therapy Centre in Germany after necessary clinical verification processes.

2. Beam tracking

The beam tracking system consists of sub-processes: 4D treatment planning (4DTP), motion detection, beam tracking parameter decision, lateral and longitudinal beam tracking. A schematic drawing of tracking components and the data flow are shown in figure 1. In advance of irradiation a reference treatment plan and sets of displacement parameters are calculated as 4DTP. During beam tracking, motion detection is continuously performed, and motion information is sent to the therapy control system (TCS). Based on the motion information beam is shifted laterally and longitudinally to track the moving targets with ion beams. In the following sections, current implementation of the sub-processes is described.

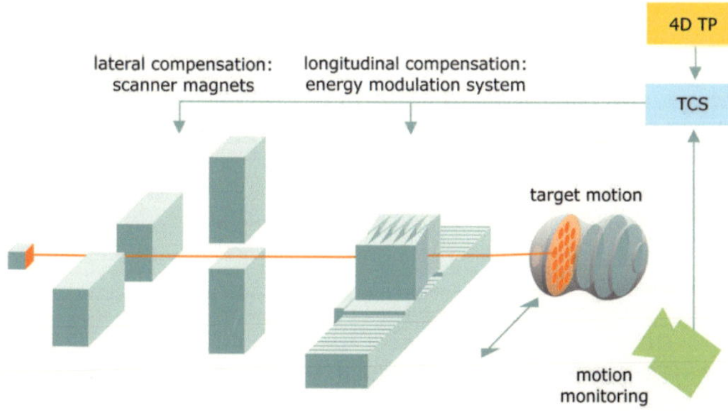

Fig. 1. Schematic drawing of the beam tracking system. Target displacement is calculated in 4D treatment planning (4DTP) and stored in the therapy control system (TCS) in advance. During a scanned beam irradiation, target motion is detected by a motion monitoring device and TCS controls scanning magnets and an energy modulation device to track the targeting spot with the ion beams continuously.

2.1 4D Treatment planning

In order to track moving targets with ion beams, information of target displacement is needed to be stored at the tracking system. In the current implementation of beam tracking, 4DCT data is used to sample internal target motion as a part of procedure in 4DTP. Basic concept and method of 4DTP for beam tracking can be found in the article (Bert & Rietzel 2007). In this method, a reference treatment plan is created as a static plan on a reference CT that is typically the 4DCT motion state at the end-exhale. The reference plan is basically a list of scanning spot information with its location in two-dimensional raster-grid coordinate (x, y), beam energy E, beam size F, beam intensity I and number of particles N. Based on the reference plan and image registration (Hartkens 1993;Shackleford et al. 2010) of the 4DCT data set, displacement of the scanning spots are calculated for each motion phase. The beam displacement along the beam path is calculated as length in water-equivalence (WE) using the 4DCT data to specify corresponding range change in WE to beam tracking system. The obtained displacement of all scanning spots and all motion phases are stored as a look-up table (LUT) in TCS. At GSI a treatment planning software TRiP98 has been upgraded to perform treatment planning in 4D (4DTRiP). The software 4DTRiP can load patient motion (either simulated proposed by Lujan (Lujan et al. 1999) or measured with motion sensors), scanned beam progress file of GSI and the Heidelberg Ion Beam Therapy Centre (HIT) for simulation of dose calculation. Detail of the recent version of the 4DTRiP software is reported in (Richter et al. 2010).

2.2 Motion detection

In the current implementation, an industrial laser distance sensor (SICK Vertriebs-GmbH, Germany) that reliably detects motion of a target phantom is used for system validation experiments. The sensor uses a triangulation method to measure the distance to the object with an accuracy of 150 µm and frequency of about 1 kHz. The relative distance is encoded continuously into an analogue voltage that is transferred to the TCS via an analogue-to-digital converter (ADC) (Janz Computer AG, Germany) installed in the TCS. The current detector was chosen because of its precision, high temporal resolution and interface with respect to phantom studies to benchmark our beam tracking system. Investigation of motion detection (e.g. detection with cameras, fluoroscopy and etc.) is currently in progress. For clinical use, however, the selection of detection system is rather defined by clinically approved devices at a clinical site. An appropriate motion detector, most likely comparable or identical to the choice made for 4DCT acquisition, is expected to be used for clinical application of beam tracking.

2.3 Therapy control system

A VME CPU board with in-house developed software is currently used as TCS. Based on the motion signal, the current motion phase is determined, and beam displacements are selected from the LUT for the corresponding raster points. Due to communication delays and response times of the beam tracking system, a prediction of target position and irradiation spots are performed at TCS. A future target position can be predicted from the current target motion with a time offset individually for lateral and longitudinal beam tracking. Since the time delays of the process are very short in milliseconds, linear extrapolation is applied for the motion prediction. For the prediction of the future irradiation point, information of the irradiation sequence is needed to be stored at TCS. Irradiation length of all individual scanning spots can be measured in advance and used for the prediction. It should be noted that the scanning spot prediction plays important role for the case of faster scanned beam delivery than the time delay of a beam tracking system. The predicted tracking parameters are updated with a typical frequency of 2 kHz and sent to lateral and longitudinal tracking sub-processes. The sub-processes react on the parameters and apply tracking just in time for the predicted motion and scan.

he tracking control system can load various experimental versions of tracking algorithms. An algorithm, for example, can overcome the discreteness of the motion phases by interpolating two sets of tracking parameters of the corresponding two discrete motion phases. In case of rotational motion a sophisticated algorithm that takes into account the changes of beam path due to motion can be applied. Figure 2 illustrates significant dose change of the proximal side of Bragg curve on different target motion states due to rotation of a target. With the algorithm for rotational motion TCS calculates the dose change occurred on the proximal side online and compensates the dose in the following scanning spots during irradiation. Detail of the implementation and experimental results of beam tracking with real-time dose compensation is described in (Lüchtenborg et al. 2011).

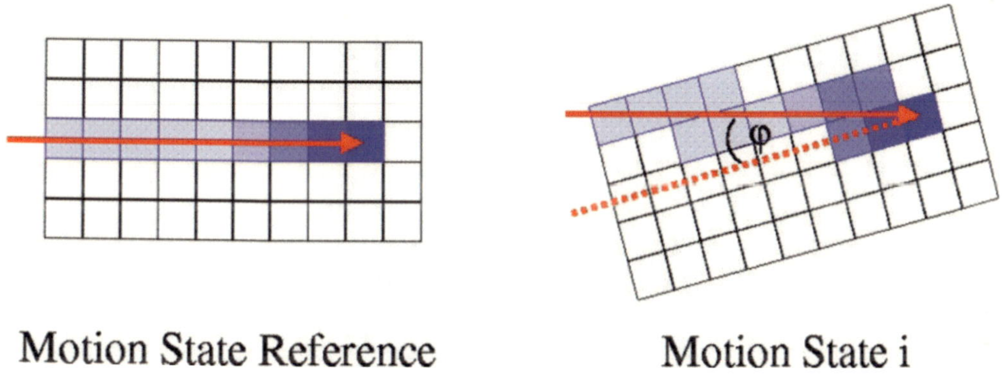

Fig. 2. Illustration of dose difference at the proximal tail of Bragg curve for different motion states with a target rotation. Dose is indicated in blue on target materials shown as pixels for a reference motion state (left) and other motion state i (right) with a rotation angle of φ. The beam tracking system is able to compensate not only the Bragg peak position (dark blue) but also the dose change occurred on the tail (light blue) due to the motion in real-time, see text in section 2.3.

2.4 Lateral tracking

The determined lateral tracking parameters at TCS are transferred to a custom made scanning magnet control unit that sets current on the scanning magnets. The scanning magnet control unit applies displacements, dx and dy, to the original raster coordinates. This process was initially implemented as a fast-feedback routine that fine-adjusts the beam position base on real-time beam position measurement (Haberer, Becher, Schardt, & Kraft 1993). For the beam tracking, the fast-feedback routine was extended to accept the displacement parameters dx and dy as position correction resulting in a real-time lateral beam adaptation to the moving target position. The fast-feedback routine typically runs with frequency of about 7 kHz. In the current implementation, the communication speed and the stability of the lateral beam tracking process have been significantly improved in comparison to the initial implementation described in (Grözinger, Bert, Haberer, Kraft, & Rietzel 2008).

2.5 Longitudinal tracking

In the current version of the beam tracking system, longitudinal tracking is performed with a range shifter that consists of two sets of polymethyl methacrylate (PMMA) wedges mounted on fast linear motors (PASIM Direktantriebe GmbH, Germany). The range shifter is installed 22 cm upstream from the iso-centre in the treatment room at GSI. Principle of the double wedge range shifter is illustrated in figure 3. Coupling of the two sets of wedges into a double wedge system results in a scan field with homogeneous absorber thickness. The range of traversing ions is modulated by moving the wedges apart or together. When the two wedges are moved apart (together), the amount of wedge material traversed by the ion beam decreases (increases) and leads to an increased (decreased) particle range in the target.

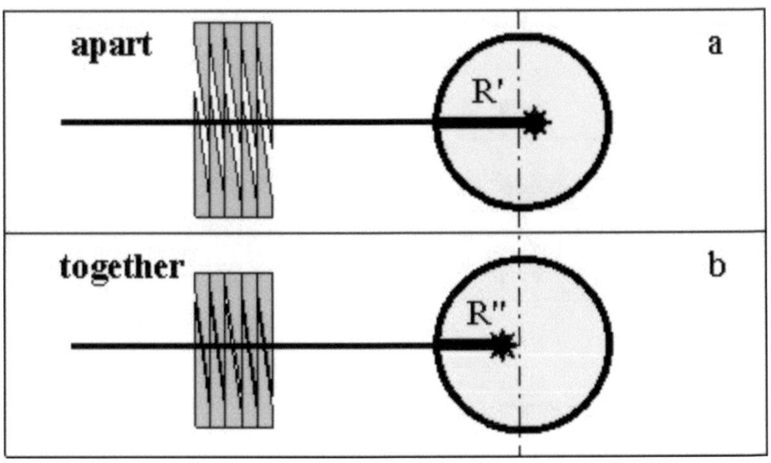

Fig. 3. Basic principle of the double wedge range shifter. Two double wedge sets are coupled to create homogeneous thickness on a scanning field. By moving the two sets apart (a) (or together (b)) ion beams penetrates more (or less) target materials.

The wedge positions are controlled by the TCS via the digital servo drives with fine-adjusted position feedback from linear encoders. With the current implementation, resolution of the sent range shift is defined by the planned maximum range shift and the resolution of the digital signal sent from TCS. For instance, a planned maximum range shift of ±10 mm in water equivalence (WE) leads to a resolution of 0.16 mm of WE for a typical configuration of the range shifter. Accuracy and its speed of the range shift depend on the specified range shift (approximately 1 mm WE deviation for 5 mm WE range shift in 10 ms), and it shows better accuracy for a slower scan speed (Saito, Bert, Chaudhri, Gemmel, Schardt, & Rietzel 2009b). The accuracy of the range adaptation is comparable to the uncertainties of CT data conversion to ion ranges (Jäkel et al. 2001;Matsufuji et al. 1998;Rietzel et al. 2007). A photo of the current implementation of the range shifter is shown in figure 4 together with motion detection device and a motion phantom with a robotic arm.

An alternative method of longitudinal beam tracking without using the double wedge system has been proposed by Chaudri et al. (Chaudhri et al. 2010). The alternative method employs a dipole magnet to shift ion beams on a static single wedge to select the appropriate thickness by deflecting the beam. Deflection of ion beams with ion beam transport magnets is much faster than the modulation time needed for the double wedge.

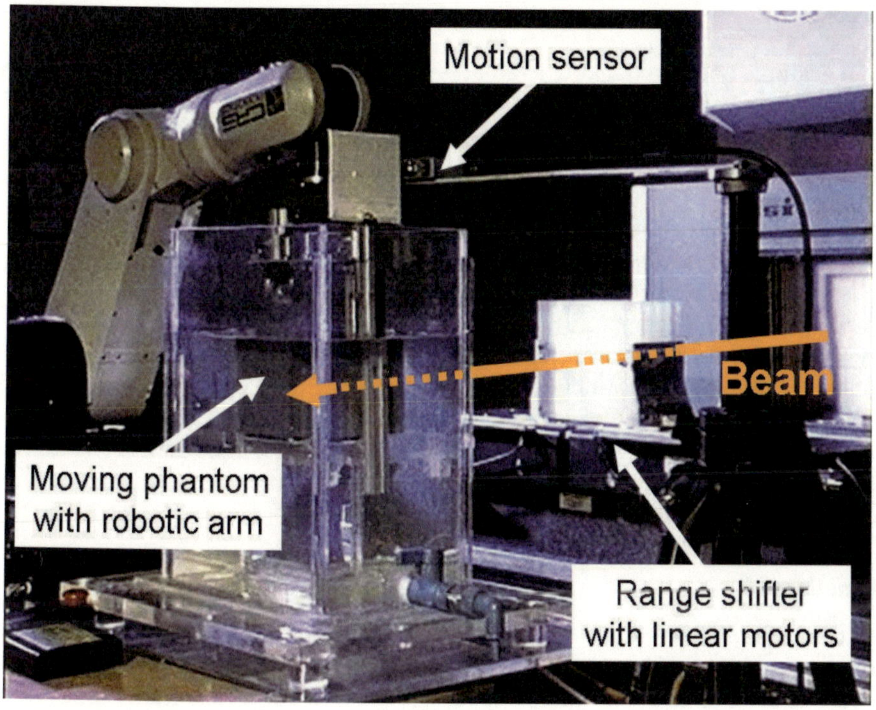

Fig. 4. A photograph of an experimental beam tracking set-up showing the range shifter, a target phantom and a motion sensor. The target phantom is moved in a water filled tank with a robotic arm in this set-up. Beam direction is indicated by a thick arrow.

Therefore the alternative solution with ion beam deflection is expected to adapt ion beam depth fast enough, so that the prediction to account the delay of the beam tracking system is not required. Investigation of the alternative system, however, requires modification of the existing beam transport line or new device to control magnet faster than the normal operation of beam transport.

3. Tracking verification

Beam tracking functionality has been experimentally tested by adapting ion beams on moving targets. In these experiments radiographic X-ray films for lateral tracking and a range telescope for longitudinal tracking were used as detectors (Bert et al. 2010;Bert, Saito, Schmidt, Chaudhri, Schardt, & Rietzel 2007). Dose quality on moving targets has also been assessed with ionization chambers, and delivered dose homogeneity has been assessed with radiographic films (Bert et al 2010). Figure 5 shows an example of beam tracking irradiation on a moving target.

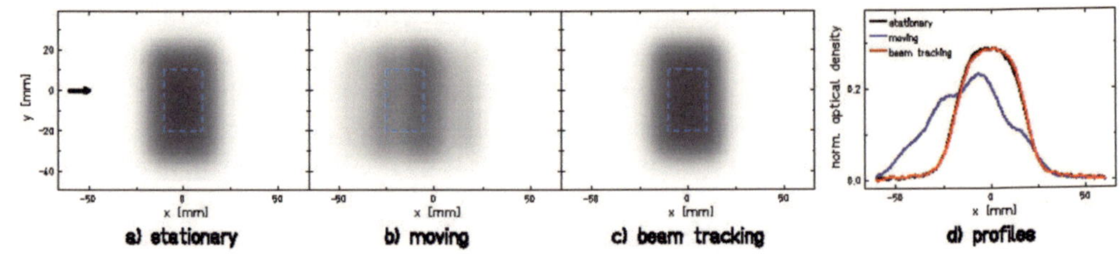

Fig. 5. An example of beam tracking irradiation on moving target. Radiographic film was either stationary (a) or in motion (b, c) irradiated without (a, b) or with (c) beam tracking. The profiles of film blackening at y=0 (indicated with an arrow (a)) are also shown (d). The profiles of the stationary case and the beam tracking case are closely overlapped.

Beam adaptation on the moving phantom has been assessed on simple phantoms with mostly simple regular motion patterns. At GSI a more complex geometry similar to a patient thorax is currently under development (Steidl et al. 2011). However, more detailed assessment for a patient geometry has to be assured for clinical usage. In the section 3.1 a method of clinical quality assurance of beam tracking is proposed. It can be used as a plan verification method of moving target treatment for individual patient's plan in clinical use, or as a daily QA process with a defined reference plan to assure tracking functionality. In the section 3.2, an online verification method of beam tracking dose delivery is described as 4D in-beam PET. The 4D-in beam PET provides the opportunity to find unexpected error on tracked dose delivery during treatment.

3.1 QA method

Beam tracking irradiation can be assured by performing a QA process with array detectors. Since the tracking parameters are closely related to the outcome of 4DTP, verification of treatment planning is also considered as a QA process for beam tracking. Section 3.1.1 describes the method of treatment plan verification for moving targets as a part of 4DTP. Section 3.1.2 reports on a measurement process to verify irradiation of beam tracking on the created treatment plan.

3.1.1 Tracking parameter tests in 4DTP

A 4D treatment plan produced as it is described in the section 2.1 can be examined in terms of dose quality in simulations. Since the beam tracking irradiation is a real-time dose delivery responding to the target motion, an assumed combination of target motion and beam delivery progress is not enough to judge the 4DTP. Therefore a number of simulations are required to test robustness of the treatment plan. In order to modulate relative timing of the target motion and beam delivery progress, various combinations of motion periods, initial motion phases, and potentially beam delivery progress are selected. Analyses of the individual dose distributions allow us to estimate the quality of delivered dose including the worst case.

For the calculation of a dose distribution in the presence of target motion, the 4DTP (reference plan and LUT), 4DCT, a simulated motion trace and the expected time cycle of beam delivery from the accelerator (beam delivery progress) are given to the 4D treatment planning software. Based on the given information, scanning spot and motion phase can be determined, and corresponding 4DCT data for the motion phase is used for dose calculation with the corresponding beam displacements for the spot that are determined from the LUT. The obtained dose distribution for each phase can be merged to the reference phase by the transformation vector maps from the image registration.

A dose-volume histogram (DVH) can be calculated by applying the reference CTV on the merged dose distribution that includes the contributions from other motion phases of the simulated motion sequence. In figure 6, an example of calculated DVHs is shown. Twelve combinations of initial motion phases and motion periods were performed in the calculation shown in the figure. DVH curves of three scenarios, no compensation, beam tracking, beam tracking including online dose compensation (see sec. 2.3) are shown. The DVHs of the no compensation case were calculated with the reference plan that was created for the CT data of the reference motion phase. The scenario of no compensation basically shows effect of the target motion. The scenarios with beam tracking apply tracking parameters based on the motion state. Depending on the level of improvement on the dose conformity in comparison to the no compensation case one can decide the beam tracking strategy, i.e. with or without dose-compensation. At this stage, one can also study the influence of transformation maps to the dose distribution as a robustness test as well as optimal selection of beam tracking parameters.

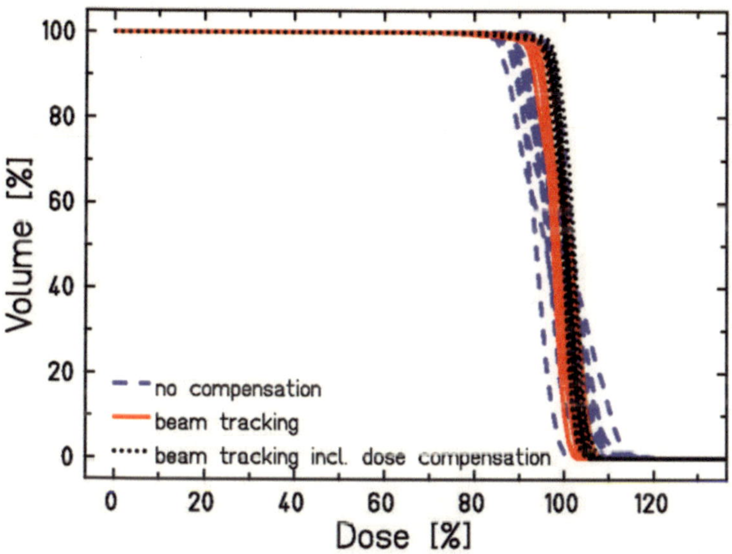

Fig. 6. DVH curves of no compensation cases (dashed blue lines), beam tracking cases (solid red lines) and beam tracking including dose-compensation, see texts in section 2.3, (dotted black lines) as an example of beam tracking parameter test in 4DTP of a lung tumour patient. In this example, a sinus motion trajectory with 4 initial motion phases and 3 different motion periods were applied for each scenario.

At GSI, a visualization tool to show deviation of the dose distribution from the prescribed dose is currently in progress (Hild et al. 2011). If the created 4DTP has insufficient dose quality due to the motion, some advanced optimization of 4DTP can be performed. The 4DTP optimization of beam tracking is under investigation. In the study of 4DTP optimization for beam tracking, various uncertainties in terms of position, time and number of particles are expected to be included.

3.1.2 QA measurements

After the verification of beam tracking parameters, the next step is to perform dose verification measurements by applying the 4D treatment plan. A dose measurement as QA for a beam tracking irradiation is performed with a virtual patient motion signal, that can be the one recorded during 4DCT acquisition or a modulated one to study the influence of e.g. different breathing parameters since the exact patient motion cannot be expected or obtained in advance of treatment. The virtual motion signal (e.g. the motion pattern investigated by Lujan et al (Lujan, Larsen, Balter, & Haken 1999)) can be produced by a motion phantom. The motion signal, dX, is sent via a motion sensor to the TCS to apply corresponding beam displacements dx, dy and dz selected from the LUT as it is performed in the corresponding patient case. During the irradiation the actual beam information (lateral position x, y, number of particles N and longitudinal displacement dz) and the motion signal dX are recorded in a time resolved manner (Saito, Bert, Chaudhri, Gemmel, Schardt, & Rietzel 2009b). As quality assurance phantom, a static water tank equipped with an array of 24 small ionization chambers (ICs) can be used as it had been performed for plan verification in therapy of static tumours at GSI (Karger et al. 1999). In the static water phantom, the dose distribution of the beam tracking case is different from that for a static target case due to the position adaptation to the virtual target, as it is illustrated in figure 7.

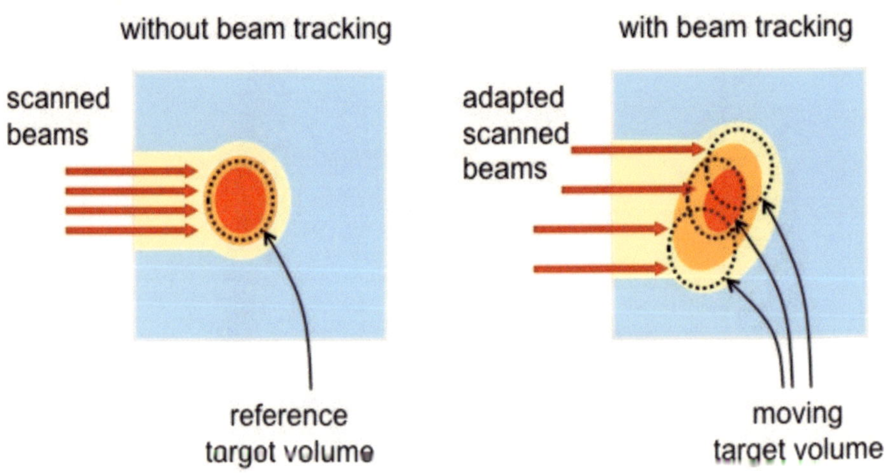

Fig. 7. Illustration of the dose distribution in a static water phantom for a static target without beam tracking (left) and for a virtually moving target with beam tracking (right). The beam tracking tracks the virtual moving target in the static water phantom, therefore the dose distribution creates a dose pattern different from the static case without beam tracking. The pattern depends on the performed beam tracking. By comparing the dose patterns of a calculated ideal beam tracking case and the measured case, the performed beam tracking can be assessed.

Hence, beam tracking induces a characteristic dose pattern in the static water phantom. The produced dose distribution can be measured and compared with the calculated distribution of the ideal beam tracking case using the recorded motion signal and ideal beam tracking parameters in LUT. This comparison assesses the performance of the beam tracking as well as the beam quality like the beam energies, focus and intensity modulation. With a defined

QA 4DTP, one can use this step as a daily QA process of beam tracking. In figure 8, an example of the dose distribution for a 4DTP is demonstrated. The upper panel of the figure 8 shows distribution without beam tracking that corresponds to the reference plan, and the lower panel corresponds to the ideal beam tracking case. In case of a failure of the beam tracking system, dose measurements on the sampled points (circles indicated on the top view in the figure 8) do not match to the dose values calculated for the ideal beam tracking case. When a QA 4DTP is selected as it is shown in the figure 8, the dose distribution has gradients on the sampled points. Therefore, even slight mis-tracking can be detected in this QA process (Saito et al. 2009a;Saito et al. 2010).

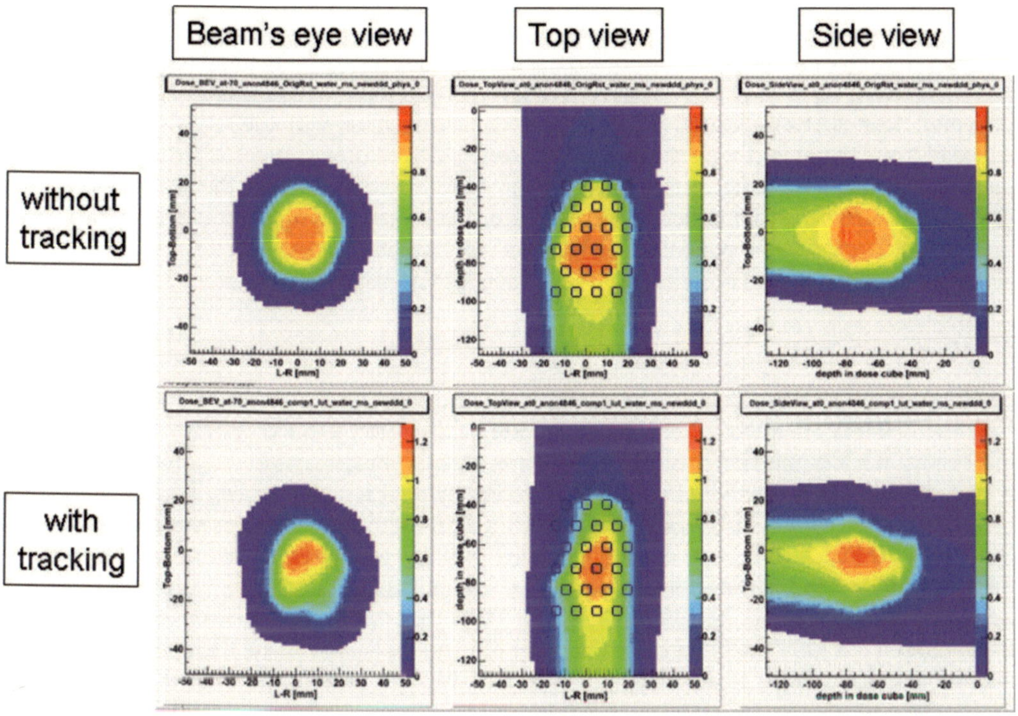

Fig. 8. An example of dose distribution cuts for a patient 4DTP without beam tracking (upper-) and with beam tracking (lower panel). Location of ionization chamber is indicated by circle on the top view. From a QA perspective, the upper panel corresponds to the dose distribution in a case of total failure of beam tracking. The lower panel corresponds to an ideal beam tracking control. (see text for more details).

After the QA measurement the dose distribution that is based on the patient 4DCT can be assessed in terms of DVH. The reconstruction of the dose distribution is based on the beam tracking data recorded during the QA measurement. The created DVH can be compared with an expected DVH range calculated with the nominal beam displacement in 4DTP that was described in the section 2.1 and 3.1.1. The DVH comparisons assesses actual beam tracking performance in a patient geometry. As the summary of the QA measurement, the workflow of the QA measurement process is shown in figure 9.

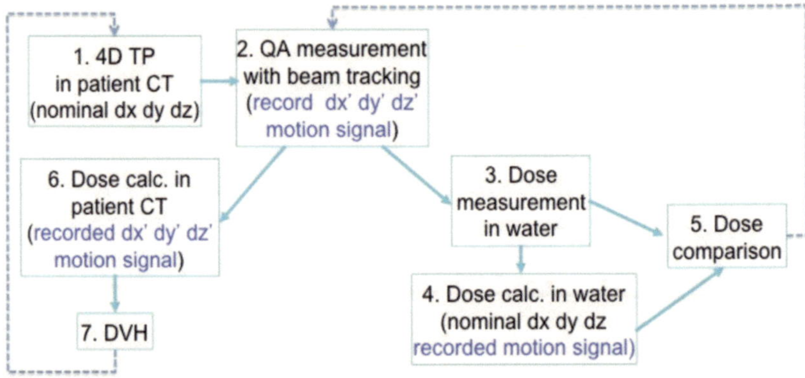

Fig. 9. A flow chart of the presented beam tracking QA process. An obtained beam tracking parameter from 4DTP (1.) can be applied in a QA measurement with a water phantom (2.). The delivered dose can be measured (3.), and dose can be calculated with the recorded motion signal (4.), and compared with the measured dose (5.). If the dose comparison does not agree, failure of the beam tracking system has to be solved followed by redoing the QA measurement (2.). If the dose delivery is confirmed (5.), the recorded beam tracking parameters can be used for dose calculation in the patient CT (6.) and its DVH can be checked (7.). If the DVH is not sufficient quality, one can process 4DTP with more optimization (1.).

3.2 4D in-beam PET

At GSI in-beam PET has been performed during ion beam therapy of static tumour to verify the dose delivery (Enghardt 2004). Since the ion beam therapy at GSI utilizes carbon ions as ion beams, nuclear reaction with the ion beams and the target materials (mainly Oxygen in water) produces beta+-activity that can be detected by the PET camera. For the case of moving target treatment with beam tracking, the in-beam PET technique can be extended to 4D. Basic concept of 4D PET is in general a time correlated positron tomography that utilizes the motion information obtained during the PET imaging. The obtained data can be sorted into subsets according the motion states. For the in-beam PET, the number of positron emission is not freely chosen but rather defined by the number of incident beams as the therapeutic beam that is optimized for tumour treatment. It leads to a limited statistics in the collection of PET data for a certain time span. In addition, the in-beam PET camera limits their solid angle coverage to spare the ion beam entrance. Therefore the 4D in-beam PET is a challenging method rather than other time correlated PET imaging techniques such as 4D PET, or gated PET. An initial feasibility study of the 4D in-beam PET has been performed and reported previously (Parodi, Saito, Chaudhri, Richter, Durante, Enghardt, Rietzel, & Bert 2009). In the feasibility study, PET data taken during the irradiation with beam tracking has been sorted for 21 motion phases, and the sorted images were co-registered into the reference coordinate of the reference phase. The results of the merged PET images were compared to the corresponding PET images of stationary cases. In the experiments motion phantom geometry and type of motion were far too simple to conclude the quality of PET image for a patient geometry, however the results encourages the further investigation of 4D in-beam PET as a real-time verification method *in vivo*. At the Helmholtz-Centre Dresden-Rossendorf in Germany an intensive investigation of 4D in-beam PET is currently conducted in cooperation with GSI and HIT. Further technical implementation of 4D in-beam PET is still needed as a real-time verification method depending on clinical needs as in vivo dose monitoring system.

4. Conclusion

The beam tracking system that adapts ion beams on moving targets has been implemented in the therapy control system at GSI. The beam tracking system requires 4D treatment planning, a motion detection device, and lateral as well as longitudinal beam tracking to adapt ion beams three-dimensionally on moving targets. The current implementation employs the scanning magnets as lateral tracking devices and a double wedge range shifter for longitudinal beam tracking. In contrast to the conventional radiotherapy with photons, ion beams are sensitive to the depth or density displacement. Therefore for ion beam tracking the longitudinal beam adaptation has to be considered as essential as the lateral adaptation. Dose delivery with the implemented beam tracking system, therefore, needs to be verified in three-dimensional dose measurements. A verification method that is dedicated for beam tracking is proposed to perform beam tracking with an artificial motion signal. That assures technical control of beam tracking by comparing the measured dose and the nominal dose delivered in a static water phantom. The QA measurement is expected to be done within a short time feasible to be a part of a daily QA. The performed beam tracking (e.g. lateral beam position and range shifter thickness) and the used motion signal can be recorded in a time correlated manner. By using beam tracking parameters in combination with a patient CT one can estimate the dose conformity on a patient-specific level. During treatment delivery, beam tracking irradiation can be monitored and verified *in vivo* with 4D in-beam PET. The 4D in-beam PET has been successfully tested with a simple motion phantom in a retrospective manner, and the further investigation to achieve a real-time application of 4D in-beam PET is foreseen. The beam tracking system is in an experimental phase at GSI, and it is expected to be applied at ion beam therapy facility if clinical approval is fulfilled.

5. Acknowledgment

The authors acknowledge fruitful discussions with and support from our colleagues at GSI-Biophysics, the Heidelberg Ion Beam Therapy Centre as well as help from Drs. Wolfgang Ott and Nikolaus Kurz regarding the details related to changes of the therapy control system at GSI. The authors appreciate the team of the *Medical Physics* groups at GSI for their support. The presented project is in part funded by Siemens AG, Healthcare Sector, Imaging & Therapy, Particle Therapy.

6. References

Badura, E., Brand, H., Essel, H.G., Haberer, T., Hardel, H., Hoffmann, J., Kurz, N., Liebold, P., Ott, W., Poppensieker, K., & Richter, M. 2000. Control system for cancer therapy with a heavy ion beam at GSI. *IEEE Transactions on Nuclear Science*, 47, (2) 170-173

Bert, C. 2006. *Bestrahlungsplanung für bewegte Zielvolumina in der Tumortherapie mit gescanntem Kohlenstoffstrahl*. PhD-Thesis TU Darmstadt.

Bert, C., Gemmel, A., Saito, N., Chaudhri, N., Schardt, D., Durante, M., Kraft, G., & Rietzel, E. 2010. Dosimetric precision of an ion beam tracking system. *Radiat Oncol*, 5, (1) 61 available from: PM:20591160

Bert, C., Gemmel, A., Saito, N., & Rietzel, E. 2009. Gated irradiation with scanned particle beams. *International journal of radiation oncology, biology, physics*, 73, (4) 1270-1275 available from: PM:19251099

Bert, C., Grözinger, S.O., & Rietzel, E. 2008. Quantification of interplay effects of scanned particle beams and moving targets. *Physics in Medicine and Biology*, 53, (9) 2253-2265 available from: http://www.iop.org/EJ/abstract/0031-9155/53/9/003

Bert, C. & Rietzel, E. 2007. 4D treatment planning for scanned ion beams. *Radiation Oncology*, 2:24

Bert, C., Saito, N., Schmidt, A., Chaudhri, N., Schardt, D., & Rietzel, E. 2007. Target motion tracking with a scanned particle beam. *Medical Physics*, 34, (12) 4768-4771 available from: http://link.aip.org/link/?MPH/34/4768/1

Chaudhri, N., Saito, N., Bert, C., Franczak, B., Steidl, P., Durante, M., Rietzel, E., & Schardt, D. 2010. Ion-optical studies for a range adaptation method in ion beam therapy using a static wedge degrader combined with magnetic beam deflection. *Phys Med Biol*, 55, (12) 3499-3513 available from: PM:20508325

Gemmel, A., Hasch, B., Ellerbrock, M., Weyrather, W.K., & Kramer, M. 2008. Biological dose optimization with multiple ion fields. *Phys Med Biol.*, 53, (23) 6991-7012

Grözinger, S.O., Bert, C., Haberer, T., Kraft, G., & Rietzel, E. 2008. Motion compensation with a scanned ion beam: a technical feasibility study. *Radiation Oncology*, 3:34

Grözinger, S.O., Rietzel, E., Li, Q., Bert, C., Haberer, T., & Kraft, G. 2006. Simulations to design an online motion compensation system for scanned particle beams. *Physics in Medicine and Biology*, 51, (14) 3517-3531 available from: PM:16825746

Haberer, T., Becher, W., Schardt, D., & Kraft, G. 1993. Magnetic scanning system for heavy ion therapy. *Nuclear Instruments and Methods in Physics Research A*, 330, 296-305

Hartkens, T. 1993. *Measuring, analysing, and visualizing brain deformation using non-rigid registration.* King's College London.

Hild, S., Durante, M., & Bert, C. 2011. Assessment of uncertainties in treatment planning for scanned ion beam therapy of moving tumors. Submitted to *Int.J.Radiat.Oncol.Biol.Phys.*

Huang, K., Hossain, S., Chuang, C., Descovich, M., Gottschalk, A., Larson, D., & Ma, L. 2008. Lung Tumor Motion Compensation during Real Time Respiratory Tracking using Gold Fiducial Markers. *International Journal of Radiation Oncology*Biology*Physics*, 72, (1, Supplement 1) S464-S465 available from: http://www.sciencedirect.com/science/article/B6T7X-4T85W5M-199/2/03a2122571c0dca1cb460021ff38bd60

Jäkel, O., Krämer, M., Karger, C.P., & Debus, J. 2001. Treatment planning for heavy ion radiotherapy: clinical implementation and application. *Physics in Medicine and Biology*, 46, (4) 1101-1116 available from: PM:11324954

Karger, C.P., Jäkel, O., & Hartmann, G.H. 1999. A system for three-dimensional dosimetric verification of treatment plans in intensity-modulated radiotherapy with heavy ions. *Medical Physics*, 26, (10) 2125-2132 available from: PM:10535629

Keall, P.J., Kini, V.R., Vedam, S.S., & Mohan, R. 2001. Motion adaptive x-ray therapy: a feasibility study. *Physics in Medicine and Biology*, 46, (1) 1-10

Kraft, G. 2000. Tumor Therapy with Heavy Charged Particles. *Progress in Particle and Nuclear Physics*, 45, (s2) s473-s544

Krämer, M., Jäkel, O., Haberer, T., Kraft, G., Schardt, D., & Weber, U. 2000. Treatment planning for heavy-ion radiotherapy: physical beam model and dose optimization. *Physics in Medicine and Biology*, 45, (11) 3299-3317

Krämer, M. & Scholz, M. 2000. Treatment planning for heavy-ion radiotherapy: calculation and optimization of biologically effective dose. *Physics in Medicine and Biology*, 45, (11) 3319-3330 available from: http://stacks.iop.org/0031-9155/45/3319

Langen, K.M. & Jones, D.T.L. 2001. Organ motion and its management. *International Journal of Radiation Oncology*Biology*Physics*, 50, (1) 265-278 available from: http://www.sciencedirect.com/science/article/B6T7X-42V69MY-15/2/6ab3ea92d5404dfa22178ca54bdce941

Li, Q., Grözinger, S.O., Haberer, T., Rietzel, E., & Kraft, G. 2004. Online compensation of target motion with scanned particle beams: simulation environment. *Physics in Medicine and Biology*, 49, (14) 3029-3046

Lu, H.M., Brett, R., Sharp, G., Safai, S., Jiang, S., Flanz, J., & Kooy, H. 2007. A respiratory-gated treatment system for proton therapy. *Medical Physics*, 34, (8) 3273-3278 available from: PM:17879790

Lüchtenborg, R., Saito, N., Durante, M., & Bert, C. 2011. Experimental verification of a real-time compensation functionality for dose changes due to target motion in scanned particle therapy. *Medical Physics*, 38, (10) 5448

Lujan, A.E., Larsen, E.W., Balter, J.M., & Haken, R.K.T. 1999. A method for incorporating organ motion due to breathing into 3D dose calculations. *Medical Physics*, 26, (5) 715-720 available from: http://link.aip.org/link/?MPH/26/715/1

Mao, W., Wiersma, R.D., & Xing, L. 2008. Fast internal marker tracking algorithm for onboard MV and kV imaging systems. *Med.Phys*, 35, (5) 1942-1949 available from: PM:18561670

Matsufuji, N., Tomura, H., Futami, Y., Yamashita, H., Higashi, A., Minohara, S., Endo, M., & Kanai, T. 1998. Relationship between CT number and electron density, scatter angle and nuclear reaction for hadron-therapy treatment planning. *Physics in Medicine and Biology*, 43, (11) 3261-3275 available from: http://stacks.iop.org/0031-9155/43/3261

Minohara, S., Kanai, T., Endo, M., Noda, K., & Kanazawa, M. 2000. Respiratory gated irradiation system for heavy-ion radiotherapy. *International Journal of Radiation Oncology*Biology*Physics*, 47, (4) 1097-1103 available from: http://www.sciencedirect.com/science/article/B6T7X-40HTYJ7-14/2/4c9348d430ff1764cc452ac73e70073e

Mori, S., Chen, G.T., & Endo, M. 2007. Effects of intrafractional motion on water equivalent pathlength in respiratory-gated heavy charged particle beam radiotherapy. *Int.J Radiat Oncol Biol.Phys.*, 69, (1) 308-317 available from: PM:17707286

Murphy, M.J. 2004. Tracking moving organs in real time. *Seminars in Radiation Oncology*, 14, (1) 91-100

Parodi, K., Saito, N., Chaudhri, N., Richter, C., Durante, M., Enghardt, W., Rietzel, E., & Bert, C. 2009. 4D in-beam positron emission tomography for verification of motion-compensated ion beam therapy. *Medical Physics*, 36, (9) 4230-4243 available from: http://link.aip.org/link/?MPH/36/4230/1

Pedroni, E., Bacher, R., Blattmann, H., Bohringer, T., Coray, A., Lomax, A., Lin, S., Munkel, G., Scheib, S., Schneider, U., & Tourovsky, A. 1995. The 200-MeV proton therapy project at the Paul Scherrer Institute: Conceptual design and practical realization. *Medical Physics*, 22, (1) 37-53 available from: http://link.aip.org/link/?MPH/22/37/1

Phillips, M.H., Pedroni, E., Blattmann, H., Boehringer, T., Coray, A., & Scheib, S. 1992. Effects of respiratory motion on dose uniformity with a charged particle scanning method. *Physics in Medicine and Biology*, 37, (1) 223-233

Richter, D., Trautmann, J., Schwarzkopf, A., Krämer, M., Gemmel, A., Jäkel, O., Durante, M., & Bert, C. 2010, *4D Treatment Planning Implementations for TRiP98*, Annual Report from GSI Helmholtzzentrum f¨ur Schwerionenforschung GmbH, Darmstadt, Germany, 2010-1. Editor: K. Grosse

Rietzel, E., Schardt, D., & Haberer, T. 2007. Range accuracy in carbon ion treatment planning based on CT-calibration with real tissue samples. *Radiat Oncol*, 2, 14 available from: PM:17381831

Saito, N., Bert, C., Chaudhri, N., Durante, M., Gemmel, A., Lüchtenborg, R., & Kraft, G. 2010, *Quality Assurance test of the real-time beam tracking system*, Annual Report from GSI Helmholtzzentrum für Schwerionenforschung GmbH, Darmstadt, Germany, 2010-01. Editor: K. Grosse

Saito, N., Bert, C., Chaudhri, N., Gemmel, A., Schardt, D., Kraft, G., Durante, M., & Rietzel, E. 2009a, Technical accuracy of a beam tracking system for scanned particle therapy of intra-fractionally moving targets, *In World Congress on Medical Physics*, 1 edn, O. Dössel & W. Schlegel, eds., Heidelberg: Springer, pp. 417-420.

Saito, N., Bert, C., Chaudhri, N., Gemmel, A., Schardt, D., & Rietzel, E. 2009b. Speed and accuracy of a beam tracking system for treatment of moving targets with scanned ion beams. *Physics in Medicine and Biology*, 54, 4849-4862

Sawant, A., Venkat, R., Srivastava, V., Carlson, D., Povzner, S., Cattell, H., & Keall, P. 2008. Management of three-dimensional intrafraction motion through real-time DMLC tracking. *Med Phys*, 35, (5) 2050-2061 available from: PM:18561681

Schmidt, A., Bert, C., ., Saito, N., ., Chaudhri, N., ., Iancu, G., ., von, N.C., & Rietzel, E. 2008. TU-EE-A2-03: Target Motion Tracking with a Scanned Particle Beam: Calculation and Experimental Validation of Biologically Effective Doses in the Presence of Motion. *Medical Physics*, 35, (6) 2911

Schulz-Ertner, D., Karger, C.P., Feuerhake, A., Nikoghosyan, A., Combs, S.E., Jäkel, O., Edler, L., Scholz, M., & Debus, J. 2007. Effectiveness of carbon ion radiotherapy in the treatment of skull-base chordomas. *Int.J.Radiat Oncol Biol.Phys.*, 68, (2) 449-457 available from: PM:17363188

Shackleford, J.A., Kandasamy, N., & Sharp, G.C. 2010. On developing B-spline registration algorithms for multi-core processors. *Physics in Medicine and Biology*, 55, (21) 6329-6351

Sihver, L., Schardt, D., & Kanai, T. 1998. Depth-dose distributions of high-energy carbon, oxygen and neon beams in water. *Jpn.J.Med.Phys.*, 18, (1) 1-21

Steidl, P., Richter, D., Schuy, C., Schubert, E., Haberer, T., Durante, M., & Bert, C. 2011. A breathing thorax phantom with independently programmable 6D tumour motion for dosimetric measurements in radiation therapy. Submitted to *Phys Med Biol*

Weber, U. & Kraft, G. 1999. Design and construction of a ripple filter for a smoothed depth dose distribution in conformal particle therapy. *Physics in Medicine and Biology*, 44, (11) 2765-2775

6

Neutron Influence in Charged Particle Therapy

Su Youwu, Li Wuyuan, Xu Junkui,
Mao Wang and Li Zongqiang
Institute of Modern Physics,
Chinese Academy of Sciences,
P. R. China

1. Introduction

In charged particle therapy, neutron will emit due to beam loss on accelerator components, beam shaping devices, treatment nozzle, and especially in the deposition of beam interactions in the patient. Those neutrons usually have high energy and strong penetrating power, so it will cause additional dose on the tumor and health tissue of the patient, which may influence the treatment effect. Thus it is essential to know the neutron influence in charged particle therapy.

Different to conventional rays, charged particle (proton and heavy ion) with certain energy has a determined "range" in matters and they loss maximum energy at the end of the "range", which is the so-called Bragg peak. During the process of cancer therapy, charged particles can be stopped at tumor site by adjusting the energy range to achieve maximum tumor destruction using the characteristics of Bragg peak. Moreover, the normal tissue surrounding a tumor will not receive much irradiation due to Bragg peak.

Based on the characteristics of charged particles in cancer therapy, many institutions develop this technique and many facilities for cancer therapy are under construction such as German Heavy Ion Research Center (GSI, Germany), National Institution of Radiological Sciences (NIRS, Japan). However secondary neutron produced due to beam loss in accelerator component may cause damage to patient body which is not well known. Thus it is important to investigate the information about neutron radiation field in charged particle cancer therapy.

2. Neutron production mechanism of ion reactions

The process of charged particles interact with matters is very complicated. The interaction of particle with matters results in the production of different types of radiation, including gamma ray and neutrons with complex energy distribution. Heavy ions loss energy mainly through ionization and excitation of atoms as it traverses matter, which is called electromagnetic cascade process. Except at low velocities, it loses a negligible amount of energy in nuclear collisions. No neutrons will be produced in electromagnetic cascade process, but it determines the energy loss of the particle in matter, and especially when the material is the patient tissue, it determines the heavy ion dose. The interaction between ions and atomic electrons can be well described by a term "stopping power" - (dE/dx).

$$-\frac{dE}{dx}=\frac{4\pi Z^2 e^4 n}{mc^2\beta^2}\left[\ln\frac{2mc^2\beta^2}{I(1-\beta^2)}\right]-\beta^2 \qquad (2\text{-}1)$$

Where Z is the atomic number of the heavy particle, e is the electron charge, n is the number of electrons per unit volume in the medium, and I is the mean excitation energy of the medium, $\beta= v/c$ is the relative velocity of the particle.

The distance traveled per unit energy loss is given by the reciprocal of the stopping power. Thus, the range R(T) of a particle with kinetic energy(T) is the integral of the reciprocal of the stopping power down to zero energy and can be written in the following form:

$$R(T)=\frac{M}{Z^2}f(\beta) \qquad (2\text{-}2)$$

It is important to note that the mean range of particles of a given speed is proportional to the mass and varies as the inverse square of their charge. The dependence of the Bethe formula on Z^2 implies that particles with the same mass and energy, but with inverse charge will have same range in matter.

Though it may not produce neutrons in the process of electromagnetic interactions with charged particles, it is the dominant way of charged particles to loss energy. A large part of heavy charged particle loss energy through interaction with extranuclear electron. Thus at different depth heavy charged particle has different energy at different depths, so the neutron produced by heavy charged particle on thick target can be regard as the results of interactions of heavy charged particle with matter in different energy region particle in different energy region and nucleus.

2.1 Nuclear interactions

Nuclear interactions include compound interactions and direct interactions.

In the process of compound interactions, the emission of the particles can be described by an evaporation process similar to the evaporation of a molecule from the surface of a liquid. The spectrum of the emitted neutrons may be described by a Maxwellian distribution:

$$\frac{dN}{dE_n}=BE_n\exp(-E_n/T) \qquad (2\text{-}3)$$

Where E_n is the energy of neutron, B is a constant, and T is the nuclear temperature. The nuclear temperature is characteristic of the target residual nucleus and its excitation energy, and has dimensions of energy. Its value lies between 2 and 8 MeV.

2.2 Heavy ion-nucleus interactions

Nuclear interactions of heavy ions as they pass through matter arise from grazing or head-on collisions. In grazing collisions, fragmentation of either the incident ion or the target nucleus occurs. Fragmentations are the major nuclear interaction. Head-on collisions are less frequent, but in such collisions, large amounts of energy are transferred compared to grazing collisions. In heavy-ion interactions, many secondary particles are created from nucleus-nucleus interactions. Nucleus-nucleus interactions have features that are different from typical hadron-nucleus interactions at either the same total energy or energy per nucleon. The cross section of nuclear collisions between two nuclei is larger than that between a single hadron and nucleus. When two high-energy nuclei interact, only the segments that interpenetrate each other undergo a significant interaction and mutual disintegration. The remainder of each nucleus is uninvolved even though each is likely to have become highly excited, as is evidenced by the fact that a substantial fragment is usually observed traveling in the same direction and at a similar speed to the incident primary ion. Even though the part of the nucleus that escapes the severe interaction becomes highly excited, it does not undergo evaporation to the extent that it breaks up into fragments with $Z < 3$[1]. It is only in a head-on collision that the projectile breaks up into many small pieces, so that no high-velocity fragment survives. The residual nucleus and the alpha particles that evaporate from the primary fragment are concentrated about the incident direction.

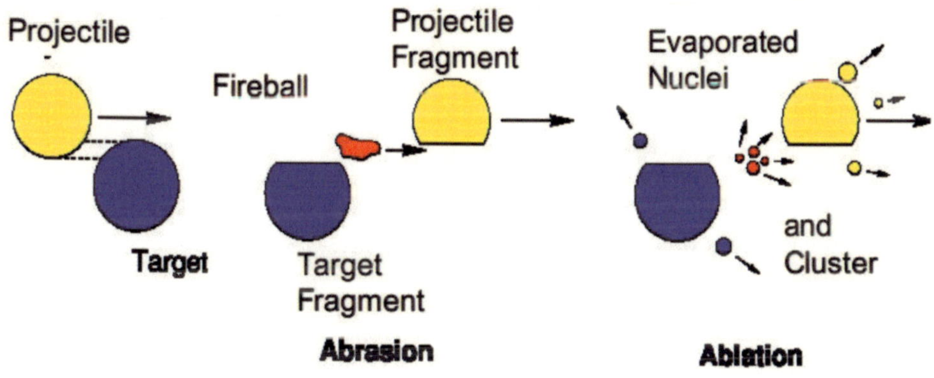

Fig. 1. Schematic illustration of fragmentation in a target

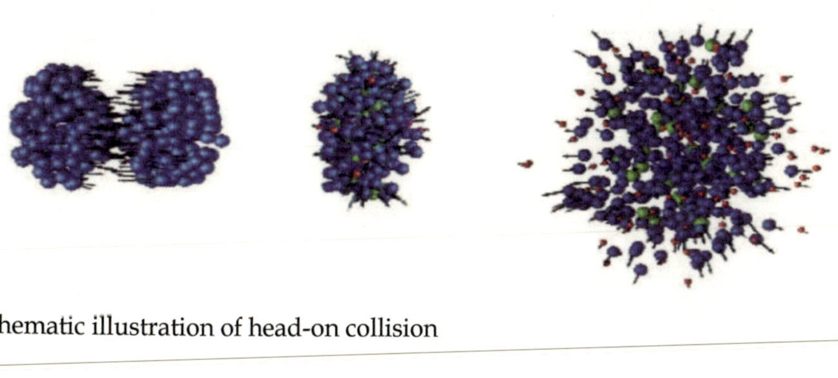

Fig. 2. Schematic illustration of head-on collision

2.3 Characteristics of neutrons emitted from charged particle interactions

Thus neutrons produced by heavy ion reaction can be divided into two categories: low energy neutrons from evaporation process and high energy neutrons from cascade process. The former was isotropic and the later was peaked in the forward direction. In the forward direction, there is a broad peak at about 2/3 of the incident particle energy per nucleon; the maximum neutron energy even can reach twice of the incident energy per nucleon, which can be explained by multi-nucleon momentum transformation.

3. Characters of neutron biological effects

In radiation protection, high energy neutrons are more important, they dominate the radiation field outside the shielding, but low energy neutrons are very important for the additional dose to patient because of their biological characters.

3.1 Neutron interactions with matter

Because neutron is uncharged particle, it can travel appreciable distances in matter without undergoing interactions. When a neutron collides with atom, it can undergo elastic or inelastic reaction. An elastic reaction is one in which the total kinetic energy of the incoming particle is conserved. In an inelastic reaction, the nucleus absorbs some energy and is left in an excited state. The neutron can also be captured or absorbed by a nucleus in reactions such as (n,p), (n,2n), (n,α) or (n,γ).

Based on the analysis above, the absorption process of neutron in matter can be expressed by the exponential decay:

$$\mathrm{I(x)} = I(0)e^{-N\sigma x} \tag{3-1}$$

Where I(x) is the neutron fluence rates at the depth of X , I(0) is the incident neutron fluence rates, N is the nuclei number , σ is the total cross section, X is the depth in matter.

3.1.1 Neutron scattering

Neutron scattering include elastic scattering and inelastic scattering. In the former situation, the total kinetic energy of the system was conserved; and in the late, the target nucleus consumes some of kinetic energy and becomes an excided nucleus.

At low energy, neutrons loss energy mainly through elastic scattering (n,n). According to law of energy conservation(neglect relative effect), the scattering neutron energy can be easily given as：

$$\mathrm{E}_n' = E_n - E_A \tag{3-2}$$

$$\mathrm{E_A} = \frac{4AE_n}{(1+A)^2}\cos^2\varphi \tag{3-3}$$

Where E_n is the incidental neutron energy, E_n' is the scattering neutron energy, A is the recoil nucleon mass, E_A is the recoil nucleon energy, φ is the emit angle of recoil nucleon.

In elastic collision, Neutron will loss energy from 0 to $\frac{4A}{(1+A)^2}E_n$; neutron loss more energy on light nucleon than on heavy nucleon, it implies that light material have higher stopping power than heavy material, for heavy nucleon such as uranium or lead target, neutron can hardly lose energy by elastic scattering. But for hydrogen nucleon, neutron can almost lose all kinetic energy at one elastic collision, so the hydrogen contents of material are of great sense to neutron moderation and measurement.

At high energy region, inelastic scattering becomes more and more important. "Elastic scattering can't absorb high energy neutrons effectively. The higher the incident energy the more forward peaked is the elastic scattering, and beyond 150MeV it is almost always in the forward direction. Since the incident particle remain at least a fraction [(A-1)/(A+1)] of the incident particle energy, it is clear that in most material with moderate mass, litter change in either energy or direction results from elastic scattering"[2].

The inelastic scattering cross section varies with energy are shown in fig. 3. H. W. Patterson had quoted the following explanation "Below 100MeV the neutron inelastic cross section increases rapidly with decreasing energy until En<25MeV where in most cases the neutron inelastic cross sections level off and then decrease suddenly……For En>100MeV the secondary neutrons may still have an effectively shorter mean free path than a higher-energy secondary, even though the inelastic cross sections are about the same, because of the increasing angular divergence with decreasing energy of secondary. It is these facts which tend to make high-energy nucleon beams attenuate approximately exponentially (after a sufficient transition region) with a mean free path which is not very sensitive to the initial energy and is not much longer than the geometric mean free path calculated from the inelastic cross sections of the elements in the shield" [3]. Thus the attenuation length λ_{atten} at high energy would be:

$$\lambda_{\text{atten}} \approx \frac{1}{N\sigma_{in}} \tag{3-4}$$

Where N is the number of atoms/cm^3 σ_{in} is the inelastic cross section.

In high energy situation inelastic scattering cross section can approximately assumed to be geometric cross section.

3.1.2 Neutron capture

The reaction of nucleus absorb a neutron and emit γ-rays (n, γ) is called neutron capture. The nucleus of mass A absorb a neutron and produce its isotopes nucleus of mass (A+1). In addition, the produced nucleus is in excited state depending on neutron binding energy and kinetic energy. Then the excited nuclei emit one or several γ photon to the ground stated. As to thermal neutron and resonance neutron, neutron capture is the main reaction. In the thermal energy rang, the capture cross section often inverse proportional to the neutron speed and almost all the stable nuclides can absorb thermal neutron except few nuclei.

The neutron capture reaction are quiet different with different nuclides at thermal and resonance energy region, but at higher energy region the neutron capture cross section variations become very gentle, fig.3 shows the neutron absorb cross section varies with target mass number. The figure can be well expressed by the following equation:

$$\sigma_a = 43A^{0.69}(mb) \tag{3-5}$$

The formula should be valid up to several hundred GeV[2], but it does not valid for some special nuclides include hydrogen. Note that the nuclides which can not apply to the formula above are those of great importance to neutron radiation protection such as H, ^{10}B Cd etc. Because hydrogen which is most widely distributed nature nuclide plays an important part in nuclear reactor moderation and cooling. Meanwhile, ^{10}B, Gd etc also very significant in neutron protection and measurement.

Experiment indicate the ratio of the neutron elastic cross section to the absorption cross section almost unchanged for mostly nuclides and energy it is 0.57 [2] .

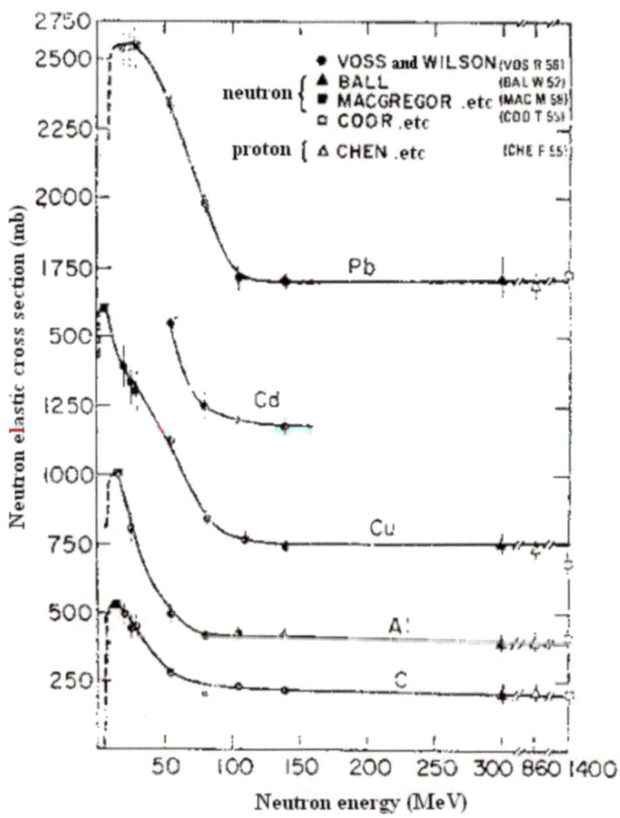

Fig. 3. Inelastic scattering cross section of neutron on different targets [3]

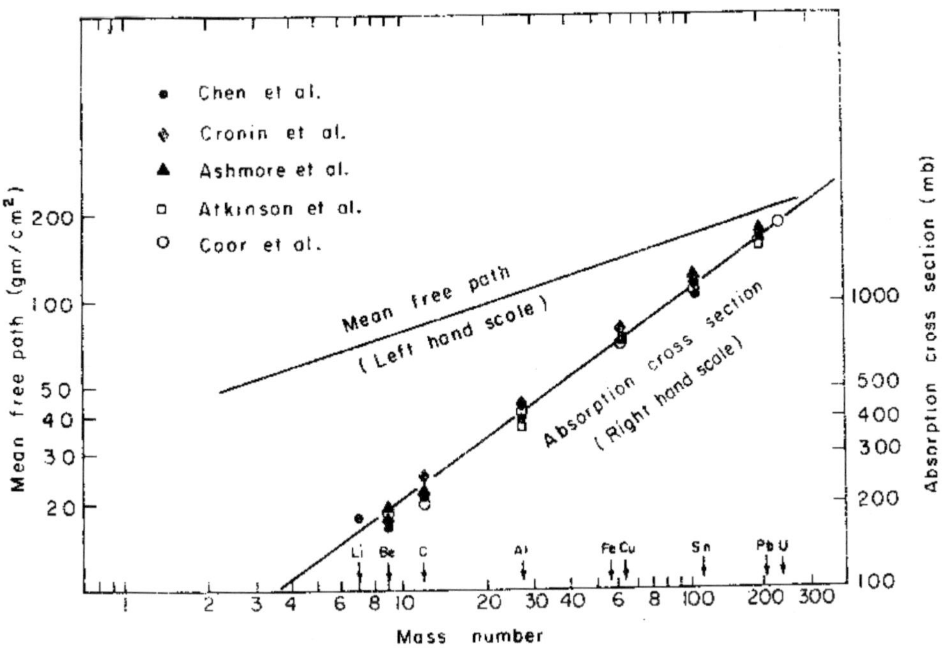

Fig. 4. Mean free path and atomic cross section as a function of mass number [2]

3.2 Human body response to neutron

3.2.1 Basic concepts of radiation dosimetry

The biological impact of radiation to human body is caused by energy deposition, thus the energy deposited per unit mass in material (absorbed dose) is the fundamental concept of radiation dosimetry. The absorbed dose unit is joule per kilogram and dedicated unit is gray (Gy). However, same absorbed dose may not always cause same biological effect for different radiation type and energy. To distinguish this difference, the International Commission on Radiological Protection (ICRP) introduced the concept of equivalent dose [4], which is defined as product of absorbed dose and radiation weighting factor:

$$H_{T,R} = w_R \bullet D_{T,R} \tag{3-6}$$

Where $D_{T,R}$ is the absorbed dose of R type radiation in tissue or organ; W_R is the radiation weighting factor.

The radiation weighting factor is a quantity which is related to relative biological effectiveness. Furthermore, the relative biological effectiveness is defined as the ratio of the doses required by two different types of radiation to cause the same level for a specified end point. And the radiation weighting factor represents damage severity. Table 1 gives the radiation weighting factor recommended by ICRP.

Radiation type and energy region	radiation weighting factor
Photon of all energy	1
Electron and μ particle : all energy region	1
Neutron energy :	
<10keV	5
10~100keV	10
100keV~2MeV	20
2MeV ~20MeV	10
>20MeV	5
proton(except recoil proton), E>2MeV	5
α particle, fission fragment, heavy nuclei	20

Table 1. Radiation weighting factor (ICRP 60 publication)

Neutron has the most complicated weighting factor, which indicates the complex of the neutron interaction with tissue. The radiation weighting factor of neutron can be expressed by the following continuous function:

$$W_R = 5 + 17\exp\left\{\frac{-[\ln(2E_n)]^2}{6}\right\} \tag{3-7}$$

The formula is not of any physics sense and just a tool of mathematics. Fig. 5 is another way to illustrate the radiation weighting factor of neutron.

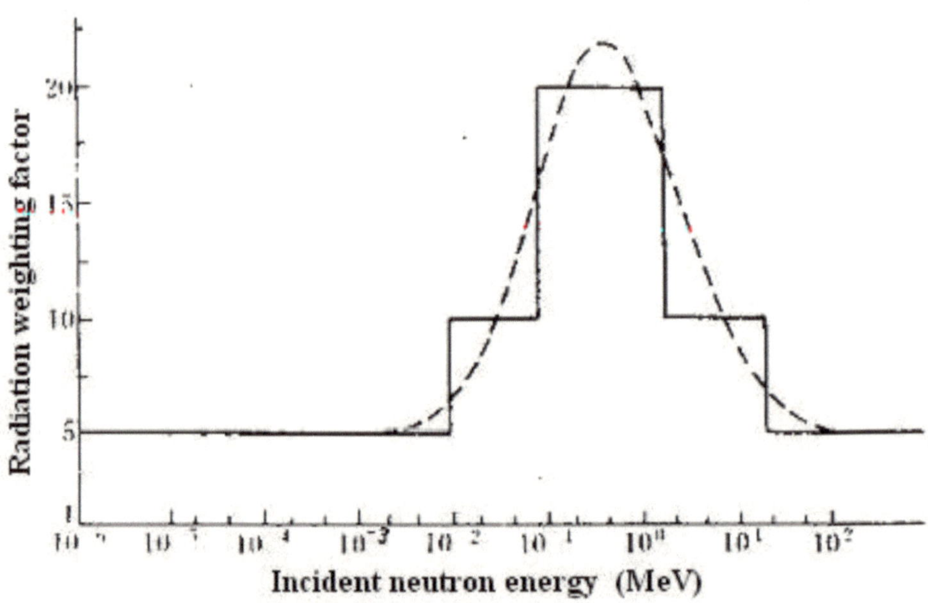

Fig. 5. Radiation weighting factor of neutron

3.2.2 Radiation response of human body

The effect of radiation on the human body can be divided into deterministic effect and stochastic effect. Deterministic effect is bounded to a certain dose; it is caused by a large number of cells death. It must have a certain threshold, below this threshold the effect will not occur. Above this threshold the severity aggravating with dose is increasing, such as skin damage, cataract and temporary or permanent infertility etc. Stochastic effect may happen in any level of radiation exposure, as it is caused by a single irradiated cell mutation. The severity degree is not related to dose, but its occurrence probability will increase with dose level， such as cancer, genetic effect and so on.

There are two ways that radiation affects human body, direct reaction and indirect reaction. The former refers to radiation reaction with Biomacromolecule such as DNA or RNA directly, and lead to cell damage. The later refers to radiation reaction with intracellular homeostasis, producing free radical and peroxide leading to cell damage. The two processes can be illustrated in Fig. 6. Two obvious characteristics should be concerned:

1. Small absorbed energy can cause very serious biological effect;

For example, 6 Gy acute radiation dose may result in acute death, but convert all the absorbed energy to heat. It can only elevate tissue temperature by 0.0014℃.

2. Transient radiation may cause forward effect
   a. Physics stage

   The ionized particles pass through atom and react with orbital electron, then deposit energy through ionization and excitation at the time of 10^{-18}seconds to 10^{-12}seconds.

   b. Physicochemical stage

   At the time of 10^{-12}seconds to10^{-9}seconds, the atomic ionizing turn to molecule ionizing, the molecules become unstable and prone to react to form free radicals

   c. Physicochemical stage

   At the time of 10^{-9}seconds to 1 second, the free radical diffuse and react with critical biological molecule, and lead to molecular damage.

   d. Biological stage

   From few seconds to years, continued molecular damage gradually develop to performance cell effect, such chromosome aberration, cell death, cell mutation etc, eventually lead to body death, long term effect or changes in offspring genetic effect.

3.2.3 Damage caused by neutron in human body

Neutrons lose energy through a lot of reactions in human body and eventually be absorbed or escaped from human body. In these processes, different types of secondary particles in a wide energy range distribution were produced. Absorbed dose in some certain position of phantom with the linear energy transfer(LET) distribution is determined by many parameters, including interaction cross section, secondary particle energy spectrum and the transport of interaction point to the location. Therefore energy spectrum at any point in human body caused by neutron external exposure is a very complex process and is strongly energy-related.

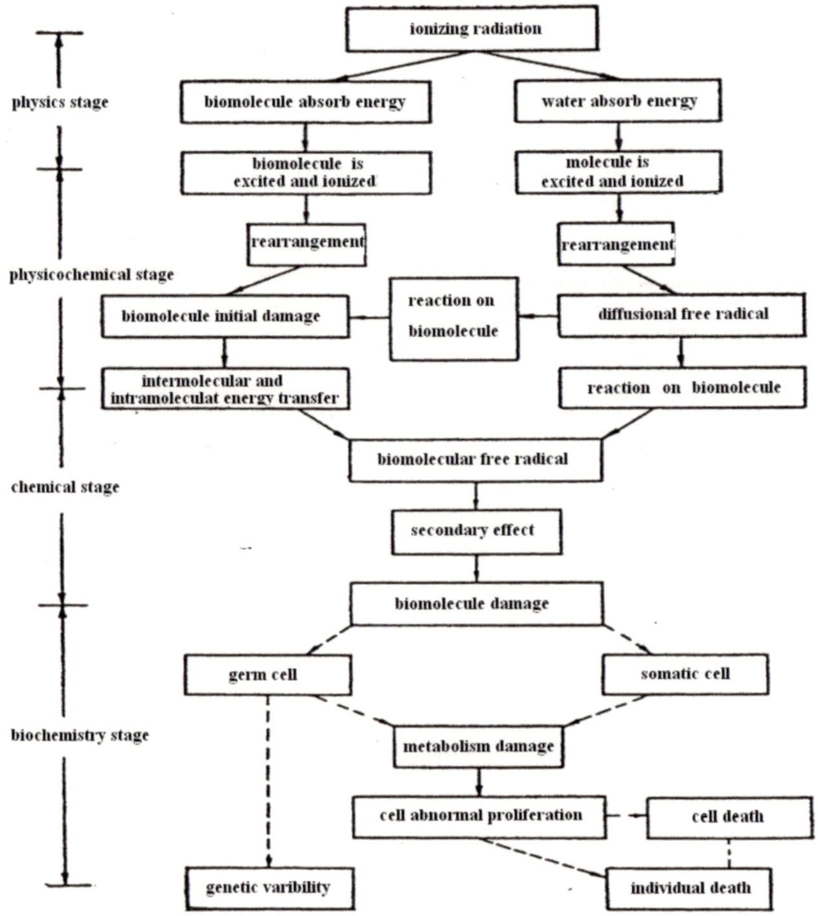

Fig. 6. Process of radiobiological damage

Secondary photon produced by interactions of neutron and tissues is very significant, especially 2.2MeV photon produced by thermal neutrons in the H(n，γ) D reaction which is of most importance in energy deposition in human body. For neutrons incident energy below 1MeV, the deep absorbed dose in human body is mainly due to secondary photon. For example, ninety percent of the absorbed dose is caused by secondary photon at the depth of 10 mm in human body in thermal and intermediate energy region. But when the neutron energy higher than 10 keV, the distribution of photon to absorb dose has sharply decreased and even down to 20% when the energy to 1MeV[5].

As the incident neutron energy increases, other radiation energy deposition play more important role. In the thermal energy region, $^{14}N(n，p)\,^{14}C$ reaction of protons(about 600keV) accounted for a large part of absorbed dose. While when the energy above 1keV, the deposition energy of recoil proton produced by hydrogen nuclei elastic scattering is more important; but when the energy is greater than a few MeV, the charged particle produced by nuclear reactions (n,D)、(n,T)、(n, α),etc becomes more and more important.

3.3 Neutron conversion factor

Conversion factor is a tool to transfer the neutron flux to dose. However, neutron flux to dose conversion factor is hard to obtain, because of the complex behavior of neutron in human body. Neutron conversion factor is the function of neutron energy and it also has relationship with neutron penetration ability, human age, gender, exposure orientation and other conditions [6]. Fig. 7 shows the conversion factor given by ICRP[6].

In the low energy case, the conversion factor change is relative stable, due to main dose distribution of the photon emit from neutron capture in the energy region. From few keV the conversion factor sharply increase and reach maximum at about 20 MeV, the recoil proton and charged particle fragment dose are of most important in this energy region. But when the neutron energy higher than 20 MeV, the conversion factor decreases, the possible reason is that charged particle equilibrium may not always be guaranteed.

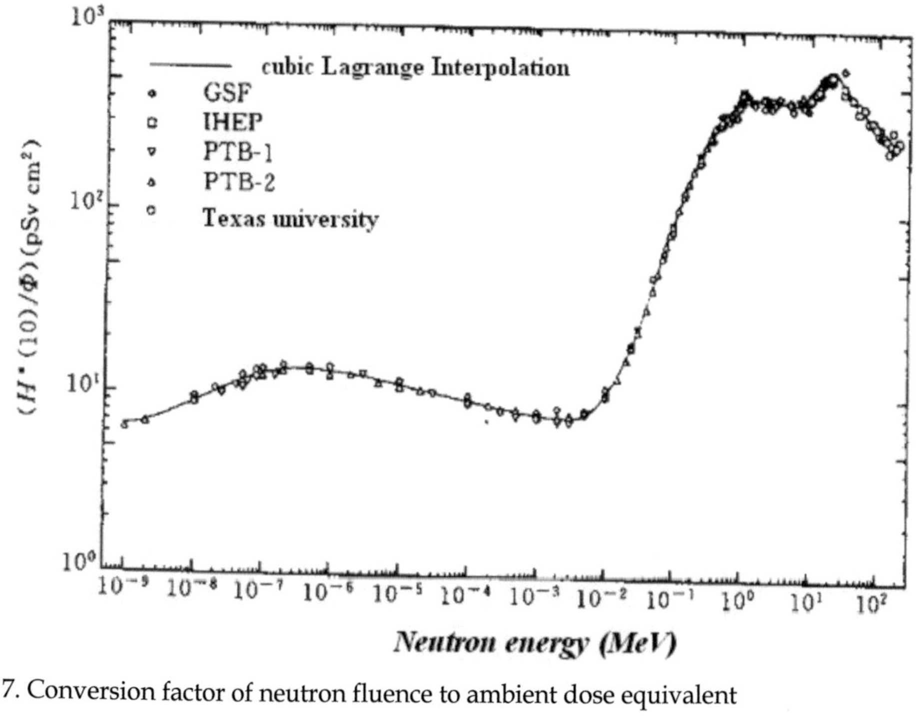

Fig. 7. Conversion factor of neutron fluence to ambient dose equivalent

4. Neutron measurements techniques in charged particle therapy

Several parameters are used to describe the neutron filed; in this chapter the principle and techniques to obtain these parameters are discussed.

4.1 Principle of neutron measurement

Since neutron is uncharged particle, it is hard to be measured directly. All neutron measurement techniques are based on neutron reaction mechanism that were discussed before.

4.1.1 Nucleus reaction method

This method is mainly based on reactions between neutron and some certain nuclide and then detecting products of these reactions to obtain the information of neutron radiation field. Usually, the reactions which can be used to detect neutron may have the following characteristics:

1. Have large neutron cross section to ensure the sensitivity of the detectors;
2. Target nucleus of high abundance in natural elements or easy to collection;
3. With large Q value which can help to discriminator the γ ray pulse amplitude

Based on this principle, the mostly used reactions are the following:

Neutron reaction with ^{10}B

Reaction equation:

$$\begin{aligned} {}_0^1n + {}_5^{10}B &\Rightarrow {}_2^4He + {}_3^7Li + 2.792MeV \\ &\Rightarrow {}_2^4He + {}_3^7Li^* + 2.31MeV \end{aligned} \quad \text{(I)}$$

$${}_3^7Li + \gamma + 480keV \quad \text{(II)}$$

The reaction product ^7Li can either be in ground state (reaction channel I) or in excited state (reaction channel II). Experiments show that, the branch ratio of reaction channel I only account 6.3%. Because the reaction energy is much greater than the kinetic energy of slow neutron, kinetic energy of reaction products mainly comes from reaction energy. Moreover the incident neutron momentum is small, the momentum of reaction products also becomes approximately equal to zero. Thus the emit direction of two products is opposite and the distribution proportion of reaction energy is determined. Therefore the kinetic energy of α particle will be 1.77 MeV and 1.47 MeV for the 7Li in ground and excited state respectively.

Neutron cross section in a wide energy range (from thermal to 1keV) is proportional to 1/v and the thermal neutron cross section is about $3840\times10^{-24}cm^2$(see Fig.8)[7]. The abundance of nature boron is about 19% and it is easy to enrich.

BF_3 proportional counter tube is most commonly used as gas ionization detector. In the center of the tube there is an anode wire which is used to collect ion. The reaction products ^7Li and α cause BF_3 ionization and these ions are amplified by the center electric field then collected by anode wire to form pulse signal. For gas detector, the detector efficiency is a crucial factor. Since the inflation pressure can be very high and 90% of the tube must be filled with BF_3. Sometimes argon gas is filled in the tube in order to improve the detector performance. The sensitivity of a BF_3 counter (neutron energy less than 100keV) is:

$$s = NV\sigma_0 \frac{v_0}{\bar{v}} \quad (4\text{-}1)$$

Where, N is the number of ^{10}B contained in a unit volume;

V is the sensitive volume of counter

v_0 is 2.2×10^5cm/s

$\overline{V}$ is the average velocity of incidental neutron

Boron can also be painted inside surface of the proportional counter tube in the form of solid. The characteristic of BF_3 gas used for gaseous detector, so avoid using BF_3 gas will bring some other advantages. However the thickness of solid boron cannot be very thick, as the reaction product cannot reach the sensitive gas and lead to decrease in efficiency. Meanwhile the reactions may occur at any depth of the boron layer, part of the kinetic energy of produced α particles will be absorbed by different thick of boron material, this will lead to generate lower pulse single and decrease the performance of γ discrimination.

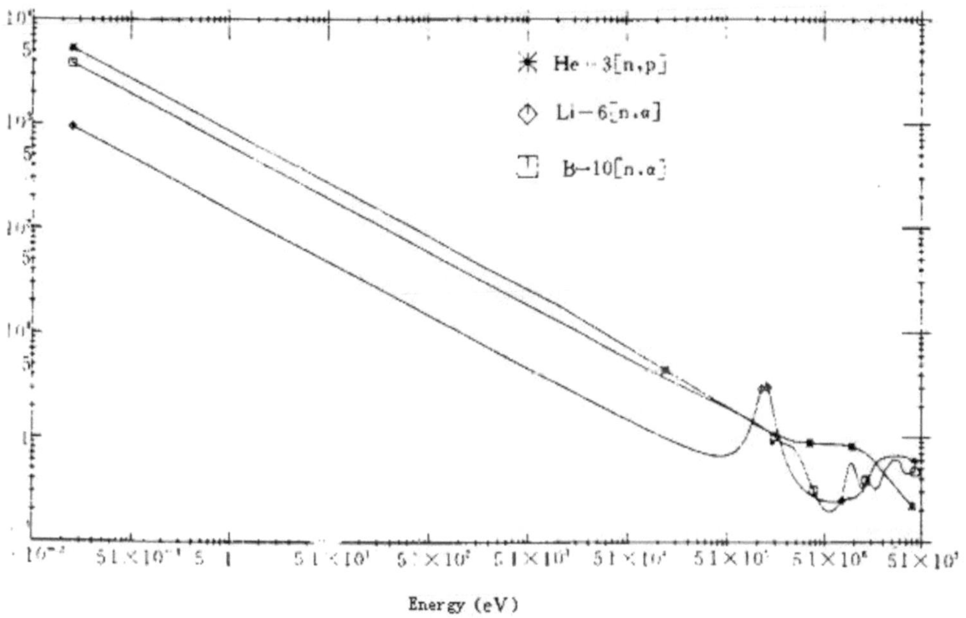

Fig. 8. Neutron cross section of ^{10}B, ^{6}Li and ^{3}He[7]

Neutron reaction with ^{6}Li

Reaction equation:

$$_0^1n + _3^6Li \Rightarrow _2^4He + _1^3T + 4.786MeV$$

In this reaction, when the incident neutron kinetic energy can be neglected, the kinetic energy of α particle and ^{3}He is 2.05MeV and 2.73MeV respectively, and they will move to opposite direction. The reaction cross section is small than $^{10}B(n, \alpha)$ (see fig. 8), but the Q value is large and easy for γ discrimination. Since there is no suitable compound for Li, it is generally used for solid-state neutron detector, including lithium iodide scintillator, lithium glass scintillator and ^{6}LiF TL detector. But the lithium glass scintillator is sensitive to γ ray, hence the size of scintillator cannot be too large for the enhancement of γ ray incentive. So the electron produced by γ ray will escape from the sensitive volume before it is absorbed. Fig. 9 shows the amplitude spectrum for γ ray and thermal neutron by a kind of lithium glass scintillator.

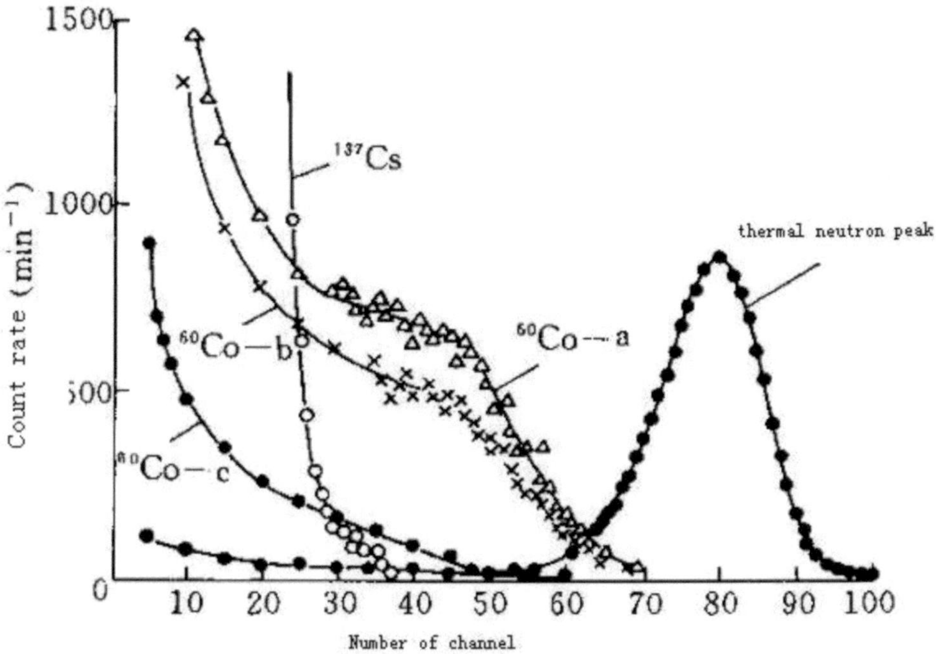

Fig. 9. Amplitude spectrum for γ ray and thermal neutron by a kind of lithium glass scintillator

Neutron reaction with ^{3}He

Reaction equation:

$$_{0}^{1}n+_{2}^{3}He\Rightarrow_{1}^{3}T+_{1}^{1}H+0.765MeV$$

In the reaction produced by thermal neutron, the kinetic energy of proton and tritium is 0.574 MeV and 0.191 MeV respectively. And the two products move in the opposite direction. The reaction cross section is $5330\times10^{-24}cm^2$ (see Fig. 8), which is the largest of the three reactions mentioned above. ^{3}He is noble gas and has no compound at present, so it is only made of gas detector such as proportional counter tube.

4.1.2 Nucleus recoil method

Nucleus recoil method uses the elastic scattering of neutron on light nucleus. Neutron field information could be obtained by detecting the recoil nuclei. According to elastic scattering principle the energy of recoil nuclei is related to its mass and scattering angle, and the momentum increases with decreasing nuclei mass. So, hydrogen is the best recoil media in neutron detection. Fig. 10 shows the recoil cross section of neutron on proton.

If the incident direction of neutron is known, the neutron energy information could be obtained by measuring the spectrum of recoil proton from a certain angle. The energy information can be expressed by:

$$\omega\ (\mathrm{E}) = \frac{\mathrm{C}}{\sigma_{\phi}(\mathrm{E})} W(aE\cos^2\varphi) \tag{4-2}$$

Here C is the normalized constant.

Thus, the conversion of recoil nucleus spectra to neutron spectra attribute to the recalibrate of energy scale(α E$\cos^2\varphi$ replace EA). This method is called differential measurement. But the weakness of this method is the low efficieny, for the thickness of recoil material and the small spanned solid angle of detector.

When the neutron energy is not too high, the scattered proton in all direction could be detected in the hydrogen-filled ionization chamber or counter, the angular distribution of the scattered proton in the center of mass is isotropic, thus the ratio of scattering cross section in certain angular(ψ) and the total scattering cross section should be $\sigma(\psi)/\sigma=1/4\pi$. The probability of recoiled proton scattering in angle ψ with energy E_A could be given:

$$\mathrm{W}(E_A) = \frac{1}{aEn} \tag{4-3}$$

Where En is the incidental neutron energy, $a = \dfrac{4A}{(1+A)^2}$

The formula shows that the recoiled proton energy could be any value in the range of 0 to aEn with equal probability when the incident neutron energy is En. So if the detector is exposed to monoenergetic neutron, the shape of recoil spectrum could be a rectangle (see Fig. 11).

If the incident neutron energy is a function of ω(E) , the recoil nucleus of energy E_A could be produced by any neutron with energy greater than E_A/a . So the producing probability of nucleus with E_A could be given as:

$$\mathrm{W}(\mathrm{E_A}) = C\int_{E_A/a}^{\infty} \sigma(E)\omega(E)\frac{dE}{E} \tag{4-4}$$

Applying differential to the formula, the measured neutron spectrum can be expressed by:

$$\omega(E) = c\frac{E}{\sigma(E)}\cdot\frac{dW}{dE_A}\bigg|_{\mathrm{E}_A=aE} \tag{4-5}$$

This method is called integration method, which is used for neutron With energy not too high and the recoil proton range is not too large. However, due to the shielding effect of the detector surface, the recoil spectrum is not a rectangle as Fig. 11 shown even the detector was exposed in a monoenergy neutron field.

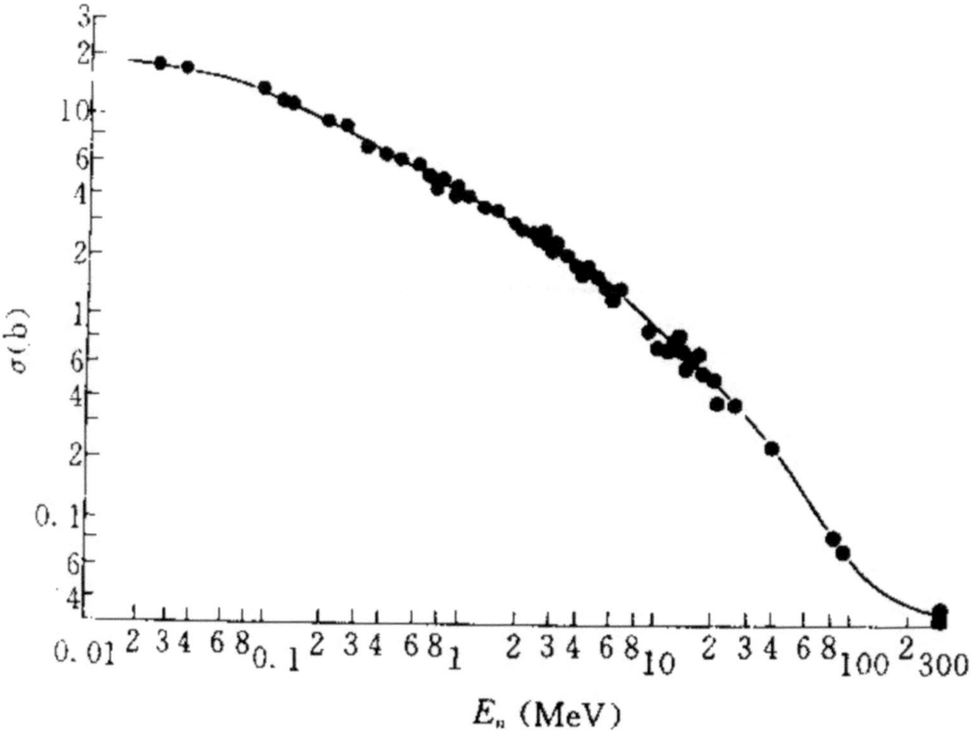

Fig. 10. Recoil cross section of neutron on proton

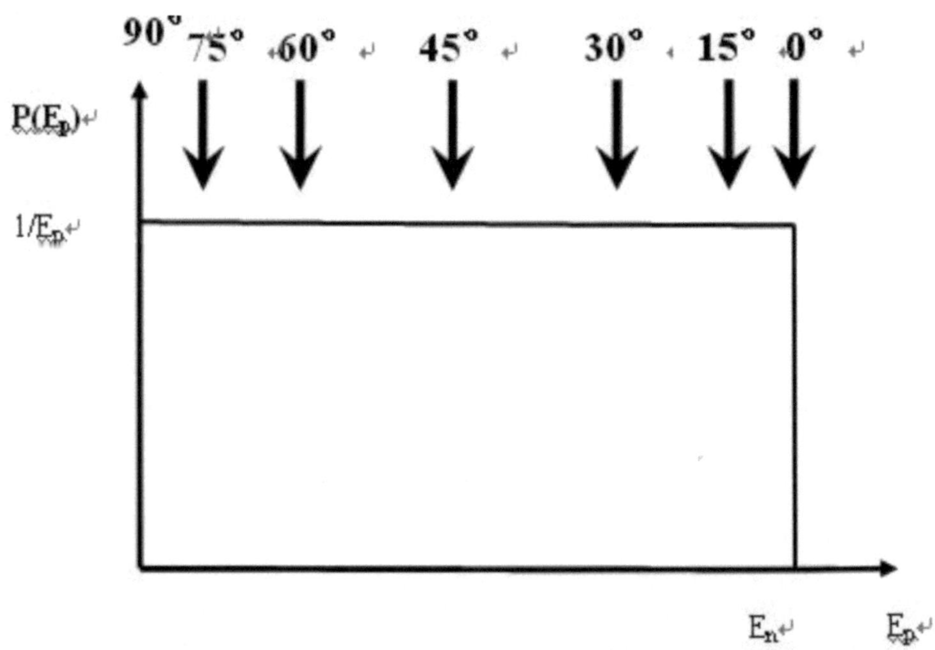

Fig. 11. Energy spectrum of recoil proton produced by mono-energetic neutron

When neutron is detected by scintillator, its efficiency is higher than gas detector. Meanwhile, due to its high stopping power, it can be used for higher energy neutron up to 20MeV measurement. But other effect of scintillator such as multiple scattering and γ ray etc will influence the shape of the neutron spectrum. Its spectrum is also not a rectangle. At the same time, the spectrometer energy resolution will become worse with the increasing volume of scintillator (the corresponding neutron energy will also increase).

4.1.3 Nucleus fission method

Nucleus fission method usually refers to detect the fission fragments which are produced by neutron captured by heavy nucleus to obtain neutron field information. Slow neutron and fast neutron can lead to heavy nucleus fission and release about 150~170MeV for each fusion, dividing into two fragments with masses nearly equal to each other. The energy released in these reactions almost equal distribute to the two fragments and each with 40~110MeV. The most advantages based on fission method is the high output pulse amplitude and is unaffected by other radiations.

4.1.4 Activation method

The information of the neutron field can be obtained by measure the neutron induced radioactivity. Some neutron reaction channel has a certain threshold, so only those neutrons with the energy higher than that threshold can induce reaction and produce certain instable nuclide. So the neutrons of some certain energy region can be obtained by measuring the products of different channel with different threshold. If we neglect the sample absorption to neutron, the radiation intensity of sample can be given as:

$$A = N\sigma\varphi(1 - e^{-T/\tau}) \qquad (4\text{-}6)$$

Where N is the number of nuclei in the sample, σ is activation cross section, φ is neutron fluence, T is the total irradiation time, τ is the mean life time of radioactive nuclei

The induced radioactivity is a function of radiation nuclide life. Generally, if the irradiation time is long enough (about 5~6 half-life), the radionuclide in sample will reach saturation ($A_{max}=N\sigma\varphi$). This value has no relationship with irradiation time, only with the number of target nucleon and the neutron flux.

For the neutron activation measurement, the ideal cross section shape of threshold detector is rectangle, below the threshold energy the cross section is 0 and above the threshold energy the cross section is a constant. However, it is difficult to find the ideal reaction channel of the characteristic mentioned above, Fig. 13 shows the excite function of some threshold detector. From Fig. 13 almost all the activation reaction cross section σ is the function of neutron energy En. So in actual measurement, the concept of average cross section is introduced.

$$\overline{\sigma} = \int \frac{\sigma(\mathrm{E}) \cdot \varphi(\mathrm{E})}{\varphi(E)} dE \qquad (4\text{-}7)$$

It can be seen from (4-7) that the average cross section is related to neutron spectrum. Yet the neutron spectrum is the parameter which is needed to be measured, so final neutron spectrum was obtained by iterative method for the original neutron spectrum which was given in advance.

For long-life products reaction, it is difficult to get saturation in limited time, so the relationship between neutron average fluence and irradiation time should be calculated. After a period of irradiation, to measure the γ ray intensity with a γ spectrometer , the flux of neutron with energy higher than the threshold could be given as:

$$\bar{\phi}=\frac{N_1 \cdot \lambda}{N \cdot \bar{\sigma}(1-e^{-\lambda t_0})[e^{-\lambda(t_1-t_0)}-e^{-\lambda(t_2-t_0)} \cdot \varepsilon \cdot \eta} \tag{4-8}$$

Where

$\bar{\sigma}$ is the average cross section in the measured energy region

λ is the decay constant of the product nucleus

N_1 is the net counts under the γ ray peak measured in the time interval t_2 to t_1

T_0 is the total irradiation time of the sample

ε is the efficiency of γ–ray spectrometer

η the branching ratio of the γ□ray measured

The selection of detector is depended to the neutron spectrum which is to be measured. At low energy region(En<20MeV) there are many activation detectors available (see Table 2). But for high energy region the choice of activation detectors are quite limited. Routti[8] et al reported that there are many reactions of neutron on Cu, many reaction channels could be used as activation detectors (see fig. 14), but there is no report about the experiment on this reaction. The attempt to measure high energy neutron was carried out at the Institute of Modern Physics, the Chinese academic of science in 2001[9]. Bi threshold detector was introduced to this experiment and the highest energy of neutron which was measured in the experiment is up to 80MeV. Fig. 15 shows the cross section of neutron activation on Bi [10].

The activation method for neutron measurement is simple, for the detector can be processed to any shape and can be used in any geometry condition. But it only gives integral neutron flux, and it requires a stable time structure of the radiation field and the off-line measurement method should also be established in advance. In addition, the choice of detectors and the irradiation time have certain requirement. For detector choice, if the half-life is too long, the yield of produced radioactive nuclides is relative low; if the half-life is too short, the produced radioactive nuclides will decay over in a short time after irradiation. As to irradiation time, the best choice is the time interval that is 5~7 times of the produced radioactive nuclides half-life, after the sample irradiated for that time it is just saturation. For short time irradiation the radioactivity in sample is too low and for long time irradiation reflects the radiation field information in the later period of the irradiation.

Almost all neutron detection method is developed based on the four basic principles mentioned above. For a certain neutron field measurement, appropriate method should be carefully selected. Table 3 shows the comparison of four methods.

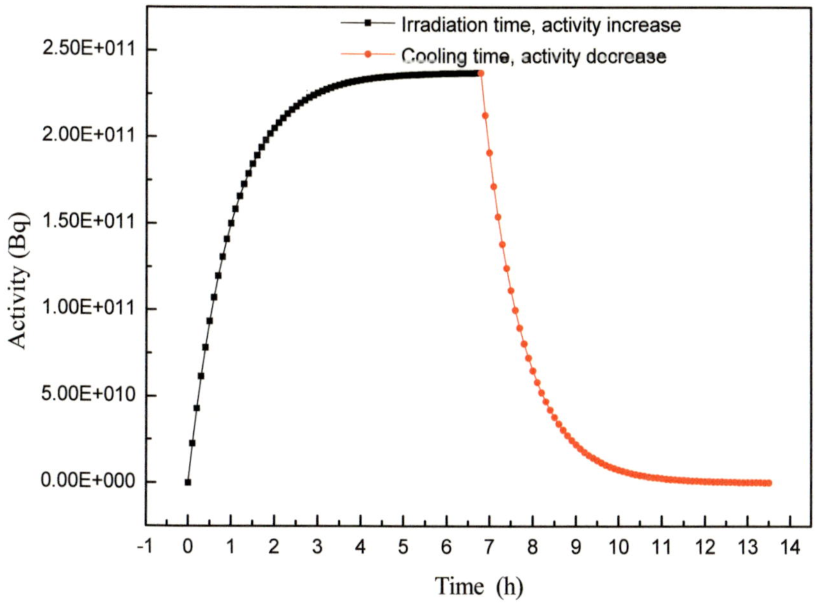

Fig. 12. Producing and decay tendency of radioactive nuclides in irradiation sample

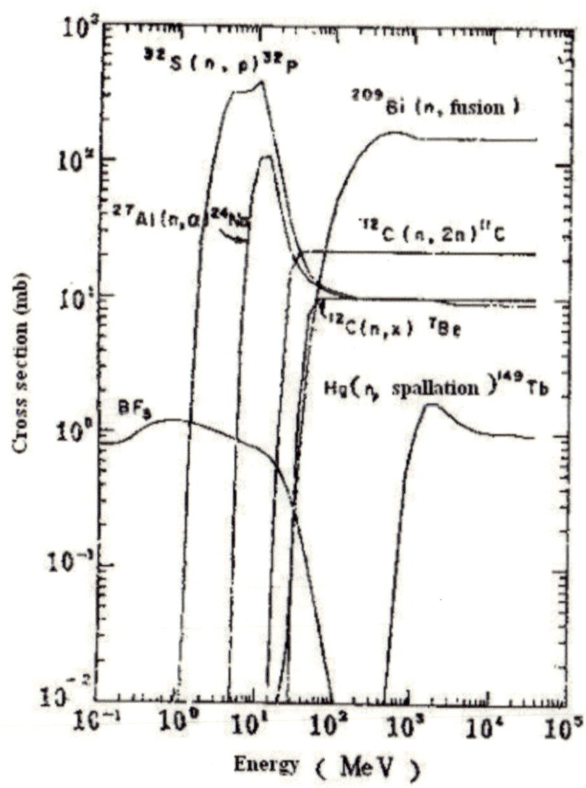

Fig. 13. Response function of neutron detector for activation method

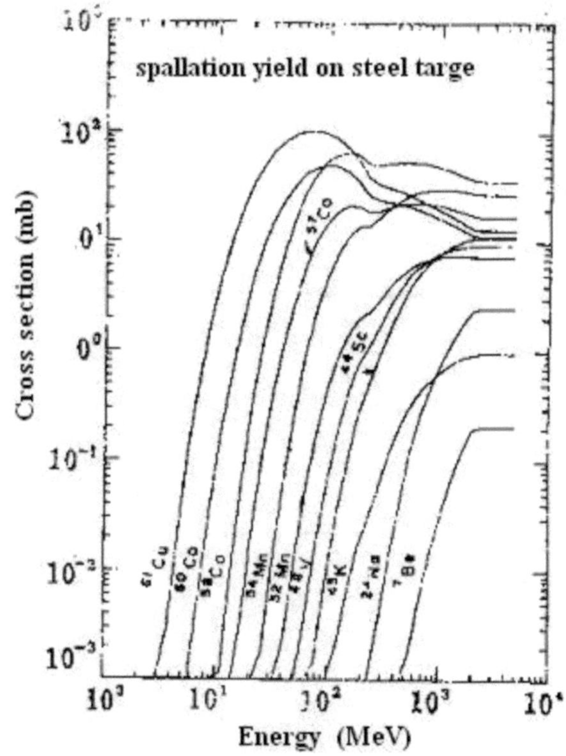

Fig. 14. Cross section of (n, Cu) reactions

Neutron energy	Detectors	Reaction channel	Reaction cross section(mb)	Half-life
Thermal neutron	^{115}In	$^{115}In(n,\gamma)^{116m}In$	119	54m
18.8ev	W	$^{186}W(n,\gamma)^{187}W$	350b	24.1m
580ev	Cu	$^{63}Cu(n,\gamma)^{64}Cu$	5.6b	12.8h
0.1MeV	Nb	$^{93}Nb(n,n')^{93m}Nb$	75.2	13.6y
1.2MeV	In	$^{115}In(n,n')^{115m}In$	170	4.5h
2.8	Zn	$^{64}Zn(n,p)^{64}Cu$	30	12.8h
4.4	Al	$^{27}Al(n,p)^{27}Mg$	4.12	9.45m
6.8	Cu	$^{63}Cu(n,\alpha)^{60}Co$	0.35	5.27y
7.2	Al	$^{27}Al(n,\alpha)^{24}Na$	0.693	15.06h
11.5	V	$^{51}V(n,\alpha)^{48}Sc$	0.08	44h
12	F	$^{19}F(n,2n)^{18}F$	0.09	1.83h
12.4	Cu	$^{63}Cu(n,2n)^{62}Cu$	0.0915	10.1m
13.5	Ni	$^{58}Ni(n,2n)^{57}Ni$	0.0056	36h
20	C	$^{12}C(n,2n)^{11}C$	0.022	20.34m

Table 2. performance of some detectors used in active reaction

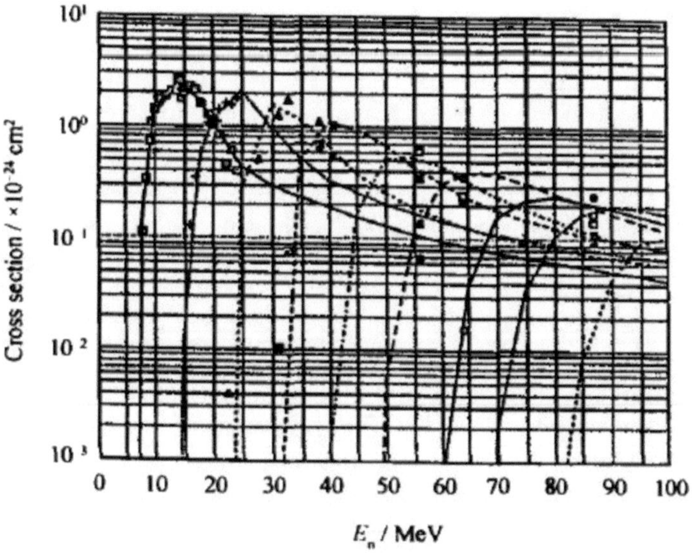

□Bi(n,2n),BNL; + Bi(n,3n),BNL; — (n,2n),ENDF/VI;
··· (n,3n),ENDF/VI; ---(n,4n),ENDF/VI; ---- (n,5n),ENDF/VI;
-·-·- (n,6n),ENDF/VI; —·(n,7n),ENDF/VI; — (n,8n),ENDF/VI;
··· (n,9n),ENDF/VI; ---(n,10n),ENDF/VI; ▲ (n,4n),INS,TIARA;
■ (n,5n),INS,TIARA; □ (n,6n),TIARA; ● (n,7n),TIARA;
O (n,8n),TIARA

Fig. 15. Neutron reaction cross section of Bi

4.2 Neutron dose measurement in particle therapy

What health physicist should be concerned in particle therapy is the additional dose to health tissue from different radiations during treatment. Neutron has strong penetration power and can affect almost whole body, and due to this it is hard to avoid neutron exposure during treatment. So neutron dose distribution in tissue is what we want to know eventually. But as discussed before, neutron absorbed dose is hard to be measured directly; usually we can measure neutron yield, fluence rate, or more precisely, neutron spectra, then using conversion factor to obtain a roughly neutron dose.

Neutron dose is hard to be measured directly, because there doesn't exit a neutron detector which has the same response as human body. However, based on the study of neutron response of Bonner sphere, it is found that larger sphere have the similar neutron energy response to that of the human body. Remmeter was then been invented, it was expected to give an approximately neutron equivalent dose response over a wide energy region (thermal to 15 MeV) when it was calibrated at only one energy beforehand.

All kind of remmeters are fitted with at least one thermal neutron detector, surrounded by moderator material, usually polyethylene or paraffin wax. Study results indicate that this kind of structure may overestimate neutron dose in the intermediate energy region, and underestimate neutron equivalent dose above 7 MeV. Thus some modifications were performed to improve the response of the remmeter to intermediate energy region. Undoubtedly A-B remmeter is the mostly successful one. In the structure of the moderator of a A-B remmeter, it contains a thermal neutron absorb layer in which holes have been drilled to allow some slow neutron pass. This improved the neutron response effectively.

For the purpose of measure higher-energy neutron dose, improvement of the moderator of the remmeter continued. Birattari[11] reported an Improved remmeters which consist of high-atomic number inserts such as lead or tungsten in the moderator. The interaction of high-energy neutrons with this inserted material causes neutron multiplication and energy

Method	Principle	Characteristics	examples
Nucleus excite	Measuring the radioactive nuclide produced by neutron reaction with some materials	• With small volume and easy operation; • With strong Anti-γ ability; • Only obtain the average fluence; • No online measurement result; • Complex technology of accumulating radioactivity measurement; • Limitation choice in detector.	$^{115}In\ (n, g)\ ^{116m}In$; $^{19}F(n,2n)^{18}F$; $^{27}Al(n,a)^{24}Na$; $^{27}Al(n,p)^{27}Mg$; $^{63}Cu(n,2n)^{62}Cu$ ……
Nucleus reaction	Measuring the producing charged particles in neutron reaction	• With strong Anti-γ ability; • Applied to thermal neutron measurement;	BF_3 count tube; ^{3}He count tube; Boron painted counter tube; Li glass scintillator
Nucleus Recoil	Measuring the emit proton by neutron scattering on proton	• Neutron-proton scattering cross section are well known; • Neutron energy information can be obtained; • γ ray influence.	Recoil proton ionizing; Organ crystal scintillator; Plastic scintillator; Recoil proton nuclear emulsion
Nucleus fission	Measuring the fusion fragments of fission reaction of neutron on ^{235}U、^{238}U、^{239}Pu etc	• With grate Q value in fusion and a obvious distinction of other; competitive reaction; • Suitable for thermal neutron measurement.	Fusion ionizing chamber

Table 3. Comparison of four methods for neutron measurement

degrading reactions such as (n, xn), thus improving the sensitivity to high-energy neutrons. By this way the survey meter can extend the neutron response to about 5 GeV. Fig. 16 shows the structure of the detector, Fig.17 shows a typical a arrangement to measure neutron dose with dosimeter.

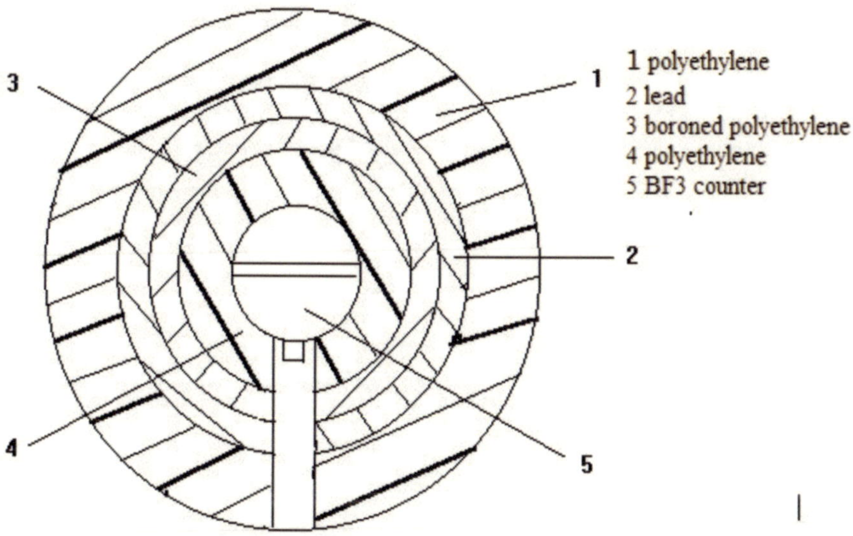

Fig. 16. Schematic illustration of the structure of an extended range neutron rem meter

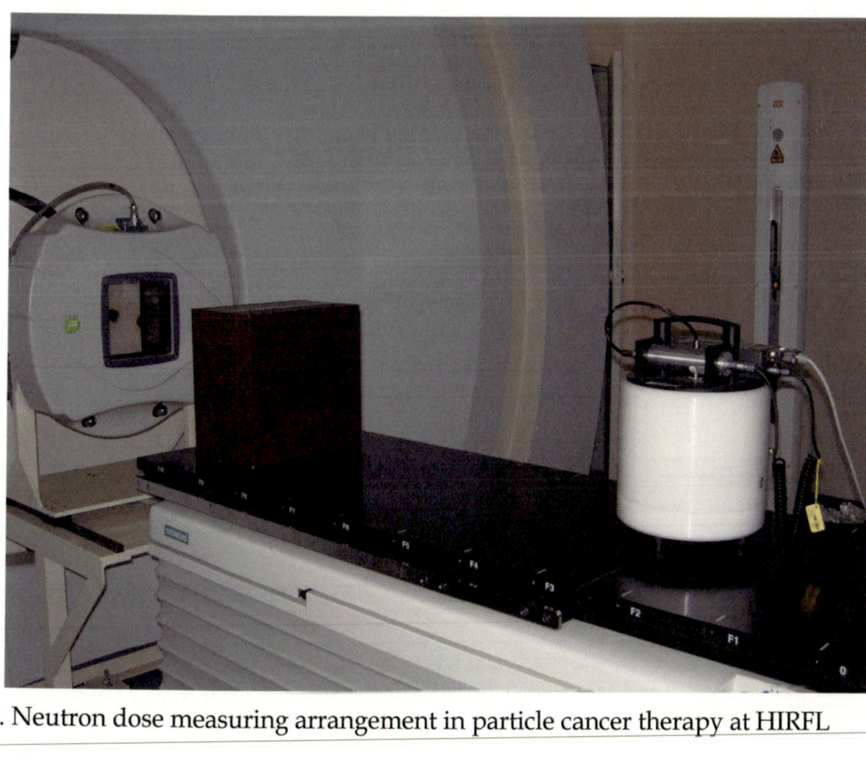

Fig. 17. Neutron dose measuring arrangement in particle cancer therapy at HIRFL

4.3 Neutron spectrum measurement

For a certain neutron field, knowing neutron dose distribution is not enough. As we described before, many factors may influence neutron dose in tissue, such as neutron energy, produce point, irradiation direction etc. thus neutron spectra measurement is required. Neutron dose distribution can be calculated by using conversion factor if neutron spectrum is known.

Activation method is an old way to measure neutron spectra, by using different threshold detectors neutron fluence with energy higher than certain energy can be measured. But at higher energy region, appropriate threshold detectors are inadequate, hence it confined the usage of this method in particle therapy.

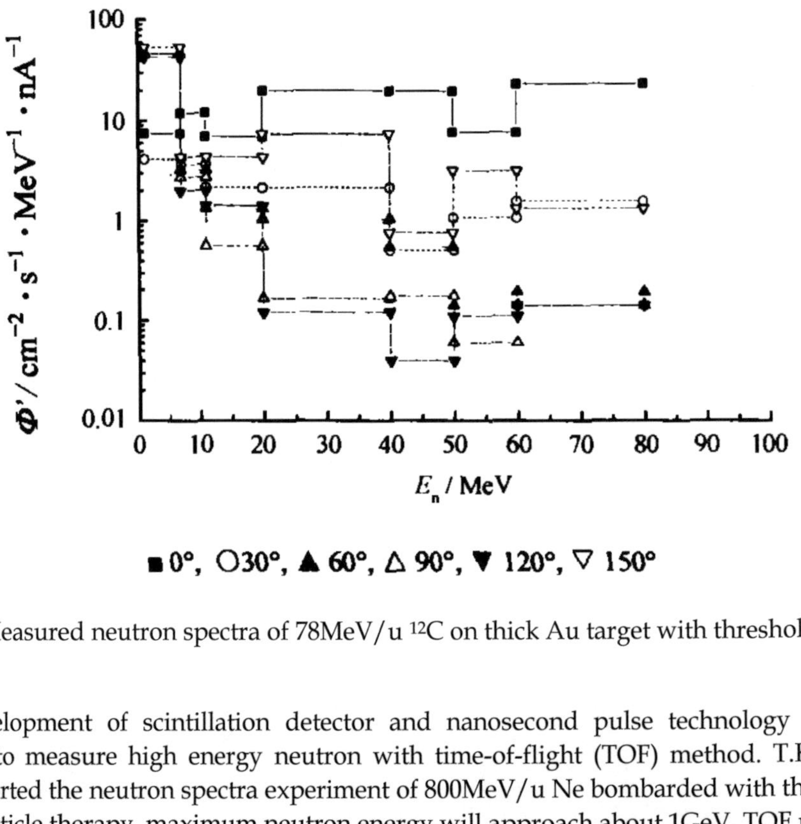

Fig. 18. Measured neutron spectra of 78MeV/u ^{12}C on thick Au target with threshold detectors

The development of scintillation detector and nanosecond pulse technology makes it possible to measure high energy neutron with time-of-flight (TOF) method. T.Kurosawa et.al reported the neutron spectra experiment of 800MeV/u Ne bombarded with thick target [12]. In particle therapy, maximum neutron energy will approach about 1GeV. TOF method is the best way to obtain neutron spectrabut it needs enough flight space to get good energy solution.

At non-relativistic energy region (E_n<20MeV), the relationship between energy E_n and flight distance L is given:

$$\text{En} = \frac{5226L^2}{t^2} \tag{4-9}$$

Where t unit is ns, L unit is m, E_n unit is MeV.

At relativistic energy region, the relationship between neutron energy and its velocity is given:

$$En = \left(\frac{m_0}{\sqrt{1-(\frac{v}{c})^2}} - m_0\right)c^2 \tag{4-10}$$

Flight distance L is easy to determine, but to obtain the flight time is more difficult, which needs to know the start time and the end time. Usually the flight time is determined by the associate particle method and the pulse bunch method, and the end single of neutron flight time is determined by the neutron detector which is located in the terminal of flying distance. Fast organic scintillator are often used as end signal detector, combined with fast photomultiplier tube due to this the system time resolution is better than 1ns.

4.4 Result and discussions

Some experimental and theoretical results were reported in recent years. Nakamura and Korusowa[13-15] had performed a systematic experiment with TOF method in 1998. In their experiment, neutron yield, energy spectrum and angular distribution for 100~800MeV/u He、C、Ne bombarded with carbon, aluminum, copper and lead target were obtained. Although the aim of these experiments are not for particle therapy, but the information obtained were quite valuable to estimate neutron dose during treatment. With the rapid progress of cancer therapy research, more attention is focused on the neutron yield in the energy region for cancer therapy. For proton it is about 250MeV and for carbon ion is about 400MeV/u. Meanwhile, the target materials more inclined to tissue equivalent such as water, polyethylene and carbon. K.Gunzert-Marx et al[16] measured the neutron distribution of 200MeV/u ^{12}C on thick water target in 2004, and D. Schardt et al[17] measured nuclear fragments and fast neutron spectrum of 200MeV/u and 400MeV/u ^{12}C on various thickness water target in 2007.

Monte-Carlo calculation is also widely used in estimating neutron field in particle therapy. A., Agosteo[18] simulated the double differential cross section(see fig 4-7) for neutron yield in particle therapy in 1996, the good agreement with experimental results provide us a good tool to evaluate neutron dose before treatment.

At the same time, neutron dose distributions are investigated in different conditions. H. Iwase et.al. measured neutron dose angular distribution in a particle therapy facility with extended neutron rem meter Wendi-II in 2007[19], pencil-like ^{12}C beam was used in their experiment.

The experimental and calculated results show that secondary neutron dose in cancer therapy is about 1% of heavy ion dose. Since neutron can affect a large volume, to obtain the same treatment dose, different tumor dimension may cause different neutron dose. Furthermore, different treatment geometry arrangement and different irradiation direction may affect neutron dose too, so it is essential to evaluate neutron dose for a certain therapy facility.

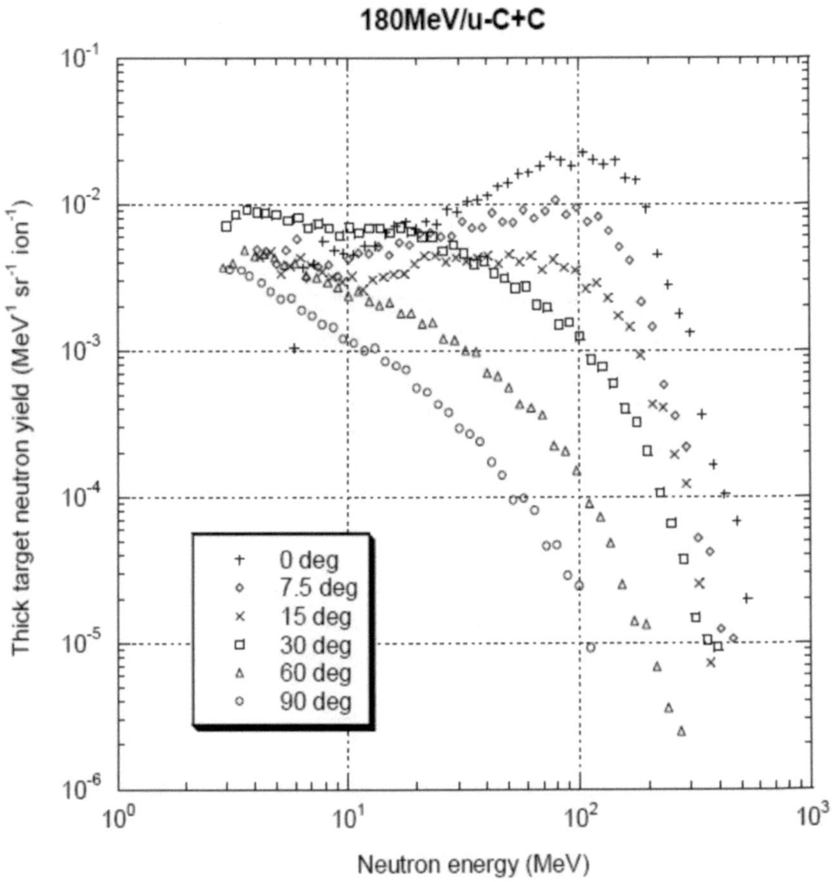

Fig. 19. Neutron spectra from 180 MeV/nucleon C ions incident on a C target

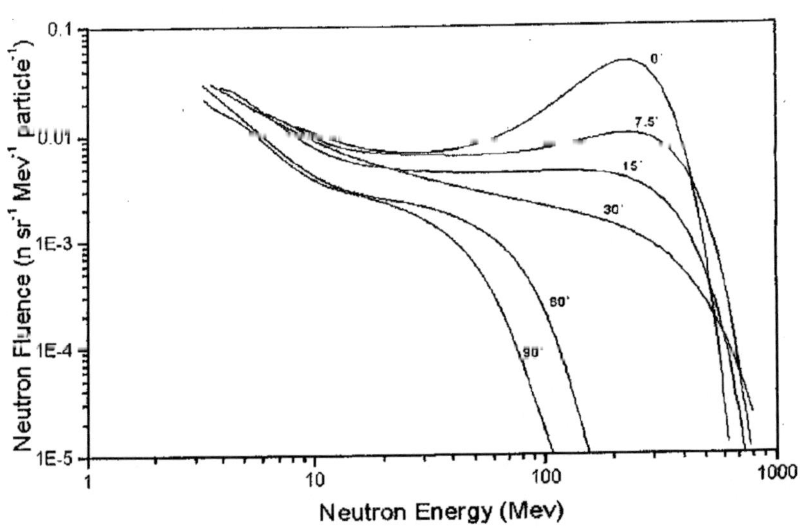

Fig. 20. Neutron angle distribution from 400 MeV/nucleon C ions incident on a Cu target

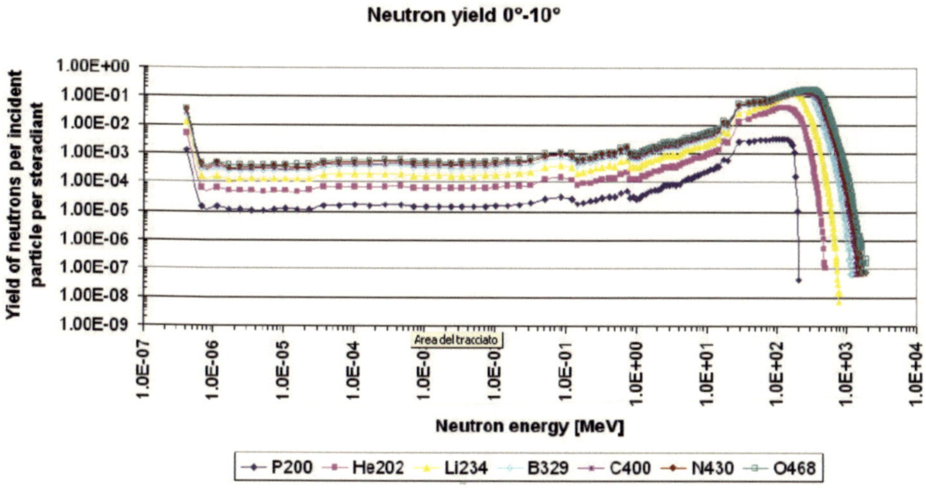

Fig. 21. Total neutron yield expressed as neutron per unit of solid angle and per incident particle in the 0° and 10° angular bin

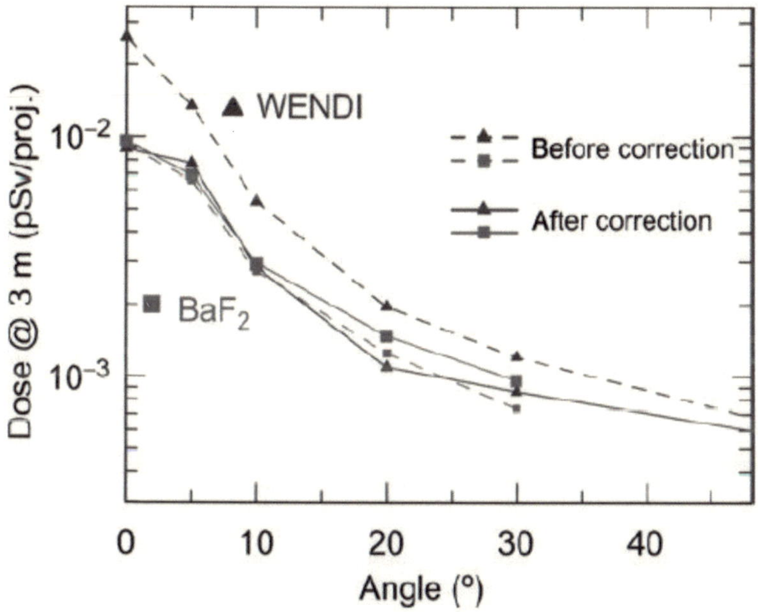

Fig. 22. Neutron angular distributions by 200 MeV/u ^{12}C on 12.8-cm water.

5. References

[1] International Commission on Radiation Units and Measurements, Basic Aspects of High Energy Particle Interactions and Radiation Dosimtery, ICRU Report 28 (International Commission on Radiation Measurements and Units, Bethesda, MD). NCRP Report 144, National Council on Radiation Protection and Measurements, Radiation Protection for Particle Accelerator Facilities.

[2] H. W. Patterson, R. H. Thomas, Accelerator Health Physics, Academic Press, INC. 1973.
[3] S. L. Lindenbaum, Shielding of high energy accelerator, Ann. Rev. Nucl. Sci. 11, 213(1961)
[4] ICRP publication 60 recommendation of the ICRP, 1990.
[5] Dietzze G. and Sibert B.R.L., Photon and neutron dose contributions and mean quality factors in phantoms of different size irradiated by monoerergetic neutrons. Radia. Res., 140, 130-133.
[6] ICRP publication 74, Conversion Coefficients for use in Radiological Protection against External Radiation, (1997).
[7] Ji Changsong, Neutron detect experimental method, Atom energy publisher, (1998, in Chinese).
[8] J. T. Routti, High energy neutron spectroscopy with activation detectors, Incorporating new methods for the analysis of Ge(Li)gamma-ray spectra and the solution of Fredholm integral equations (Ph. D. thesis), Lawrence Radiation Laboratory report UCRL-18514. 1969, 4.
[9] Li Guisheng, et.al., Experimental verification of a new threshold detector and neutron measurement in heavy ion target area, Nuclear techniques, Vol 26， No.11(2003, in Chinese).
[10] M. Imamura，S. Shibata，Y. Uwamino et a1 · JAERI TIARA Annual Report, April 1994'—March 1995.
[11] Birattari C, Ferrrari A, Nuccetelli C, et al. Nucl Inst and Meth, 1990, A297:250.
[12] T.Kurosawa et.al., Phys.Rev. C62, 044615 (2000).
[13] T. Nakamura et.al., the 4th International Meetings of SATIF, ORNL, Knoxville, USA, Sep., (1998).
[14] T. Kurosawa, et.al., Measurements of Secondary Neutrons Produced from Thick Targets Bombarded by High-Energy Helium and Carbon Ions Nucl. Sci. Eng. 132, 30(1999).
[15] T. Kurosawa et.al., Journal of Nucl. Sci. Tech., Vol.36, No.1, 41 (1999).
[16] Konstanze Gunzert-Marx_, Dieter Schardt and Reinhard S. Simon， Fast neutrons produced by nuclear fragmentation in treatment irradiations with 12C beam Radiation Protection Dosimetry (2004), Vol. 110, Nos 1-4, pp. 595---600
[17] D. Schardt, H. Iwase, R. S. Simon and K. Gunzert-Marx, "Experimental investigation of secondary fast neutrons produced in carbon ion radiotherapy," in Proc. of the International Workshop on Fast Neutron Detectors and Applications, 3-6 April 2006, 7566 University of Cape
[18] Agosteo, S., Campi, F., and Caresana, Double-differential spectra of secondary particles from hadrons on tissue equivalent targets, Radiat. Prot. Dosim. 132 (1), 29-41.
[19] H. Iwase1, et.al., Experimental and theoretical study of the neutron dose produced by carbon therapy beam, Radiation Protection Dosimetry (2007), Vol. 126, No. 1 - 4, pp. 615 - 618

7

The Stopping Power of Matter for Positive Ions

Helmut Paul
Johannes Kepler University Linz, Austria

1. Introduction

When a fast positive ion travels through matter, it excites and ionizes atomic electrons, losing energy. For a quantitative understanding of radiotherapy by means of positive ions, it is necessary to know the energy loss per unit distance of matter transversed, *S*, which is alternatively called stopping power or stopping force or linear energy transfer (LET)[1]. To avoid a trivial dependence of the linear stopping power *S* upon the density ρ, one often uses the mass stopping power S/ρ instead. In the following, we discuss experimental and theoretical stopping power data. Using our large collection[2] (Paul, 2011a) of experimental stopping data for ions from $_1$H to $_{92}$U, the reliability of various stopping theories and stopping tables is estimated by comparing them statistically to these data. We consider here only the electronic (not the "nuclear") energy loss of ions in charge equilibrium.

We treat both gaseous and condensed targets (i.e., targets gaseous or condensed at normal temperature and pressure), and we treat them separately. Solid targets are assumed to be amorphous or polycrystalline. We treat elements, compounds and mixtures.

1.1 Tables and programs

The tables and computer programs used here are listed in Table 1. Program PASS (on which the tables in ICRU Report 73 are based) and the program by Lindhard and Sørensen (1996) (LS) are based on first principles only. The same is true for CasP (Grande & Schiwietz, 2004) and HISTOP (Arista & Lifshitz, 2004), except that they use empirical values (Schiwietz& Grande, 2001) for the ionic charge. The programs by Janni, by Hubert et al. and by Ziegler, and the program MSTAR are semi-empirical. Program LET is not further considered here since it is not independent, but based on Ziegler's programs.

To represent stopping for heavy ions at the highest energies correctly, it is necessary to use the non-perturbational LS theory which is fully relativistic and, in addition, assumes

[1] While "stopping power" considers the energy reducing force of the material, the term "linear energy transfer (LET)" aims at the energy transferred to the surroundings by secondary electrons. If the energy transferred is restricted to electron energies below a certain threshold, this is then the "restricted linear energy transfer"

[2] See the "matrix" in (Paul, 2001a) for the availability of stopping data for various ions and targets.

Name, reference	Z_1	Z_2	(Specific) energy range	Remarks
ATIMA (Geissel et al., 2011)	1 - 92	1 – 92	≥10 MeV/u (as used here)	Based[3] on Lindhard-Sørensen above 30 MeV/u
BEST (Berger, Bichsel, 1994)	1 - 92	1 – 92; 180 compounds[4]	≥ 0.5 MeV/u	Bethe theory with corrections; bare ions
CasP v. 5.0 (Grande & Schiwietz, 2004)	1 - 92	1 – 92, any compound[5]	0.0001 – 200 MeV/u	Default settings used here for target ionization[6]
HISTOP (Arista and Lifshitz, 2004)	many	6	0.01 – 30 MeV/u	HISTOP for the valence electrons, SCA for the K shell of carbon
Hubert et al. (1990)	2 - 103	36 solid elements	2.5 – 500 MeV/u	
ICRU Report 49[7] (1993)	1, 2	25 elements, 48 compounds or mixtures	0.001-10000 MeV (p); 0.001-1000 MeV (α)	Programs NIST PSTAR, NIST ASTAR
ICRU Report 73 (2005)	3 – 18, 26	25 elements, 31 compounds	0.025 – 1000 MeV/u	Based on PASS
Janni (1982)	1	1 – 92; 63 compounds	0.001 – 10000 MeV	
LET (Zajic et al. (1999), Zajic (2001)	1 – 92	19 materials	0.2 – 1000000 MeV/u	Based on Ziegler's TRIM/SRIM programs (before 1999)
MSTAR (Paul, 2003)	3 - 18	31 elements, 48 compounds or mixtures	0.00025 – 250 MeV/u	Based on alpha stopping powers of ASTAR
PASS (Sigmund & Schinner, 2002)	many	many	Above 0.025 MeV/u	Binary Theory. Used for ICRU 73
SRIM[8] 2003 (Ziegler, 2004)	1 – 92	1 – 92, many compounds	1.1 eV – 10 GeV/u	SRIM stopping was not changed since 2003
Ziegler et al. (1985)	1 - 92	1 – 92; many other targets	0.1 – 100000 keV/u	First program to treat all ions, all targets

Table 1. Tables and computer programs for the stopping power of positive ions. "u" is the unified atomic mass unit, also called dalton.

[3] Below 10 MeV/u, the values are based on an old version of SRIM (Ziegler et al., 1985). Between 10 and 30 MeV/u, the values are interpolated between SRIM and LS.
[4] Additional compounds may be calculated by entering a chemical formula
[5] Compounds are calculated according to chemical formula, assuming Bragg additivity.
[6] Target and projectile ionization must be calculated separately, and added.
[7] At high energy, the ICRU table was calculated using BEST
[8] SRIM was called TRIM in earlier times

projectile nuclei of finite size. For convenience, we have employed the program ATIMA (Geissel, Scheidenberger, et al., 2011) which is based on the LS program above 30 MeV/u and which includes shell, Barkas and Fermi-density effect corrections and in addition, a correction for projectile mean charge. But the use of the LS program will hardly be necessary for radiation therapy, since even for oxygen ions at 690 MeV/u, there is no difference between LS theory and Bethe theory (eq. 1) (Scheidenberger et al. 1994).

At high (but not too high) energies, where the ion has lost all electrons, the stopping power can be calculated by the relativistic Bethe theory without corrections (Bethe, 1932; ICRU Report 49):

$$S/\rho = (0.307075 \mathrm{MeVcm}^2 g^{-1}) \frac{Z_1^2}{\beta^2} \frac{Z_2}{A_2} L(\beta) \tag{1}$$

where Z_1 and v are charge number and velocity of the ion; Z_2 and A_2 are charge number and mass number of the target; $\beta = v/c$ (c = speed of light); and the stopping number L is given by

$$L(\beta) = \ln \frac{2mv^2}{I(1-\beta^2)} - \beta^2 \tag{2}$$

where m is the rest mass of the electron, and I is the mean ionization energy of the target. In this simple case, I is the only non-trivial constant that describes the stopping power. It can be deduced from optical or from stopping data. An earlier claim (Smith et al., 2006) that the results of these two methods may be in conflict, has been disproved (Paul et al., 2009a).

Lists of mean ionization energies I can be found in ICRU Report 49. The high energy parts of the stopping tables in this report were calculated using program BEST (Berger & Bichsel 1994). This program is also useful to calculate the stopping power eq. (1); normally, it uses the same I values as ICRU 49, but it also permits to enter a different value. BEST also includes the shell, Barkas, Bloch and Fermi-density effect corrections (see ICRU 49) not shown in eq. (2). It assumes a bare nucleus and is therefore not useful below about 1 MeV/nucleon.

At lower energies, the ion will carry electrons, and equilibrium between capture and loss of electrons will develop, leading to a certain mean charge of the ion, lower than $Z_1 e$. Also, the Bethe eq. (1) must then be extended by the corrections mentioned.

1.2 Mixtures and compounds

For a mixture or, assuming Bragg's additivity rule (Bragg & Kleeman, 1905), for a compound, the mass stopping power is obtained by a linear combination of the constituent stopping powers (ICRU Report 49):

$$\frac{S}{\rho} = \sum_j w_j \left(\frac{S}{\rho} \right)_j \tag{3}$$

where w_j is the fraction by weight, and $(S/\rho)_j$ is the mass stopping power of the jth constituent. The corresponding relation for the mean ionization energy is

$$\ln I = \left(\sum w_j \frac{Z_{2j}}{A_{2j}} \ln I_j \right) \Big/ \left\langle \frac{Z_2}{A_2} \right\rangle \tag{4}$$

where

$$\left\langle \frac{Z_2}{A_2} \right\rangle = \sum w_j \frac{Z_{2j}}{A_{2j}} . \tag{5}$$

A list of mean ionization energies I and other properties for 48 compounds and mixtures of interest to particle therapy can also be found in ICRU Report 49. To calculate the stopping power, I values different from those in the main list for elements were used, in an attempt to correct for the influence of binding and phase effects (see Table 2.11 of ICRU Report 49).

Some of the I-values in ICRU 49 are probably outdated, and a commission of the ICRU is working to improve the values for water and graphite. Comparisons with newer values of ionization energies are shown by Paul and Berger (1995), and by Paul et al. (2007a). The particular case of water is discussed in sect. 5 below.

BEST will also calculate the stopping of any compound defined by a chemical formula, and in particular, for 180 numbered compounds and mixtures[9] identified by three-digit ID numbers.

The file compound.dat in the SRIM program contains information for many compounds, including those covered by ICRU Report 49. Compound.dat also includes corrections for a deviation[10] from Bragg additivity (Ziegler & Manoyan, 1988) that becomes noticeable below 1 MeV/nucleon. In addition, it contains instructions on how to add more compounds to the SRIM program. To produce the data in table 6 below, we have added the properties of many compounds contained in our data base. Properties of compounds are also given by Janni (1988), and by Moyers et al. (2010).

1.3 Statistical comparisons

For statistical comparisons between experimental data and tables, we use our program "Judge", v. 3.19 (Paul & Schinner, 2001). This program calculates the normalized differences

$$\delta = (S_{ex} - S_{tab}) / S_{ex} \tag{6}$$

for every data point. Here, S_{ex} is the experimental value, and S_{tab} the corresponding table value for the same ion, same target and same energy. In every range of specific energy, i.e., energy per nucleon, it then determines the average normalized difference:

$$\Lambda = \langle \delta \rangle \tag{7}$$

[9] For only 48 of these, where experimental low-energy stopping data were available, stopping tables are given in ICRU Report 49. The properties of all the 180 substances can be found in program NIST ESTAR for electron stopping powers.

[10] These corrections are only applied for H and He ions. The absolute values of the non-zero Bragg corrections amount to about 3 %, on the average (Paul & Schinner, 2006). An attempt to test the accuracy of those corrections statistically is shown in the same paper.

and its standard deviation

$$\sigma = \sqrt{\langle \delta^2 \rangle - \langle \delta \rangle^2} \tag{8}$$

The averages are unweighted, except that obviously discrepant data are rejected (Paul, 2011a). A small Δ usually signifies good agreement between table and experimental data; in such a case, σ is related to the mean experimental accuracy, and σ may be taken as a measure of the accuracy of the table, as determined from experiment.

2. Hydrogen and helium ions

2.1 Hydrogen and helium ions in elements

In Tables 2 and 3, the reliability of various stopping power tables for H and He ions in solid elements is given in terms of Δ ± σ. Here, *E* is the energy of the ion. These tables were originally published by Paul & Schinner (2005), but many new data have since been added. This has not changed the results much, but it adds to the reliability.

E/A_1 (MeV) *Number of points*	0.01-0.1 *1357*	0.1 – 1 *2492*	1 – 10 *1212*	10 - 100 *225*	0.01 -100 *5286*
Janni, 1982	2.1 ± 11	-1.1 ± 7.1	-0.9 ± 3.6	-0.3 ± 0.5	-0.2 ± 7.7
Ziegler et al., 1985	-1.3 ± 11	-3.1 ± 7.8	-0.4 ± 4.2	0.4 ± 2.2	-1.9 ± 8.2
ICRU, 1993	0.8 ± 11	-0.7 ± 7.0	-0.3 ± 4.0	-0.1 ± 0.5	-0.2 ± 7.5
SRIM, 2003	0.6 ± 10.3	-0.9 ± 6.7	-0.6 ± 3.7	-0.2 ± 0.6	-0.4 ± 7.2

Table 2. Mean normalized deviations Δ ± σ (in %) for H ions in 17 solid elements covered by the ICRU Table, compared to various tables.

E/A_1 (MeV) *Number of points*	0.01-0.1 *1036*	0.1 – 1 *1913*	1 – 10 *400*	10 - 100 *11*	0.01 -100 *3360*
Ziegler et al., 1985	3.2 ± 8.7	0.6 ± 5.6	-0.8 ± 3.3	0.8 ± 2.4	1.2 ± 6.7
ICRU, 1993	2.6 ± 8.3	0.2 ± 5.6	0.1 ± 3.3	0.9 ± 0.9	0.9 ± 6.4
SRIM, 2003	3.5 ± 8.2	0.6 ± 5.2	-0.3 ± 3.1	0.2 ± 0.9	1.4 ± 6.3

Table 3. Mean normalized deviations Δ ± σ (in %) for He ions in 16 solid elements covered by the ICRU Table.

One can see that σ always decreases with increasing energy, due to the higher accuracy of measurements at high energy. The numbers of experimental points averaged is also shown, to give an idea of the accuracy. To provide a fair comparison with the smaller number of targets in the ICRU table, we compare only with the targets of that table, even though we have many more targets in our files (Paul, 2011a). We see that generally, σ has decreased and hence, the overall agreement has improved in time, with the exception of (Ziegler et al., 1985); but this was the first table capable of treating all ions and all targets.

Table 4 gives results for H ions in elemental gases. Here, we exclude measurements for low energy H ions in helium (Golser & Semrad, 1991; Schiefermüller et al., 1993; Raiola et al., 2001). Due to the threshold effect (Fermi & Teller, 1947) these data would produce a very large Δ and thus obscure any other discrepancy. Except for the tables by Ziegler et al. (1985) (due to large discrepancies for H and He targets), the gas measurements appear here more reliable than those on solids.

E/A$_1$ (MeV)	0.001 - 0.01	0.01 - 0.1	0.1 - 1.0	1 - 10	10 - 100	0.001 - 100
No. of points	*124*	*335*	*535*	*303*	*11*	*1308*
Janni, 1982	-0.9 ± 9.2	-0.0 ± 4.6	0.5 ± 3.9	0.9 ± 3.2	3.2 ± 0.6	0.4 ± 4.7
Ziegler et al., 1985	22 ± 14	22 ± 11	0.4 ± 6.8	-1.1 ± 1.7	-1.0 ± 0.5	7.7 ± 14
ICRU, 1993	-0.6 ± 6.7	-1.2 ± 5.0	-1.2 ± 3.7	-0.8 ± 1.6	-0.2 ± 0.5	-1.0 ± 4.1
SRIM, 2003	2.1 ± 5.2	-0.1 ± 4.7	-0.4 ± 3.6	-0.2 ± 1.6	0.2 ± 0.3	-0.1 ± 3.9

Table 4. Mean normalized difference Δ ± σ (in %) for H ions in all elemental gases except F, Cl, Rn

Table 5 shows results for He ions in elemental gases. Again, the agreement with the data is much better than for solids, and we can observe a gradual improvement in time.

E/A$_1$ (MeV)	0.001 - 0.01	0.01 - 0.1	0.1 - 1.0	1 – 10	0 - 10
No. of points	*5*	*267*	*863*	*238*	*1373*
Ziegler et al., 1985	7.2 ± 13	2.5 ± 5.9	3.0 ± 4.9	-0.5 ± 2.5	2.3 ± 5.0
ICRU, 1993	0.5 ± 6.8	-1.0 ± 4.2	0.1 ± 4.2	0.7 ± 2.3	0.0 ± 4.0
SRIM, 2003	-5.4 ± 6.1	0.3 ± 3.9	0.1 ± 3.8	-0.2 ± 2.2	0.1 ± 3.7

Table 5. Mean normalized difference Δ ± σ (in %) for He ions in all elemental gases except F, Cl, Rn

2.2 Hydrogen and helium ions in compounds

Data for compounds have been treated in (Paul & Schinner, 2006). In our data base (Paul, 2011a), we have data for 150 different compounds. Table 6 shows results for hydrogen and helium ions in these compounds, compared to SRIM. Because of the different low energy limit chosen[11], some of the results appear somewhat better than for elements. Again, the errors σ tend to be smaller for gases than for solids.

Ions	Targets	E/A$_1$ (MeV)	0.025-0.25	0.25 – 2.5	2.5 - 30	0.025 – 30
H	cond.	*No. of points*	*412*	*946*	*232*	*1590*
		Δ ± σ	-1.3 ± 8.2	1.4 ± 6.3	-0.1 ± 4.0	0.5 ± 6.7
	gas	*No. of points*	*508*	*378*	*24*	*910*
		Δ ± σ	-0.9 ± 4.3	0.1 ± 3.3	-0.9 ± 2.1	-0.5 ± 3.9
He	cond.	*No. of points*	*472*	*1460*	*14*	*1946*
		Δ ± σ	0.4 ± 6.8	-0.5 ± 4.3	-2.0 ± 3.1	-0.3 ± 5.1
	gas	*No. of points*	*997*	*1742*	*0*	*2739*
		Δ ± σ	-2.6 ± 7.2	1.1 ± 2.9		-0.3 ± 5.2

Table 6. Mean normalized deviations Δ ± σ (in %) for H and He ions in condensed or gaseous compounds, as compared to SRIM (2003).

[11] This is to avoid large deviations due to the threshold effect in LiF (Markin et al., 2009)

Table 7 shows a comparison between SRIM 2003 and ICRU Report 49, for the smaller number of compounds covered by the latter table, for H and He ions together (Paul & Schinner, 2006). For this restricted number of targets, ICRU Report 49 is clearly better than SRIM.

E/A_1 (MeV)	0 – 0.03	0.03 – 0.3	0.3 – 3.0	3 – 30	0 – 30
Number of points	*116*	*1036*	*1237*	*135*	*2524*
ICRU, 1993	0.2 ± 8.9	1.4 ± 5.9	1.3 ± 5.2	1.0 ± 4.4	1.3 ± 5.7
SRIM, 2003	-7.8 ± 12	-1.0 ± 6.4	0.4 ± 5.6	-0.6 ± 4.0	-0.6 ± 6.6

Table 7. Mean normalized deviations $\Delta \pm \sigma$ (in %) for H and He ions in 23 (solid or gaseous) compounds covered by ICRU Report 49 (1993)

Moyers et al. (2010) have recently measured the linear stopping powers for protons at 135, 175, and 225 MeV in many compounds of interest to particle therapy, relative to a water target. They compared their results to the Janni (1988) or LET tables (Zajic, 2001), finding agreement within 1 to 3 %. As examples, Fig. 1 shows a few results by Moyers et al., compared to the Janni, BEST and SRIM tables. The BEST calculation uses the *I*-values of ICRU Report 49, except that I = 78 eV was taken for water (cf. Sect. 5 below). It should be noted that in this energy region, corrections to eq. (2) are small[12], hence eq. (2) would also suffice in place of BEST.

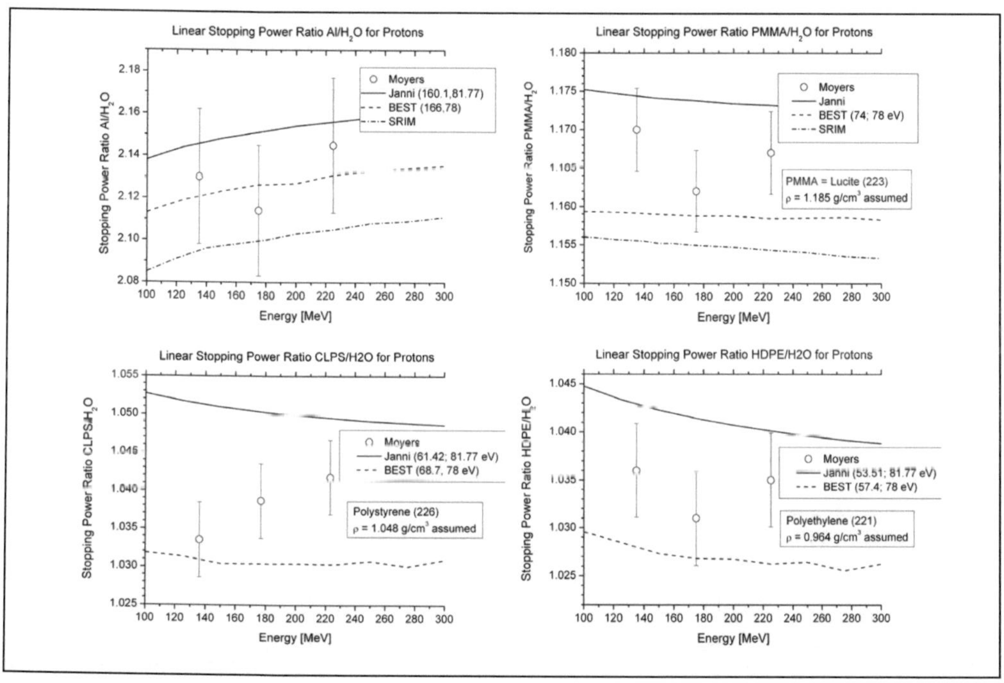

Fig. 1. The linear stopping power of Al, polymethyl methacrylate (PMMA), clear polystyrene (CLPS) and high density polyethylene (HDPE), for protons relative to water, compared to the tables Janni, BEST, and SRIM. The 3-digit ID numbers from ICRU 49 are shown in parentheses. For the curves, the *I* values for both substances are shown in parentheses, where available. Experimental data are from Moyers et al. (2010) and Moyers (2011).

[12] In the case of p in Al, e.g., corrections are below 0.2 %.

Inspection of Fig. 1 shows that the curves are essentially determined by the *I* values. In particular, the Janni curves are always above the BEST curves because of the rather high *I* value for water and the rather low *I* values for the other substances. Evidently, BEST agrees best with the Al measurements. For the compounds, BEST appears slightly low; this might point to slight errors of the I values used.

2.3 Application to particle therapy

Inspection of Tables 2 and 4 shows that, for protons in elements, in the range 10 – 100 MeV, the value of Δ is negligible for the ICRU and SRIM tables, and σ is 0.5 %, on the average. Hence, in this energy range important for therapy, the ICRU and SRIM tables can be expected to be accurate to 0.5 %. And the same accuracy may be expected up to 1000 MeV, if the ICRU or SRIM tables are extended[13] using the pure Bethe theory eq. (1), since the corrections to Bethe are minimal (cf. Fig. 8 below). The same holds for the Janni table for elemental solids (not for gases).

For protons in compounds, the highest energy range (Table 6) goes only up to 30 MeV, and σ is larger (2 – 4 %). Hence, the predictive quality of SRIM appears worse for compounds. On the other hand, since Bragg additivity holds at high energy, the stopping power of compounds at high energy may be calculated using eq. (3), and in this way, the accuracy could be improved somewhat.

3. Ions from $_3$Li to $_{18}$Ar

In tables 8 to 10, MSTAR v.3 (Paul, 2003), SRIM (2003), and ICRU Report 73 (2005) are compared to experimental data. To provide a fair comparison between MSTAR and SRIM, we compare both tables to the same data; not all of these are covered by ICRU 73. These comparisons are based upon our earlier analyses (Paul, 2006) but contain many newer data. This has hardly changed the results, but it adds credibility.

E/A_1 (MeV)	0.025 - 0.1	0.1- 1	1 – 10	10 - 100	100-1000	0.025-1000
No. of points	*1426*	*3821*	*1370*	*190*	*11*	*6818*
MSTAR	2.3 ± 9.6	0.3 ± 6.5	1.1 ± 4.9	0.2 ± 2.1	0.7 ± 1.4	0.9 ± 7.0
SRIM, 2003	1.3 ± 8.8	-0.5 ± 5.0	-0.1 ± 4.8	1.5 ± 2.8	-0.1 ± 1.6	-0.1 ± 6.4
ICRU Rep. 73	-11.7 ± 20	-6.3 ± 11	-2.9 ± 5.8	-0.9 ± 2.9	-0.8 ± 1.9	-6.6 ± 13

Table 8. Mean normalized deviations Δ ± σ (in %) for ions from $_3$Li to $_{18}$Ar in all the elemental solids covered by MSTAR. The number of points refers to MSTAR and SRIM; for ICRU 73, it is slightly smaller since that table does not cover B, Zr, Gd, and Ta targets.

Table 8 shows the reliability of the tables in terms of Δ ± σ for ions from $_3$Li to $_{18}$Ar in solid elements. Similarly, Table 9 gives the reliability of the same tables for the 10 compounds for which we have data. Finally, Table 10 shows results for all gases covered by MSTAR and ICRU Report 73 for which we have data. We find that MSTAR and SRIM describe the data about equally well, and that ICRU 73 is too high at low energy, on the average. Fig. 2 shows an extreme example: the stopping power of Ag for Li ions, where ICRU 73 is too high, and

[13] In the case of ICRU, this simply means using the ICRU table up to 1000 MeV.

E/A_1 (MeV)	0.025 – 0.1	0.1 – 1	1- 10	10 - 100	0.025-100
No. of points	*180*	*775*	*554*	*16*	*1525*
MSTAR	4.8 ± 10.1	0.8 ± 6.4	5.4 ± 4.3	0.8 ± 2.4	3.0 ± 6.7
SRIM, 2003	-2.2 ± 9.4	-0.5 ± 6.0	0.1 ± 5.0	-1.5 ± 2.5	-0.5 ± 6.2
ICRU Rep. 73	-11 ± 11	-2.6 ± 7.4	-1.1 ± 5.0	-0.8 ± 1.7	-3.1 ± 7.9

Table 9. Mean normalized deviations $\Delta \pm \sigma$ (in %) for ions from $_3$Li to $_{18}$Ar in 10 condensed compounds[14]. The number of points refers to MSTAR and SRIM; for ICRU 73, it is slightly smaller since that table does not cover polypropylene and toluene.

E/A_1 (MeV)	0.025 – 0.1	0.1 – 1	1- 10	10 - 100	0.025-100
No. of points	*163*	*190*	*574*	*189*	*1116*
MSTAR	-2.5 ± 10.4	-2.1 ± 12	0.1 ± 3.8	0.7 ± 2.4	-0.5 ± 7.2
SRIM, 2003	3.2 ± 10.1	-7.6 ± 12	-1.0 ± 5.9	-2.2 ± 3.9	-1.7 ± 8.2
ICRU Rep. 73	-50 ± 28	-3.1 ± 16	-1.9 ± 10.3	-0.1 ± 3.8	-8.8 ± 23

Table 10. Mean normalized deviations $\Delta \pm \sigma$ (in %) for ions from $_3$Li to $_{18}$Ar in all (elemental and compound) gases covered by MSTAR and ICRU 73 for which we have data.

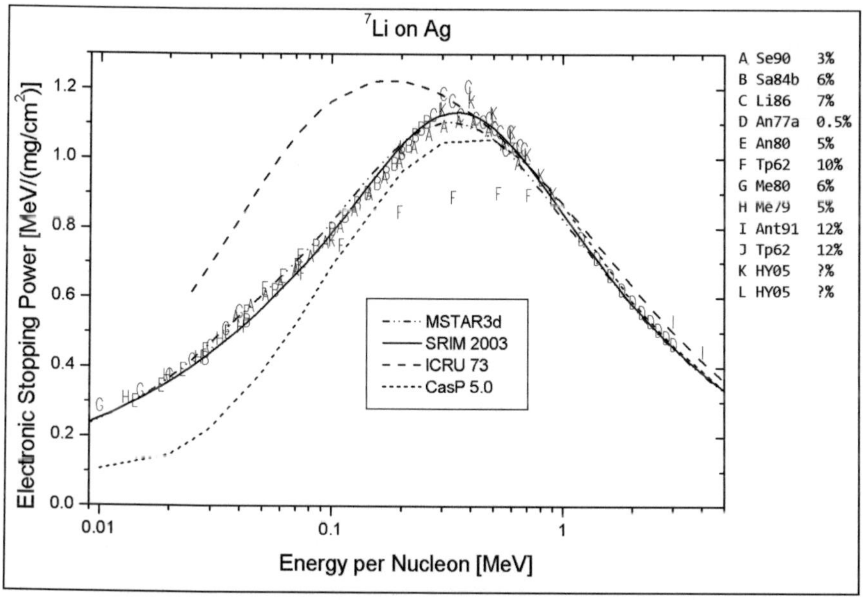

Fig. 2. Electronic stopping power as a function of specific energy for Li ions in Ag, compared to various tables. Experimental points are marked by letters; the references corresponding to the reference codes given in the margin can be found in (Paul, 2011a).

CasP is too low at low energy. Table 10 shows that the overall agreement is here not better for gases than for solids. The agreement of ICRU 73 with the data for gases at low energy is noticeably worse than for solids. This could be related to the fact that PASS uses the same ionic charge for gases as for solids.

[14] Aluminum oxide, kapton polyimide, polycarbonate (makrolon), polyethylene, polyethylene terephthalate (mylar), polypropylene, polyvinyl chloride, silicon dioxide, toluene, water (liquid)

3.1 In particular: Carbon ions

We consider carbon ions especially because of their importance for medical therapy. As an example, Fig. 3 shows stopping powers for carbon ions in carbon. Here, there is good agreement between the experimental data and the MSTAR, SRIM, HISTOP and ICRU 73 tables in most energy regions, while CasP is too low at low energy[15].

Table 11 shows the reliability of MSTAR, SRIM and ICRU 73 for C ions in elemental solids. The overall agreement for MSTAR and SRIM is slightly better than in Table 7 for all ions (Li to Ar), but the highest energy range goes only up to 100 MeV/nucleon. Here, the accuracy in the highest range (10 - 100 MeV/nucleon) is only about 3 % for MSTAR and SRIM, much worse than for protons.

E/A_1 (MeV)	0.025 - 0.1	0.1- 1	1 – 10	10 - 100	0.025-100
No. of points	*202*	*632*	*229*	*8*	*1071*
MSTAR	-1.6 ± 9.6	0.6 ± 5.8	0.9 ± 5.1	0.0 ± 2.8	0.2 ± 6.6
SRIM, 2003	0.4 ± 8.3	-0.5 ± 5.3	-0.6 ± 5.2	1.0 ± 3.0	-0.3 ± 6.0
ICRU Rep. 73	-13.0 ± 12	-9.2 ± 11	-2.6 ± 5.8	-0.6 ± 3.8	-8.5 ± 11

Table 11. Mean normalized deviations Δ ± σ (in %) for C ions in 15 elemental solids covered by MSTAR. The number of points refers to MSTAR and SRIM; for ICRU 73, it is slightly smaller since that table does not cover Gd and Ta targets.

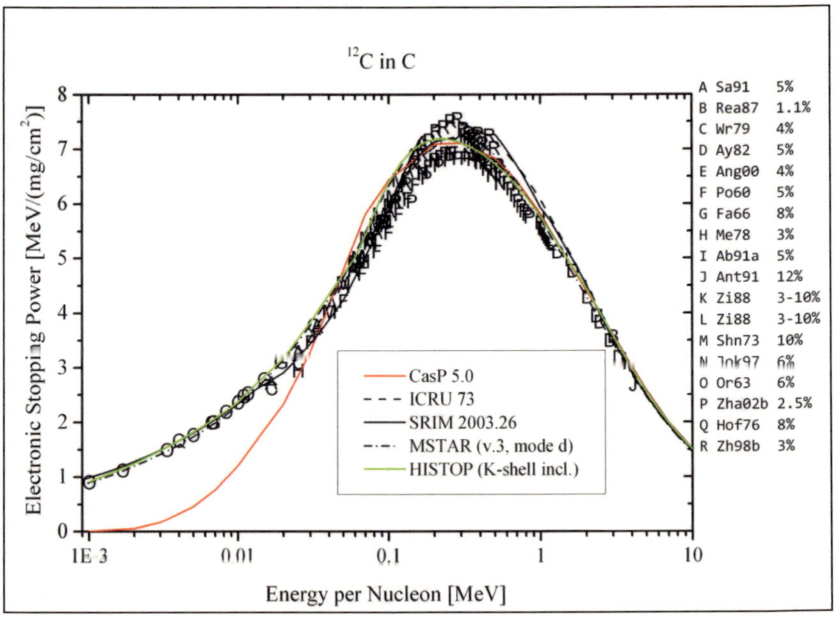

Fig. 3. Like Fig. 2, for C ions in amorphous carbon[16].

[15] This discrepancy has not changed much from CasP v. 3.1 to v. 5.0

[16] The CasP calculation was done using oscillator strengths for carbon, and adding projectile ionization to target ionization.

4. Ions from $_{19}K$ to $_{92}U$

In Table 12 (Paul, 2010), the reliability of stopping tables for ions $_{19}K$ to $_{92}U$ in elemental solids is given numerically. We find that, at the highest energy, only the non-perturbational Lindhard-Sørensen theory (calculated using ATIMA) is correct. Between 2.5 and 100 MeV/nucleon, the Hubert table is best. SRIM is fairly good everywhere, except near the maximum (2.5 - 30 MeV/n). By detailed analysis, it can be shown, that on the average, for heavy ions in solid elemental targets, SRIM is 6 % high in heavy targets and 5 % low in light targets at the maximum, as has been noted already by Randhawa & Virk (1996). For examples, see the graphs for U in Au (Fig. 4) and for Pb in C (Fig. 5).

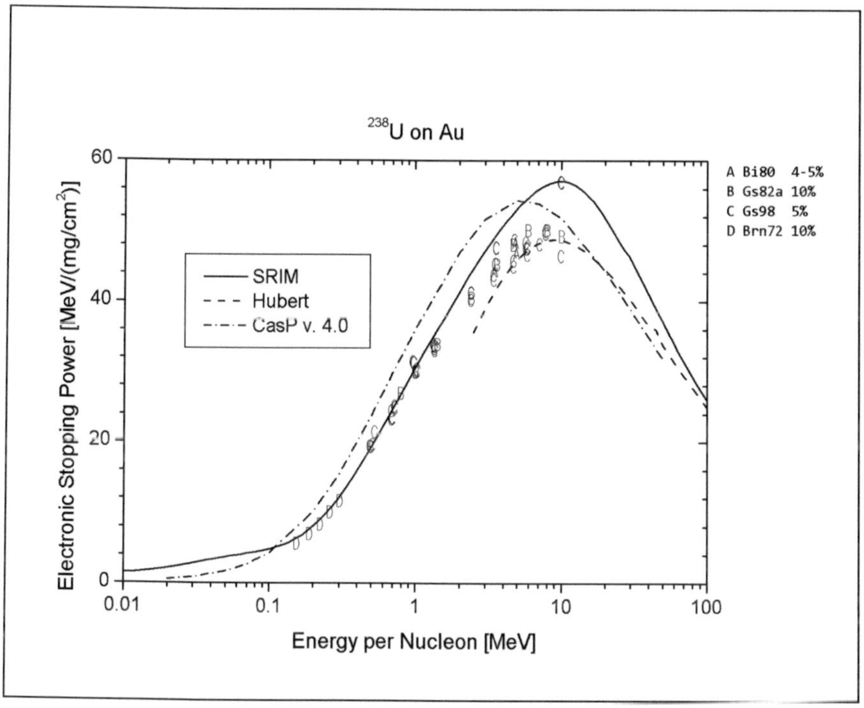

Fig. 4. Electronic stopping power as a function of specific energy for U ions in Au. The data points are indicated by letters; corresponding references can be found in (Paul, 2011a).

Fig. 6 (Paul, 2011b) shows the stopping power for U ions at 10 MeV/nucleon in elements, versus target atomic number Z_2. One can see the well known positive solid-gas difference due to the high collision frequency of fast ions in solids (Geissel et al., 1982; Paul, 2009b) which is well described by CasP 4.0 (due to the different ionic charge states used by CasP for solids and gases) but not by SRIM[17]; SRIM is too high for heavy ions in gaseous elements (see Table 13). For gaseous compounds, SRIM is also too high, especially for the heaviest ions at the maximum (see, e.g., the graph for U ions in Butane in (Paul, 2011a) and Table 14).

[17] The „Gas Tgt" button in SRIM does, however, describe the *negative* solid-gas difference due to polarization screening in the solid, found at low energy, see ref. (Paul 2009b).

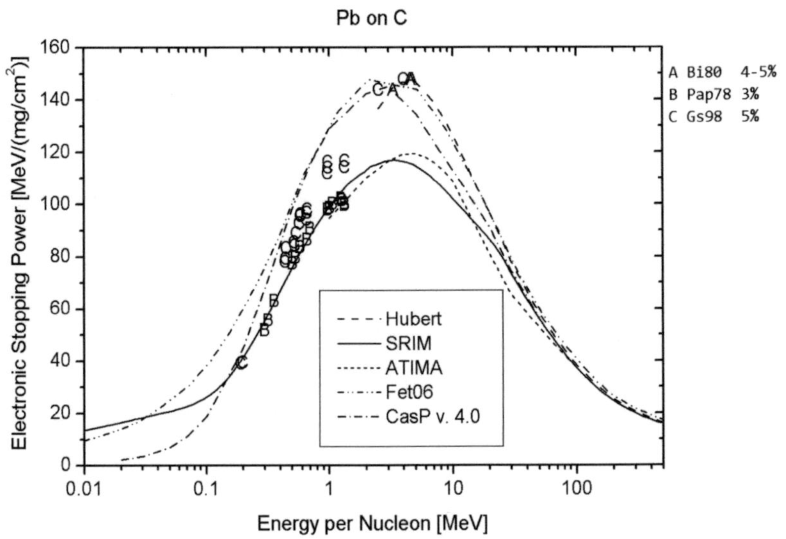

Fig. 5. Electronic stopping power of carbon for Pb ions, versus specific energy. Measured points are indicated by letters; the corresponding references are found in (Paul, 2011a). The curve designations are explained in Table 1, except for Fet06 (Fettouhi et al.). This curve is based on PASS, but incorporating a realistic mean charge of the ion.

E/A_1 (MeV)	0.025-0.25	0.25 – 2.5	2.5 - 30	30 - 100	100 - 500	500 - 1000	Total range
No. of pts.	*655*	*3025*	*1058*	*65*	*43*	*13*	*4859*
SRIM	2.0 ± 19	1.8 ± 6.6	-2.0 ± 9.0	-0.3 ± 3.4	5.0 ± 2.4	7.5 ± 2.2	1.0 ± 9.9
No of pts.			*934*	*65*	*43*		*1042*
Hubert			0.8 ± 5.1	1.1 ± 3.2	4.6 ± 2.5		1.0 ± 5.0
No. of pts.				*65*	*43*	*13*	*121*
ATIMA				2.3 ± 4.0	1.2 ± 1.5	0.9 ± 0.8	1.7 ± 3.1

Table 12. Mean normalized deviations Δ ± σ of experimental data for 31 ions from $_{19}$K to $_{92}$U in all 54 solid elemental targets for which we have data, in various ranges of specific energy.

E/A_1 (MeV)	0.25 – 2.5	2.5 - 30	30 - 100	Total Range
Number of points	*276*	*459*	*38*	*773*
SRIM	1.4 ± 6.9	-6.0 ± 10.3	-7.2 ± 5.9	-3.4 ± 9.7

Table 13. Δ ± σ (in %) for SRIM, for ions from $_{19}$K to $_{92}$U in all elemental gas targets for which we have experimental data in (Paul, 2011a).

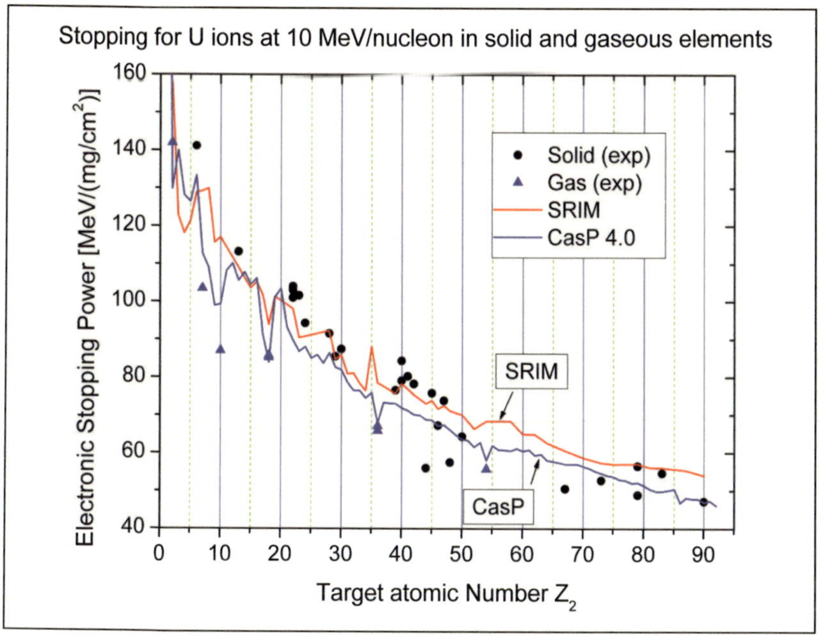

Fig. 6. The stopping power of elements for U ions at 10 MeV/nucleon, as a function of target atomic number. The graph shows the well-known positive solid-gas difference.

E/A_1 (MeV)	0.025-0.25	0.25 – 2.5	2.5 – 30	30 – 100	Total Range
No. of points	*21*	*112*	*195*	*15*	*343*
SRIM	-1.6 ± 9.5	-4.8 ± 5.3	-9.9 ± 7.7	-0.9 ± 7.3	-7.4 ± 7.8

Table 14. $\Delta \pm \sigma$ (in %) for SRIM, for ions from $_{19}$K to $_{92}$U in all gaseous compounds for which we have data in (Paul, 2011a): butane, CF_4, methane, CO_2, and C_3F_8 (Freon-218).

Table 15 shows a statistical comparison for solid compounds. The deviation between SRIM and experiments is larger than for elements, and SRIM is too low, on the average. An example is the case for Ni ions in SiC (see the figure in (Paul, 2011a)).

E/A_1 (MeV)	0.025-0.25	0.25-2.5	2.5-30	30-100	Total Range
No. of points	*239*	*211*	*86*	*10*	*546*
SRIM	8.1 ± 12	4.6 ± 9.2	8.5 ± 9.8	5.9 ± 5.8	6.8 ± 10.7

Table 15. $\Delta \pm \sigma$ (in %) for ions from $_{19}$K to $_{92}$U in all solid compounds (Al_2O_3, Formvar, Havar, Mylar, NE111 Plastic Scintillator, Polycarbonate, Polyethylene naphthalate, Polypropylene, Polystyrene, SiC, Silicon Nitride, ZrO_2) for which we have data in (Paul, 2011a) and which are calculable, compared to SRIM.

5. Water as a target

Water as a target is especially important for medical physics. Fig. 7 gives an overview of experimental and tabulated values of the stopping power of solid and liquid water for protons. Fig. 8 shows the same data again, but divided by the values of ICRU 49 (to make small differences apparent), and only the high energy part which is most important for radiation physics. Because corrections to the simple Bethe formula, eqs. (1 & 2), are smaller than 0.68 % beyond 10 MeV, the value of the stopping power is essentially given by the value of the mean ionization energy in this entire region.

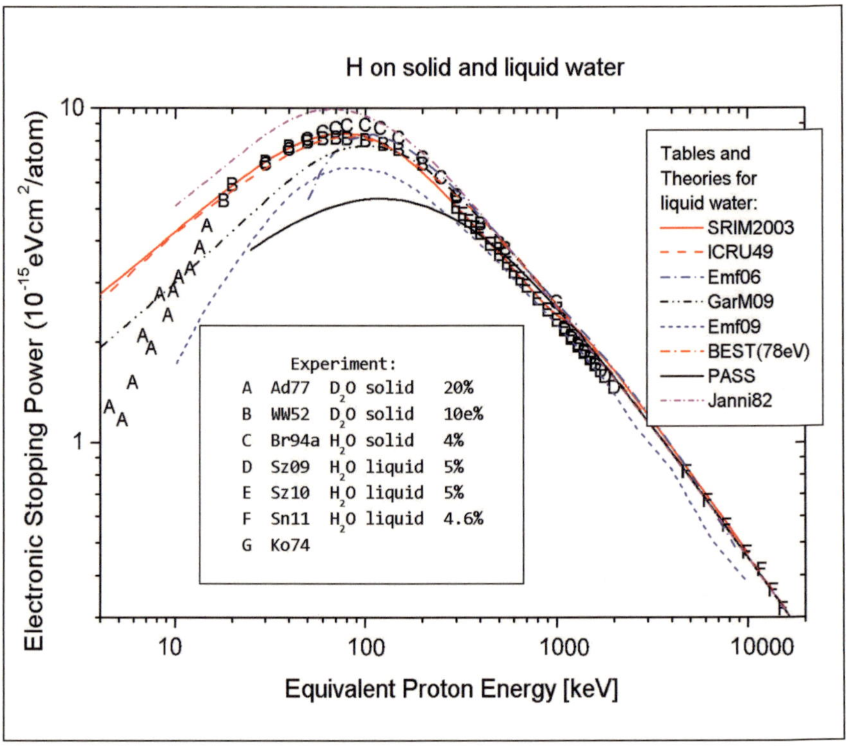

Fig. 7. Electronic stopping power of solid and liquid water for protons, versus energy. The file designations for experiments are explained in Paul (2011a). The table and theory designations are explained in Table 1, except for the following: Emf06 (Emfietzoglou et al, 2006), Emf09 (Emfietzoglou et al., 2009), GarM09 (Garcia-Molina et al., 2009), PASS (Sigmund & Schinner, 2002; Sigmund, 2010)

Table 16 gives an overview of calculated and measured values of the mean ionization energy of liquid water[18] (Paul et al., 2007a). On the basis of the data available in 1984, the value I = 75.0 was chosen in ICRU 37 (1984) and again in ICRU 49 (1993). But evidently, all the more recent determinations indicate a larger value.

Recently, there have been measurements of the stopping power of liquid water for protons by two groups: the Kyoto group (Shimizu et al., 2009, 2010) using a liquid water jet in vacuum, and the Jyväskylä group (Siiskonen, et al., 2011) using a thin water sheet (enclosed within two thin copper sheets) in transmission. The results are shown as points D, E and F in Figs. 7 and 8.

[18] We assume that the I-values for solid and liquid water are equal.

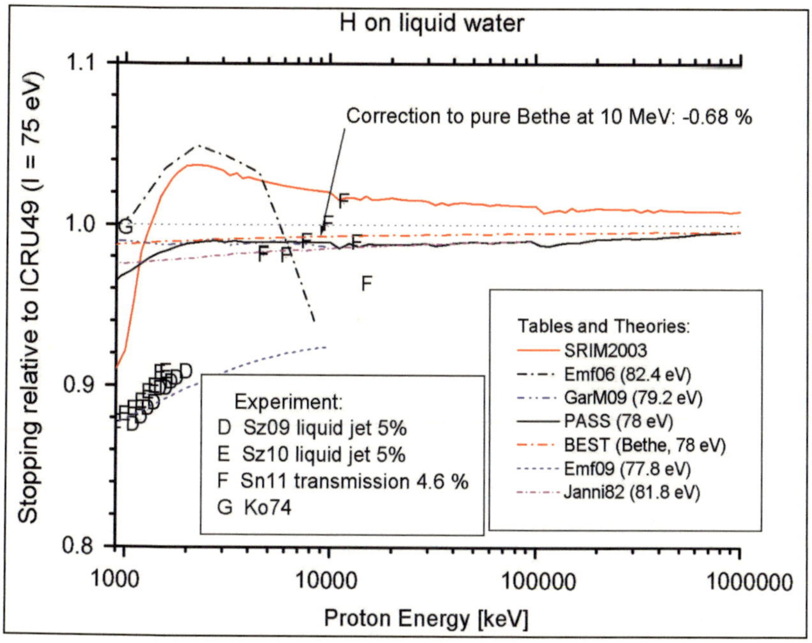

Fig. 8. Stopping power of liquid water for protons, normalized by the table ICRU 49. The designations for tables and for experimental points are as in Fig. 7. *I*-values are shown in parentheses.

I (eV)	Reference	Method or remark
75.4 ± 1.9[19]	Thompson, 1952	Range, 340 – 200 MeV p, assuming I_{Cu} = 322 eV
74.6 ± 2.7	Nordin et al., 1979	Stopping power, 60 MeV pions
75	Ritchie et al., 1978	Dielectric response function
75.4	Ashley, 1982	Dielectric response function
81.77	Janni, 1982	Averaging data for H and O
79.7 ± 2	Bichsel et al., 1992	Ionization curves, 70 MeV p
81.8	Dingfelder et al., 1998	Dielectric response function
80.0	Bichsel et al., 2000	C ions, 290 MeV/u
77	Kramer et al., 2000	Depth dose curves for C ions
78.4	Kumazaki et al., 2007	Depth dose curves for protons
78	Schardt et al., 2008	Bragg curves for H, He, Li, C, and O ions
75.0 ± 3	Chosen in ICRU 37, 49	
78.0 ± 2	Chosen in Sigmund et al. (2009)	Replaces the value 67.2 eV in ICRU Report 73

Table 16. *I* values for liquid water.

This brings up a problem (Paul, 2010). The Bethe equation (i.e., BEST) is generally reliable (Paul & Schinner, 2005) and depends only on *I* and on the shell correction in this energy region, and the latter correction is quite small here. The GarM09 curve and the PASS curve are very close to BEST, about 1 % below ICRU 49 (i.e., unity) due to the higher *I*-value. Hence it appears that the Emf09 curve (and also the Shimizu measurements) may be low by about 10 %.

[19] The data were analyzed by Berger (ICRU Report 37)

It appears that at present, the most precise measurements of the mean ionization energy of water are the range measurements made in Darmstadt (Schardt et al., 2008), leading to $I = 78$ eV. And this value has also been assumed for the corrected table of ICRU 73 (Sigmund et al., 2009). Evidently, the recent Jyväskylä measurements are in very good agreement with BEST and PASS (using I = 78 eV): the points F yield an average of 0.986 ± 0.005, very close to the relative value of BEST: 0.993. But the Jyväskylä results alone would not yield a precise *I* value; it is the range measurements (Schardt et al., 2008) that give a clear distinction between various values.

6. Some remarks concerning the physics of radiation therapy

In radiation therapy, water is used as tissue reference medium (Schardt et al., 2010). Rules for the application of proton therapy have been defined in ref. (ICRU, 2007).

For the dosimetry of fast heavy ions, following a recommendation of the International Atomic Energy Agency (Andreo, 2000), air filled ionization chambers should be used. To convert the absorbed dose in air thus determined to the dose in water (Paul, Geithner and Jäkel, 2007a, 2007b), the first approximation is to use the ratio of mass stopping powers

$$\frac{(S(E)/\rho)_w}{(S(E)/\rho)_{air}} \tag{9}$$

where $(S(E)/\rho)_m$ denotes the mass stopping power of medium *m* evaluated at the energy *E*.

This ratio is essentially determined by the mean ionization energies *I* of water and air. It should be sufficiently accurate from about 5 MeV/nucleon up to (but not beyond) the primary ion energy.

To obtain a more accurate correspondence between the measurement in air and its application to water, it is necessary to use Monte Carlo calculations to take into account all physical processes, especially the effect of fragments produced by nuclear reactions. One defines the 'stopping power ratio' (Andreo, 2000) (as opposed to the simple ratio of stopping powers defined above), i.e., the fluence-weighted average ratio of stopping powers

$$s_{w,air} = \frac{\sum_i \int_0^\infty \Phi_{E,i,w} (S_i(E)/\rho)_w \, dE}{\sum_i \int_0^\infty \Phi_{E,i,w} (S_i(E)/\rho)_{air} \, dE} \tag{10}$$

where $S_i(E)/\rho$ is the mass stopping power for a (primary or secondary) particle i with energy E in water or air, and $\Phi_{E,i,w}$ is the particle spectrum differential in energy, at a particular depth in water, for particles of type i.

For carbon ions of 400 MeV/nucleon and assuming an increased mean ionization energy $I_{water} = 80.8$ eV for water, it was shown (Paul et al., 2007b) that $s_{w,air}$ still fits[20] into the range $s_{w,air} = 1.13 \pm 0.02$ adopted for heavy ion beams by the IAEA Code of Practice (Andreo, 2000). This would probably hold also for the more realistic value $I_{water} = 78$ eV adopted by ICRU 73.

[20] For lower ion energies, the limit 1.15 might be exceeded, however.

7. Conclusion

For a quantitative understanding of radiotherapy by positive ions, one needs information about stopping powers. In this chapter, we discuss stopping power tables and programs, and we compare them statistically to our large collection of experimental data. In this way, the reliability of various tables can be estimated. We describe it by $\Delta \pm \sigma$, where Δ is the average normalized difference between experimental and tabulated values, and σ is its standard deviation. A small Δ usually signifies good agreement; in this case, σ may be taken as a measure of the accuracy of the table. We treat both condensed and gaseous targets, and we consider elements, compounds and mixtures. We give an overview of relevant tables and programs, and of the basic formulas of Bethe theory.

We find that σ always decreases with increasing energy, and that in general, the agreement between tables and experimental data has improved in time. For H ions in elements, in the highest range of specific energy (10 - 100 MeV/nucleon), we find that σ = 0.5 %, on the average. For H and He ions in elements, σ is always smaller than 1 % in that energy range, except for He ions in elemental gases, where we have data only up to 10 MeV/nucleon. The SRIM tables and the tables from ICRU Report 49 are equally good in general, but the SRIM tables describe many more targets. For H and He ions, the gas measurements appear more reliable than those on solids. For compounds, results are similar to those for elements, except that experimental data go only up to 30 MeV/nucleon, so that σ is larger (2 - 4 %) in the highest energy range.

For ions from $_{3}$Li to $_{18}$Ar in elemental solid targets compared to SRIM and MSTAR, we find that σ is about 1.5 % in the highest specific energy range (100 - 1000 MeV/nucleon) and that the energy-dependent accuracy is comparable in condensed compounds, except that data go only up to 100 MeV/nucleon in that case. ICRU Report 73 is too high at low energy, particularly for gases. For ions from $_{3}$Li to $_{18}$Ar, the overall agreement is not better for gases than for solids.

For carbon ions in particular, the overall agreement is slightly better than for all ions (Li to Ar), but the accuracy in the highest energy range (10 - 100 MeV/nucleon) is only about 3 %, much worse than for protons.

For ions from $_{19}$K to $_{92}$U, the ATIMA, Hubert and SRIM tables are best in different ranges of specific energy. The positive gas-solid difference due to the high collision frequency of fast ions in solids is well described by the CasP program but not by SRIM.

Precise values of the mean ionization energy of a substance, I, deduced from range measurements, could often be more useful at high energy than measurements of stopping power. It is shown that the value I = 75 eV that has long been accepted for water, should be increased to I = 78 eV, following the very precise range measurements of Schardt et al. Recent stopping power measurements for water at Jyväskyläa are in good agreement with this value, but the measurements by the Kyoto group are probably too low by about 10 %. The I-value of water is also discussed in relation with the validity of the IAEA Code of Practice for heavy ions.

8. Acknowledgments

The help of N. Arista, P. Grande, D. Petrascheck, J. Rosina, P. Sigmund, A.H. Sørensen, H. Weick and W. Zulehner in various phases of this work is gratefully acknowledged.

9. References

Andreo, P., et al. (2000). Absorbed Dose Determination in External Beam Radiotherapy: An International Code of Practice for Dosimetry Based on Standards of absorbed Dose to Water, *Tech. Report Series No. 398*, International Atomic Energy Agency, Vienna

Arista, N.R., Lifschitz, A.F.(2004). Non-Linear Approach to the Energy Loss of Ions in Solids, in *Advances in Quantum Chemistry*, ed. by J. Sabin, Vol. 45, pp. 47-77, Elsevier Inc.

Ashley, J.C. (1982). Stopping Power of Liquid Water for low energy electrons. *Radiat. Res.*, Vol. 89, No. 1, pp. 25-31

Berger, M.J., Bichsel, H. (1994). BEST, BEthe STopping power program. Unpublished.

Bethe, H. (1932). Bremsformel für Elektronen relativistischer Geschwindigkeit, *Z. Phys.* Vol. 76, pp. 293-299

Bichsel, H., Hiraoka, T. (1992). Energy loss of 70 MeV protons in elements. *Nucl. Instrum. Methods B*, Vol. 66, No. 3, pp. 345-351

Bichsel, H., Hiraoka, T., Omata, K.. (2000). Aspects of fast ion dosimetry. *Radiat. Res.* Vol. 153, no. 2, pp. 208-219

Bragg, W.H., Kleeman, R. (1905). Range of ALPHA-particles of radium. *Phil. Mag.* Vol. 10, pp. 318-340

Emfietzoglou, D., Nikjoo, H., Pathak, A. (2006) Electronic cross sections for proton transport in liquid water based on optical-data models. *Nucl. Instrum. Methods B*, Vol. 249, no. 1-2, pp. 26-28

Emfietzoglou, D., Garcia-Molina, R., Kyriakou, I., Abril, I., Nikjoo, H. (2009). A dielectric response study of the electronic stopping power of liquid water for energetic protons and a new *I* value for water. *Phys. Med. Biol.* Vol. 54, no. 11, pp. 3451-3472

Fermi, E., Teller, E. (1947). The capture of negative mesotrons in matter. *Phys. Rev.* Vol. 72, pp. 399-408

Fettouhi, A., Geissel, H., Schinner, A., Sigmund, P. (2006). Stopping of high-Z ions at intermediate velocities. *Nucl. Instrum. Methods B* Vol. 245, pp. 22-27

Garcia-Molina, R., Abril, I., Denton, C.D., Heredia-Avalos, S., Kyriakou, I., Emfietzoglou, D. (2009). Calculated depth-dose distributions for H^+ and He^+ beams in liquid water. *Nucl. Instrum. Meth. B*, Vol. 267, no. 16, pp. 2647-2652

Geissel, H., et al. (1982). Observation of a gas-solid difference in the stopping powers of (1-10) MeV/u heavy ions. *Phys. Letters,* Vol. 88A, pp. 26-28

Geissel, H., Scheidenberger, Ch., Malzacher, P., Kunzendorf, J., Weick, H. (2011). ATIMA. Available from http://www-linux.gsi.de/~weick/atima/

Golser, R., Semrad, D. (1991). Observation of a striking departure from velocity proportionality in low-energy electronic stopping. *Phys. Rev. Lett.*, Vol. 66, no. 14, pp. 1831-1833

Grande, P.L., Schiwietz, G., CasP (2004). In: Advances in Quantum Chemistry, Cabrera-Trujillo, R., Sabin, J.R. (Eds.).Vol. 45, pp. 7-46, Elsevier. Program available from http://www.helmholtz-berlin.de/people/gregor-schiwietz/casp_en.html

Hubert, F., Bimbot, R. & Gauvin, H. (1990). Range and stopping-power tables for 2.5-500 MeV/nucleon heavy ions in solids, *Atomic Data Nucl. Data Tables*, Vol. 46, no. 1, pp. 1-213

ICRU Report 37 (1984), Stopping Powers for Electrons and Positrons, International Commission on Radiation Units and Measurements, Bethesda, MD, USA

ICRU Report 49 (1993), Stopping Powers and Ranges for Protons and Alpha Particles, International Commission on Radiation Units and Measurements, Bethesda, MD, USA

ICRU Report 73 (2005), Stopping of Ions Heavier Than Helium, *J. ICRU*, Vol. 5, no. 1, pp. 1-253

ICRU Report 78 (2007). Prescribing, Recording, and Reporting Proton-Beam Therapy. *J. ICRU,* Vol. 7, no. 2

Janni, J.F. (1982). Proton Range-Energy Tables. *Atomic Data Nucl. Data Tables,* Vol. 27, No. 1, pp. 147-529

Krämer, M., Jäkel, O., Haberer, T., Schardt, D., Weber, U. (2000). Treatment planning for heavy-ion radiotherapy : physical beam model and dose optimization. *Phys. Med. Biol.* Vol. 45, no.11, pp. 3299-3317

Lindhard, J., Sørensen, A.H. (1996). Relativistic theory of stopping for heavy ions, *Phys. Rev. A* Vol. 53, no. 4, pp. 2443-2456

Markin, S.N., Primetzhofer, D., Bauer, P. (2009). Vanishing Electronic Energy Loss of Very Slow Light Ions in Insulators with Large Band Gaps. *Phys. Rev. Lett.* Vol. 103, pp. 113201, 1-4

Moyers, M.F., et al. (2010). Ion stopping powers and CT numbers. *Medical Dosimetry,* Vol. 35, No. 3, pp. 179-194

Moyers, M.F. (2011). Personal communication

NIST ASTAR. Available from http://physics.nist.gov/PhysRefData/Star/Text/ASTAR.html

NIST ESTAR. Available from http://physics.nist.gov/PhysRefData/Star/Text/ESTAR.html

NIST PSTAR. Available from http://physics.nist.gov/PhysRefData/Star/Text/PSTAR.html

Nordin, J.A., Henkelman, R.M. (1979). Measurement of stopping power ratios for 60 MeV positive or negative pions. *Phys. Med. Biol.,* Vol. 24, no. 4, 781-792

Paul, H., Berger, M.J. (1995), Stopping Powers, Ranges and Straggling, In: *Atomic and molecular data for radiotherapy and radiation research,* pp. 415-545, IAEA-TECDOC-799

Paul, H., Schinner, A. (2001). An empirical approach to the stopping power of solids and gases for ions from Li to Ar. *Nucl. Instrum. Methods B* Vol. 179, pp. 299-315

Paul, H. (2003). MSTAR version 3. Available from <http://www.exphys.uni-linz.ac.at/stopping/MSTARWWW/MSTAR312.zip>

Paul, H., Schinner, A. (2005). Judging the reliability of stopping power tables and programs for protons and alpha particles using statistical methods. *Nucl. Instrum. Methods B* Vol. 227, pp. 461-470

Paul, H. (2006). A comparison of recent stopping power tables for light and medium-heavy ions with experimental data, and applications to radiotherapy dosimetry. *Nucl. Instrum. Methods B* Vol. 247, no. 2, pp. 166-172

Paul, H., Schinner, A. (2006). Statistical analysis of stopping data for protons and alphas in compounds. *Nucl. Instrum. Methods B* Vol. 249, pp. 1-5

Paul, H., Geithner, O., Jäkel, O. (2007a). The influence of stopping powers upon dosimetry for radiation therapy with energetic ions. *Adv. Quantum Chemistry,* John R. Sabin, Erkki Brandäs (Eds.) Vol. 52, pp. 289-306

Paul, H., Geithner, O., Jäkel, O. (2007b). The ratio of stopping powers of water and air for dosimetry applications in tumor therapy. *Nucl. Instrum. Methods B* Vol. 256, pp. 561-564

Paul, H., Grande, P., Smith, D.Y. (2009a). Optical oscillator strengths, mean ionization energy, shell corrections and experimental values for stopping power. *Nucl. Instrum. Methods B* Vol. 267, pp. 2471-2474

Paul, H. (2009b). Some new results on stopping power for fast ions. *AIP Conf. Proc. 1099,* pp. 251-254

Paul, H. (2010). Recent results in stopping power for positive ions, and some critical comments. *Nucl. Instrum. Methods B* Vol. 268, pp. 3421-3425

Paul, H. (2011a). Stopping Power for Light Ions, Available from <http://www.exphys.uni-linz.ac.at/stopping/>

Paul, H. (2011b). Comparing experimental stopping power data for positive ions with stopping tables, using statistical analysis. *Nucl. Instrum. Meth.*B. doi :10.1016/j.nimb.2011.07.026

Randhawa, G.S., Virk, H.S. (1996). Stopping power and range of heavy ions in solids: a comparative study. *Radiation Measurements* Vol. 26, no. 4, pp. 541-560

Raiola, F., et al. (2001). Stopping power of low-energy deuterons in ^{3}He gas. *Eur. Phys. J. A* Vol. 10, pp. 487-491

R. H. Ritchie and W. Brandt, Projectile-charge dependence of stopping powers. *Phys. Rev.* A, 1978, Vol. 17, no. 6, 2102-2106

D. Schardt, P. Steidl, M. Krämer, U. Weber, K. Parodi, and S. Brons (2008). Precision Bragg-curve measurements for light-ion beams in water. GSI Scientific Report 2007, 2008-1, pp. 373-373 (GSI-Gesellschaft f. Schwerionenforschung, Darmstadt, Germany). Available from http://www.gsi.de/informationen/wti/library/scientificreport2007/PAPERS/RADIATION-BIOPHYSICS-19.pdf.

Scheidenberger, C., et al. (1994). Direct observation of Systematic Deviations from the Bethe Stopping Theory for Relativistic Heavy ions. *Phys. Rev. Lett.* Vol. 73, pp. 50-53

Schiefermüller, A., Golser, R., Stohl, R., Semrad, D. (1993). Energy loss of hydrogen projectiles in gases. *Phys. Rev. A* Vol. 48, no. 6, pp. 4467-4475

Schiwietz, G., Grande, P.L. (2001). Improved charge-state formulas. *Nucl. Instrum. Methods* B 175-177 pp. 125-131

Shimizu, M., Kaneda, M., Hayakawa, T., Tsuchida, H., Itoh, A. (2009). Stopping cross sections of liquid water for MeV energy protons. *Nucl. Instrum. Methods* B 267 no. 16, pp. 2667-2670

Shimizu, M., Hayakawa, T., Kaneda, M., Tsuchida, H., Itoh, A. (2010). Stopping cross sections of liquid water. *Vacuum* Vol. 84 no. 8, pp. 1002-1004

Sigmund, P., Schinner, A. (2002) Binary theory of electronic stopping. *Nucl. Instrum. Methods* B 195, no. 1-2, pp. 64-90

Sigmund, P., Schinner, A., Paul, H. (2009). Errata and Addenda for ICRU Report 73, *J. ICRU* Vol. 5, no. 1

Sigmund, P. (2010). Personal communication to H. Paul

Siiskonen, T., et al. (2011). Energy loss measurement of protons in liquid water. *Phys. Med. Biol.* Vol. 56, pp. 2367-2374

Smith, D.Y., Inokuti, M., Karstens, W., Shiles, E. (2006). Mean excitation energy for the stopping power of light elements, *Nucl. Instrum. Methods* B Vol. 250, No. 1-2, pp 1-5

Thompson, T.J. (1952). Report UCRL-1910, Univ. of California Radiation Lab., Berkeley, CA, USA

Zajic, V., Thieberger, P. (1999). Heavy Ion Linear Energy Transfer Measurements during Single Event Upset Testing of Electronic Devices, *IEEE Transactions on Nuclear Science* Vol. 46, pp. 59-69

Zajic, V. (2001). LET: Energy-LET-Range calculator v. 1.24. New York: Brookhaven National Laboratory. Available from http://tvdg10.phy.bnl.gov/let.html

Ziegler, J.F., Biersack, J.P., Littmark, U. (1985). *The Stopping and Ranges of Ions in Matter*, Vol.1, Pergamon, New York. ISBN 0-08-021603-X

Ziegler, J.F., Manoyan, J.M. (1988). The stopping of ions in compounds. *Nucl. Instrum. Methods B* Vol. 35, no. 3-4, pp. 215-228

Ziegler, J.F. (2004). SRIM 2003. Available from http://www.srim.org

Ziegler, J.F., Biersack, J.P., Ziegler, M.D. (2008). *SRIM - The Stopping and Range of Ions in Matter*, SRIM Co., Chester, MD, 21619 USA.

Part 3

Brachytherapy and Intraoperative Radiation Treatments

8

Prostate Seed Brachytherapy – Methods to Improve Implant Characteristics

Bruce Libby, Matthew D. Orton, Haidy Lee,
Mark E. Smolkin, Stanley H. Benedict and Bernard F. Schneider
University of Virginia Health System,
Charlottesville, Virginia,
USA

1. Introduction

Prostate cancer is diagnosed in over 230,000 men each year in the United States (Jemel, et al, 2006), and with the use of screening prostate specific antigen (PSA) the majority is diagnosed with locally confined prostate cancer. Many of these patients are good candidates for prostate brachytherapy. With the development of transperineal implantation using trans-rectal ultrasound guidance the number of patients undergoing permanent radioactive seed implants for prostate cancer has increased significantly over the past ten years. (Cooperberg, et al, 2004)

A variety of techniques have been used to deliver transperineal prostate brachytherapy, using Pd-103, I-125 and, more recently, Cs -131(Sommerkamp, et al 1988, Reed, et al, 2007, Spadlinger, et al, 2006, Meigooni, et al, 2004, Yue, et al, 2005). These sources are typically implanted as individual seeds, or as seeds that are stranded together or held together using plastic linking devices (Sommerkamp, et al 1988, Reed, et al, 2007, Spadlinger, et al, 2006). Stranded or linked seeds have the advantage that they are less likely to migrate from their implant position than individually implanted seeds (Sommerkamp, et al 1988, Reed, et al, 2007, Spadlinger, et al, 2006). Additionally, implants can be either performed via pre-plans, in which a planning ultrasound is performed before the surgery, or with intra-operative planning (Matzkin, et al, 2003).

While a linear brachytherapy source in the form of a coiled ^{103}Pd wire was developed, it never became available clinically (Meigooni, et al, 2004). The possibility of using a continuous linear source led to the consideration of continuously linked seeds in an implant rather than the normal use of seeds separated by spacers. The elimination of spacers would allow the same number of seeds to be implanted into the prostate with fewer needles. The number of needles used in prostate seed implants has been shown to be correlated with acute urinary morbidity after seed implantation. (Eapen, et al, 2004, Ohashi, et al, 2006, Bucci, et al, 2002, Lee, et al, 2000, Buskirk, et al, 2004, Bottomley, et al, 2007, Keyer, et al, 2009, Thomas, et al, 2008). Also, the use of fewer needles should cause less trauma to the base of the penis, which may help preserve erectile function (Macdonald, et al, 2005, Steggerda, et al, 2010).

In eliminating spacers between seeds and reducing the number of needles used for implantation, it is important that implant quality is not adversely affected. In this study a comparison of post-implant dosimetry in patients treated with conventional linked seed implants with spacers (**+S**) to those treated with a novel technique using linked seed implants without spacers between the sources (**-S**) is performed, as well as a comparison of the numbers of needles used for each technique.

2. Strategy to provide monotherapy palladium-103 implants

The day 0 post implant dosimetry of 101 consecutive patients who received monotherapy Palladium-103 implants was retrospectively reviewed. To avoid selection bias the dosimetric data from the final 48 **+S,** and the first 53 **-S** patients implanted at our institution were analyzed. Prior to this study the implant team had performed more than 800 permanent seed prostate implants. The ultrasound images taken at the time of the implant in the operating room were transferred to the Variseed ™ 7.0 software (Varian, Palo Alto, CA). The radiation oncologist then developed a treatment plan to deliver a minimum dose of 125 Gy to the prostate with a margin of approximately 5 mm anteriorly and laterally, with a smaller posterior margin. All patients were implanted using Bard® BrachyStar® seed implant needles containing Theragenics® TheraSeed® palladium-103 sources with an average activity of 2.08 (Range 1.78-2.31) linked with 0.5 mm seed-to-seed SourceLink™ links or 5 mm SourceLink™ spacer links, for the **-S** and **+S** cohorts respectively, assembled using a SourceLink™ loader. A photograph of the linked seeds without the use of spacers is shown in Figure 1, along with a schematic diagram of the linked seeds. All patients had a CT scan for post-implant dosimetry within 3 hours following the implant. A Foley catheter was in place during the implant procedure, and for the post-implant CT scan. A single radiation oncologist planned, treated, and performed post-implant contouring of the prostate.

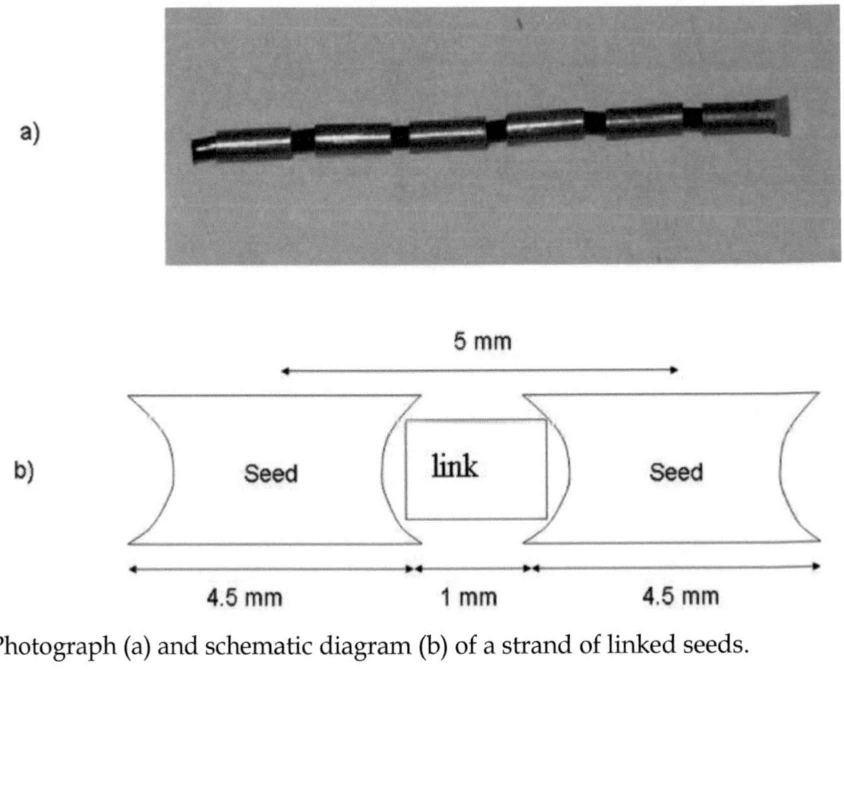

Fig. 1. Photograph (a) and schematic diagram (b) of a strand of linked seeds.

Forty-eight patients (+S) were implanted using linked seeds with a single spacer between each seed, except at the apex of the gland where 2 seeds were frequently placed without a spacer between them. Another 53 patients (-S) had implants in which all seeds were linked without spacers between them. Other than use of spacers, both groups of patients were planned and implanted in the same manner.

Using the Variseed ™ 7.0 software (Varian, Palo Alto, CA), dose-volume histograms were compiled for each patient and were used to determine the prostate V800, V400, V350, V300, V250, V200, V150, V100, V90 and V80, prostate D100, D90 and D80, and urethral D90, D30, and D10 using a point source approximation (Pd-103 M200 Corrected [NIST 99]).

The automatic seed finder was utilized to locate the Pd-103 seeds in the CT data sets, which were subsequently reviewed by the physician to assure that location of the seeds was correctly identified and that the number of seeds identified matched the number that was implanted. The post implant prostatic D90 is defined as the dose covering 90% of the prostate volume. The V100 is defined as the volume of the prostate that receives 100% of prescribed dose. The post implant dosimetric analysis was performed according to the American Brachytherapy Society's guidelines for permanent prostate brachytherapy (Nag, et al, 2000).

Statistical Methods: The dosimetric variables, as well as the average number of needles and seeds used for each group, were statistically compared using Student's t-tests to determine if there were significant differences between the two types of implants. Demographic and clinical factors including prostate volume, patient age, clinical stage, pre-implant PSA, and Gleason score were evaluated using Student's t-tests or chi-square tests of association to determine whether the two cohorts were similar. All statistical analyses were performed using SAS® Version 9.1 (Cary, NC, USA). All tests were performed using a Type I error rate of 0.05.

3.1 Demographics, disease and treatment characteristics

The clinical characteristics of our study cohorts are presented in Table 1. Statistical analysis showed no significant differences between age, mean pre-implant PSA, Gleason's score or pre-implant prostate volume. The prostate volumes were similar for the two groups (39.3 cm^3 for +S and 36.7 cm^3 for -S).

3.2 Number of needles and seeds

Comparison between the two cohorts (Table 1) showed that an average of 100.8 seeds and 23.1 needles were used for implants without spacers, while an average of 94 seeds and 31.5 needles were required when implanting with spacers. This difference in numbers of needles used was statistically significant ($p<0.001$), but there was no statistically significant difference in the number of seeds in the two types of implants ($p=0.16$).

3.3 Dosimetric parameters and outcomes

Detailed dosimetric analysis results are found in Table 2. The mean prostatic D90 for the +S cohort was 99.2 Gy, slightly higher than the 95.5 Gy calculated for the -S cohort, but this difference was not statistically significant (p=0.22). The prostate D80 and D100 values were also not significantly different between the two cohorts (Figure 2).

Variable	+S*	-S†	p-Value
Mean Age (years)	66.2	63.5	0.068
Mean Preimplant PSA‡ (ng/ml)	6.3	6.0	0.93
Mean Combined Gleason Score	6.2	6.2	0.88
T-stage, number (%)			
T1c	38 (79.2)	45 (84.9)	
T2a	9 (18.8)	7 (13.2)	
T2b	0 (0)	0(0)	0.45
T2c	1 (2.1)	1 (1.9)	
Mean Prostate Volume (cm^3)	39.3	36.7	0.31
Mean Urethral Volume (cm^3)	1.4	1.4	0.97
Mean Number of Needles	31.5	23.1	<0.001
Mean number of Seeds	94	100.8	0.16
Mean number of seeds per needle	3.0	4.4	<0.001

*+S: Patients receiving implantation with spacers. † -S: Patients receiving implantation with linked seed implants without spacers between the sources. ‡PSA = Prostate Specific Antigen.

Table 1. Demographic and Clinical Cohort Comparison

Dosimetric Variable	**+S***	**-S†**	**p-Value**
Prostate, Mean V800(%)	2.4	2.0	0.029
Prostate, Mean V400(%)	7.1	6.1	0.005
Prostate, Mean V200(%)	30.6	27.4	0.058
Prostate, Mean V150(%)	56.3	52.0	0.077
Prostate, Mean V100(%)	88.1	85.8	0.19
Prostate, Mean V90(%)	92.2	90.8	0.28
Prostate, Mean D100 (Gy)	44.2	45.4	0.63
Prostate, Mean D90 (Gy)	99.2	95.5	0.22
Prostate, Mean D80 (Gy)	116.8	112.3	0.15
Urethra, Mean V100(%)	41.5	51.5	0.038
Urethra, Mean V100(%)	121.4	119.7	0.70
Urethra, Mean V100(%)	139.8	137.6	0.70

*+S = Patients receiving implantation with spacers between the linked sources.

† -S = Patients receiving implantation with linked seed implants without spacers

Table 2. Summary of Dosimetric Analysis

The urethral D90 was significantly higher (p=0.038) in the group without spacers (51.5 Gy) compared to the group with spacers (41.5 Gy) shown in Figure 3.

The mean V100 for the **-S** cohort was 85.8% compared to a mean V100 of 88.1% for the **+S** cohort (p= 0.19). Also there were no statistically significant differences in the mean V80 and V90 for each of the cohorts.

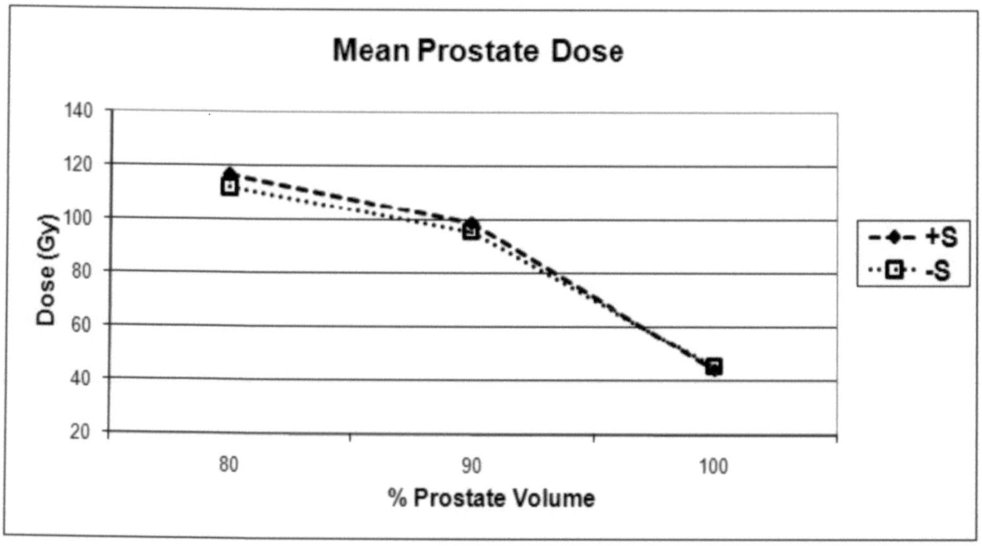

Fig. 2. Mean Prostate dose for the implants performed with (+S) and without (-S) spacers.

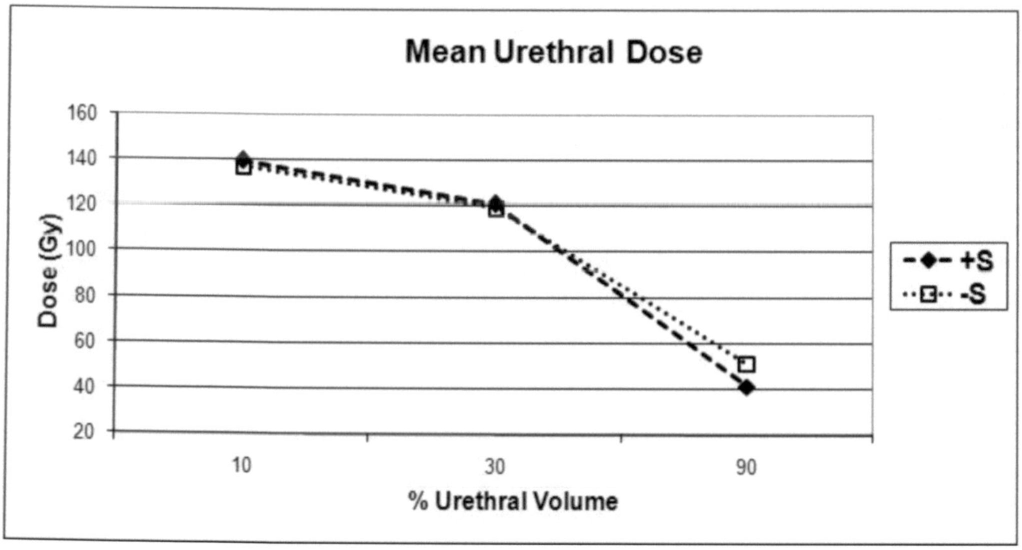

Fig. 3. Mean urethral dose for the implants performed with (+S) and without (-S) spacers.

The mean V150, V200,V250, V300, V350, V400 and V800 were also calculated and compared between the two cohorts, as shown in Figure 4.

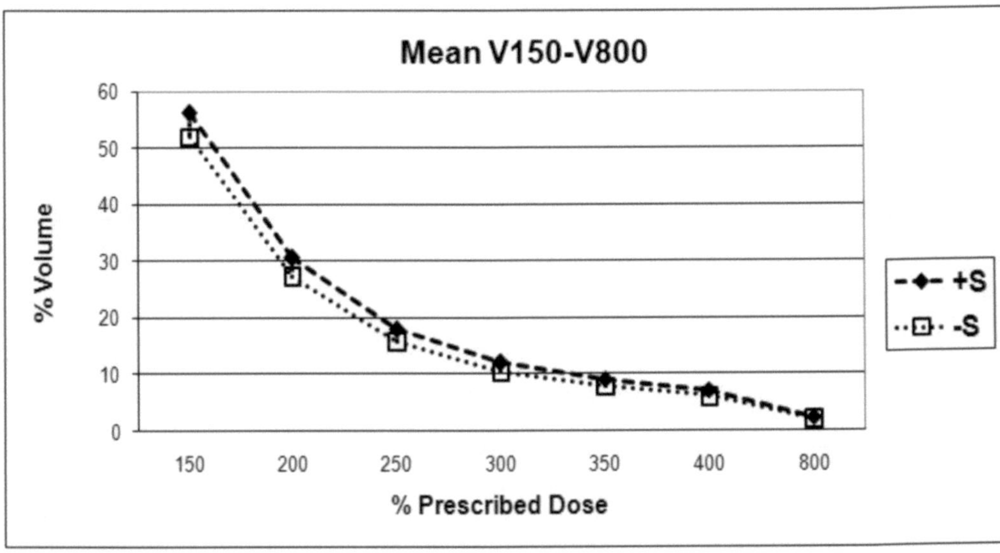

Fig. 4. Mean Prostate V150-800 for the implants performed with (+S) and without (-S) spacers.

The mean V150 and V200 for the **+S** were 56.3% and 30.6% of the prostatic volume respectively compared to 52.0% and 27.4% of the prostatic volume respectively for the **-S** group. While the differences in the V150 and V200 are surprisingly higher for the **+S** as compared to the **-S** cohort these differences failed to be statistically significant but were trending (p=0.077 and p=0.058 respectively). This trend of higher V values for the **+S** cohort as compared to the **-S** group continued. The **+S** mean V250, V300,V350, V400 and V800 are 17.9%, 12% , 9.1%, 7.1% and 2.4% of the total prostatic volume respectively and are elevated compared to these same values for the **-S** cohort which are, 15.8%, 10.5%, 7.8%, 6.1% and 2% of the total prostatic volume respectively (Figure 3). Statistical analysis demonstrated that for each of these measurements that the percentage of volume of the prostate receiving 2.5-8 times the prescribed dose was significantly higher for the **+S** cohort. A plot of total implanted activity as a function of prostate volume for both the +S and -S cohorts is shown in Figure 5. This plot shows that the total implanted activity is the same for implants using spacers or not, which confirms the dosimetric data.

4. Review of implants

The development of the RadioCoil™ linear ^{103}Pd source prompted studying the use of Pd-103 seeds linked in a continuous linear fashion, in order to decrease the number of implant needles and possibly decrease toxicity. This study was not intended to assess toxicity, but rather to compare post-implant dosimetry for patients receiving implants using conventional spacers with those whose implants used seeds continuously linked without spacers. Interestingly there are few dosimetric differences between the two groups.

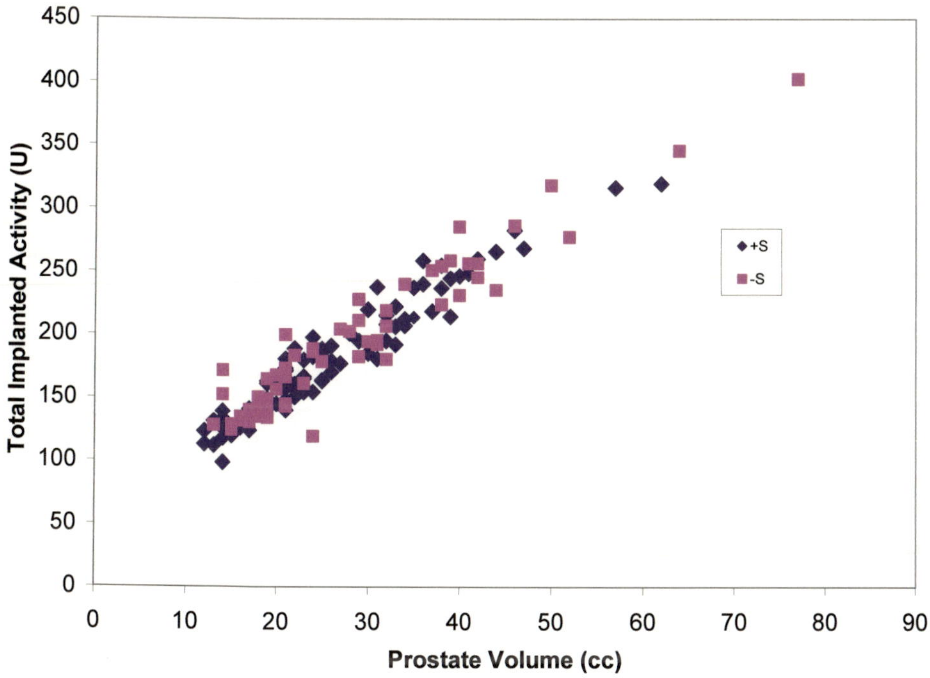

Fig. 5. Total Implanted activity as a function of prostate volume for the two arms of the study

Our data show a significant decrease in the volume of prostate tissue receiving greater that 100% of the prescribed dose (V250-V800) in the **-S** patients. This result was not expected. One of the main concerns about using continuous sources is the creation of "hot spots". However, our data show that this is not the case. All of the mean V values calculated for the **-S** cohort were less than those of the **+S** cohort.

The elimination of spacers between seeds allowed a significant decrease in the total number of needles used during prostate implants. This may prove to be clinically significant since needle trauma likely plays a role in post implant morbidity. Acute urinary toxicity is the predominant side effect of prostate brachytherapy and acute urinary retention (AUR) is a well recognized and described early toxicity (Mallick, et al, 2003). Most obstructive symptoms occur quickly after the implant procedure before the dose deposited by the sources to the surrounding tissue can reach a significant level (Bucci, et al, 2009). This suggests that prostatic trauma from the procedure is the predominant factor in obstructive urinary symptoms. Studies have been conducted to try to identify factors that predict post implant AUR.

There have been many studies reporting on AUR rates after prostate brachytherapy, with the number of needles used, the number of seeds implanted, hormonal manipulation, pre-implant prostate volume, the level of post-implant prostatic edema, and diabetes all linked to increased AUR rates(Ohashi, et al 2006, Bucci, et al, 2002, Lee, et al 2000, Buskirk, et al, 2004, Bottomley, et al, 2007, Keyer, et al, 2009, Thomas, et al 2008, Macdonald, et al, 2005). While there are no large randomized controlled trials, many of the references cited provide evidence that prostate gland trauma caused by needle insertion during brachytherapy implant plays some role in the development of post-operative AUR. Therefore, decreasing the number of needles required to perform quality implants could play a role in decreasing post implant AUR. It remains to be seen if reduced AUR is seen in patients undergoing **-S** compared to **+S**. It is important to note that decreasing the number of needles used for prostate implantation can predispose to unwanted hot or cold regions if the needle position is not placed correctly. Thus, this technique should only be undertaken by experienced brachytherapy teams.

The use of a retrospective, rather than a randomized, method to study possible dosimetric differences between the +S and –S arms of the study could have led to errors associated with all non-randomized trials. The use of Day 30, rather than or in addition to Day 0 CT scans, may have shown either differences in the data not present in the Day 0 or allowed the tracking of the dosimetric parameters as a function of time.

5. Conclusion

This study was designed as an analysis of prostate seed implants comparing the dosimetric characteristics of implants performed both with (**+S**) or without (**-S**) the use of spacers between the seeds. The results presented in this study show that **+S** and **-S** implants were dosimetrically similar, with a significant reduction in the number of needles required for the treatments without the use spacers (**-S**). Analysis of the total implanted activity as a function of prostate volume are essentially identical between the two arms of the study, which helps to clarify the similarity in dosimetric characteristics of the **-S** and **+S** implants. Studies have shown that the number of needles used for brachytherapy correlates to AUR; therefore, we expect a decreased rate of AUR with this technique. It remains to be evaluated what significance implants with **-S** has on urinary quality of life.

6. References

Bottomley D, Ash D, Al-Qaisieh B, et al. Side effects of permanent I125 prostate seed implants in 667 patients treated in Leeds. *Radiotherapy and Oncology* 2007;82:46-49.

Bucci J, Morris JW, Keyes M, et al. Predictive factors of urinary retention following prostate brachytherapy. *Int J Radiat Oncol Biol Phys* 2002;53:91-98.

Buskirk SJ, Pinkstaff DM, Petrou SP, et al. Acute urinary retention after transperineal template- guided prostate biopsy. *Int J Radiat Oncol Biol Phys* 2004;59:1360-1366.

Cooperberg M, Lubeck D, Meng M, et al. The changing face of low risk prostate cancer: Trends in clinical presentation and primary management. *J Clin Oncol* 2004;22:2141-2149.

Eapen L, Kayser C, Deshaies Y, et al. Correlating the degree of needle trauma during prostate brachytherapy and the development of acute urinary toxicity. *Int J Radiat Oncol Biol Phys* 2004;59:1392-1394.

Jemal A, Siegel R, Ward E, et al. Cancer statistics. *2006 CA Cancer J Clin* 2006;56:106-130.

Keyes, M, Miller, S, Moravan, V, et al. Predictive Factors for Acute and Late Urinary Toxicity After Permanent Prostate Brachytherapy: Long-Term Outcome in 712 Consecutive Patients, *Int J Radiat Oncol Biol Phys* 2009; 73:1023-1032.

Lee N, Wuu C, Brody R, et al. Factors predicting for postimplantation urinary retention after permanent prostate brachytherapy. *Int J Radiat Oncol Biol Phys* 2000;48:1457-1460.

Macdonald AG, Keyes M, Kruk A, et al. Predictive factors for erectile dysfunction in men with prostate cancer after brachytherapy: is dose to the penile bulb important? *Int J Radiat Oncol Biol Phys* 2005;63:155-163.

Mallick S, Azzouzi R, Cormier L, et al. Urinary morbidity after ^{125}I brachytherapy of the prostate. *BJU Int* 2003;92:555-558.

Matzkin, H, Kaver, I, Stenger, A, et al, Iodine-125 Brachytherapy for localized prostate cancer and urinary morbidity: A prospective comparison of two seed implant methods-preplanning and intraoperative planning,*Urology* 2003;3:497-502.

Meigooni AS, Zhang H, Clark JR, et al. Dosimetric characteristic of the new RadioCoil™ ^{103}Pd wire line source for use in permanent brachytherapy implants. *Medical Physics* 2004;31:3095-3105.

Nag S, Bice W, DeWyngaert K, et al. The American Brachytherapy Society recommendations for permanent prostate brachytherapy post implant dosimetric analysis. *Int J Radiat Oncol Biol Phys* 2000;46:221-230.

Ohashi T, Yorozu A, Toya K, et al. Predicative factors of acute urinary retention requiring catheterization following ^{125}I prostate brachytherapy. *Jpn J Clin Oncol* 2006;36:285-289.

Reed DR, Wallner KE, Merrick GS, et al. A prospective randomized comparison of stranded vs. loose ^{125}I seeds for prostate brachytherapy. *Brachytherapy* 2007;6:129-134.

Sommerkamp H, Ruppercht M, Wannenmacher M. Seed loss in interstitial radiotherapy of prostatic carcinoma with I-125. *Int J Radiat Oncol Biol Phys* 1988;14:389-392.

Spadingler I, Hilts M, Keyes M, et al. Prostate brachytherapy postimplant dosimetry: a comparison of suture-embedded and loose seed implants. *Brachytherapy* 2006;5:165-173.

Steggerda, M, van der Poel, H, and Moonen, L, Minimizing the number of implant needles for 125I brachytherapy: An investigation of possibilities and implications, *Brachytherapy* 2010;9: 319-327.

Thomas C, Keyes M, Liu M, and Moravan, V, Segmental Urethral Dosimetry and Urinary Toxicity in Patients with no Urinary Symptons Before Permanent Prostate Brachytherapy. *Int J Radiat Oncol Biol Phys* 2008; 72:447-455

Yue N, Heron DE, Komanduri K, et al. Prescription dose in permanent (131)Cs seed prostate implants. *Medical Physics* 2005;32:2496-2502.

9

Intra-Operative Radiotherapy with Electron Beam

Ernesto Lamanna[1], Alessandro Gallo[1],
Filippo Russo[1], Rosa Brancaccio[2],
Antonella Soriani[3] and Lidia Strigari[3]
[1]*Magna Graecia University, Medicine Faculty,*
[2]*Bologna University, Physics Department,*
[3]*Istituto Regina Elena, Medical Physics Department,*
Italy

1. Introduction

Surgery is in many cases the most effective therapy to eradicate tumors in the human body. It is applied effectively in cases of tumors with low production of metastasis. However since the early decades of the last century a large fraction of recurrence after the operation have been observed. The number of cases with the onset of recurrence greatly decreases when the region which underwent surgical resection is treated with radiotherapy. External beam radiation therapy with photons is currently the most widely used. The conformal techniques and the intensity modulated radiotherapy reduce but do not completely eliminate damage to healthy tissues traversed by the radiation. In cases where the tumor is located very close to radiosensitive normal tissues or in cases of cancer for which fractioned treatments are ineffective, external beam radiation becomes difficult to apply. The technique of intra-operative radiotherapy (IORT) is effective in such cases as it allows direct visualization of the region to be irradiated after the removal of the lesion and it allows healthy tissue to be protected. The IORT technique consists in the delivering of a single high dose of radiation to the target volume, by shielding the healthy tissue, during the operation.

In the early years of application of IORT, beams of low energy photons (Comas & Prio, 1907; Beck, 1909) were used. The penetrating ability, however, produced damage to underlying tissues traversed by the radiation. This approach was then abandoned until beams of electrons instead of photons were used in Japan (Abe & Takahashi, 1981; Abe, 1989).

Fig. 1 shows the trends of the absorbed dose in water as a function of depth normalized to the value of maximum dose. Two beams generated by a Varian Clinac 2100 DH were used: A 6-MV photon beam and an electron beam of energy 6 MeV. It can be observed that the release of the maximum dose occurs at depths very similar for both electrons and photons. However, while the dose of electrons is released in a few cm from the entry point, the dose of photons is released at greater depths. The electrons are the most suitable particles to give the required dose of radiation directly to the tissues displayed during surgery, thus protecting the underlying healthy tissues.

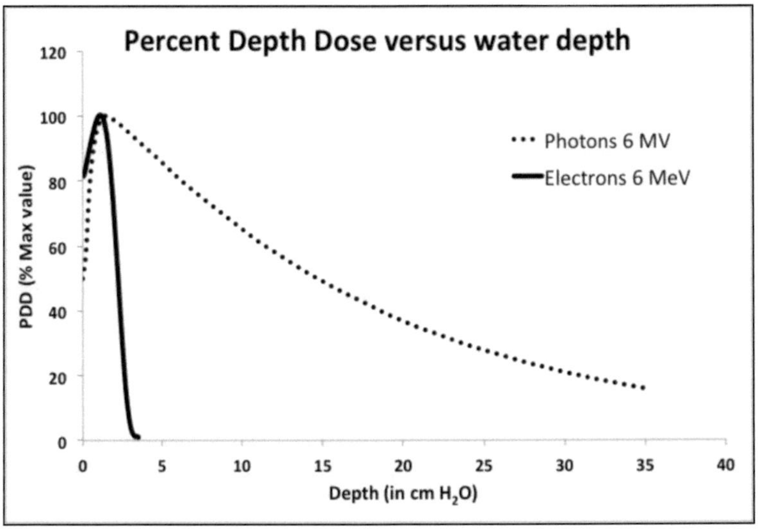

Fig. 1. Per cent Depth Dose (PDD) as a function of depth in water. The photon behaviour is compared to the electron delivered dose. In a few cm the electron dose is absorbed while a greater depth is needed to absorb the photon dose.

The results obtained with electrons have contributed to the diffusion of IORT in the world and to the development of dedicated mobile accelerators to be used in operating rooms. The three compact Linac (linear accelerator), Mobetron (IntraOp Medical Inc., 2011), NOVAC7 (Ronsivalle et al., 2001), LIAC (Soriani et al., 2010), were the most used in the first decade of 2000. These machines generate electron beams of variable energy range 3-12 MeV. The ability of high dose rate (up to 20 Gy/min) allows the delivering of the required dose in a few minutes. The experience accumulated over the past ten years for the research of more versatile systems has produced new proposals on the market (NRT, 2011).

In the next section we will describe such systems. Two were completely designed and built in Italy.

A detailed description of the technique and a presentation of results according to disease-site may be found in the recent textbook "Intraoperative Irradiation" (Gunderson et al., 2011).

However, two aspects of IORT technique with dedicated accelerators have not yet been fully defined. The first is the difficulty of making a treatment plan similar to that prepared for treatment with external beam. The second is the difficulty of using dosimeters recommended by the standard protocols for radiotherapy, due to the high dose per pulse (3-12 cGy/pulse).

In the following paragraphs we present some results and methods intended to overcome these difficulties and provide a good basis for future developments.

2. Dedicated LINAC

Until some years ago, it was not possible to perform intraoperative radiotherapy with electrons in a conventional operating theatre. In fact, the accelerators were located within a bunker of radiotherapy departments and patients had to be transferred under anesthesia from the operating theatre to the accelerator room for treatment. In some cases, the entire surgical procedure was performed in a modified accelerator treatment room.

In the last ten years there has been an increasing interest in the IORT technique due to the development of mobile accelerators which produce only electron beams. This type of machine can be introduced directly into the operating room without any other special fixed shielding systems. Several types of mobile accelerator are now available on the market (Mobetron, Novac7, Liac®). These mobile machines have solved logistical and radioprotection problems which are related to their use in conventional operating theatres. These machines have allowed a widespread use of this approach. In the following paragraphs, we will describe the most important features of the three mobile accelerators mentioned previously.

2.1 Mobetron

The Mobetron® (Mobetron is a registered trademark of IntraOp Medical, Inc.), has been designed and configured for intraoperative radiotherapy.

The Mobetron is a lightweight X-band linear accelerator mounted on a C-arm gantry. The gantry is attached to a stand that contains the accelerator cooling system and a transportation system. A mobile modulator rack, a lightweight operator control console, and connecting cables complete the Mobetron system. The Mobetron may be adjusted for two configurations: accelerator horizontal with a low center of gravity for transportation and storage; and accelerator vertical for treatment. In transport configuration, the Mobetron is compact; its dimensions are such that it may fit on to many elevators. The unit can be removed from the operating theatre for maintenance and annual calibrations. The control system contains the dosimetry readout parameters, accelerator controls, machine interlock status, and a color video output of the treatment viewing system.

The Mobetron produces electron beams of nominal energies 4, 6, 9, and 12 MeV.

At 12 MeV, R_{50} value (i.e. the value of the depth in water at which the dose is 50% of its maximal value) is 47.7 mm.

While conventional medical linear accelerators operate in the S band (10 cm wavelength, 3 GHz frequency), the Mobetron operates in the X band (3 cm wavelength, 10 GHz frequency) and this allows transportability and positioning flexibility. In fact, the diameter of the accelerator structures is therefore reduced by a factor of three.

The gantry is in the configuration of a C-arm, but with some additional flexibility of movement. The gantry may be rotated ±45° downward in the transverse plane. In addition, the gantry may be tilted ±30° in the radial plane. Also, the gantry may be moved in and out, and from side to side in the horizontal plane, ±5 cm. The gantry tilt and horizontal movements are unique features not found in conventional accelerators used for intraoperative radiotherapy. The axis of rotation is 99 cm above the floor and the nominal electron source to treatment surface distance (SSD) is 50 cm (i.e to the end of the treatment applicators). Gantry rotation and tilt movements are controlled by the hand held pendant and are variable from 0 to 1° per second. Horizontal movements are controlled in a similar manner and vary from 0 to 2 mm per second. The gantry design includes an integral beamstop to intercept photon contamination generated in the accelerator, collimation system, and the patient. The gantry with accelerator, cooling system, beamstop, and transportation system has a mass of 1250 Kg. Mobetron transportation is accomplished by using a modified pallet jack, located at the rear of the gantry stand. Wheels attached to the front of the gantry support legs and wheels integral to the pallet jack provide a stable support for transportation.

The Mobetron uses two X-band linear accelerators in tandem. One-third of the radio frequency power is injected into the first accelerator, producing electron energy of 4 MeV. The remaining two-thirds of the power may be absorbed in a water load and/or injected into the second accelerator guide. Adjusting the phase to change the amount of power that enters the second guide, as opposed to the water load, varies the energy. As the power in the second guide is changed, the phase of the microwaves in the second guide is simultaneously adjusted to maintain optimal resonance in the accelerator structure. This allows energy control between 4 and 12 MeV without using a bending magnet, which corresponds to a therapeutic range of 4 cm. The injector system, together with a prebuncher and beam alignment system, control the electrons to occupy a very narrow energy spectrum, reducing radiation leakage. Since the bending magnet is a major source of radiation leakage in conventional accelerator designs, this design feature also contributes to a significant reduction of photon leakage.

Applicator sizes ranged from 3 to 10 cm diameter for flat applicators, and 3 to 6 cm diameter for 30° beveled applicators. The Mobetron uses a soft-docking system in which the treatment applicator is connected to a special rigid clamp system attached to the surgical bed and the gantry is optically guided to the docking position above the applicator. During irradiation, the docking is interlocked for both alignments of the treatment head with the applicator and for treatment distance.

The dose-rate ranges from 2.5 to 10 Gy/min at SSD of 50 cm with an applicator of 10 cm diameter.

As the unit is designed to operate only in the electron mode, beam currents are low, producing less inherent radiation leakage. Together with the compact beamstop opposite the electron beam, the overall design allows the system to be used in rooms with no additional shielding.

2.2 Novac

The first model, named Novac7 (Hitesys SpA (LT) Italy 1997), is a dedicated accelerator with four nominal electron energy levels: 3,5,7,9 MeV. At 9 MeV, R_{50} value is 31 mm. This value is related to the mean energy of electrons, about 7.2 MeV, on the surface of the water equivalent phantom.

The most important Novac7 dosimetric characteristic is the very high dose-per-pulse, ranging from 2.5 to 12 cGy/pulse, values up to 100 times greater than the doses per pulse produced by a conventional accelerator.

Novac7 has both a mobile and a fixed unit. The mobile unit is a stand structure on a motorized base, which supports the accelerator and modulator. The stand structure has the form of an articulated arm with four rotational joints, allowing movements similar to those of human arms. The base permits the entire structure to move without modifying the head orientation.

The beam collimation is performed through poly methyl methacrylate applicators consisting of two separated sections: the upper is fastened to the accelerator's head, with the lower in contact with the patient. These sections are aligned and finally hard-docked together before dose delivery. The applicators set consists of cylindrical tubes with a wall thickness of 5 mm, diameter ranging from 4 to 10 cm, and face angles of 0–45°. The length of the applicators varies according to the diameter: 80 cm for diameters up to 8 cm and 100 cm for those up to 10 cm. Using an applicator of 10 cm diameter and the maximum energy, the depth corresponding to 85% of maximum dose, measured in water, is about 18 mm on the geometrical central axis. When beveled applications are used the dose distribution is asymmetric and with high gradients of dose.

The Novac7 does not employ scattering filters which in conventional machines are the main source of stray radiation, but for this reason it is complicated to modulate accelerator dose rate, which is high compared to conventional accelerators. The total bremmstrahlung photon dose for conventional accelerators is at least 2-3% of the dose at the depth of R_{max}, mainly due to head scatter. For Novac7 9MeV nominal energy this value is 0.2% of the dose value at R_{max}.

The novelty of the Novac7 system (NRT, 2011) concerns the accelerating structure. The accelerating structure consists of a β graded SW 2998 MHz on-axis coupled linac operating in π/2 mode with 11 accelerating cavities and is 50 cm long, powered by a 2.6 MW magnetron: it is a compact accelerating structure in which the beam focusing is automatically achieved without using external magnetic lenses and the losses are kept at low energy so getting a negligible diffused X radiation.

Model	Old Novac7 (Hitesys)	New Novac7 (NRT)
Nominal Energy	3 – 5 – 7 – 9 MeV	4 – 6 – 8 – 10 MeV
Beam current	1.5 mA	1.5 mA
Frequency of emission	5 Hz	9 Hz
Scattering foil	No	No
Dose rate	9 > e < 21 Gy/min	>6 e < 39 Gy/min
Field Diameter	4,5,6,7,8,10	3,4,5,6,7,8,10
X-ray contamination	< 0.2 %	< 0.2 %
Power dissipation	<1kW	<1kW

Table 1. Novac system evolution

The single cavity shapes were optimized in order to maximize the efficiency and to reduce the dark currents, which could be a serious problem for the operation of the system at very low currents. In particular the beam hole diameter was reduced from 8 to 6 mm and the cavity nose shape was modified. The shunt impedance was increased of 15%: in this way adding only four cavities to the Novac7 structure it is possible to increase the maximum energy with the same power.

The main characteristics are reported in Table 1.

2.3 Liac

Liac® 10 MeV (SORDINA SpA Italy) is an intraoperative radiotherapy system, which produces electrons with energies of 4, 6, 8 and 10 MeV with a dose rate between 5 and 20Gy/min and a pulse frequency between 5 and 20 Hz.

At 10 MeV, R_{50} value (i.e. the value of the depth in water at which the dose is 50% of its maximal value) is 38 mm.

The Liac® system consists of a mobile radiant unit and a operator control rack, connected by a 10 meter cable, which during the irradiation supplies the radiant unit with electrical power and transmits the treatment parameters. During the IORT session the Liac® is not connected to the local electrical system but is fed by the UPS (Uninterruptable Power Supply) hosted in the control rack. The weight of the Liac® radiant unit is less than 400Kg, so that there are no installation problems in any surgical suite; a battery system lets this unit move independently in the operative block. An innovative robotic system allows the LIAC to be extremely mobile and strongly simplifies hard-docking procedures. The LIAC head has three degrees of freedom: it can be moved up and down for a maximum excursion of 100 cm, it has a roll angle of ± 60° and a pitch angle between + 30° and – 15°.

The standing wave S-band linear accelerating structure, specifically designed for Liac , is 850 mm long and consists of 17 autofocusing cavities; it is supplied with a 3.1MW Magnetron, with 2.5 µs pulse length and produces an electron beam of 12 MeV maximum energy. The pulse repetition can vary from 1 up to 20 p.p.s (pulse per second).

The output beam has a 3 mm diameter and is collimated by a sterilizable cylindrical perspex applicator 60 cm long, different diameters and terminal beveled angles. The dose homogeneity on the surface to be treated is generally guaranteed by a 100 µm brass foil scattering filter inserted in front of the titanium window. This technique allows the optimization of the accelerator performances keeping the level of stray radiation below the required limits. A new type of dosimetric system has been implemented to monitor the beam. It is based on a properly designed resonant cavity. The signal, proportional to the absorbed dose, is picked up from the cavity, acquired and displayed in real time on the control rack, with a good signal-noise ratio. This dosimeter is not affected by saturation phenomena and temperature, pressure and humidity are independent.

Due to the need to minimize the length of the Perspex applicator, a scattering foil, made of brass (50- 150 µm thickness), was introduced in the beam to produce a homogenous profile. This technique allows the optimization of the accelerator performances, keeping the level of stray radiation below the required limits.

Sordina's 10 MeV model has been on the market since 2001 and, to meet customer demands, a new Liac® model able to accelerate electrons up to 12 MeV was developed in the last few years.

The Liac® 12 MeV accelerating system is a newly designed linac operating in the π/2 mode at 2998 MHz. The electron energy is set by varying the radiofrequency power from 1.2 up to 3 MW. The new machine setup provides four clinical energy points: 6, 8, 10 and 12 MeV. At 12 MeV, R_{50} value (i.e. the value of the depth in water at which the dose is 50% of its maximal value) is 48 mm.

The 12 MeV LINAC is 92.5 cm long (19 accelerating cavities) and its total weight, including electron gun and ionic vacuum pumps, is less than 30 kg. Radiofrequency power is supplied by an E2V magnetron MG6090.

The particular design of the LIAC 12 MeV accelerator head guarantees a minimal head leakage radiation, much lower than target scatter radiation. The accelerating waveguide has no external solenoid for electron beam radial focusing, but electrostatic focusing is used instead. This radial focusing system decreases the tail of electron beam distribution hitting the copper waveguide, reducing bremsstrahlung radiation, and focalizes the electrons along the beam line. It was manufactured in accordance with Italian regulations of radioprotection.

Furthermore there is no bending magnet and the metallic elements which the electron beam crosses along its path are a titanium window, 55 µm thick, an aluminum scattering foil, 820 µm thick and four ionization chamber electrodes, in total 20 µm thick. The total head leakage is less than the scatter radiation by a factor 10. The choice of using an aluminum scattering foil (820 µm) for the 12 MeV instead of a brass one (75 µm), as in the previous model (10 MeV), was made after an experimental study where the optimization parameters had the following characteristics: limited applicator length (for manageability reasons), beam flatness (within ± 5% evaluated at 80% of the dose profile), a controlled environmental x-ray radiation and a low neutron contamination.

It is worth noting that by reducing the applicator diameter (from 100 mm to 40 mm) the dose per pulse increases correspondingly. This type of collimation together with the absence of a bending magnet and the planning choice of light material make this equipment workable in operating rooms, keeping the stray radiation at a low level. Pulse Repetition Frequency (PRF) can be varied from 1 to 60 Hz. The PRF is set by the manufacturer according to the various e-beam energies to keep the dose rate around 10 Gy/min with an applicator diameter of 100 mm. However, up to 30 Gy/min higher or lower dose rates are readily obtained. A newly designed system has been used to guarantee the LIAC output reliability. The current injected by the electron gun can be adjusted (± 5% maximum) by an automatic dose control board (ADCB) to keep constant the read-out of the two monitor chambers so that the ratio cGy/MU is kept reliably constant. Radiofrequency power is supplied by E2V magnetron MG6090. Electron energy is set by varying the radiofrequency power from 1.2 up to 3 MW. The new machine setup provides four clinical energy points: 6; 8; 10 and 12 MeV.

The PMMA applicators are 60 cm long and 0.5 cm thick and fully gas sterilizable; various diameters (from 30 to 100 mm typically) and bevel angles are available. The distance from the scattering foil to the end of the applicator or SSD is 713 mm.

This passive beam shaping technique allows good uniformity and flatness of the radiation field and a very low x-ray contamination. Furthermore, the electron beam interaction with the PMMA applicator generates low energy electrons which deposit the dose in the region very close to the surface: this fact explains the higher value of the skin dose with respect to (External Beam Radiation Therapy) EBRT linac. Surface dose is greater than 85% with 4 MeV e-beam and reaches about 94% with 12 MeV e-beam. The main characteristics are reported in Table 2.

Model	LIAC 10 MeV	LIAC 12 MeV
Nominal Energy	**4 – 6 – 8 – 10**	**6 – 8 – 10 – 12**
Beam current	1.5 mA	1.5 mA
Frequency of emission	1 – 60 Hz (variabile)	1 – 60 Hz (variabile)
Scattering foil	75 micron brass	850 micron aluminum
Dose rate	2-30 Gy/min	3-40 Gy/min
Field Diameter	3,4,5,6,7,8,10 & 12 opz	3,4,5,6,7,8,10 & 12 opz
X-ray contamination	< 0.5 %	< 0.5 %
Power dissipation	2 kW	2 kW

Table 2. Liac® system characteristics

3. Dosimetry of the beam

The IORT technique using the dedicated Linac machines requires special dosimetrical determinations, which are sometimes different in comparison to conventional external-beam radiotherapy. The main reason stems from the fact that a single high dose of radiation is delivered to a selectively defined volume of tissue, whose extension and depth are directly determined in the operating theatre. Particularly, the dosimetric data must allow the calculation of the Monitor Unit (MU) necessary to deliver the dose prescribed to the target volume. A further difference between IORT and external radiotherapy is related to the use of specific applicators that contribute to the determination of the physical-geometrical characteristics of the electron beams (quality, topology, homogeneity, etc.). All definitions are reported in the main international guideline (Istituto Superiore di Sanità [ISS], 2003; Palta et al. 1995; AAPM, 2006; Beddar et al., 2006).

A square applicator 10×10 cm^2 or a circular applicator of diameter 10 cm with a plane basis is recommended for measurements in reference conditions and for each energy. This choice should allow, in most cases, to have a SSD = 100 cm or a nominal SSD when the length of the applicator does not allow it. The depth of R_{max} (i.e. the depth at which the maximum dose is obtained) is recommended as reference depth for the dosimetry, both in reference and in no reference conditions.

The use of ionization chambers for the calibration of the beam in terms of dose per MU may be ineffective with dedicated machines because of the high density of electric charge produced in the chamber volume per radiation pulse. In particular the correction factor for ion recombination (K_{sat}) can be largely overestimated if the correction methods recommended by the international protocols are used. (AAPM, 1999; IAEA, 2001)

Italian guidelines (ISS, 2003) recommend the use of the absolute dosimetric system of Fricke for the measurement of the absorbed dose in water in reference conditions.

If other dosimetry systems are used, it is in any case required that all measurements can be traceable to national and international standards of the quantity "absorbed dose to water". This goal can be achieved through the calibration of the dosimeters at a Primary Metrological Institute or by a recognized Calibration Centre.

The Fricke dosimeter is a chemical dosimeter based on a solution of iron sulphate (Olszanski A. et al., 2002) and it consists of a glass-sealed ampoule (8.7 mm in diameter and 28 mm in length, thickness of glass 0.5 mm) filled with a ferrous sulphate acqueous solution. Dose assessment is performed through optical absorption measurements with a spectrophotometer at a wavelength of 304 nm. The perturbation introduced by the glass walls of the vial should be taken into account. The calibration in terms of dose to water is made using a 60Co gamma ray field in a Primary Standard Laboratory. The stated uncertainty in dose measurement has been evaluated to be 1.5% (1s).

An alternative dosimetric system with sensitivity independent from the dose-rate, from the beam energy and from the angle of incidence of the electron beam is the alanine dosimetry.

This type of dosimeter consists of a blend of alanine (95% by weight) and polyethylene (5% by weight) pressed into pellets of 4.9 mm in diameter and 2 mm in length (1.2 g/cm^3 mass density) (De Angelis et al., 2006). The sample is measured with a spectrometer using the Electron Paramagnetic Resonance (EPR). Typically a set of five alanine pellets is used for each point of dose measurement. Each set is inserted into a quartz tube that is positioned in the microwave cavity for measurement. The alanine dosimeters should be calibrated in terms of dose to water in a 60Co gamma ray field against a Primary Standard Laboratory. The combined uncertainty in the measure is 1% (1s) for a test dose of 10 Gy.

Dose measurements performed using Fricke and alanine dosimeters have shown a good agreement, generally within 1% for plane-base applicators (De Angelis et al., 2006).

However, since the ionization chamber is the online absolute dose measurement device accepted as a reference dosimeter in clinical dosimetry, several authors have proposed corrections to take into account the free-electron fraction component which causes the overestimated value. The correction is introduced through the estimation of a saturation factor K_{sat}.

All the approaches are based on three improved theoretical models proposed by Boag (Boag et al., 1996). These authors did not provide particular criteria for choosing any of their three different expressions of k_{sat}.

Italian researches propose two different experimental approaches. Di Martino (Di Martino et al., 2005) suggested a new equation for k_{sat} in high dose-per-pulse beams based on the first Boag's formulas and experimentally derived the free-electron fraction p. The evaluation of some chamber-specific parameters is needed for calculation of ion recombination correction factor. The intercalibration with a second dose-per-pulse independent dosimeter is needed. Laitano (Laitano et al., 2006) proposed the evaluation of k_{sat} starting from Boag's two voltages analysis (TVA), suggesting that the third of three Boag's models is more adequate. The latter approach has the advantage of being able to avoid any chamber calibration using, however, the calculated value of p as a function of chamber characteristics and experimental conditions.

In the general dosimetric characterization of the electron beams produced by an accelerator dedicated to IORT, the absolute dose in the point of the clinical prescription (buildup depth in water on the clinical axis) for the beveled applicators must also be determined. We refer to this type of dosimetry as "in non reference conditions".

Dosimetric characteristics of the electron beams requires the knowledge of:

- PDD (Percentage Depth Dose) measured along the clinical axis of the beam (which is different from the geometrical axis in the case of base-beveled applicators, with the indication of the values of the main parameters: R_{max}, practical range (R_p), depth in water at which the dose is reduced to 90% and 50% of maximum dose (R_{90}, R_{50}), surface dose and per cent dose due to the photon contamination of the beam (tail of bremsstrahlung radiation);
- Beam profiles measured in two orthogonal directions at the depths of R_{max}, of R90, of R80 and of R_{50};

Values of the output expressed as dose per Monitor Units (MU) (cGy/MU), measured in a point at the reference depth on the clinical axis of the beam;

Such characteristics must be measured for every applicator, energy and SSD of clinical interest.

For low energy electron beams (4-10 MeV), as suggested by international protocol, a parallel-plate ionization chamber should be used. Ross, Markus and advanced Markus types (PTW Freiburg Germany) are the most utilized (PTW, 2011). But the response of these types of chamber is angle dependent so their use with a beveled applicator is not possible.

Because of its high spatial resolution and independence on beam direction, a small cylindrical chamber, such as CC01 (Wellhofer, 2011) or Pin point (PTW), seems to be particularly suitable to perform absolute dose measurements for electron beams collimated with inclined applicators (Karaj et al 2007, Iaccarino et al. 2011).

Due to the presence of electrons scattered by the applicator, the electron fields obtained with IORT-dedicated applicators are characterized by a wider energy spectrum and a wider angular distribution than electron beams collimated with conventional systems. This implies a higher surface dose (especially for the lower nominal energies) and less steep dose gradients (especially for the higher nominal energies).

For this reason, a dosimetry system must be selected characterized by a minimum dependence of the response from the beam energy and from the angle of incidence of electrons. It is recommended that, when measuring the dose distribution, the same dose-rate (MU/min) should be used during the determination of the output and during the treatment of the patients. It is also recommended to investigate and determine the percentage of the radiation scattered through the applicator's walls, as a function of the beam energy and of the distance from the walls and from the base of the applicator.

For PDDs and Profiles diodes or diamond detectors should be used in water phantom. Relative absorbed dose measurements, i.e. percentage depth doses (PDDs) and off-axis profiles (OAPs) could be performed using a p-type diode in the water phantom. The signal from the diode detector is stated to be approximately proportional to the absorbed dose to water. The uncertainty in the diode position is typically less than 0.2 mm and PDDs and OAPs are acquired with a spatial resolution of 1.0 mm.

For the Output measurements, in no reference conditions and in particular in the case of small and or beveled applicators, a solid phantom and radiographic or radiochromic films (Fiandra et al., 2008), or TLD (Palta et al., 1995) may be employed.

Because of its high spatial resolution, weak energy dependence, and near tissue equivalence, Gafchromic films have been selected as the reference detector for relative measurements of the output factor, for applicators of various dimensions length, diameter and angle of incidence.

4. Decision parameters during surgery

The final goal of IORT is an enhanced control in loco-regional tumor. IORT is feasible for intra-abdominal, retroperitoneal, pelvic and other malignancies. More recently, clinical experiences have shown that IORT may improve local control and disease-free survival, especially when used in adjuvant setting, combined with external beam irradiation in the stomach, pancreas, colon-rectum cancer and soft tissue sarcoma.

The rationale for the use of this segmental radiation therapy in place of whole-breast irradiation (WBI) is based on the results of some long-term studies reporting that local relapses after conservative surgery and External Beam Radio Therapy (EBRT) occur at the original tumor site at a rate of 80% or more, with few exceptions. This has been also confirmed by the results of the Milan III trial, which compared quadrantectomy alone with the same conservative surgery plus EBRT on the whole breast, have confirmed that 85% of local relapses were in, or close to, the previous index quadrant (Ivaldi et al., 2008). Another important advantage of the IORT is the avoidance of interaction with systemic therapy that may determine delays in the initiation or in performing conventional EBRT when CT, and particularly anthracyclin-based cycles, are given.

Widespread applications of IORT at various disease sites are feasible due to improvements in technology. By increasing the maximum energy of the linear accelerators of IORT and the total radiation dose it is possible to improve the therapeutic ratio and the tumor local control without increasing morbidity. Moreover, IORT is used as an adjuvant therapy, i.e., given as a boost after conventional fractionated radiotherapy.

IORT is generally given as a single fraction of 10 to 20 Gray (Gy). The single fraction is considered biologically equivalent from two to three times that of conventional fractionation for tumors while it is radiobiologically inferior due to impaired cellular repair of surrounding normal tissues, thus the protection of these tissues is mandatory. In fact the benefit in local recurrence needs to be carefully considered against the complications arising from the addition of IORT. This means that the appropriate entrance of the beam should be selected as well as the appropriate applicator diameter/kind and beam energy. This determines the Output factors (OFs) to be used and the monitor units evaluated using on an appropriate formula. In order to determine a dose-response effect a dosimeter could be placed within the field to be representative of the tumor bed or close to an Organ at Risk (OAR) near the applicator or under an additional shield.

The dose to be prescribed to target depends on the kind of disease, as well as on the microscopic spread expected in the tumor bed. This is particularly important when the resection margins are positive and depends on tumor stage and on radio-resistance of cells.

5. Simulation of treatment plans

Treatment planning is the process in which a team of radiation oncologists, surgeons and physicists plan the most appropriate radiation treatment for the patient using an external beam or by means of an internal beam such as IORT.

Currently, it is very complicated to plan the radiotherapy process beforehand, this is because the team of specialists must choose at the very moment of operation the cone dimension, its positioning, the bevel angle and the electron beam energy according to its medical and surgical experience and the information gathered during surgery. One of the main limitations of IORT is due to the removal of diseased tissue from the patient during surgery which will change the geometric shape. In this way it is difficult to carry out a feasible dosimetry calculation using pre-operative images.

This aspect highlights two main problems for the planning of the intervention:

- Before the operation: it is difficult to estimate the dose to be received;
- After the operation: since such images are not available, the results cannot be assessed.

Today, treatment planning is entirely focused on the use of specific software. This software uses algorithms that exploit the patient's images, by means of the computed tomography (CT), for extremely realistic simulations.

Systems like Novac11 and Radiance are now a single integrated platform, which include not only the dedicated mobile accelerator, but also simulation software (NRT, 2011).

Radiance (Radiance, 2011) is the only available Radiotherapy Treatment Planning System (RTPS) that has been specifically designed for Intraoperative Radiotherapy (IORT) and the only one that works in such a field of radiotherapy. Radiance allows the specialists to simulate the whole IORT workflow. The operator can simulate the full delivery of the dose loading and visualizing CT images of the patient. The best choice of the parameters needed may be taken (applicator geometry, accelerator energy, number of MU …) to maximize the dose on the tumor or tumor bed and to minimize the contribution to the healthy tissues.

A new step in the IORT procedure has been introduced, the preplanning phase. In this pre-simulation, it is possible to modify the different parameters of the procedure to evaluate the outcome without stress in the real treatment decision-making process. Thus, it improves the safety of the global procedure. Radiance is a state-of-the-art development unrivalled today worldwide. It uses advanced visualization, simulation and dosimetry algorithms that are far ahead, concerning performance, of any current software. DICOM images are loaded and their volumetric representation built online. The computation of both graphics and dosimetry computation algorithms is almost in real time. The capability of simulation of different tissue densities opens a new era in the planning of IORT. At the same time the measurements required for commissioning a linear accelerator have been reduced considerably.

The dose calculation is carried out by means of two dosimetry computation engines which are available in radiance:

- An adapted and validated fast implementation of the Pencil Beam algorithm used in external radiotherapy.
- A parallel implementation of Monte Carlo algorithm.

In the future built-in tools that make use of advanced simulation software, as described, will be increasingly used. This software will be able to improve not only the quality of the radiosurgery procedure, and therefore the results on patients, but also the interaction of the various specialists involved (radiation oncologists, surgeons and Physicists) through simulations of different scenarios.

6. Use of scintillating fibers for dosimetry

Several types of dosimeters have been developed to measure the precise dose and the percentage depth dose (PDD) using detectors such as ion chambers, silicon-diode detectors, diamond detectors, liquid ion chambers and radiographic films. They are partially ineffective, considering requests for new approaches to therapy such as: small sensitive volume, dose rate independence, measurements in real-time, use of high doses, no dependence on environmental conditions, such as temperature, humidity and pressure. For this reason new dosimetric systems have been studied and proposed in recent years (Aoyama et al., 1997; Staub et al., 2004; Archambault et al., 2005; Consorti et al. 2005).

These include dosimeters based on scintillating fibers. Difficulties arising in the conventional techniques have been overcome by providing features such as: small sensitive volume, high spatial resolution, water or tissue equivalent material, linear response, independence from energy and from dose rate, independence from environmental conditions such as temperature, pressure and humidity, high resistance to radiation exposure, real-time response.

One of the first works that showed the potential of scintillating fibers to measure the depth-dose for electron beams was published in 1997 (Takahiko et al., 1997). Over the past 10 years several proposals have appeared in literature with the use of scintillating fibers. Some of these have used small volumes of scintillating fibers coupled to light guides for reading the response to light (Archambault & Arsenault, 2005 ; Frelin & Fontbonne, 2006 ; Bongsoo et al., 2008 ; Staub, 2004). In others, detector elements have been incorporated in a water phantom equivalent in order to have a 3D representation (Fontbonne et al., 2002; Guillot et al., 2010; Lacroix et al., 2008). In all these approaches, the technique of elimination of Cerenkov light was needed. This contribution is very small and negligible compared to the scintillation light in the scintillating fibers. However the Cerenkov light becomes dominant, in the light guides, connecting the scintillating fibers to the electronic readout, when the guides are subject to radiation (Archambault et al., 2006).

The approach described in this paragraph is different because it is very similar to that adopted in X-ray diagnostics (Bartesaghi et al., 2007a, 2007b). The dose absorbed along the beam is read by scintillating fibers assembled in homogeneous planes orthogonal to the beam axis. The light produced is integrated and collected for the read out at the end of each fiber, giving the projection of the signal in one dimension. The fibers assembled in each homogeneous plane are ready at one end. The approach is similar to the reading of one-dimensional Radon transformation of the dose absorbed. In a previous work (Lamanna et al., 2009) we studied the basic elements constituting the device. In this chapter we present the results of the study of the therapeutic beam of electrons (the system can be used for other particles, photons, protons etc.) with a homogeneous layer of plastic scintillating fibers.

6.1 Dosimeter characteristics

According to the clinical dosimetry, the absorbed dose is calculated in the biological tissue. For this reason, dosimeters built with equivalent tissue are more suitable. Water is the source material for phantoms taken as equivalent tissue in all electron beam dosimetry protocols. This choice comes from the fact that approximately 2/3 of the human body is made up water. The dose distribution curves in water and in tissues are very similar, because water is an excellent image of diffusion and absorption properties of the human body. However, in many cases the employment of water as a test phantom is not very useful. Thus it can be replaced by solid materials with similar physical properties like polystyrene and perspex (PMMA). In theory, water tissue-equivalent materials should have effective atomic number Z_{eff} (electron number per gram), and a density equal to water. In clinical applications, however, a material with electron density as close as possible to water is fairly suitable (McLaughlin & Chalkey, 1965).

For these reasons the system DOSIORT (IORT Dosimeter) is conceived as a box made of a tissue-equivalent material (polystyrene) with a density $\rho = 1.05$ g/cm^3. Inside the box there are six sensitive layers spaced 4 mm apart and set perpendicularly to the Z direction of the incident beam, as shown in Fig 2.

Each layer is composed of a grid consisting of two bundles of 190 scintillating optical fibers having a square cross section of 0.5×0.5 mm^2 and crossing one over another so as to define a XY layer with an area of approximately 10×10 cm^2. We use 380 BCF-60 scintillating optical fibers produced by Saint Gobain Crystals.

The light emitted in the fibers is read-out by two photodiode (Phd) arrays Hamamatsu S8865-128 for each bundle. The Phd are read sequentially using the Hamamatsu driver C9118 and the signal is digitized and processed by a computer with dedicated software.

The read-out system is therefore composed of 24 arrays. A dedicated electronic system is able to acquire, process and display the reconstructed electron beam image in real time (within a few seconds).

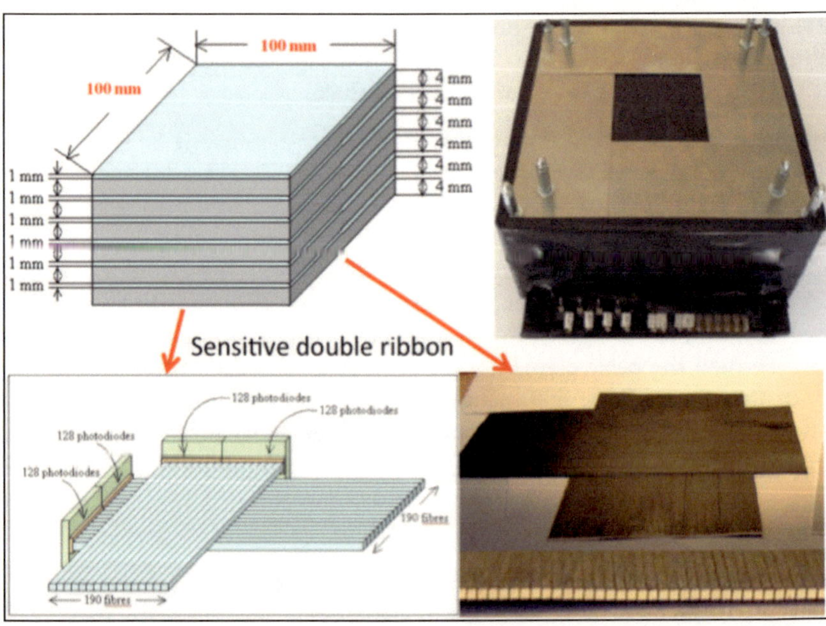

Fig. 2. Scintillating fiber dosimeter. On the left the characteristics of the layers are shown. On the right the photo of the dosimeter and of one double ribbon are shown.

The system is thus a solid phantom having a density approaching 1 g/cm^3, with sensitive layers of scintillating fibers set at fixed positions in a calorimetric configuration for the containment of electrons of energy 4-12 MeV. The prototype is able to define the physical and geometrical characteristics of the electron beam (energy, isotropy, homogeneity, etc) and to measure the parameters needed to select the energy, the intensity and the Monitor Units (MU) for the exposure: the Percentage Depth Dose (PDD); the Beam profiles; the Isodose curves; the values of dose per MU (cGy/MU).

Another important thing that must be considered is spatial resolution. The spatial resolution is a key feature for next-generation dosimeters. This feature will become more important in the next few years, in fact next-generation accelerators have a beam size up to 3 mm. DOSIORT responds fully to the demands of new approaches to radiotherapy. Its innovative detection system, made up of optical fibers which provide high spatial resolution and photodiodes with a sensitive area of 0.3×0.6 mm^2 with pitch of 0.4 mm, gives a high spatial resolution approximately 0.5 mm.

The materials used to make the detector and the operations needed before the data taking have been studied and described in details in previous papers (Lamanna et al., 2009a, 2009b). The sequence of the most relevant steps before beginning are: electronic noise measurement to be subtracted; choice of the integration time to ensure a dynamic range large enough to have a linear response from the detector electronic; evaluation of calibration factor for each Phd response by exposing the fibers homogeneously to the same beam.

In this paragraph we include the results obtained from testing the system with a photon beam of 6 MV and an electron beam of energy 9 MeV generated through a Varian Clinac 2100 DHX. The data are compared to the measurements obtained using the PTW Freiburg TM31010 ionization chamber and the PTW MP3 water phantom, when they are available.

The data taken through DOSIORT may be selected and elaborated considering the full detector or only part of the system. There are three different configurations for applying the dosimeter: using one layer (1D) at fixed depth and rotating it around the beam axis , using one double layers (2D) at different depths and using all six double layers (3D) at different depths. All the configurations are able to get the results in real time but the first system gives a more accurate measurement of the dose at fixed depth and it provides a series of measurements at different angles, the second one gives the results in depth more quickly but is less accurate in the reconstruction of the dose outside the FOV (field of view) region, the third is able to give a 3D estimation of the beam in depth into a single measure. However, all configurations require the acquisition of a number of measurements in a time that is in any case about a tenth of that needed to obtain the results with the ionization chambers or other traditional dosimeters. The choice is connected to the level of accuracy needed in the measurements.

6.2 Results using one layer of DOSIORT

The response of the detector was tested through exposure to a beam orthogonal to the layer surface using only one layer. The photon beam of 6 MV was selected with a FOV of 8x8 cm^2. The system was rotated manually around the beam axis and the projected data were collected every 5 degrees for a total of 37 positions from -87.5 to 92.5 degrees.

The projection of the dose in arbitrary units at different angles and positions along the rotating axes is shown in Fig. 3,a.

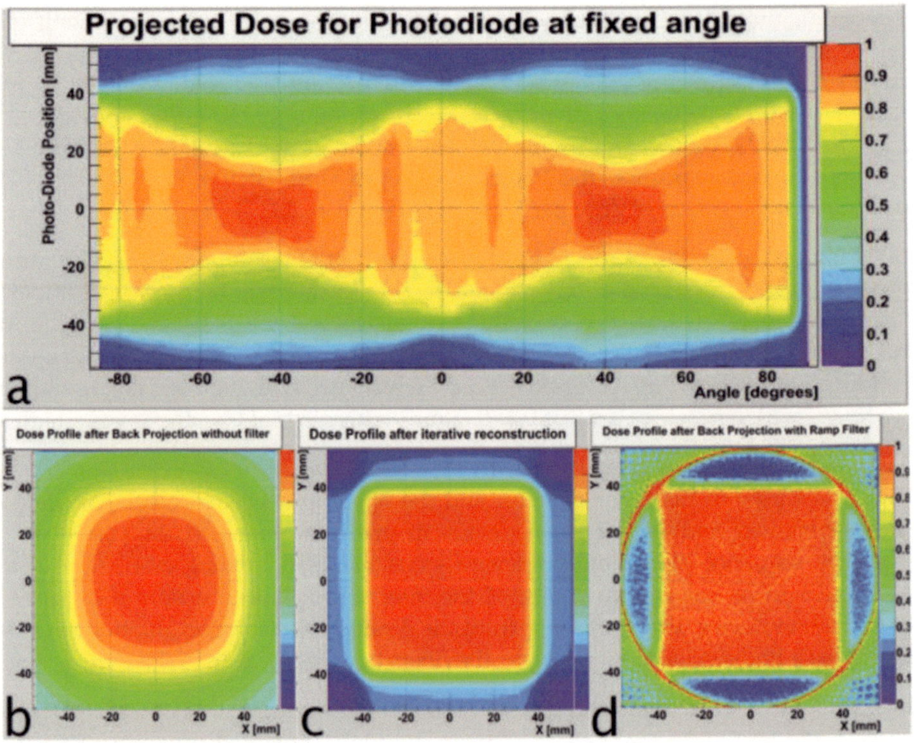

Fig. 3. a) Projected dose at 37 rotating angles around the beam axis; b) Back Projection without filter; c) Iterative reconstruction; d) Filtered Back Projection with ramp filter.

The plot shows the typical behaviour of a square FOV with maximum values around the diagonal at -45 and 45 degrees.

The 37 projections were used to reconstruct the transverse profile of the dose. The fibers collect the light generated along their axis, thus each acquired profile corresponds to the projection of the beam delivered along an axis. This system is not dissimilar to image reconstruction problems in tomography, where several projections have to be composed to trace them back to the original image. For this reason the first reconstruction method chosen was the back-projection algorithm widely used in tomography. Nevertheless using a back-projection approach without a filter we obtained a poor profile (Fig.3,b) while with the introduction of a ramp filter (Fig. 3,d) a better reconstructed image can be obtained. We additionally developed a dedicated algorithm based on the principle of the tomographic iterative methods.

The iterative method uses only two orthogonal projections. The choice of the two projections is very important to determine a good result in the reconstruction. We have selected the projections at 0 and 90 degrees. The idea is to sum the data collected by each fiber along an axis with the corresponding fiber for each of the two different angles, weighing the projection contribution on the basis of its concordance with the other projections results.

The image obtained in this way is a weighted sum of the contributions of two projections. Then the difference between the sum along an axis of the reconstructed image and the acquired values is calculated. This parameter is taken as the error to minimize iterating the method. Each value difference is projected back on the reconstructed image to correct the values along the fiber and then a new difference between reconstructed and acquired data is calculated. This method, described in (Brancaccio et al., 2004), is able to converge into seven/eight steps and in less than a second. The reconstructed image seems to be more precise (Fig. 3,c). A more accurate comparison is shown in Fig. 4 where the reconstructed dose profiles are superimposed on the profile obtained using the ionization chamber around the centre of the beam.

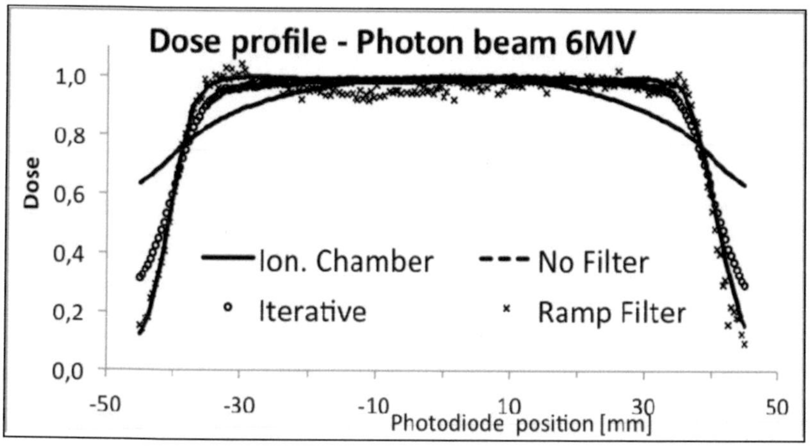

Fig. 4. Dose profile measured using Ionizing chamber superimposed on the reconstructed dose profiles using the iterative and the back projection with and without filters.

The reconstruction using the back projection with ramp filter reproduces quite accurately the dose profile from the maximum to about 10%. Some fluctuations are visible in the flat region. The iterative approach describes quite well the profile in the FOV used [-40 to +40 mm] where the dose varies from 100% to 50%.

6.3 2D Results using one double layer of DOSIORT

The energy 9 MeV was selected for the test of 1 double layer (XY) of DOSIORT. The FOV was set at 4×4 cm^2 in order to study the energy response containing the doses absorbed.

The data were taken in different acquisition tests. In each test we changed the geometry of the setup superimposing over the previous configuration a sheet of polystyrene of 4.2 mm in water equivalent thickness. In this way we simulated a homogeneous phantom with measurements at different depths.

The double layer 2D was used for the reconstruction of the XY map of dose absorption at different depths. The XY projections collected for each acquisition, after correction for noise and calibration, were used to reconstruct the dose through the iterative method used for 1D detector and explained in (Brancaccio, 2004).

In Fig. 5 isodose curves at four fixed depths are shown for electrons of 9 MeV. The performance is similar to the reconstructions obtained in medical imaging using only two projections. The results are acceptable in a region coinciding with the FOV, [-20 to +20 mm] in the pictures. Outside this region the reproduction is poor.

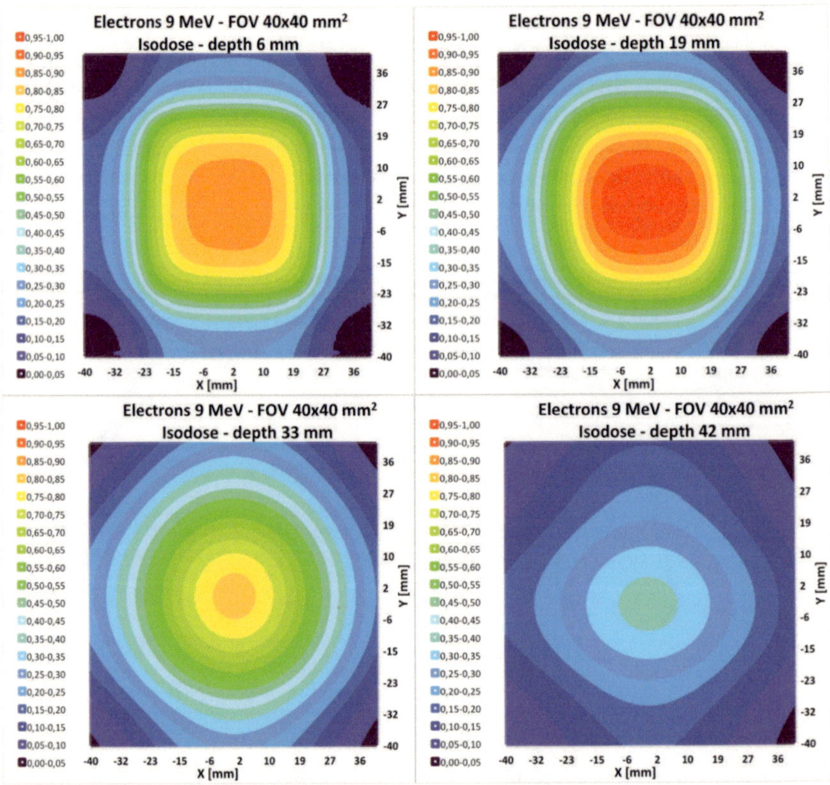

Fig. 5. Isodose at four fixed depths for electrons of energy 9 MeV. A FOV 40×40 mm² was selected.

The reconstructed doses may be used to visualize the depth dose profiles, selecting one central slice for each fixed step. The performance obtained is shown in Fig. 6.

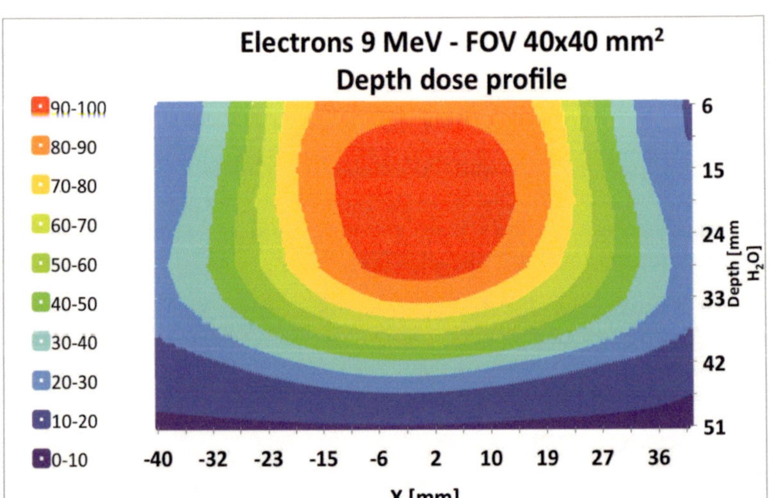

Fig. 6. Depth dose profile for electrons of energy 9 MeV. A (FOV) 40×40 mm² was selected.

Also in this picture the region inside the FOV is well represented.

6.4 3D Results using DOSIORT

The full detector DOSIORT may be used to measure the dose in three dimensions.

The system, optimized through Monte Carlo simulation as explained in (Lamanna et al., 2009a), is able to contain the full shower produced with electrons of energy 6 to 9 MeV. Greater energies may be measured using the same technique as in the previous paragraph: by superimposing sheets of polystyrene over the detector. The thickness of the sheets must be selected to position the build-up inside the detector. The isodose curves measured in each double layer for a beam of electrons of energy 9 MeV, using a FOV of 40×40 mm^2 without external sheets are shown in Fig. 7. The dose is reconstructed using the iterative method described previously. The curves are well described in the FOV region. Outside some artefacts connected to the reconstruction method are visible.

The central X slice of the isodoses shown in Fig. 7 are represented on the left side of Fig. 8. The dose profile in depth is well reconstructed. The comparison with the ionization chamber results is done in the right part of the same figure, using the average of the dose in a central surface of area 2×2 mm^2, corresponding to 4×4 scintillating fibers. The measurement through DOSIORT reproduces the PDD curve obtained through the ionization chamber.

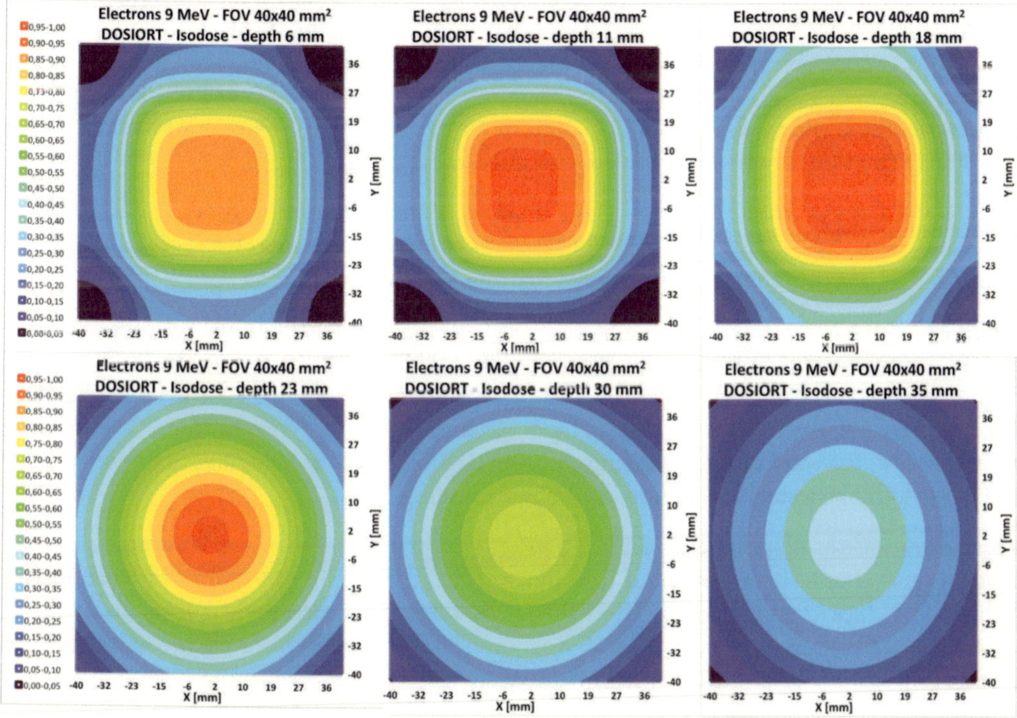

Fig. 7. Isodose curves reconstructed through the iterative method described in the test using data taken with 6 double layers of DOSIORT.

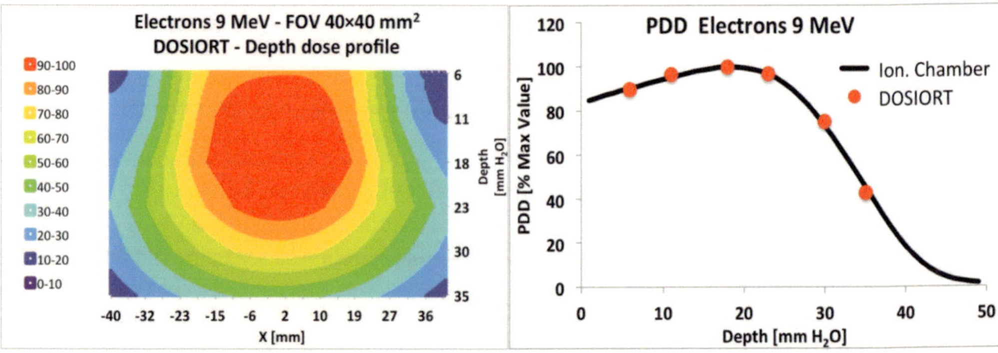

Fig. 8. Depth dose X profile on the left and PDD curve on the right: DOSIORT compared to the results of the Ionization Chamber.

7. Current activities and future clinical perspectives

Many efforts have been made by physicists and physicians in order to improve this therapeutic approach. In the main, medical physicists have conducted many studies to determine the dosimetric characteristics of the beams produced by these dedicated devices, in order to overcome the limits of the dosimetric systems, such as ionization chambers. In particular, dosimetric data and the OFs of flat and bevelled applicators for all available applicator/angle combinations have been investigated using Monte Carlo simulation. The aim was to predict with higher accuracy the dose distributions delivered to the target and Organs at Risks, and to use this information in modern treatment planning systems, which are mandatory to improve the knowledge of dose-effect relationships.

In fact, even if the Monte Carlo method requires longer computing time, it is capable of accurately calculating the dose distribution under almost all circumstances and can be employed as a benchmark of conventional treatment planning systems.

Another future clinical perspective is based on the deep investigation of radiobiology of IORT. In fact, the current radiobiological models should be applied up to 16-18 Gy, i.e. for higher doses the validity of Linear Quadratic model should be proven. Because IORT is used as an adjuvant therapy, i.e. given as a boost after conventional fractionated radiotherapy, the modality to combine the expected effects of fractionated and single dose fraction should be tested and verified in a clinical setting.

Moreover, the development of on-line systems to monitor the three dimensional dose distributions should be encouraged in order to reduce the time of verification and quality assurance before treatment delivery; as well as during the acceptance tests in order to reduce the time of the physical characterization of these machines for each combination of energy and applicator diameter or collimator systems.

Finally, the availability of large field sizes (larger than 10 cm), as well as of beam modifiers should prove to be more efficacious for larger targets such as sarcomas, while sparing normal tissues.

8. Conclusion

The implementation of the new Linac for the production of electron beams dedicated to IORT in the last 10 years has allowed a diffusion of the therapeutic approach in a large number of health facilities by making the approach easier to use. Its use has facilitated the possibility to carry out clinical trials in international contexts for different types of tumor. Accelerators available on the market are today excellent for use in operating rooms. The improvements of the IORT technique pass through the proposal of means that include devices and methods to cover completely the therapeutic intervention. In particular through introduction of tools that help operators to select parameters required for this radio-surgical technique. Among these the most important are: a specific treatment planning system, and an efficient distribution map of the beam dosimetry. Today effort is devoted to introducing improvements in these two fields.

GMV-RADIANCE (GMV, 2011) has recently introduced on the market a Proposal for IORT treatment planning. This package is included in the method ELIOT (Electron Beam Intraoperative Radiotherapy) of the NRT (NRT, 2011) and it is really promising.

The Dosimetry of the beam is assessed using conventional systems with some difficulties to provide all the necessary measurements. In this chapter a method based on scintillating fibers is described. The results of the tests performed using a Varian electron beam are promising. The system allows, quickly and in detail, the measurements of the dosimetric distributions of the beam. The evolution of the system will be the engineering of the prototype to improve the electronic read-out and its stability.

9. Acknowledgment

We are grateful to the Italian INFN (National Institute of Nuclear Physics) for supporting the study of scintillating fiber dosimeters and the '"Hospital of Cosenza" for allowing the use of the accelerator Varian to test the dosimeter.

10. Abbreviations used in this chapter

AAPM: American Association of Physicists in Medicine;
ADCB: Automatic Dose Control Board;
DICOM: Digital Imaging and COmmunications in Medicine;
EBRT: External Beam Radiation Therapy;
FOV: Field Of View;
IORT: Intra-Operative RadioTherapy;
IAEA: International Atomic Energy Agency;
INFN: National Institute of Nuclear Physics;
ISS: Istituto Superiore di Sanità;
K_{sat}: Correction factor for ion recombination in ionization chamber;
Linac: Linear Accelerator;
MU: Monitor Unit;
NRT: New Radiant Technology;
OFs: Output factors;
PDD: Per cent Depth Dose;
Phd: Photodiode;

PMMA: Poly(Methyl Methacrylate);
PRF: Pulse Repetition Frequency;
R_{max}: Depth in water at which the dose has the maximal value;
R_p: Practical Range;
R_{50}: Depth in water at which the dose is 50% of its maximal value;
R_{90}: Depth in water at which the dose is reduced to 90% of its maximal value;
SSD: Source to treatment Surface Distance;
TLD: Thermo Luminescent Dosimeter;
TVA: Two Voltages Analysis;
UPS: Uninterruptable Power Supply;
Z_{eff}: Electron number per gram;

11. References

AAPM (1999) American Association of Physicists in Medicine. Radiation Therapy Committee Task Group No. 51. Protocol for Clinical Dosimetry of High-Energy Photon and Electron Beams, Available from http://www.aapm.org/pubs/reports/

AAPM (2006) Radiation Therapy Committee Task Group No. 72. Intraoperative radiation therapy using mobile electron linear accelerators. Available from http://www.aapm.org/pubs/reports/

Abe, M. ; Takahashi M. (1981) Intraoperative Radiotherapy: the Japanese Experience. *Int.J. Radiat.Oncol.Biol. Phys.* Vol.7, pp.863-868

Abe, M. (1989). History of Intraoperative radiation therapy. In:Debelbower RR, Abe M (eds) *Intraoperative radiation therapy.* CRC, Boca Raton; :1-10.

Aoyama, T.; Koyama, S.; Tsuzaka, M.; & Maekoshi, H. (1997). A depth-dose measuring device using a multichannel scintillating fibre array for electron beam therapy; *Med. Phys.*, Vol. 24, no. 8, pp: 1235–1239.

Archambault, L.; Arsenault, J.; et al. (2005) Plastic scintillation dosimetry: Optimal selection of scintillating fibers and scintillators. *Medical Physics,* Vol. 32, pp.2271-2278.

Archambault, L. et al. (2006) Measurement accuracy and Cerenkov removal for high performance, high spatial resolution scintillation dosimetry. *Medical Physics,* Vol.33, pp.128-135.

Bartesaghi, G. et al. (2007a) A real time scintillating fiber dosimeter for gamma and neutron monitoring on radiotherapy accelerators. *Nuclear Instruments and Methods in Physics Research Section A: Accelerators, Spectrometers, Detectors and Associated Equipment,* Vol.572 (1 SPEC. ISS.), pp.228-230. DOI: 10.1016/j.nima.2006.10.323

Bartesaghi, G. et al. (2007b) A scintillating fiber dosimeter for radiotherapy. *Nuclear Instruments and Methods in Physics Research Section A: Accelerators, Spectrometers, Detectors and Associated Equipment,* Vol.581 (1-2 SPEC. ISS.), pp.80-83

Beck C. (1909). On external roentgen treatment of internal structures (eventration treatment). *New York Medical Journal,* Vol.LXXXIX No.13, pp.621-622

Beddar, A.S.; Biggs, P.J.; Chang, S.; Ezzell G.A.; Faddegon B.A.; Hensley F.W.; Mills M.D. (2006) Intraoperative radiation therapy using mobile electron linear accelerators: report of AAPM Radiation Therapy Committee Task Group No. 72. *Med Phys. May;* Vol.33(5), pp.1476-89.

Boag J.W. ; Hochhauser, E. ; & Balk, O.A. (1996) The effect of free-electron collection on the recombination correction to inization measurements of pulsed radiation. *Phys Med. Biol.* Vol. 41(5), pp.885-897

Bongsoo L. et al. (2008) Measurement of Two-Dimensional Photon Beam Distributions Using a Fiber-Optic Radiation Sensor for Small Field Radiation Therapy. *Ieee Transaction on Nuclear Science*, Vol.55, pp.2632-2636

Brancaccio, R. et al. (2004) Study of an appropriate reconstruction algorithm for an innovative electron beam imaging system for dosimetry in IORT (Intra Operative Radiation Therapy). *Nuclear Science Symposium Conference Record, 2004 IEEE*, Rome (Italy), (Oct.16-22, 2004) pp.3531–3535.

Comas, C.; Prio, A. (1907). Irradiation roentgen intra-abdominale ,après intervention chirurgicale dans un cas de cancer de l'uterus. *Congres International d'Electrologie .Imprenta Francesca Badia*, pp.5-14 Barcelona, Espagna

Consorti, R. ; Petrucci, A. ; Fortunato, F. ; Soriani, A.; Marzi, S. ; Iaccarino, G. ; V. Landoni, V. ; & M. Benassi, M. (2005) In vivo dosimetry with mosfets: Dosimetric characterization and first clinical results in intraoperative radiotherapy. *Int. J. Radiat. Oncol. Biol. Phys.*, Vol. 63, no. 3, pp : 952–960.

De Angelis C., A. Soriani,M. Benassi and S. Onori (2006) On measuring the output of an IORT mobile dedicated accelerator; *Radiat Prot Dosimetry* Vol.120, pp.221-225

Di Martino, F.; Giannelli, M.; Traino A.C.; & Lazzeri M. (2005). Ion recombination correction for very high dose-per-pulse high energy electrons beams. *Med. Phys.* Vol.32, pp.2204–2210.

Fiandra, C.; Ragona, R.; Ricardi, U.; Anglesio, S.; Giglioli, F.R. (2008). Absolute and relative dose measurements with Gafchromic EBT film for high energy electron beams with different doses per pulse. *Med Phys.* Vol.35(12), pp.5463-70.

Fontbonne, J.M.; et al. (2002) Scintillating Fiber Dosimeter for Radiation Therapy Accelerator. *Ieee Transaction on Nuclear Science,* Vol.49, pp.2223-2227.

Frelin, A.-M.; Fontbonne, J.-M.; et al. (2006) A New Scintillating Fiber Dosimeter Using a Single Optical Fiber and a CCD Camera. *Ieee Transaction on Nuclear Science,* Vol.53, pp.1113-1117.

Guillot, M. et al. (2010) Toward 3D dosimetry of intensity modulated radiation therapy treatments with plastic scintillation detectors. *Journal of Physics: ConferenceSeries,* Vol.250, pp.1-5. doi:10.1088/1742-6596/250/1/012006.

Gunderson, L.L.; Willett, C.G.; Calvo F. A. & Harrison, L.B. (Eds) (2011) *Intraoperative Irradiation, Techniques and Results.* Humana Press, Springer Science+Business Media. available from http://www.springer.com/medicine/oncology/book/978-1-61779-014-0 .

Iaccarino, G.; Strigari, L.; D'Andrea, M.; Bellesi, L.; Felici, G.; Ciccotelli, A.; Benassi, M.; Soriani, A.; (2011). Monte Carlo simulation of electron beams generated by a 12 MeV dedicated mobile IORT accelerator. *Phys Med Biol.* (2011 Jul 21); Vol.56(14), pp.4579-96. Epub 2011 Jul 1.

IAEA (2001) International Atomic Energy Agency. Absorbed dose in external beam radiotherapy: An international code of practice for dosimetry based on standards of absorbed dose to water. *IAEA Technical Report Series* No. 398.

IntraOp Medical, Inc. (2011) Santa Clara, CA . Available at http://www.intraopmedical.com

Istituto Superiore di Sanità [ISS] (2003), Guidelines for quality assurance in intra-operative radiation therapy *ISTISAN. Report* No. 03/1 EN

Ivaldi, G.B.; Leonardi, M.C.; Orecchia, R.; Zerini, D.; Morra, A.; Galimberti, V.; Gatti, G.; Luini, A.; Veronesi, P.; Ciocca, M.; Sangalli, C.; Fodor, C.; Veronesi, U. (2008) Preliminary results of electron intraoperative therapy boost and hypofractionated external beam radiotherapy after breast-conserving surgery in premenopausal women. *Int J Radiat Oncol Biol Phys* Vol.72(2), pp.485-93

Karaj E., Righi S., Di Martino F. (2007) Absolute dose measurements by means of a small cylindrical ionization chamber for very high dose per pulse high energy electron beams. *Med Phys.* Vol.34(3), pp.952-959

Lacroix, F. et al. (2008) Clinical prototype of a plastic water-equivalent scintillating fiber dosimeter array for QA application. *Medical Physics,* Vol.35, pp.3682-3690.

Laitano R. F. et al., (2006) Charge collection efficiency in ionization chambers exposed to electron beams with high dose per pulse. *Phys. Med. Biol.* Vol.51(24), pp.6419–6436

Lamanna, E. et al. (2009a) Dosimetry of High Intensity Electron Beams Produced by Dedicated Accelerators in Intra-Operative Radiation Therapy (IORT). *Ieee Transaction on Nuclear Science,* Vol. 56, pp : 3844 - 3940.

Lamanna, E. et al. (2009b) Calorimetric approach for 3D dosimetry of high intensity therapeutic electron beams. *Nuclear Physics B - Proceedings Supplements* Vol.197 (1), pp.24-29

McLaughlin, L.; Chalkey, L. (1965) Low atomic numbered dye systems for ionising radiation measurements. *Photo.Sci.Eng.* Vol.9, pp.159-165.

NRT (2011). SpA - New Radiant Technology Italy. Available from http://www.newrt.it/

Olszanski, A.; Klassen, N.V.; C. K. Ross, C.K.; & Shortt K.R. (2002) The IRS Fricke Dosimetry *System Ionizing Radiation Standards Institute for National Measurement Standards National Research Council* Ottawa, Ontario

Palta, J.R.; Biggs, P.J.; Hazle J.D.; Huq, M.S.; Dahl, R.A.; Ochran T.G.; Soen, J.; R. R. Dobelbower, R.R.; & McCullough E.C. (1995) Intraoperative electron beam radiation therapy: Technique, dosimetry and dose specification: Report of task force 48 of the radiation therapy committee, American Association of Physicists in Medicine. *Int. J. Radiat. Oncol., Biol., Phys.* Vol.33, pp.725–746.

PTW (2011) available at http://www.ptw.de/advanced_markus_electron_chambe.html

Radiance (2011) available at http://www.gmv.com/healthcare/news_radiance/

Ronsivalle, C.; Picardi, L.; Vignati, A.; Tata, A.; Benassi, M. (2001). Accelerators development for intraoperative radiation therapy. *Proceedings of the 2001 Conference on Particle Accelerator.* Chicago, IL , USA Vol. 4, pp.2494-2496

Soriani A., Felici G., Fantini M. et al. (2010) Radiation protection measurements around a 12 MeV mobile dedicated IORT accelerator. *Med Phys.* ; Vol.37(3), pp.995-1003

Staub, D. (2004) Real-time radio-transparent dosimeter for X-ray imaging system. *Nuclear Instruments and Methods in Physics Research* A, Vol.525, pp:303-307

Takahiko, A.; Shuji, K.; Masatoshi, T.; Hisashi M. (1997) A depth-dose measuring device using a multichannel scintillating fiber. *Medical Physics* Vol.8, pp.1235-1239.

Wellhofer (2011) Available at http://www.elimpex.com/companies/

10

Intraoperative Radiotherapy for Early Breast Cancer

Masataka Sawaki
*Department of Breast Oncology,
Aichi Cancer Center Hospital,
Japan*

1. Introduction

The standard treatment for early breast cancer is breast-conserving therapy (BCT) with whole breast external irradiation therapy (WBI), and local control plays crucial role on survival (Clarke et al., 2005). It has been established that there is no significant difference in disease-free or overall survival rates between treatment by mastectomy or by lumpectomy with WBI for women with early breast cancer (Fisher et al., 2002; Veronesi et al., 2002). WBI actually provides statistically significant local control and survival data out to 15 years in favor of WBI compared to none (Clarke, et al., 2005). In actuality many women are still encouraged to proceed to mastectomy, because of the lack of access to radiotherapy centers or the long course of treatment of WBI. On the other hand, local recurrences after BCT with or without WBI arise most in the same quadrant as the primary cancer (Veronesi et al., 2001). The main objective of radiotherapy after BCT is considered to be the destruction of residual cancer cells in the operative field. Partial breast irradiation (PBI) has been tested in clinical trials for selected patients, and these studies have shown adequate local control, minimal toxicity, and good cosmetic appearance (Njeh et al., 2010). Intraoperative radiotherapy (IORT) is one of these PBI methods, which has recently been used in early stage breast cancer. Partial breast radiation therapy administered around the tumor bed has been comparable to WBI in selected patients (Antonucci et al., 2009; Benitez et al., 2007; Vicini et al., 2001; Vicini et al., 2003). Many phase II or III trials evaluating adjuvant IORT are actively accruing patients in the United States (NSABP-B39), Europe, United Kingdom, and Australia (Holmes et al., 2007). The standard treatment for early breast cancer is BCT with WBI, and outside the setting of a clinical trial, use of IORT as well as accelerated partial breast irradiation (APBI) is not yet recommended (Njeh, et al., 2010; Skandarajah et al., 2009). Then, particularly in practice setting, patients' selection is critical to the successful application of PBI (Polgár et al., 2010; Vicini et al., 2011).

This review concentrates on the eligibility, methods, outcome, and the point at issue of IORT for early breast cancer. With regard to terms, I defined IORT as the delivery of single fractional dose irradiation directly to the tumor bed during operation, and PBI as irradiation confined to the tumor bed either during operation or after surgery.

2. Eligibility for IORT – Who is suitable for IORT?

Many attempts have been made to identify subgroups of patients who might avoid radiotherapy after BCT, but these are not distinct factors. In general, the risk factors for local recurrence after BCT are larger tumor size, higher tumor grade, younger age, lymph node-positive, and close surgical margin (Clarke, et al., 2005; Fisher, 1997; Park et al., 2000). Then, we review clinical questions of which patients can be considered for IORT. Some points are discussed separately as below.

2.1 Age

Cumulative incidence of recurrence of tumor in the ispilateral breast after WBI with a boost is different among age, younger women have higher incidence (Bartelink et al., 2007). Age ≥ 50 years has been selected in most prospective trials, and studies have shown that elderly patients treated with WBI (Antonini et al., 2007; Bartelink, et al., 2007) or MammoSite® (Chao et al., 2007) were low risk. Few women younger than 50 has been treated with PBI in prospective single arm studies.

2.2 Tumor size

A maximum tumor size of 2cm has been selected in most prospective trials. T2 tumors (>2cm, ≤5cm) or T0 (ductal carcinoma *in situ*; DCIS) tumors are cautionary recommended (Smith et al., 2009). Patients with T3 or T4 tumors should not receive PBI. An extensive intraductal component should be treated with caution. Patients with multicentric tumors; i.e. presence of foci of cancer in different quadrants, should not receive PBI because of the extent of disease. Patients of clinically unifocal or multifocal tumors with a total tumor size no greater than 2 cm could be suitable for PBI (Smith, et al., 2009).

2.3 Nodal status

Node positive is one of the risk factor for ispilateral breast cancer (Clarke, et al., 2005). Then, majority of patients who have been treated in prospective single arm APBI trials had pathologically node negative disease. Patients who do not undergo surgical nodal assessment or who have pathologic evidence of nodal involvements should not receive PBI.

2.4 Pathology

One area of concern in the use of IORT is the management of positive surgical margins as positivity is discovered at the final histology, a few days after surgery and IORT. Attention should be paid to ensure negative margins on final pathology (Beal et al., 2007), although margin positivity does not always influence the rate of local recurrences if effective radiotherapy is delivered (Chism et al., 2006; Mariani et al., 1998; Veronesi et al., 2010). Intraoperative frozen sections may be used to reduce positive margins (Fukamachi et al., 2010). Patients with close but negative margins (<2 mm) may be treated with caution (Smith, et al., 2009). Tumors with higher tumor grade is not suitable for PBI, because it is one of risk factors for local recurrence and most prospective trials have not considered as eligibility criteria. As for the tumor characteristics biologically, such as HER2-type or basal-type have been shown to involve higher risk for local recurrence than luminal A or B-types (Kyndi et al., 2008; Nguyen et al., 2008; Veronesi, et al., 2010), then tailored local-regional treatment for early-stage breast cancer has been reported to be mandatory now (Solin, 2010).

2.5 Oncology

There has been only limited study of IORT including PBI in patients receiving neoadjuvant or concurrent chemotherapy. For patients who will receive adjuvant chemotherapy, it is recommended that APBI be performed first and that there should be an interval of at least 2 to 3 weeks between completion of APBI and initiation of chemotherapy (Smith, et al., 2009). Thus, IORT including PBI allows radiotherapy to be given without delaying administration of chemotherapy or hormonal therapy. But careful observation is needed; a retrospective analysis from MammoSite registry single arm trial reported an association between initiation of adjuvant chemotherapy within 3 weeks of the last MammoSite treatment and an increased risk of both radiation recall skin retraction and suboptimal cosmetics (Haffty et al., 2008). There is no data when adjuvant endocrine therapy with APBI should be started.

We show recommendation for APBI in Table 1.

Factors	Suitable group
Age	≥60
BRCA1/2 mutation	Not present
Tumor size	≤2 cm
T stage	T1
Margins	Negative by at least 2 mm
Grade	Any
LVSI	No
ER statis	Positive
Multicentricity	Unicentric only
Mutifocally	Clinically unifocal with total size ≤2 cm
Histology	Invasive ductal or other favorable subtypes
Pure DCIS	Not allowed
EIC	Not allowed
Associated LCIS	Allowed
Nodal status	pN0
Nodal surgery	SN Bx or ALND
Neoadjuvant Therapy	Not allowed

Table 1. Suitable patient group recommendation selections for APBI outside of clinical trials; ASTRO Consensus Statement (Smith, et al., 2009)

3. Radiation methods

Several radiation methods are commonly in use for IORT. We show various radiation techniques for IORT in Table 2.

Author	Year (publish)	No. of cases	Median follow up (months)	Technique	Ispilateral breast cancer recurrence (%)
Sawaki M	2011	32	26	Electrons	0
Veronesi U	2010	1,822	36.1	Electrons	1.3
Vaidya JS	2010	854	48	Photons	1.2
Lemanski C	2010	42	30	Electrons	4.8
Mussari S	2006	47	48	Electrons	0

Table 2. Clinical studies using full-dose Intraoperative radiation therapy (IORT)

3.1 Electrons Intraoperative Therapy (ELIOT)

3.1.1 As a single fractional dose

ELIOT is one of these PBI methods, which has recently been used in early stage breast cancer, mainly at the European Institute of Oncology (Italian: *Istituto Europeo di Oncologia;* IEO) in Milan since 1999 (Intra et al., 2006; Luini et al., 2005; Veronesi et al., 2005; Veronesi, et al., 2010; Veronesi et al., 2001). They have been promoted to prospective trials to investigate tolerance to increased IORT doses and ultimately to introduce the use of 21 Gy in the context of breast conserving surgery. A single dose of 21 Gy at 90% isodose has been shown to be feasible in European breast cancer patients and biologically equivalent to a full dose of conventional WBI (Intra, et al., 2006; Luini, et al., 2005; Veronesi, et al., 2005; Veronesi, et al., 2001). The main advantages are follows. 1) To be able to deliver the radiation before tumor cells have a chance to proliferate under surgical intervention have a rich vascularization, which makes them more sensitive to the action of the radiation. 2) To be able to deliver under direct visualization at the time of surgery. It has the potential for accurate dose delivery by permitting delivery of the radiation dose directly to the surgical margins. 3) IORT could minimize some potential side effects, since skin and the subcutaneous tissue can be displaced. 4) The spread of irradiation to lung and heart is reduced significantly. 5) IORT allows radiotherapy to be given without delaying administration of chemotherapy or hormonal therapy. 6) There is potential for decreasing healthcare cost, because it is one fraction as opposed to 25 fractions. For elderly patients, it is feasible and corresponded to acceptable quality index criteria (Lemanski et al., 2010). In Asian breast cancer patients 21 Gy was recommended as a result of phase I study, which had been used a scheme of dose-escalation from 19, 20, and 21 Gy. (Sawaki et al., 2009), and subsequent phase II study indicated feasibility at 21 Gy (Sawaki et al., 2011). In this technique a disk should be needed to reduce the spread of irradiation to lung and heart. The disk is located between gland and pectoralis muscle. The aluminum and lead disk has been used in Italy (Intra, et al., 2006, Mussari et al., 2006). As for an another kind of disk, two layers disk; a first layer (source side) of polymethyl methacrylate (PMMA) and a second layer of copper was designed and selected from metals such as aluminum, copper and lead after testing for their shielding capabilities and the range of the backscatter (Oshima et al., 2009).

IORT extends the primary operation only for an additional 15 minutes plus approximately 30 minutes of a radiotherapy physicist's time to prepare the device(Sawaki, et al., 2009), although conventional WBI radiotherapy usually requires 5 weeks of outpatient treatment.

The largest randomized clinical trial to date is now in progress at the Milan Institute. The goal of the trial is to determine the equivalence of local recurrence rates between quadrantectomy with conventional WBI and that with IORT. To date, they remain investigational until information on its long-term efficacy and safety becomes available (Buchholz, 2009). In the trial at the Milan Institute, 21 Gy, which is recommended through more than 1000 IORT procedures (Intra, et al., 2006), was used for the IORT arm. In the most update data, local recurrence rate was 1.3% (24/1,822) (Veronesi, et al., 2010). In addition, IORT can achieve early initiation of radiotherapy (RT). Delay in the initiation of RT is associated with a decrease in local recurrence rate (Huang, 2003).

3.1.2 As a boost

With regard to another concept of full-dose intraoperative radiotherapy, an anticipated boost during operation has been studied (Reitsamer et al., 2002; Reitsamer et al., 2006). A single dose of 9 Gy was applied to the 90% reference isodose with energies ranging from 4 to 15 MeV, using round applicator tubes 4–8 cm in diameter. After wound healing, the patients received additional 51 - 56 Gy external boost radiation (EBRT) to the whole breast (Reitsamer, et al., 2002). The advantages are follows. 1) To complete skin sparing, 2) the precise application of the boost directly to the tumor bed with a homogeneous tissue radiation and 3) to reduce postoperative radiation time for 7-10 days (time of postoperative boost radiotherapy) (Reitsamer, et al., 2002; Reitsamer, et al., 2006).

3.1.3 Nipple sparing mastectomy

Nipple sparing mastectomy can be applied for treatment of breast cancer when mastectomy is indicated. To reduce the risk of retro areolar recurrence, radio-surgical treatment combining subcutaneous mastectomy with intraoperative radiotherapy (ELIOT) is proposed (Petit et al., 2009; Petit et al., 2009). The IORT with electrons of 16 Gy in one shot was delivered on the nipple areolar area. Local recurrence rate was not higher than the usual rate observed in the literature, although longer follow up is needed.

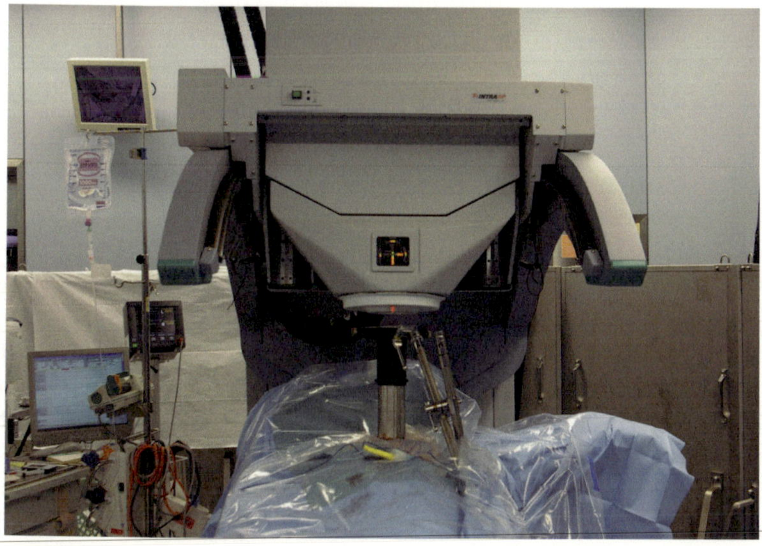

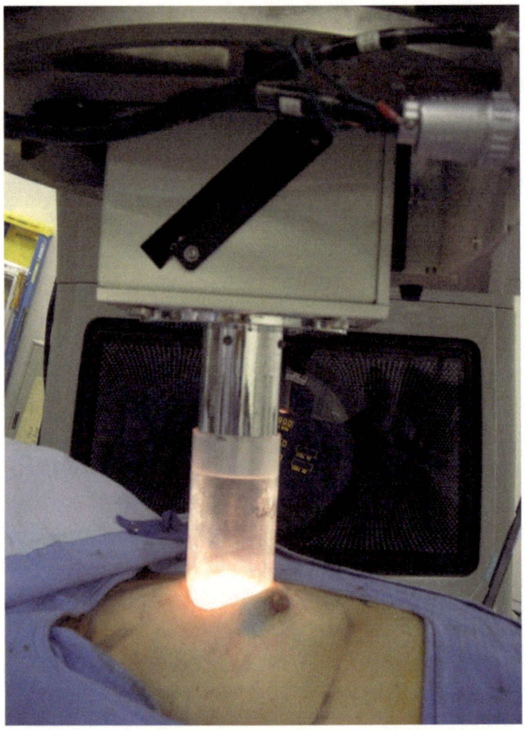

Fig. 1. Intraoperative radiotherapy (electrons)

3.2 Targeted intraoperative radiotherapy (TARGIT)

This device is inserted intraopetatively into the tumor cavity after excision of the tumor and emits X-rays from within the breast (Vaidya et al., 2010). The authors used a miniature electron-beam-driven X-ray source called Intrabeam®, which emits low energetic X-rays with 50 kV from the point source. In large randomized clinical trial, TARGIT trial for selected patients with early breast cancer, a single dose of radiotherapy delivered at the time of surgery by use of targeted intraoperative radiotherapy is considered as an alternative to external beam radiotherapy delivered over several weeks (Vaidya, et al., 2010), although it needs longer follow up to conclude the no inferiority to the WBI (Reitsamer et al., 2010).

4. Conclusion

In conclusion, IORT is an option applied for breast conserving therapy in the selected patients. TARGIT trial has been considered as an alternative to external beam radiotherapy delivered over several weeks (Vaidya, et al., 2010). And also ELIOT appears a promising feature in early breast cancer treated with breast conserving surgery, reducing the exposure of normal tissues to radiations and shortening the radiation course from 6 weeks to one single session (Veronesi, et al., 2010). These clinical studies have shown adequate local control, minimal toxicity, and good cosmetic appearance, although a longer follow up is needed for the evaluation of the late side effects. In practice setting, careful management is needed because patients' selection is critical to the successful application of IORT.

5. Conflict of Interest

The author states that I have no conflict of interest.

6. Acronyms and abbreviations

ALND; Axillary lymph node dissection
APBI; Accelerated partial breast irradiation
ASTRO; American Society for Radiation Oncology
BCT; Breast conserving therapy
DCIS; Ductal carcinoma *in situ*
EBRT; External boost radiation
EIC; Extensive intraductal component
ELIOT; Electrons intraoperative therapy
ER; Estrogen receptor
IORT; Intraoperative radiotherapy
LCIS; lobular carcinoma *in situ*
LVSI; Lymph-vascular space invasion
PBI; Partial breast irradiation
PMMA; polymethyl methacrylate
RT; Radiotherapy
SN Bx; Sentinel lymph node biopsy
TARGIT; Targeted intraoperative radiotherapy
WBI; Whole breast external irradiation therapy

7. References

Antonini N, Jones H, Horiot JC et al (2007) Effect of age and radiation dose on local control after breast conserving treatment: EORTC trial 22881-10882. *Radiother Oncol* 82:265-271

Antonucci JV, Wallace M, Goldstein NS et al (2009) Differences in Patterns of Failure in Patients Treated With Accelerated Partial Breast Irradiation Versus Whole-Breast Irradiation: A Matched-Pair Analysis With 10-Year Follow-Up. *International Journal of Radiation OncologyBiologyPhysics* 74:447-452

Bartelink H, Horiot JC, Poortmans PM et al (2007) Impact of a higher radiation dose on local control and survival in breast-conserving therapy of early breast cancer: 10-year results of the randomized boost versus no boost EORTC 22881-10882 trial. *J Clin Oncol* 25:3259-3265

Beal K, McCormick B, Zelefsky M et al (2007) Single-Fraction Intraoperative Radiotherapy for Breast Cancer: Early Cosmetic Results. *International Journal of Radiation OncologyBiologyPhysics* 69:19-24

Benitez P, Keisch M, Vicini F et al (2007) Five-year results: the initial clinical trial of Mammosite balloon brachytherapy for partial breast irradiation in early-stage breast cancer. *The American Journal of Surgery* 194:456-462

Buchholz TA (2009) Radiation therapy for early-stage breast cancer after breast-conserving surgery. *N Engl J Med* 360:63-70

Chao KK, Vicini FA, Wallace M et al (2007) Analysis of treatment efficacy, cosmesis, and toxicity using the MammoSite breast brachytherapy catheter to deliver accelerated partial-breast irradiation: the william beaumont hospital experience. *Int J Radiat Oncol Biol Phys* 69:32-40

Chism D, Freedman G, Li T et al (2006) Re-excision of margins before breast radiation—diagnostic or therapeutic? *International Journal of Radiation OncologyBiologyPhysics* 65:1416-1421

Clarke M, Collins R, Darby S et al (2005) Effects of radiotherapy and of differences in the extent of surgery for early breast cancer on local recurrence and 15-year survival: an overview of the randomised trials. *Lancet* 366:2087-2106

Fisher B, Anderson S, Bryant J et al (2002) Twenty-year follow-up of a randomized trial comparing total mastectomy, lumpectomy, and lumpectomy plus irradiation for the treatment of invasive breast cancer. *N Engl J Med* 347:1233-1241

Fisher ER (1997) Lumpectomy margins and much more. *Cancer* 79:1453-1458; discussion 1459-1460

Fukamachi K, Ishida T, Usami S et al (2010) Total-Circumference Intraoperative Frozen Section Analysis Reduces Margin-Positive Rate in Breast-Conservation Surgery. *Jpn J Clin Oncol* 40:513-520

Haffty BG, Vicini FA, Beitsch P et al (2008) Timing of Chemotherapy after MammoSite radiation therapy system breast brachytherapy: analysis of the American Society of Breast Surgeons MammoSite breast brachytherapy registry trial. *Int J Radiat Oncol Biol Phys* 72:1441-1448

Holmes D, Baum M, Joseph D (2007) The TARGIT trial: targeted intraoperative radiation therapy versus conventional postoperative whole-breast radiotherapy after breast-conserving surgery for the management of early-stage invasive breast cancer (a trial update). *The American Journal of Surgery* 194:507-510

Huang J (2003) Does Delay in Starting Treatment Affect the Outcomes of Radiotherapy? A Systematic Review. *J Clin Oncol* 21:555-563

Intra M, Luini A, Gatti G et al (2006) Surgical technique of intraoperative radiation therapy with electrons (ELIOT) in breast cancer: A lesson learned by over 1000 procedures. *Surgery* 140:467-471

Kyndi M, Sorensen FB, Knudsen H et al (2008) Estrogen Receptor, Progesterone Receptor, HER-2, and Response to Postmastectomy Radiotherapy in High-Risk Breast Cancer: The Danish Breast Cancer Cooperative Group. *J Clin Oncol* 26:1419-1426

Lemanski C, Azria D, Gourgon-Bourgade S et al (2010) Intraoperative radiotherapy in early-stage breast cancer: results of the montpellier phase II trial. *Int J Radiat Oncol Biol Phys* 76:698-703

Luini A, Orecchia R, Gatti G et al (2005) The Pilot Trial on Intraoperative Radiotherapy with Electrons (ELIOT): Update on the Results. *Breast Cancer Res Treat* 93:55-59

Mariani L, Salvadori B, Marubini E et al (1998) Ten year results of a randomised trial comparing two conservative treatment strategies for small size breast cancer. *Eur J Cancer* 34:1156-1162

Mussari S, Sabino Della Sala W, Busana L et al (2006) Full-dose intraoperative radiotherapy with electrons in breast cancer. First report on late toxicity and cosmetic results from a single-institution experience. *Strahlentherapie und Onkologie : Organ der Deutschen Rontgengesellschaft [et al]* 182:589-595

Nguyen PL, Taghian AG, Katz MS et al (2008) Breast Cancer Subtype Approximated by Estrogen Receptor, Progesterone Receptor, and HER-2 Is Associated With Local and Distant Recurrence After Breast-Conserving Therapy. *J Clin Oncol* 26:2373-2378

Njeh CF, Saunders MW, Langton CM (2010) Accelerated Partial Breast Irradiation (APBI): A review of available techniques. *Radiat Oncol* 5

Oshima T, Aoyama Y, Shimozato T et al (2009) An experimental attenuation plate to improve the dose distribution in intraoperative electron beam radiotherapy for breast cancer. *Phys Med Biol* 54:3491-3500

Park CC, Mitsumori M, Nixon A et al (2000) Outcome at 8 years after breast-conserving surgery and radiation therapy for invasive breast cancer: influence of margin status and systemic therapy on local recurrence. *J Clin Oncol* 18:1668-1675

Petit JY, Veronesi U, Orecchia R et al (2009) Nipple sparing mastectomy with nipple areola intraoperative radiotherapy: one thousand and one cases of a five years experience at the European institute of oncology of Milan (EIO). *Breast Cancer Res Treat* 117:333-338

Petit JY, Veronesi U, Rey P et al (2009) Nipple-sparing mastectomy: risk of nipple-areolar recurrences in a series of 579 cases. *Breast Cancer Res Treat* 114:97-101

Polgár C, Limbergen EV, Pötter R et al (2010) Patient selection for accelerated partial-breast irradiation (APBI) after breast-conserving surgery: Recommendations of the Groupe Européen de Curiethérapie-European Society for Therapeutic Radiology and Oncology (GEC-ESTRO) breast cancer working group based on clinical evidence (2009). *Radiother Oncol* 94:264-273

Reitsamer R, Fastner G, Kopp M et al (2010) Intraoperative radiotherapy for early breast cancer. *Lancet* 376:1141; author reply 1143-1144

Reitsamer R, Peintinger F, Sedlmayer F et al (2002) Intraoperative radiotherapy given as a boost after breast-conserving surgery in breast cancer patients. *Eur J Cancer* 38:1607-1610

Reitsamer R, Sedlmayer F, Kopp M et al (2006) The Salzburg concept of intraoperative radiotherapy for breast cancer: results and considerations. *Int J Cancer* 118:2882-2887

Sawaki M, Sato S, Kikumori T et al (2009) A phase I study of intraoperative radiotherapy for early breast cancer in Japan. *World J Surg* 33:2587-2592

Sawaki M, Sato S, Noda S et al (2011) Phase I/II study of intraoperative radiotherapy for early breast cancer in Japan. *Breast Cancer*:2011 Jul 16. [Epub ahead of print]

Skandarajah AR, Lynch AC, Mackay JR et al (2009) The Role of Intraoperative Radiotherapy in Solid Tumors. *Ann Surg Oncol* 16:735-744

Smith BD, Arthur DW, Buchholz TA et al (2009) Accelerated partial breast irradiation consensus statement from the American Society for Radiation Oncology (ASTRO). *J Am Coll Surg* 209:269-277

Solin LJ (2010) Tailored Local-Regional Treatment for Early-Stage Breast Cancer. *Clinical Breast Cancer* 10:343-344

Vaidya JS, Joseph DJ, Tobias JS et al (2010) Targeted intraoperative radiotherapy versus whole breast radiotherapy for breast cancer (TARGIT-A trial): an international, prospective, randomised, non-inferiority phase 3 trial. *Lancet* 376:91-102

Veronesi U, Cascinelli N, Mariani L et al (2002) Twenty-year follow-up of a randomized study comparing breast-conserving surgery with radical mastectomy for early breast cancer. *N Engl J Med* 347:1227-1232

Veronesi U, Marubini E, Mariani L et al (2001) Radiotherapy after breast-conserving surgery in small breast carcinoma: long-term results of a randomized trial. *Ann Oncol* 12:997-1003

Veronesi U, Orecchia R, Luini A et al (2005) Full-Dose Intraoperative Radiotherapy With Electrons During Breast-Conserving Surgery. *Ann Surg* 242:101-106

Veronesi U, Orecchia R, Luini A et al (2010) Intraoperative radiotherapy during breast conserving surgery: a study on 1,822 cases treated with electrons. *Breast Cancer Res Treat* 124:141-151

Veronesi U, Orecchia R, Luini A et al (2001) A preliminary report of intraoperative radiotherapy (IORT) in limited-stage breast cancers that are conservatively treated. *Eur J Cancer* 37:2178-2183

Vicini F, Arthur D, Wazer D et al (2011) Limitations of the American Society of Therapeutic Radiology and Oncology Consensus Panel guidelines on the use of accelerated partial breast irradiation. *Int J Radiat Oncol Biol Phys* 79:977-984

Vicini FA, Baglan KL, Kestin LL et al (2001) Accelerated treatment of breast cancer. *J Clin Oncol* 19:1993-2001

Vicini FA, Kestin L, Chen P et al (2003) Limited-Field Radiation Therapy in the Management of Early-Stage Breast Cancer. *JNCI Journal of the National Cancer Institute* 95:1205-1210

Part 4

Scope of Radiation Therapy for Specific Diseases

11

Enhancing Therapeutic Radiation Responses in Multiple Myeloma

Kelley Salem and Apollina Goel
Free Radical and Radiation Biology Program, Department of Radiation Oncology, University of Iowa, Iowa City, USA

1. Introduction

Multiple myeloma (MM) is hematologic malignancy characterized by the accumulation of malignant plasma cells in the bone marrow. The annual incidence of newly diagnosed MM cases in the United States is 3 to 4 per 100,000 people and accounts for approximately 1% of all malignant diseases (Jemal et al., 2011). MM is diagnosed at an advanced stage in 95% of patients and the median age at diagnosis is 65 years. It is a progressive malignancy that begins with monoclonal gammopathy of undetermined significance (MGUS), progresses to asymptomatic or smoldering myeloma and then symptomatic MM. MGUS is a disorder that exhibits clonal proliferation of plasma cells and can eventually evolve into MM or other B-cell disorders (Landgren et al., 2011). Clinically, patients with symptomatic myeloma have 10% or more malignant plasma cells in bone marrow, abnormal levels of serum free light chain, osteolytic bone disease, and show damage to other tissues or organs. Smoldering myeloma has the same plasma cell and M-protein characteristics of symptomatic but lacks evidence of organ damage. A rare type of MM, nonsecretory myeloma, has no detectable M-protein and accounts for only 1-5% of MM cases. Solitary plasmacytoma is a plasma cell neoplasm that has a single bone or extramedullary lesion (Mendenhall et al., 2003). MM is characterized by significant heterogeneity at the molecular level (Herve et al., 2011) and the bone marrow microenvironment plays an active role in supporting tumor growth, angiogenesis, bone disease, and drug resistance (Anderson and Carrasco, 2011). The disease initially responds to alkylating agents, corticosteroids, and thalidomide but eventually becomes refractory (Sirohi and Powles, 2004). High dose melphalan combined with peripheral blood stem cell transplant has improved the response rate in myeloma patients, but is not curative (Fassas and Tricot, 2001). To date, MM remains uniformly fatal with a median survival of approximately 50 months after diagnosis.

MM is extremely susceptible to radiation treatment and targeted radiotherapy including bone-seeking radiopharmaceuticals, monoclonal antibodies conjugated to radionuclides (radioimmunotherapy), and radiotargeted gene therapy using recombinant oncolytic viruses (radiovirotherapy) now offers a new paradigm to target this systemic malignancy. Combining targeted radiotherapy with radiation-sensitizing chemotherapeutic drugs provides additional benefit by improving treatment efficacy and extends the clinical use of radiotherapy in MM beyond palliative care or myeloablative preconditioning regimens. In this chapter, we will discuss recent advances in the field of targeted radiotherapy and chemotherapeutic drugs that have been utilized to increase radiation responses in MM patients.

2. Conventional radiotherapy in MM

Radiation therapy is a powerful treatment modality for MM (Bosch and Frias, 1988; Mill, 1975) where ionizing radiation generates free radicals that cause DNA damage, leading to the death of tumor cells. Approximately 80% of myeloma patients present with skeletal-related problems such as diffuse osteopenia, focal lytic lesions, pathological fractures, and bone pain; all these clinical manifestations are associated with myeloma bone disease that compromises quality of life and contributes towards morbidity and mortality (Kyle, 1975). Conventional external beam radiation therapy (EBRT), based on an outside-in approach, is used in MM for ablation of bony lesions and utilizes nuclear medicine methods that deliver radiation as either a local or a wide-field beam (Cole, 1989; Friedland, 1999; Price et al., 1986). EBRT has been combined with vertebroplasty and kyphoplasty for palliation of bone pain caused by vertebral compression fractures in MM patients (Hirsch et al., 2011). Radiotherapy is effective in the treatment of solitary plasmacytomas that manifest either as soft tissue disease (extramedullary tumors) or have bone involvement (osseous tumors) (Bolek et al., 1996; Kilciksiz et al., 2008; Krause et al., 2011; Lewanski et al., 1999; Tsang et al., 2001). The availability of new intensity-modulated radiation treatment (IMRT) techniques such as helical tomotherapy (HT) (Chargari et al., 2009), 3D conformal radiotherapy (3D-CRT) (Chargari et al., 2011) has enabled specific delivery of radiation to plasmacytomas with minimum normal tissue toxicity. The combination of localized fractionated radiotherapy with novel chemotherapeutic agents such as thalidomide (Marchand et al., 2008) and bortezomib (Berges et al., 2008) has provided good clinical outcomes with reduced radiotoxicity to normal tissues.

For systemic diffused myeloma disease, hemibody irradiation has been utilized, however, this method is associated with significant toxicity (Biswal, 2004; Hu and Yahalom, 2000). In MM patients, double hemibody irradiation has been combined with granulocyte-macrophage and granulocyte colony-stimulating factors (GM-CSF, G-CSF) to reduce toxic side effects of radiation on hematopoiesis (Troussard et al., 1995). Total body irradiation (TBI) has provided improved long-term survival rates for certain MM patient cohorts (Rostom, 1988). To alleviate TBI induced pulmonary complications, fractionation regimens of radiotherapy have also been evaluated in MM (Soejima et al., 2007) with improved *in vitro* clonogenic cell death of MM cell lines (Gluck et al., 1994). For hematological malignancies such as B-cell lymphoma and MM, curative radiation doses are estimated in the 20-30 Gy range, but without stem cell transplantation, a 2 Gy of radiation dose can result in hematologic toxicity (Brahme and Agren, 1987; Fletcher, 1976). Hence, clinical utilization of radiotherapy as a definitive therapeutic approach in MM has been mainly limited to a conditioning regimen prior to autologous or allogeneic stem-cell transplantation (Moehler and Goldschmidt, 2011; Snowden et al., 2011). However, in a study comparing melphalan plus TBI with melphalan alone for conditioning regimens before autologous stem cell transplantation (ASCT), melphalan alone showed less toxicity and was found to be as effective as melphalan plus TBI (Moreau et al., 2002). Technological advances such as IMRT,

HT and linear accelerator-based intensity-modulated total marrow irradiation now enable the delivery of systemic radiotherapy to myeloma cells with higher and more tumoricidal doses of radiation with potential curative benefit (Wong et al., 2006; Wong et al., 2009; Yeginer et al., 2011).

3. Novel targeted radiotherapeutic agents in MM

A new generation of targeted radiotherapeutic methods such as radioimmunotherapy (Chatterjee et al., 2006; Goel, 2006), radiovirotherapy (Dingli et al., 2004; Goel et al., 2007), and bone-seeking radiopharmaceuticals have been tested for systemic radiotherapy of MM. Since these agents deliver radiation to myeloma cells either by directly targeting the cancer cells (radioimmunotherapy and radiovirotherapy) or bone (skeletal-targeted radiotherapy), they deliver radiation from the inside-out thereby minimizing normal tissue toxicity with increased tumor cell death, resulting in an overall increase in therapeutic efficacy.

3.1 Radioimmunotherapy (RIT)

Radioimmunotherapy (RIT) combines the advantages of antibody specificity, by binding to a tumor-associated antigen, with the cytotoxicity of radionuclides, resulting in targeted radiation therapy. RIT is a systemic treatment that has shown promising clinical remission rates in metastatic cancers such as non-Hodgkin lymphoma (NHL) and MM (Chatterjee et al., 2006; Mayes et al., 2011). Several monoclonal antibodies (MAbs) targeting the myeloma cell or the bone marrow microenvironment have been tested in preclinical and clinical studies (van de Donk et al., 2011); these MAbs are potentially amenable to RIT. RIT with MAb targeting the CD20 marker such as Zevalin (^{90}Y-iritumomab Tiuxetan) or Bexxar (^{131}I-tositumomab) has provided clinical benefit in B-cell lymphomas (Ahmed et al., 2010). Another CD20- targeting monoclonal antibody, rituximab, is being studied in patients with lymphocytic leukemia and other hematological diseases (NCT00669318) (Barcellini and Zanella, 2011). The monoclonal antibody, daratumumab, targets CD38^{+} MM *in vitro* and has shown promising results in selectively killing MM cells *in vivo* (de Weers et al., 2011). van der Veer et al. demonstrated a synergistic effect when tumor cells were pretreated with lenalidomide prior to treatment with daratumumab (van der Veer et al., 2011).

Most RIT developed and tested in clinical trials utilizes beta-particle emitting radionuclides in which short-range beta emitters such as iodine-131 and copper-67 are used to target small tumor cell clusters (Wun et al., 2001). Long-range beta emitters such as yttrium-90 are used to target larger tumor masses, tumor areas that remain inaccessible to RIT agents due to poor vascularity, and tumor cells that lack antigen expression by utilizing bystander radiation toxicity (Bethge and Sandmaier, 2005). RIT with alpha-emitters, such as bismuth-212, bismuth-213, astatine-211, actinium-225, lead-212 offer the advantage of a short path length with a high linear energy transfer of radiation, resulting in more specific tumor cell killing with less damage to the surrounding healthy tissues (Brechbiel, 2007). However, RIT with α-particles is fraught with challenges such as limited availability, radiolysis, suboptimal specificity of radiolabeled conjugates, and heterogeneous dose deposition in tumors (Cherel et al., 2006). In MM cell lines, Supiot et al demonstrated superior tumor cell killing by anti-CD138 (syndecan-1) B-B4 MAb labeled with bismuth-213 as compared to iodine-131-labeled antibody suggesting that alpha-RIT might be more suitable for treating single cell tumors such as MM (Supiot et al., 2002). Besides B-B4, the MAb MA5, which recognizes mucin-1 expressed by both normal and malignant plasma cells, has been coupled to bismuth-213 to target myeloma cells (Couturier et al., 1999; Supiot et al., 2005). In ASCT conditioning regimens, RIT is a good alternative to TBI as it results in less radiotoxicity for normal organ systems and delivers more radiation to tumors as reflected in improved cure rates (Gustavsson et al., 2003). MM patients have higher microvessel density than control subjects at bone marrow biopsy (Bhatti et al., 2006; Rajkumar et al., 2000). A preclinical study using bevacizumab, a humanized anti-VEGF MAb radiolabeled with Bi-213, showed promising results for prostate cancer treatment (Abbas Rizvi et al., 2008). As bevacizumab is now undergoing phase I/II clinical trials (Somlo et al., 2011), an alpha-RIT with this antibody may hold some clinical benefit for myeloma patients. It can be speculated that myeloablative conditioning protocols involving RIT with or without chemotherapy followed by ASCT may hold clinical benefit in MM. Also, approaches like pre-targeted RIT that separates delivery of the targeting molecule from radionuclide delivery can offer dose escalation (DeNardo et al., 2006) and radiolabeled high affinity antibody fragments (Goel and Batra, 2001; Goel et al., 2000) remain yet to be developed and tested in MM. The physical properties of few radionuclides tested in preclinical and clinical trials for cancer therapy are listed in Table 1.

Isotope	Radiation	Physical half-life	Mean particle energy (Mev)	Maximum energy (Mev)	Tisssue range (mm)
Iodine-131	β^-, γ	8 days	0.19	0.6	2.3
Yttrium-90	β^-	2.7 days	0.9	2.3	11.3
Rhenium-188	β^-	17 h	0.8	2.1	10.4
Rhenium-186	β^-, γ	88.8 h	0.35	1.1	2.4
Lutetium-177	β^-, γ	6.7 days	0.15	0.5	1.6
Copper-67	β^-, γ	61.8 h	0.14	0.58	2.1
Samarium-153	β^-, γ	1.9 d	0.23	0.81	0.6
Bismuth-213	α	46 min	8.3	8.4	0.09
Astatine-211	α	7.2 h	6.8	7.4	0.08
Actinium-225	α	10 days	8.4	5.8	0.08

Table 1. Physical characteristics of few isotopes studied in nuclear medicine for cancer therapy.

3.2 Radiovirotherapy

Oncolytic viruses have natural or engineered tropism for tumor cells which permits specific targeting and destruction of cancer cells (virotherapy) (Parato et al., 2005; Stief and McCart, 2008). In MM, studies with measles virus (MV) (Peng et al., 2001), vaccinia virus (Deng et al., 2008; Kawa and Arakawa, 1987), vesicular stomatitis virus (VSV) (Goel et al., 2007), and coxsackievirus A21 (Au et al., 2007; Hadac et al., 2011), have demonstrated *in vitro* and *in vivo* killing of tumor cells. Attenuated MV, which is an Edmonston vaccine lineage derivative (MV-Edm), has entered clinic trials for recurrent ovarian cancer, recurrent glioblastoma multiforme, and MM (Msaouel et al., 2009; Myers et al., 2007). However, intravenous administration of MV may be less effective in patients who have been previously vaccinated with the measles vaccine as these patients' antiviral antibodies may neutralize the oncolytic MV (Liu et al., 2010; Ong et al., 2007). Liu et al. performed a study demonstrating the feasibility and efficacy of using irradiated, MV-infected myeloma cells as carriers in mice (Liu et al., 2010). Using cells as viral carriers prevents neutralization by the humoral immune response; using myeloma cells ensures that the carriers are shuttled to the bone marrow and virus is delivered to the tumor site.

Oncolytic viruses have been used for radiotargeted gene therapy whereby radionuclides can be localized at tumor sites by inducing tumor cells to express sodium-iodide symporter (hNIS) gene (radiovirotherapy) (Chung, 2002). Such "designer oncolytic viruses" that express the human NIS gene have been engineered and tested in MM (Dingli et al., 2004; Goel et al., 2007). By using radionuclides, such as iodine-123, iodine-124, or technicium-99m, combined with detection with either a γ camera, positron emission tomography (PET), or single photon emission computed tomography (SPECT)/computed tomography (CT), NIS can be used as a reporter gene to non-invasively monitor viral localization and spread. Furthermore, NIS can be used as a therapeutic transgene by allowing intracellular uptake of isotopes such as iodine-131 which can cause direct radiation damage to tumor cells, thereby enhancing the therapeutic efficacy of radiovirotherapy. Currently, a phase I clinical trial of MV-NIS given with or without cyclophosphamide for treatment of patients with recurrent or refractory MM (NCT00450814) is ongoing at Mayo Clinic (Msaouel et al., 2009). Combining MV-NIS with other therapeutic radioisotopes such as rhenium-186, rhenium-188, or astatine-211 may be worth exploring in MM. Ongoing preclinical studies have shown that using cellular virus-delivery vehicles (i.e. mesenchymal progenitor cells, monocytes, T cells) can facilitate viral delivery to tumor cells (Munguia et al., 2008; Russell and Peng, 2008; Willmon et al., 2009). Irradiated 5TGM1 myeloma cells transfected with VSV-GFP have been shown to deliver VSV to sites of myeloma tumor growth in an orthotopic human myeloma model (Munguia et al., 2008). Since intravenous delivery of radiotargeted gene therapy is prerequisite for targeting systemic myeloma tumor sites, selection of the optimal cell carrier for radiovirotherapy is expected to improve the tumor remission rate in MM.

3.3 Skeletal-Targeted Radiotherapy (STR)

Bone seeking radionuclide therapy enables the delivery of precisely focused radiation to the major bone marrow sites where myeloma cells reside and reduces the radiation exposure of healthy organs. Samarium-153-ethylene diamine tetramethylene phosphonate (153-Sm-EDTMP or ^{153}Samarium lexidronam) is an US Food and Drug Administration (FDA)-approved radiopharmaceutical that demonstrates good therapeutic ratio for a dose of 1 mCi/kg for palliation of pain in cancer patients with osseous metastases (Lamb and Faulds, 1997; Lewington, 2005). Reversible myelosuppression is the only significant toxic effect of 153-Sm-EDTMP; retreatment with 153-Sm-EDTMP is considered a safe, feasible, and efficacious for palliative treatment of bone metastasis (Serafini, 2000). 153-Sm-EDTMP is taken up by portions of the skeleton undergoing active remodeling resulting in rapid clearance from the blood (Bayouth et al., 1994). 153-Sm-EDTMP has also been used for total marrow irradiation in myeloablative clinical protocols for MM (Abruzzese et al., 2008; Anderson et al., 2002; Dispenzieri et al., 2003; Dispenzieri et al., 2010; Dispenzieri et al., 2005; Macfarlane et al., 2002). In non-transplant situations, 153-Sm-EDTMP treatment reduced pain in more than 70% of patients with osteoblastic metastases (Serafini, 2001).

Several other investigational radiopharmaceuticals such as rhenium-186-hydroxyethylidenediphosphonic acid (^{186}Re-HEDP) (Lam et al., 2004) and 166-Holmium-1,4,7,10-tetraazacyclododecane-1,4,7,10-tetramethylene-phosphonate (^{166}Ho-DOTMP) (Breitz et al., 2006) have been developed for targeted radiotherapy of bone malignancies (Jansen et al., 2010). In one preclinical study, EDTMP labeled with 166-dysprosium/166-Ho was used to establish an *in vivo* generator system for myeloablative radiotherapy/chemotherapy protocols in MM (Pedraza-Lopez et al., 2004).

4. Chemo-radiotherapy for MM

Chemotherapy alone has been proven to be insufficient treatment for patients with MM; however, when combined with radiation or stem cell transplantation, chemotherapy can improve the rate of remission (Galli et al., 2005). High-dose chemotherapy combined with ASCT is considered a standard part of initial therapy for patients with MM. Over 75% of myeloma patients are over 50 years old at diagnosis; the majority of these patients do not qualify to receive aggressive therapeutic protocols involving ASCT due to their advanced age (Gautier and Cohen, 1994; Palumbo and Gay, 2009; Turesson et al., 2010). Chemotherapy can be combined with radiotherapy (chemo-radiotherapy) permitting chemotherapy and/or radiation to be offered at reduced dosages; such regimens may also inhibit the emergence of therapy resistant disease frequently seen with prolonged usage of high dosing regimens (Greenstein et al., 2002). In MM, tumor microenvironment has been shown to induce myeloma-cell drug resistance (Shain and Dalton, 2009).

Chemotherapy has been combined with STR in MM in a study in which sequential therapy with 153-Sm-EDTMP, melphalan, and bone marrow transplant resulted in less radiotoxicity in non-hematopoietic organs as compared to TBI in preclinical studies (Turner et al., 1993). Recently, STR using 153-Sm-EDTMP has been combined with high-dose melphalan and ASCT was used as a myeloma-conditioning regimen and found to be safe and well tolerated (Dispenzieri et al., 2010). Similarly, in primary refractory myeloma patients, conditioning with 166-Ho-DOTMP plus melphalan was found to be both safe and efficacious as compared to melphalan alone (Clapp, 2004; Giralt et al., 2003).

Radiation therapy has been shown to induce apoptotic cell death of endothelial cells (Garcia-Barros et al., 2003) and recently treatment of MM with regimens combining a designer anti-angiogenic drug and radiotherapy showed promising preclinical results for the treatment of focal MM (Jia et al., 2010). Bisphosphonates have also been studied in combination with other treatments, like thalidomide, to target myeloma cells in patients with osteolytic lesions (Ciepluch et al., 2002). Zoledronic acid and pamidronate are anti-catabolic nitrogen-containing bisphosphonates used in MM therapy (Pozzi and Raje, 2011). Combining bisphosphonates with radiotherapy for treatment of myeloma bone disease has been suggested as a method for improved myeloma control (Ural and Avcu, 2007; Yeh and Berenson, 2006).

Our increased understanding of the role of endogenous and therapy-induced oxidative stress, which results from an imbalance in the production of reactive oxygen species (ROS) and cellular antioxidant defenses, offers a biochemical rationale for designing novel ways to induce oxidative stress-mediated killing of cancer cells while sparing healthy tissues (Gius and Spitz, 2006; Goel et al., 2011; Spitz et al., 2004). Below are few cytotoxic agents that have been shown to induce ROS-mediated anti-myeloma activity. It is reasonable to hypothesize that combining radiation with such chemotherapeutic agents, that partially act by altering the redox parameters, may lead to increased anti-myeloma cellular activity.

Dexamethasone (Dex), a synthetic glucocorticoid, is an agent commonly used to treat MM. Although most newly diagnosed patients are sensitive, prolonged Dex therapy results in the development of Dex resistance and treatment failure (Alexanian et al., 1992; Greenstein et al., 2002). We have recently proposed a novel combination of Dex plus radiation for treatment of MM in which the combination of 153-Sm-EDTMP radiotherapy and Dex should selectively enhance killing of myeloma cells (Bera et al., 2010). Normal BM hematopoiesis would be protected via a mechanism that involves the selective increase of certain types of oxidative stress in myeloma cells (Bera et al., 2010).

Proteasome inhibitors - Bortezomib (BTZ, also known as Velcade/PS-341) is boronic acid inhibitor of the catalytic site of the 20S proteasome and is first in the class to be approved by the FDA for clinical use (Terpos et al., 2008). BTZ induces myeloma cell apoptosis in its supportive bone marrow microenvironment by disrupting multiple signaling pathways affecting cell cycle and survival related proteins like NF-κB, p53, and Bax among others (Mitsiades et al., 2002). BTZ inhibits NF-κB activation by stopping IκB degradation (Goel et al., 2005; Hussein, 2002). BTZ was approved for the treatment of relapsed/refractory MM patients in 2003 and data suggest that the initial combination of BTZ with immunomodulatory drugs (IMiD) can increase the response rate in MM patients (Blade and Rosinol, 2008). Studies have shown that cellular upregulation of target enzymes is a common mode of resistance to several types of chemotherapeutic drugs (Schimke et al., 1984). In one study, high levels of acquired BTZ resistance were seen after *in vitro* selection using stepwise increases in BTZ concentrations which were achieved by selective overexpression of a structurally altered β5 proteasome subunit (Oerlemans et al., 2008). In an effort to overcome BTZ resistance, novel proteasome inhibitors are being developed that act through mechanisms distinct from BTZ (Ruschak et al., 2011). Our group has shown that BTZ/PS-341 can sensitize myeloma cells to conventional radiotherapy by both intrinsic and extrinsic apoptotic pathways (Goel et al., 2005). BTZ acts as a "radiation modifier" in MM predominantly by attenuating endogenous and IR-induced NF-κB activity; indeed, several relevant molecularly targeted drugs are being tested and developed in combination with ionizing radiation to specifically target and eliminates the tumor cells while simultaneously decreasing radiotoxicity toward normal tissues (Begg et al., 2011). Using the orthotopic, syngeneic 5TGM1 myeloma model, we demonstrated that the combination of BTZ with 153-Sm-EDTMP resulted in increased survival time without a corresponding increase in the myelosuppressive effects of 153-Sm-EDTMP (Goel et al., 2006). In a phase I trial, combining 153-Sm-EDTMP with BTZ was well-tolerated and showed clinical activity in patients with relapsed or refractory MM (Berenson et al., 2009).

Studies have shown that BTZ induces apoptosis in cancer cells by increasing ROS generation in mitochondria (Ling et al., 2003; Yu et al., 2004) and endoplasmic reticulum (Fribley et al., 2004). Besides radiation, proteasome inhibitors have been combined with ROS-generating chemotherapeutic drugs like histone deacetylase (HDAC) inhibitors (Feng et al., 2008; Feng et al., 2007b; Heider et al., 2008; Miller et al., 2007; Pei et al., 2004), non-steroidal anti-inflammatory drugs (Minami et al., 2005) and rituximab (Bellosillo et al., 2001; Wang et al.,

2008) with improved elimination of cancer cells. In MM, combined treatment of BTZ with the Bcl-2 inhibitor (Pei et al., 2003) or HDAC inhibitors (sodium butyrate, suberoylanilide hydroxamic acid, PXD101) (Feng et al., 2007b; Pei et al., 2004) have shown synergistic myeloma cell killing by oxidative injury. Recently, BTZ was shown to induce Nrf-2-mediated antioxidant responses by upregulating glutamate cysteine ligase and heme-oxygenase I HMOX1 (Nerini-Molteni et al., 2008). Clinically, BTZ has been combined with 153-Sm-EDTMP with good clinical outcomes in MM with reduced radiotoxicity to normal tissues (Berenson et al., 2009; Berges et al., 2008).

Non-steroidal anti-inflammatory drugs (NSAIDs) are non-selective cyclooxygenase inhibitors (for Cox-1 and Cox-2) that were developed as anti-inflammatory agents (Rigas and Sun, 2008). Nitric oxide-donating sulfosalycylic acid was used mainly for chemo-preventive effects (Rigas and Kashfi, 2004). Recent studies have shown that NSAIDs induce apoptosis in a variety of tumor cell lines including hematological malignancies (Bernard et al., 2008; Robak et al., 2008). Drugs like SC-58125 and SDX-101 without Cox-2 inhibitory activity induces cytotoxicity, overcomes drug resistance and enhances the activity of dexamethasone in MM (Feng et al., 2007a). SDX-308 (an indole-pyran analog of SDX-101) also shows anti-myeloma effect (Feng and Lentzsch, 2007; Lentzsch et al., 2007), inhibits RANKL-stimulated NF-κB activation and osteoclast formation (Feng et al., 2007a), and β-catenin/T-cell factor pathway (Feng and Lentzsch, 2007; Yasui et al., 2007). Sulindac also shows anti-myeloma effects by accumulation of p53, Bax, and Bak in mitochondria mediated by p38 MAPK activation downstream of ROS production (Seo et al., 2007).

Arsenic trioxide (ATO)- Preclinical data shows ATO activity in B-cell lymphoma and MM (Bahlis et al., 2002; Gartenhaus et al., 2002; Grad et al., 2001). In myeloma cell lines ATO-mediated oxidative stress has been shown to upregulate proapoptotic Bcl-2 family proteins with release of cytochrome c and apoptosis (Karp and Lancet, 2005; Santucci et al., 2003). However, gene expression studies in myeloma cell lines have suggested ATO may result in a protective antioxidant response by upregulating genes such as HMOX1 and metallothionein-2A (Zhou et al., 2005). Studies have shown that the sulfhydryl oxidizing action of ATO exerts cytotoxic effects by elevating oxidative stress and by inhibiting the proper function of the glutathione/ glutathione peroxidase system (Dalton, 2002; Hussein, 2003). In support of this mechanism, myeloma cell lines with lower antioxidant capacity were found to be sensitive to ATO induced apoptosis (Zhu et al., 2000). Agents that deplete cellular glutathione, such as green tea, ascorbic acid, PI3K/Akt inhibitor, and buthionine sulfoximine have been shown to enhance ATO-induced apoptosis (Bachleitner-Hofmann et al., 2001; Gartenhaus et al., 2002; Grad et al., 2001; Nakazato et al., 2005a; Ramos et al., 2005). ATO has been combined with trolox (an analogue of α-tocopherol) with increased apoptosis in acute promyelocytic leukemia and myeloma cell lines (Diaz et al., 2005; Diaz et al., 2007). In acute myeloid leukemia cell lines ATO has been combined with polyunsaturated fatty acid docosahexaenoic acid (forming toxic lipid peroxidation products) with increased oxidative cell death (Bachleitner-Hofmann et al., 2001; Gartenhaus et al., 2002; Grad et al., 2001; Nakazato et al., 2005a; Ramos et al., 2005; Sturlan et al., 2003). Recently, ATO and 2-methoxyestradiol have been combined with BTZ to enhance BTZ-induced toxicity in myeloma cell lines via inhibition of β-catenin protein accumulation (Zhou et al., 2008). The use of ATO in B-cell lymphoma and MM clinical trials has however resulted in modest success. In MM patients that are refractory to conventional salvage therapy, ATO produced responses in 3/14 patients and prolonged stable disease in a fourth patient (Munshi et al., 2002). ATO has also been combined with BTZ is patients with relapsed/refractory MM and objective responses were observed in 6/22 patients (Hofmeister et al., 2008). However, ATO when combined with DVd (Doxil, vincristine, and dexamethasone) in 11 newly diagnosed myeloma patients failed to improve the response rate compared to DVd alone (Hofmeister et al., 2008). Overall, ATO has shown promising preclinical and clinical responses in malignant B-cells.

Motexafin gadolinium (MGd) is a metallotexaphyrin that acts by generating ROS and depletion of reducing metabolites such as protein thiols, thioredoxin, NADPH, ascorbate, and glutathione besides other mechanisms of action (Evens, 2004). It is a broad-spectrum anti-cancer agent that in under clinical trials as a single agent and in conjunction with radiotherapy and chemotherapy (Evens, 2004). MGd induces ROS-mediated toxicity in chemotherapy-sensitive and -resistant myeloma cell lines and in primary myeloma cells (Evens et al., 2005b). In B-cell lymphoma cell lines, MGd has been shown to sensitize cells to IR (Magda et al., 2001), disrupt intracellular zinc homeostasis by inducing metal response element-binding transcription factor-1 (MTF-1)-regulated and HIF-1-regulated genes (Lecane et al., 2005), and inhibit HMOX1 activity (Evans et al., 2007). MGd has shown single agent activity in very heavily pretreated B-chronic lymphocytic leukemia /small lymphocytic lymphoma patients, and showed complete remissions in combination with zevalin for relapsed B-cell NHL (Evens et al., 2005a). MGd is yet to be combined with radiotherapy and chemotherapy in clinical treatment of MM.

Myeloma and lymphoma cells harbor Ras mutations and respond to cell death by the **farnesyltransferase inhibitor (FTI)** (Karp and Lancet, 2005; Santucci et al., 2003). Manumycin (Man)-A induces apoptosis in B-cell tumors by inhibiting prenylation (Frassanito et al., 2002) and also by generating ROS that inhibits Ras/MEK/ERK and Ras/PI3K/Akt pathways by cleaving MEK and Akt (Sears et al., 2008). R115777 (zarnestra/tipifarnib) is a FTI that has shown promising clinical results in acute myeloid leukemia, myelodysplastic syndromes, chronic myelogenous leukemia, and MM and with anti-tumor effects noted independent of Ras mutations (Martinelli et al., 2008). In breast and thyroid cancer cells, similar to Man-A, R115777 was shown to perturb the redox balance and induce caspase independent DNA damage and apoptosis (Pan et al., 2005). With the recent understanding of the role ROS in FTI-induced tumor cytotoxicity, further studies may show a more promising role of these agents in Ras harboring B-cell malignancies.

Imexon, a cyanoaziridine, directly impairs mitochondria function via decreasing levels of cellular thiols, and by inducing oxidative damage of mitochondrial DNA (Dvorakova et al., 2000; Dvorakova et al., 2001; Salmon and Hersh, 1994). Imexon has shown activity in MM (Dvorakova et al., 2000; Salmon and Hersh, 1994; Samulitis et al., 2006) and promyelocytic leukemia (Dvorakova et al., 2001), and large cell lymphoma cell lines (Hersh et al., 1993). In myeloma cell lines, imexon treatment is associated with decreased levels of cellular thiols (cysteine and glutathione) with partial rescue of cytotoxicity by N-acetylcysteine and theonyltrifluoroacetone (inhibitor of mitochondrial complex II) (Dvorakova et al., 2000; Dvorakova et al., 2001). Also, resistance to imexon has been correlated with increased Cu/Zn superoxide dismutase 1 expression in myeloma cell lines (Samulitis et al., 2006). In an imexon-resistant myeloma cell line and peripheral blood mononuclear cells from normal volunteers and advanced cancer patients, imexon treatment resulted in adaptive response by up-regulation of thioredoxin reductase-1, glutaredoxin-2, and peroxiredoxin-3 that is thought to be mediated by increased AP-1 binding and nuclear levels of NF-E2-related factor 2, Nrf2 (Baker et al., 2007). In a myeloma cell line, imexon showed synergistic cytotoxicity with chemotherapeutic drugs such as cisplatin, dacarbazine, and melphalan (DNA-alkylating agents), cytarabine, fluorouracil, and gemcitabine (pyrimidine-based anti-metabolites), docetaxel (taxane), dexamethasone (glucocorticoid), and BTZ (Baker et al., 2007; Scott et al., 2007). In advanced cancer patients imexon showed clinical activity with decrease in plasma thiols and resulted in partial response in a heavily pretreated patient with B-cell NHL (Dragovich et al., 2007). The preclinical and clinical findings suggest that combining imexon with alkylating agents and pyrimidine-based anti-metabolites could result in a ROS-mediated increase in therapeutic responses in MM patients.

Naturally occurring compounds - Several natural compounds have been shown to induce cytotoxicity in B-cell lymphoma and MM cells via increased oxidative stress. Studies have shown that procarbazine (a plant sesquiterpene lactone) induces myeloma cell apoptosis by mechanisms that involves ROS (Wang et al., 2006) or by inhibiting NF-κB and caspase-dependent and -independent pathways (Suvannasankha et al., 2008). Procarbazine generates H_2O_2 during oxidation to its azo derivative (Berneis et al., 1963), has been incorporated in a combination chemotherapy called MMPP (ranimustine, melphalan, procarbazine and prednisolone) however this regimen did not show superior chemotherapy over MMCP (with cyclophosphamide) in MM (Nagura et al., 1997). It has been shown that these compounds can induce Nrf2/antioxidant response element pathway and antioxidant enzymes resulting in increased resistance to oxidative damage (Umemura et al., 2008). Resveratrol, a polyphenolic compound (stilbenes) has been shown to be both chemo-preventive and possess anti-tumor effects, presumably by altering intracellular redox reactions that regulate the activity of Nrf2 (Aggarwal et al., 2004). The anti-tumor effect of resveratrol has been hypothesized to occur in a ROS-dependent pathway (Dong et al., 2008; Juan et al., 2008; Sekhar et al., 2002). Resveratrol induces apoptosis in B-cell lymphoma (Faber and Chiles, 2006; Faber et al., 2006; Shimizu et al., 2006) and MM by several mechanisms (Bhardwaj et al., 2007; Boissy et al., 2005; Sun et al., 2006) and synergizes with radiotherapy (Baatout et al., 2004) and paclitaxel chemotherapy (Jazirehi and Bonavida, 2004). Curcumin (diferuloylmethane), a phytochemical compound of turmeric induces apoptosis by inhibition of NF-κB and STAT3 activation in myeloma (Bharti et al., 2003a; Bharti et al., 2003b; Bharti et al., 2004) and B-cell lymphoma cell lines (Hussain et al., 2008; Mackenzie et al., 2008). In B-cell lymphoma, curcumin has been shown to down modulate Syk cell activity and Akt activation (Gururajan et al., 2007), and a ROS-mediated lysosomal rupture and caspase activation with ascorbic acid-mediated enhancement of curcumin's action has been reported (Skommer et al., 2006). Catechin (epigallocatechin-3-gallate), green tea polyphenol has been shown to increase ROS levels with an increase in apoptosis in MM and lymphoma cells with enhanced killing when combined with ATO (Nakazato et al., 2005b) or etoposide (Nakazato et al., 2005a). Chaetocin (a thiodioxopiperazine produced by fungi) is a competitive and selective substrate for thioredoxin reductase-1 (Tibodeau et al., 2008) and induces myeloma cell apoptosis by oxidative stress (Isham et al., 2007). Parthenolide (a sesquiterpene lactone from herb feverfew) induces ROS-mediated apoptosis in MM cells (Wang et al., 2006) targets both myeloma and bone marrow microenvironment by caspase-dependent and -independent mechanisms (Suvannasankha et al., 2008) with anti-angiogenic effects (Kong et al., 2008).

5. Conclusion

Although major advances have been made in the treatment of MM, this disease remains incurable (Jemal et al., 2011). Myeloma tumors are considered to be inherently radiosensitive; thus the importance of radiation therapy as a part of a comprehensive treatment approach is expected to provide a clinical benefit in MM protocols. Modern radiotherapy now offers new methods and techniques to deliver high doses of radiation with enhanced anatomical precision to cancerous sites. Targeted radiotherapy using monoclonal antibodies conjugated to radionuclides, radiotargeted gene therapy using recombinant oncolytic viruses (radiovirotherapy), and bone-seeking radiopharmaceuticals now offer a new paradigm to target this systemic malignancy. Furthermore, increased understanding of the dysregulation of cancer signaling pathway(s) have lead to novel preclinical and clinical chemo-radiotherapy protocols that may offer improved response rates for MM patients.

6. Acknowledgement

The authors thank Cedar Ridge Medical Writing for their editorial services. This work was supported by the National Institutes of Health Grants [CA127958 and P30CA086862 (AG), and T32CA078586 (KS)].

7. List of abbreviations

MM, multiple myeloma
MGUS, monoclonal gammopathy of undetermined significance
NHL, non-Hodgkin lymphoma
TBI, total body irradiation
EBRT, external beam radiation therapy
IMRT, intensity-modulated radiation treatment
HT, helical tomotherapy
CRT, conformal radiotherapy
RIT, radioimmunotherapy
ASCT, autologous stem cell transplantation
MAbs, monoclonal antibodies
MV, measles virus
VSV, vesicular stomatitis virus
NIS, sodium-iodide symporter gene
SPECT, single photon emission computed tomography
CT, computed tomography
PET, positron emission tomography
STR, skeletal-targeted radiotherapy
BTZ, bortezomib
Dex, dexamethasone
NSAIDs, non-steroidal anti-inflammatory drugs
ATO, arsenic trioxide
MGd, motexafin gadolinium
FTI, farnesyltransferase inhibitor

EDTMP, ethylene diamine tetramethylene phosphonate
HEDP, hydroxyethylidenediphosphonic acid
NF-κB, nuclear factor- κB
Nrf2, NF-E2-related factor 2
IMiD, immunomodulatory drugs
ROS, reactive oxygen species

8. References

Abbas Rizvi, S.M., Song, E.Y., Raja, C., Beretov, J., Morgenstern, A., Apostolidis, C., Russell, P.J., Kearsley, J.H., Abbas, K., and Allen, B.J. (2008). Preparation and testing of bevacizumab radioimmunoconjugates with Bismuth-213 and Bismuth-205/Bismuth-206. Cancer Biol Ther *7*, 1547-1554.

Abruzzese, E., Iuliano, F., Trawinska, M.M., and Di Maio, M. (2008). 153Sm: its use in multiple myeloma and report of a clinical experience. Expert Opin Investig Drugs *17*, 1379-1387.

Aggarwal, B.B., Bhardwaj, A., Aggarwal, R.S., Seeram, N.P., Shishodia, S., and Takada, Y. (2004). Role of resveratrol in prevention and therapy of cancer: preclinical and clinical studies. Anticancer Res *24*, 2783-2840.

Ahmed, S., Winter, J.N., Gordon, L.I., and Evens, A.M. (2010). Radioimmunotherapy for the treatment of non-Hodgkin lymphoma: current status and future applications. Leuk Lymphoma *51*, 1163-1177.

Alexanian, R., Dimopoulos, M.A., Delasalle, K., and Barlogie, B. (1992). Primary dexamethasone treatment of multiple myeloma. Blood *80*, 887-890.

Anderson, K.C., and Carrasco, R.D. (2011). Pathogenesis of myeloma. Annu Rev Pathol *6*, 249-274.

Anderson, P.M., Wiseman, G.A., Dispenzieri, A., Arndt, C.A., Hartmann, L.C., Smithson, W.A., Mullan, B.P., and Bruland, O.S. (2002). High-dose samarium-153 ethylene diamine tetramethylene phosphonate: low toxicity of skeletal irradiation in patients with osteosarcoma and bone metastases. J Clin Oncol *20*, 189-196.

Au, G.G., Lincz, L.F., Enno, A., and Shafren, D.R. (2007). Oncolytic Coxsackievirus A21 as a novel therapy for multiple myeloma. Br J Haematol *137*, 133-141.

Baatout, S., Derradji, H., Jacquet, P., Ooms, D., Michaux, A., and Mergeay, M. (2004). Enhanced radiation-induced apoptosis of cancer cell lines after treatment with resveratrol. Int J Mol Med *13*, 895-902.

Bachleitner-Hofmann, T., Gisslinger, B., Grumbeck, E., and Gisslinger, H. (2001). Arsenic trioxide and ascorbic acid: synergy with potential implications for the treatment of acute myeloid leukaemia? Br J Haematol *112*, 783-786.

Bahlis, N.J., McCafferty-Grad, J., Jordan-McMurry, I., Neil, J., Reis, I., Kharfan-Dabaja, M., Eckman, J., Goodman, M., Fernandez, H.F., Boise, L.H., *et al.* (2002). Feasibility and correlates of arsenic trioxide combined with ascorbic acid-mediated depletion of intracellular glutathione for the treatment of relapsed/refractory multiple myeloma. Clin Cancer Res *8*, 3658-3668.

Baker, A.F., Landowski, T., Dorr, R., Tate, W.R., Gard, J.M., Tavenner, B.E., Dragovich, T., Coon, A., and Powis, G. (2007). The antitumor agent imexon activates antioxidant gene expression: evidence for an oxidative stress response. Clin Cancer Res *13*, 3388-3394.

Barcellini, W., and Zanella, A. (2011). Rituximab therapy for autoimmune haematological diseases. Eur J Intern Med *22*, 220-229.

Bayouth, J.E., Macey, D.J., Kasi, L.P., and Fossella, F.V. (1994). Dosimetry and toxicity of samarium-153-EDTMP administered for bone pain due to skeletal metastases. J Nucl Med *35*, 63-69.

Begg, A.C., Stewart, F.A., and Vens, C. (2011). Strategies to improve radiotherapy with targeted drugs. Nat Rev Cancer *11*, 239-253.

Bellosillo, B., Villamor, N., Lopez-Guillermo, A., Marce, S., Esteve, J., Campo, E., Colomer, D., and Montserrat, E. (2001). Complement-mediated cell death induced by rituximab in B-cell lymphoproliferative disorders is mediated in vitro by a caspase-independent mechanism involving the generation of reactive oxygen species. Blood *98*, 2771-2777.

Bera, S., Greiner, S., Choudhury, A., Dispenzieri, A., Spitz, D.R., Russell, S.J., and Goel, A. (2010). Dexamethasone-induced oxidative stress enhances myeloma cell radiosensitization while sparing normal bone marrow hematopoiesis. Neoplasia *12*, 980-992.

Berenson, J.R., Yellin, O., Patel, R., Duvivier, H., Nassir, Y., Mapes, R., Abaya, C.D., and Swift, R.A. (2009). A phase I study of samarium lexidronam/bortezomib combination therapy for the treatment of relapsed or refractory multiple myeloma. Clin Cancer Res *15*, 1069-1075.

Berges, O., Decaudin, D., Servois, V., and Kirova, Y.M. (2008). Concurrent radiation therapy and bortezomib in myeloma patient. Radiother Oncol *86*, 290-292.

Bernard, M.P., Bancos, S., Sime, P.J., and Phipps, R.P. (2008). Targeting cyclooxygenase-2 in hematological malignancies: rationale and promise. Curr Pharm Des *14*, 2051-2060.

Berneis, K., Kofler, M., Bollag, W., Kaiser, A., and Langemann, A. (1963). The degradation of deoxyribonucleic acid by new tumour inhibiting compounds: the intermediate formation of hydrogen peroxide. Experientia *19*, 132-133.

Bethge, W.A., and Sandmaier, B.M. (2005). Targeted cancer therapy using radiolabeled monoclonal antibodies. Technol Cancer Res Treat *4*, 393-405.

Bhardwaj, A., Sethi, G., Vadhan-Raj, S., Bueso-Ramos, C., Takada, Y., Gaur, U., Nair, A.S., Shishodia, S., and Aggarwal, B.B. (2007). Resveratrol inhibits proliferation, induces apoptosis, and overcomes chemoresistance through down-regulation of STAT3 and nuclear factor-kappaB-regulated antiapoptotic and cell survival gene products in human multiple myeloma cells. Blood *109*, 2293-2302.

Bharti, A.C., Donato, N., and Aggarwal, B.B. (2003a). Curcumin (diferuloylmethane) inhibits constitutive and IL-6-inducible STAT3 phosphorylation in human multiple myeloma cells. J Immunol *171*, 3863-3871.

Bharti, A.C., Donato, N., Singh, S., and Aggarwal, B.B. (2003b). Curcumin (diferuloylmethane) down-regulates the constitutive activation of nuclear factor-kappa B and IkappaBalpha kinase in human multiple myeloma cells, leading to suppression of proliferation and induction of apoptosis. Blood *101*, 1053-1062.

Bharti, A.C., Shishodia, S., Reuben, J.M., Weber, D., Alexanian, R., Raj-Vadhan, S., Estrov, Z., Talpaz, M., and Aggarwal, B.B. (2004). Nuclear factor-kappaB and STAT3 are constitutively active in CD138+ cells derived from multiple myeloma patients, and suppression of these transcription factors leads to apoptosis. Blood *103*, 3175-3184.

Bhatti, S.S., Kumar, L., Dinda, A.K., and Dawar, R. (2006). Prognostic value of bone marrow angiogenesis in multiple myeloma: use of light microscopy as well as computerized image analyzer in the assessment of microvessel density and total vascular area in multiple myeloma and its correlation with various clinical, histological, and laboratory parameters. Am J Hematol *81*, 649-656.

Biswal, B.M. (2004). Assessment of the usefulness of hemibody irradiation in painful bone metastasis. J Indian Med Assoc *102*, 133-134, 136-137.

Blade, J., and Rosinol, L. (2008). Advances in therapy of multiple myeloma. Curr Opin Oncol *20*, 697-704.

Boissy, P., Andersen, T.L., Abdallah, B.M., Kassem, M., Plesner, T., and Delaisse, J.M. (2005). Resveratrol inhibits myeloma cell growth, prevents osteoclast formation, and promotes osteoblast differentiation. Cancer research *65*, 9943-9952.

Bolek, T.W., Marcus, R.B., and Mendenhall, N.P. (1996). Solitary plasmacytoma of bone and soft tissue. Int J Radiat Oncol Biol Phys *36*, 329-333.

Bosch, A., and Frias, Z. (1988). Radiotherapy in the treatment of multiple myeloma. Int J Radiat Oncol Biol Phys *15*, 1363-1369.

Brahme, A., and Agren, A.K. (1987). Optimal dose distribution for eradication of heterogeneous tumours. Acta Oncol *26*, 377-385.

Brechbiel, M.W. (2007). Targeted alpha-therapy: past, present, future? Dalton Trans, 4918-4928.

Breitz, H.B., Wendt, R.E., 3rd, Stabin, M.S., Shen, S., Erwin, W.D., Rajendran, J.G., Eary, J.F., Durack, L., Delpassand, E., Martin, W., *et al.* (2006). 166Ho-DOTMP radiation-absorbed dose estimation for skeletal targeted radiotherapy. J Nucl Med *47*, 534-542.

Chargari, C., Hijal, T., Bouscary, D., Caussa, L., Dendale, R., Zefkili, S., Fourquet, A., and Kirova, Y.M. (2011). The role of helical tomotherapy in the treatment of bone plasmacytoma. Med Dosim.

Chargari, C., Kirova, Y.M., Zefkili, S., Caussa, L., Amessis, M., Dendale, R., Campana, F., and Fourquet, A. (2009). Solitary plasmocytoma: improvement in critical organs sparing by means of helical tomotherapy. Eur J Haematol *83*, 66-71.

Chatterjee, M., Chakraborty, T., and Tassone, P. (2006). Multiple myeloma: monoclonal antibodies-based immunotherapeutic strategies and targeted radiotherapy. Eur J Cancer *42*, 1640-1652.

Cherel, M., Davodeau, F., Kraeber-Bodere, F., and Chatal, J.F. (2006). Current status and perspectives in alpha radioimmunotherapy. Q J Nucl Med Mol Imaging *50*, 322-329.

Chung, J.K. (2002). Sodium iodide symporter: its role in nuclear medicine. J Nucl Med *43*, 1188-1200.

Ciepluch, H., Baran, W., and Hellmann, A. (2002). Combination of pamidronate and thalidomide in the therapy of treatment resistant multiple myeloma. Med Sci Monit *8*, PI31-36.

Clapp, K. (2004). In focus: (166) Ho-DOTMP in the pretransplant treatment of multiple myeloma. Clin Adv Hematol Oncol *2*, 753-754.

Cole, D.J. (1989). A randomized trial of a single treatment versus conventional fractionation in the palliative radiotherapy of painful bone metastases. Clin Oncol (R Coll Radiol) *1*, 59-62.

Couturier, O., Faivre-Chauvet, A., Filippovich, I.V., Thedrez, P., Sai-Maurel, C., Bardies, M., Mishra, A.K., Gauvrit, M., Blain, G., Apostolidis, C., *et al.* (1999). Validation of 213Bi-alpha radioimmunotherapy for multiple myeloma. Clin Cancer Res *5*, 3165s-3170s.

Dalton, W.S. (2002). Targeting the mitochondria: an exciting new approach to myeloma therapy. Commentary re: N. J. Bahlis et al., Feasibility and correlates of arsenic trioxide combined with ascorbic acid-mediated depletion of intracellular glutathione for the treatment of relapsed/refractory multiple myeloma. Clin. Cancer Res., 8: 3658-3668, 2002. Clin Cancer Res *8*, 3643-3645.

de Weers, M., Tai, Y.T., van der Veer, M.S., Bakker, J.M., Vink, T., Jacobs, D.C., Oomen, L.A., Peipp, M., Valerius, T., Slootstra, J.W., *et al.* (2011). Daratumumab, a novel therapeutic human CD38 monoclonal antibody, induces killing of multiple myeloma and other hematological tumors. J Immunol *186*, 1840-1848.

DeNardo, G.L., Sysko, V.V., and DeNardo, S.J. (2006). Cure of incurable lymphoma. Int J Radiat Oncol Biol Phys *66*, S46-56.

Deng, H., Tang, N., Stief, A.E., Mehta, N., Baig, E., Head, R., Sleep, G., Yang, X.Z., McKerlie, C., Trudel, S., *et al.* (2008). Oncolytic virotherapy for multiple myeloma using a tumour-specific double-deleted vaccinia virus. Leukemia *22*, 2261-2264.

Diaz, Z., Colombo, M., Mann, K.K., Su, H., Smith, K.N., Bohle, D.S., Schipper, H.M., and Miller, W.H., Jr. (2005). Trolox selectively enhances arsenic-mediated oxidative stress and apoptosis in APL and other malignant cell lines. Blood *105*, 1237-1245.

Diaz, Z., Laurenzana, A., Mann, K.K., Bismar, T.A., Schipper, H.M., and Miller, W.H., Jr. (2007). Trolox enhances the anti-lymphoma effects of arsenic trioxide, while protecting against liver toxicity. Leukemia *21*, 2117-2127.

Dingli, D., Peng, K.W., Harvey, M.E., Greipp, P.R., O'Connor, M.K., Cattaneo, R., Morris, J.C., and Russell, S.J. (2004). Image-guided radiovirotherapy for multiple myeloma using a recombinant measles virus expressing the thyroidal sodium iodide symporter. Blood *103*, 1641-1646.

Dispenzieri, A., Wiseman, G.A., Lacy, M.Q., Geyer, S., Litzow, M.R., Tefferi, A., Inwards, D.J., Micallef, I.N., Ansell, S., Gastineau, D.A., *et al.* (2003). A phase II study of high dose 153-samarium EDTMP (153-sm EDMTP) and melphalan for peripheral stem cell transplantation (PBSCT) in multiple myeloma (MM). Blood *102*, 982a-982a.

Dispenzieri, A., Wiseman, G.A., Lacy, M.Q., Hayman, S.R., Kumar, S.K., Buadi, F., Dingli, D., Laumann, K.M., Allred, J., Geyer, S.M., *et al.* (2010). A Phase II study of (153)Sm-EDTMP and high-dose melphalan as a peripheral blood stem cell conditioning regimen in patients with multiple myeloma. Am J Hematol *85*, 409-413.

Dispenzieri, A., Wiseman, G.A., Lacy, M.Q., Litzow, M.R., Anderson, P.M., Gastineau, D.A., Tefferi, A., Inwards, D.J., Micallef, I.N., Ansell, S.M., *et al.* (2005). A phase I study of 153Sm-EDTMP with fixed high-dose melphalan as a peripheral blood stem cell conditioning regimen in patients with multiple myeloma. Leukemia *19*, 118-125.

Dong, L.F., Low, P., Dyason, J.C., Wang, X.F., Prochazka, L., Witting, P.K., Freeman, R., Swettenham, E., Valis, K., Liu, J., *et al.* (2008). Alpha-tocopheryl succinate induces apoptosis by targeting ubiquinone-binding sites in mitochondrial respiratory complex II. Oncogene *27*, 4324-4335.

Dragovich, T., Gordon, M., Mendelson, D., Wong, L., Modiano, M., Chow, H.H., Samulitis, B., O'Day, S., Grenier, K., Hersh, E., *et al.* (2007). Phase I trial of imexon in patients with advanced malignancy. J Clin Oncol *25*, 1779-1784.

Dvorakova, K., Payne, C.M., Tome, M.E., Briehl, M.M., McClure, T., and Dorr, R.T. (2000). Induction of oxidative stress and apoptosis in myeloma cells by the aziridine-containing agent imexon. Biochem Pharmacol *60*, 749-758.

Dvorakova, K., Waltmire, C.N., Payne, C.M., Tome, M.E., Briehl, M.M., and Dorr, R.T. (2001). Induction of mitochondrial changes in myeloma cells by imexon. Blood *97*, 3544-3551.

Evans, J.P., Xu, F., Sirisawad, M., Miller, R., Naumovski, L., and de Montellano, P.R. (2007). Motexafin gadolinium-induced cell death correlates with heme oxygenase-1 expression and inhibition of P450 reductase-dependent activities. Mol Pharmacol *71*, 193-200.

Evens, A.M. (2004). Motexafin gadolinium: a redox-active tumor selective agent for the treatment of cancer. Curr Opin Oncol *16*, 576-580.

Evens, A.M., Balasubramanian, L., and Gordon, L.I. (2005a). Motexafin gadolinium induces oxidative stress and apoptosis in hematologic malignancies. Curr Treat Options Oncol *6*, 289-296.

Evens, A.M., Lecane, P., Magda, D., Prachand, S., Singhal, S., Nelson, J., Miller, R.A., Gartenhaus, R.B., and Gordon, L.I. (2005b). Motexafin gadolinium generates reactive oxygen species and induces apoptosis in sensitive and highly resistant multiple myeloma cells. Blood *105*, 1265-1273.

Faber, A.C., and Chiles, T.C. (2006). Resveratrol induces apoptosis in transformed follicular lymphoma OCI-LY8 cells: evidence for a novel mechanism involving inhibition of BCL6 signaling. Int J Oncol *29*, 1561-1566.

Faber, A.C., Dufort, F.J., Blair, D., Wagner, D., Roberts, M.F., and Chiles, T.C. (2006). Inhibition of phosphatidylinositol 3-kinase-mediated glucose metabolism coincides with resveratrol-induced cell cycle arrest in human diffuse large B-cell lymphomas. Biochem Pharmacol *72*, 1246-1256.

Fassas, A., and Tricot, G. (2001). Results of high-dose treatment with autologous stem cell support in patients with multiple myeloma. Semin Hematol *38*, 231-242.

Feng, R., Anderson, G., Xiao, G., Elliott, G., Leoni, L., Mapara, M.Y., Roodman, G.D., and Lentzsch, S. (2007a). SDX-308, a nonsteroidal anti-inflammatory agent, inhibits NF-kappaB activity, resulting in strong inhibition of osteoclast formation/activity and multiple myeloma cell growth. Blood *109*, 2130-2138.

Feng, R., and Lentzsch, S. (2007). Treatment of multiple myeloma with SDX-308. Drug News Perspect *20*, 431-435.

Feng, R., Ma, H., Hassig, C.A., Payne, J.E., Smith, N.D., Mapara, M.Y., Hager, J.H., and Lentzsch, S. (2008). KD5170, a novel mercaptoketone-based histone deacetylase inhibitor, exerts antimyeloma effects by DNA damage and mitochondrial signaling. Mol Cancer Ther *7*, 1494-1505.

Feng, R., Oton, A., Mapara, M.Y., Anderson, G., Belani, C., and Lentzsch, S. (2007b). The histone deacetylase inhibitor, PXD101, potentiates bortezomib-induced anti-multiple myeloma effect by induction of oxidative stress and DNA damage. Br J Haematol *139*, 385-397.

Fletcher, G.H. (1976). Indications for combination of irradiation and surgery. J Radiol Electrol Med Nucl *57*, 379-390.

Frassanito, M.A., Cusmai, A., Piccoli, C., and Dammacco, F. (2002). Manumycin inhibits farnesyltransferase and induces apoptosis of drug-resistant interleukin 6-producing myeloma cells. Br J Haematol *118*, 157-165.

Fribley, A., Zeng, Q., and Wang, C.Y. (2004). Proteasome inhibitor PS-341 induces apoptosis through induction of endoplasmic reticulum stress-reactive oxygen species in head and neck squamous cell carcinoma cells. Molecular and cellular biology *24*, 9695-9704.

Friedland, J. (1999). Local and systemic radiation for palliation of metastatic disease. Urol Clin North Am *26*, 391-402, x.

Galli, M., Nicolucci, A., Valentini, M., Belfiglio, M., Delaini, F., Crippa, C., Barbui, A.M., Giussani, U., Rambaldi, A., and Barbui, T. (2005). Feasibility and outcome of tandem stem cell autotransplants in multiple myeloma. Haematologica *90*, 1643-1649.

Garcia-Barros, M., Paris, F., Cordon-Cardo, C., Lyden, D., Rafii, S., Haimovitz-Friedman, A., Fuks, Z., and Kolesnick, R. (2003). Tumor response to radiotherapy regulated by endothelial cell apoptosis. Science *300*, 1155-1159.

Gartenhaus, R.B., Prachand, S.N., Paniaqua, M., Li, Y., and Gordon, L.I. (2002). Arsenic trioxide cytotoxicity in steroid and chemotherapy-resistant myeloma cell lines: enhancement of apoptosis by manipulation of cellular redox state. Clin Cancer Res *8*, 566-572.

Gautier, M., and Cohen, H.J. (1994). Multiple myeloma in the elderly. J Am Geriatr Soc *42*, 653-664.

Giralt, S., Bensinger, W., Goodman, M., Podoloff, D., Eary, J., Wendt, R., Alexanian, R., Weber, D., Maloney, D., Holmberg, L., *et al.* (2003). 166Ho-DOTMP plus melphalan followed by peripheral blood stem cell transplantation in patients with multiple myeloma: results of two phase 1/2 trials. Blood *102*, 2684-2691.

Gius, D., and Spitz, D.R. (2006). Redox signaling in cancer biology. Antioxid Redox Signal *8*, 1249-1252.

Gluck, S., Van Dyk, J., and Messner, H.A. (1994). Radiosensitivity of human clonogenic myeloma cells and normal bone marrow precursors: effect of different dose rates and fractionation. Int J Radiat Oncol Biol Phys *28*, 877-882.

Goel, A., and Batra, S.K. (2001). Antibody constructs for radioimmunodiagnosis and treatment of human pancreatic cancer. Teratog Carcinog Mutagen *21*, 45-57.

Goel, A., Carlson, S.K., Classic, K.L., Greiner, S., Naik, S., Power, A.T., Bell, J.C., and Russell, S.J. (2007). Radioiodide imaging and radiovirotherapy of multiple myeloma using VSV(Delta51)-NIS, an attenuated vesicular stomatitis virus encoding the sodium iodide symporter gene. Blood *110*, 2342-2350.

Goel, A., Colcher, D., Baranowska-Kortylewicz, J., Augustine, S., Booth, B.J., Pavlinkova, G., and Batra, S.K. (2000). Genetically engineered tetravalent single-chain Fv of the pancarcinoma monoclonal antibody CC49: improved biodistribution and potential for therapeutic application. Cancer Res *60*, 6964-6971.

Goel, A., Dispenzieri, A., Geyer, S.M., Greiner, S., Peng, K.W., and Russell, S.J. (2006). Synergistic activity of the proteasome inhibitor PS-341 with non-myeloablative 153-

Sm-EDTMP skeletally targeted radiotherapy in an orthotopic model of multiple myeloma. Blood *107*, 4063-4070.

Goel, A., Dispenzieri, A., Greipp, P.R., Witzig, T.E., Mesa, R.A., and Russell, S.J. (2005). PS-341-mediated selective targeting of multiple myeloma cells by synergistic increase in ionizing radiation-induced apoptosis. Exp Hematol *33*, 784-795.

Goel, A., Spitz, D.R., and Weiner, G.J. (2011). Manipulation of cellular redox metabolism for improving therapeutic responses in B-cell lymphoma and multiple myeloma. J Cell Biochem.

Goel, A.a.R., SJ (2006). Enhancing the therapeutic index of radiation in multiple myeloma. Drug Discovery Today *3*, 515-522.

Grad, J.M., Bahlis, N.J., Reis, I., Oshiro, M.M., Dalton, W.S., and Boise, L.H. (2001). Ascorbic acid enhances arsenic trioxide-induced cytotoxicity in multiple myeloma cells. Blood *98*, 805-813.

Greenstein, S., Ghias, K., Krett, N.L., and Rosen, S.T. (2002). Mechanisms of glucocorticoid-mediated apoptosis in hematological malignancies. Clin Cancer Res *8*, 1681-1694.

Gururajan, M., Dasu, T., Shahidain, S., Jennings, C.D., Robertson, D.A., Rangnekar, V.M., and Bondada, S. (2007). Spleen tyrosine kinase (Syk), a novel target of curcumin, is required for B lymphoma growth. J Immunol *178*, 111-121.

Gustavsson, A., Osterman, B., and Cavallin-Stahl, E. (2003). A systematic overview of radiation therapy effects in Hodgkin's lymphoma. Acta Oncol *42*, 589-604.

Hadac, E.M., Kelly, E.J., and Russell, S.J. (2011). Myeloma xenograft destruction by a nonviral vector delivering oncolytic infectious nucleic acid. Mol Ther *19*, 1041-1047.

Heider, U., von Metzler, I., Kaiser, M., Rosche, M., Sterz, J., Rotzer, S., Rademacher, J., Jakob, C., Fleissner, C., Kuckelkorn, U., *et al.* (2008). Synergistic interaction of the histone deacetylase inhibitor SAHA with the proteasome inhibitor bortezomib in mantle cell lymphoma. Eur J Haematol *80*, 133-142.

Hersh, E.M., Grogan, T.M., Funk, C.Y., and Taylor, C.W. (1993). Suppression of human lymphoma development in the severe combined immune-deficient mouse by imexon therapy. J Immunother Emphasis Tumor Immunol *13*, 77-83.

Herve, A.L., Florence, M., Philippe, M., Michel, A., Thierry, F., Kenneth, A., Jean-Luc, H., Nikhil, M., and Stephane, M. (2011). Molecular heterogeneity of multiple myeloma: pathogenesis, prognosis, and therapeutic implications. J Clin Oncol *29*, 1893-1897.

Hirsch, A.E., Jha, R.M., Yoo, A.J., Saxena, A., Ozonoff, A., Growney, M.J., and Hirsch, J.A. (2011). The use of vertebral augmentation and external beam radiation therapy in the multimodal management of malignant vertebral compression fractures. Pain Physician *14*, 447-458.

Hofmeister, C.C., Jansak, B., Denlinger, N., Kraut, E.H., Benson, D.M., and Farag, S.S. (2008). Phase II clinical trial of arsenic trioxide with liposomal doxorubicin, vincristine, and dexamethasone in newly diagnosed multiple myeloma. Leuk Res *32*, 1295-1298.

Hu, K., and Yahalom, J. (2000). Radiotherapy in the management of plasma cell tumors. Oncology (Williston Park) *14*, 101-108, 111; discussion 111-102, 115.

Hussain, A.R., Ahmed, M., Al-Jomah, N.A., Khan, A.S., Manogaran, P., Sultana, M., Abubaker, J., Platanias, L.C., Al-Kuraya, K.S., and Uddin, S. (2008). Curcumin suppresses constitutive activation of nuclear factor-kappa B and requires functional

Bax to induce apoptosis in Burkitt's lymphoma cell lines. Mol Cancer Ther *7*, 3318-3329.

Hussein, M.A. (2002). Nontraditional cytotoxic therapies for relapsed/refractory multiple myeloma. Oncologist *7 Suppl 1*, 20-29.

Hussein, M.A. (2003). Trials of arsenic trioxide in multiple myeloma. Cancer Control *10*, 370-374.

Isham, C.R., Tibodeau, J.D., Jin, W., Xu, R., Timm, M.M., and Bible, K.C. (2007). Chaetocin: a promising new antimyeloma agent with in vitro and in vivo activity mediated via imposition of oxidative stress. Blood *109*, 2579-2588.

Jansen, D.R., Krijger, G.C., Kolar, Z.I., Zonnenberg, B.A., and Zeevaart, J.R. (2010). Targeted radiotherapy of bone malignancies. Curr Drug Discov Technol *7*, 233-246.

Jazirehi, A.R., and Bonavida, B. (2004). Resveratrol modifies the expression of apoptotic regulatory proteins and sensitizes non-Hodgkin's lymphoma and multiple myeloma cell lines to paclitaxel-induced apoptosis. Mol Cancer Ther *3*, 71-84.

Jemal, A., Bray, F., Center, M.M., Ferlay, J., Ward, E., and Forman, D. (2011). Global cancer statistics. CA Cancer J Clin *61*, 69-90.

Jia, D., Koonce, N.A., Halakatti, R., Li, X., Yaccoby, S., Swain, F.L., Suva, L.J., Hennings, L., Berridge, M.S., Apana, S.M., *et al.* (2010). Repression of multiple myeloma growth and preservation of bone with combined radiotherapy and anti-angiogenic agent. Radiat Res *173*, 809-817.

Juan, M.E., Wenzel, U., Daniel, H., and Planas, J.M. (2008). Resveratrol induces apoptosis through ROS-dependent mitochondria pathway in HT-29 human colorectal carcinoma cells. J Agric Food Chem *56*, 4813-4818.

Karp, J.E., and Lancet, J.E. (2005). Development of the farnesyltransferase inhibitor tipifarnib for therapy of hematologic malignancies. Future Oncol *1*, 719-731.

Kawa, A., and Arakawa, S. (1987). The effect of attenuated vaccinia virus AS strain on multiple myeloma; a case report. Jpn J Exp Med *57*, 79-81.

Kilciksiz, S., Celik, O.K., Pak, Y., Demiral, A.N., Pehlivan, M., Orhan, O., Tokatli, F., Agaoglu, F., Zincircioglu, B., Atasoy, B.M., *et al.* (2008). Clinical and prognostic features of plasmacytomas: a multicenter study of Turkish Oncology Group-Sarcoma Working Party. Am J Hematol *83*, 702-707.

Kong, F., Chen, Z., Li, Q., Tian, X., Zhao, J., Yu, K., You, Y., and Zou, P. (2008). Inhibitory effects of parthenolide on the angiogenesis induced by human multiple myeloma cells and the mechanism. J Huazhong Univ Sci Technolog Med Sci *28*, 525-530.

Krause, S., Hillengass, J., Goldschmidt, H., Debus, J., and Neuhof, D. (2011). Radiotherapy of solitary plasmacytoma. Ann Hematol *90*, 1093-1097.

Kyle, R.A. (1975). Multiple myeloma: review of 869 cases. Mayo Clin Proc *50*, 29-40.

Lam, M.G., de Klerk, J.M., and van Rijk, P.P. (2004). 186Re-HEDP for metastatic bone pain in breast cancer patients. Eur J Nucl Med Mol Imaging *31 Suppl 1*, S162-170.

Lamb, H.M., and Faulds, D. (1997). Samarium 153Sm lexidronam. Drugs Aging *11*, 413-418; discussion 419.

Landgren, O., Kyle, R.A., and Rajkumar, S.V. (2011). From myeloma precursor disease to multiple myeloma: new diagnostic concepts and opportunities for early intervention. Clin Cancer Res *17*, 1243-1252.

Lecane, P.S., Karaman, M.W., Sirisawad, M., Naumovski, L., Miller, R.A., Hacia, J.G., and Magda, D. (2005). Motexafin gadolinium and zinc induce oxidative stress responses and apoptosis in B-cell lymphoma lines. Cancer research *65*, 11676-11688.

Lentzsch, S., Elliott, G., and Roodman, G.D. (2007). SDX-308 and SDX-101, non-steroidal anti-inflammatory drugs, as therapeutic candidates for treating hematologic malignancies including myeloma. Arch Pharm (Weinheim) *340*, 511-516.

Lewanski, C.R., Bates, T., Bowen, J., and Ashford, R.F. (1999). Solitary bone plasmacytoma: management of isolated local relapse following radiotherapy. Clin Oncol (R Coll Radiol) *11*, 348-351.

Lewington, V.J. (2005). Bone-seeking radionuclides for therapy. J Nucl Med *46 Suppl 1*, 38S-47S.

Ling, Y.H., Liebes, L., Zou, Y., and Perez-Soler, R. (2003). Reactive oxygen species generation and mitochondrial dysfunction in the apoptotic response to Bortezomib, a novel proteasome inhibitor, in human H460 non-small cell lung cancer cells. The Journal of biological chemistry *278*, 33714-33723.

Liu, C., Russell, S.J., and Peng, K.W. (2010). Systemic therapy of disseminated myeloma in passively immunized mice using measles virus-infected cell carriers. Mol Ther *18*, 1155-1164.

Macfarlane, D.J., Durrant, S., Bartlett, M.L., Allison, R., and Morton, A.J. (2002). 153Sm EDTMP for bone marrow ablation prior to stem cell transplantation for haematological malignancies. Nucl Med Commun *23*, 1099-1106.

Mackenzie, G.G., Queisser, N., Wolfson, M.L., Fraga, C.G., Adamo, A.M., and Oteiza, P.I. (2008). Curcumin induces cell-arrest and apoptosis in association with the inhibition of constitutively active NF-kappaB and STAT3 pathways in Hodgkin's lymphoma cells. Int J Cancer *123*, 56-65.

Magda, D., Lepp, C., Gerasimchuk, N., Lee, I., Sessler, J.L., Lin, A., Biaglow, J.E., and Miller, R.A. (2001). Redox cycling by motexafin gadolinium enhances cellular response to ionizing radiation by forming reactive oxygen species. Int J Radiat Oncol Biol Phys *51*, 1025-1036.

Marchand, V., Decaudin, D., Servois, V., and Kirova, Y.M. (2008). Concurrent radiation therapy and lenalidomide in myeloma patient. Radiother Oncol *87*, 152-153.

Martinelli, G., Iacobucci, I., Paolini, S., and Ottaviani, E. (2008). Farnesyltransferase inhibition in hematologic malignancies: the clinical experience with tipifarnib. Clin Adv Hematol Oncol *6*, 303-310.

Mayes, S., Brown, N., and Illidge, T.M. (2011). New antibody drug treatments for lymphoma. Expert Opin Biol Ther *11*, 623-640.

Mendenhall, W.M., Mendenhall, C.M., and Mendenhall, N.P. (2003). Solitary plasmacytoma of bone and soft tissues. Am J Otolaryngol *24*, 395-399.

Mill, W.B. (1975). Radiation therapy in multiple myeloma. Radiology *115*, 175-178.

Miller, C.P., Ban, K., Dujka, M.E., McConkey, D.J., Munsell, M., Palladino, M., and Chandra, J. (2007). NPI-0052, a novel proteasome inhibitor, induces caspase-8 and ROS-dependent apoptosis alone and in combination with HDAC inhibitors in leukemia cells. Blood *110*, 267-277.

Minami, T., Adachi, M., Kawamura, R., Zhang, Y., Shinomura, Y., and Imai, K. (2005). Sulindac enhances the proteasome inhibitor bortezomib-mediated oxidative stress and anticancer activity. Clin Cancer Res *11*, 5248-5256.

Mitsiades, N., Mitsiades, C.S., Poulaki, V., Chauhan, D., Fanourakis, G., Gu, X., Bailey, C., Joseph, M., Libermann, T.A., Treon, S.P., *et al.* (2002). Molecular sequelae of proteasome inhibition in human multiple myeloma cells. Proc Natl Acad Sci U S A *99*, 14374-14379.

Moehler, T., and Goldschmidt, H. (2011). Therapy of relapsed and refractory multiple myeloma. Recent Results Cancer Res *183*, 239-271.

Moreau, P., Facon, T., Attal, M., Hulin, C., Michallet, M., Maloisel, F., Sotto, J.J., Guilhot, F., Marit, G., Doyen, C., *et al.* (2002). Comparison of 200 mg/m(2) melphalan and 8 Gy total body irradiation plus 140 mg/m(2) melphalan as conditioning regimens for peripheral blood stem cell transplantation in patients with newly diagnosed multiple myeloma: final analysis of the Intergroupe Francophone du Myelome 9502 randomized trial. Blood *99*, 731-735.

Msaouel, P., Dispenzieri, A., and Galanis, E. (2009). Clinical testing of engineered oncolytic measles virus strains in the treatment of cancer: an overview. Curr Opin Mol Ther *11*, 43-53.

Munguia, A., Ota, T., Miest, T., and Russell, S.J. (2008). Cell carriers to deliver oncolytic viruses to sites of myeloma tumor growth. Gene Ther *15*, 797-806.

Munshi, N.C., Tricot, G., Desikan, R., Badros, A., Zangari, M., Toor, A., Morris, C., Anaissie, E., and Barlogie, B. (2002). Clinical activity of arsenic trioxide for the treatment of multiple myeloma. Leukemia *16*, 1835-1837.

Myers, R.M., Greiner, S.M., Harvey, M.E., Griesmann, G., Kuffel, M.J., Buhrow, S.A., Reid, J.M., Federspiel, M., Ames, M.M., Dingli, D., *et al.* (2007). Preclinical pharmacology and toxicology of intravenous MV-NIS, an oncolytic measles virus administered with or without cyclophosphamide. Clin Pharmacol Ther *82*, 700-710.

Nagura, E., Ichikawa, A., Kamiya, O., Kato, R., Utsumi, M., Tanaka, M., Takeyama, H., Shimizu, K., Kobayashi, M., Naito, K., *et al.* (1997). A randomized study comparing VMCP and MMPP in the treatment of multiple myeloma. Cancer Chemother Pharmacol *39*, 279-285.

Nakazato, T., Ito, K., Ikeda, Y., and Kizaki, M. (2005a). Green tea component, catechin, induces apoptosis of human malignant B cells via production of reactive oxygen species. Clin Cancer Res *11*, 6040-6049.

Nakazato, T., Ito, K., Miyakawa, Y., Kinjo, K., Yamada, T., Hozumi, N., Ikeda, Y., and Kizaki, M. (2005b). Catechin, a green tea component, rapidly induces apoptosis of myeloid leukemic cells via modulation of reactive oxygen species production in vitro and inhibits tumor growth in vivo. Haematologica *90*, 317-325.

Nerini-Molteni, S., Ferrarini, M., Cozza, S., Caligaris-Cappio, F., and Sitia, R. (2008). Redox homeostasis modulates the sensitivity of myeloma cells to bortezomib. Br J Haematol *141*, 494-503.

Oerlemans, R., Franke, N.E., Assaraf, Y.G., Cloos, J., van Zantwijk, I., Berkers, C.R., Scheffer, G.L., Debipersad, K., Vojtekova, K., Lemos, C., *et al.* (2008). Molecular basis of bortezomib resistance: proteasome subunit beta5 (PSMB5) gene mutation and overexpression of PSMB5 protein. Blood *112*, 2489-2499.

Ong, H.T., Hasegawa, K., Dietz, A.B., Russell, S.J., and Peng, K.W. (2007). Evaluation of T cells as carriers for systemic measles virotherapy in the presence of antiviral antibodies. Gene Ther *14*, 324-333.

Palumbo, A., and Gay, F. (2009). How to treat elderly patients with multiple myeloma: combination of therapy or sequencing. Hematology Am Soc Hematol Educ Program, 566-577.

Pan, J., She, M., Xu, Z.X., Sun, L., and Yeung, S.C. (2005). Farnesyltransferase inhibitors induce DNA damage via reactive oxygen species in human cancer cells. Cancer research *65*, 3671-3681.

Parato, K.A., Senger, D., Forsyth, P.A., and Bell, J.C. (2005). Recent progress in the battle between oncolytic viruses and tumours. Nat Rev Cancer *5*, 965-976.

Pedraza-Lopez, M., Ferro-Flores, G., Arteaga de Murphy, C., Morales-Ramirez, P., Piedras-Ross, J., Murphy-Stack, E., and Hernandez-Oviedo, O. (2004). Cytotoxic and genotoxic effect of the [166Dy]Dy/166Ho-EDTMP in vivo generator system in mice. Nucl Med Biol *31*, 1079-1085.

Pei, X.Y., Dai, Y., and Grant, S. (2003). The proteasome inhibitor bortezomib promotes mitochondrial injury and apoptosis induced by the small molecule Bcl-2 inhibitor HA14-1 in multiple myeloma cells. Leukemia *17*, 2036-2045.

Pei, X.Y., Dai, Y., and Grant, S. (2004). Synergistic induction of oxidative injury and apoptosis in human multiple myeloma cells by the proteasome inhibitor bortezomib and histone deacetylase inhibitors. Clin Cancer Res *10*, 3839-3852.

Peng, K.W., Ahmann, G.J., Pham, L., Greipp, P.R., Cattaneo, R., and Russell, S.J. (2001). Systemic therapy of myeloma xenografts by an attenuated measles virus. Blood *98*, 2002-2007.

Pozzi, S., and Raje, N. (2011). The role of bisphosphonates in multiple myeloma: mechanisms, side effects, and the future. Oncologist *16*, 651-662.

Price, P., Hoskin, P.J., Easton, D., Austin, D., Palmer, S.G., and Yarnold, J.R. (1986). Prospective randomised trial of single and multifraction radiotherapy schedules in the treatment of painful bony metastases. Radiother Oncol *6*, 247-255.

Rajkumar, S.V., Leong, T., Roche, P.C., Fonseca, R., Dispenzieri, A., Lacy, M.Q., Lust, J.A., Witzig, T.E., Kyle, R.A., Gertz, M.A., *et al.* (2000). Prognostic value of bone marrow angiogenesis in multiple myeloma. Clin Cancer Res *6*, 3111-3116.

Ramos, A.M., Fernandez, C., Amran, D., Sancho, P., de Blas, E., and Aller, P. (2005). Pharmacologic inhibitors of PI3K/Akt potentiate the apoptotic action of the antileukemic drug arsenic trioxide via glutathione depletion and increased peroxide accumulation in myeloid leukemia cells. Blood *105*, 4013-4020.

Rigas, B., and Kashfi, K. (2004). Nitric-oxide-donating NSAIDs as agents for cancer prevention. Trends Mol Med *10*, 324-330.

Rigas, B., and Sun, Y. (2008). Induction of oxidative stress as a mechanism of action of chemopreventive agents against cancer. Br J Cancer *98*, 1157-1160.

Robak, P., Smolewski, P., and Robak, T. (2008). The role of non-steroidal anti-inflammatory drugs in the risk of development and treatment of hematologic malignancies. Leuk Lymphoma *49*, 1452-1462.

Rostom, A.Y. (1988). A review of the place of radiotherapy in myeloma with emphasis on whole body irradiation. Hematol Oncol *6*, 193-198.

Ruschak, A.M., Slassi, M., Kay, L.E., and Schimmer, A.D. (2011). Novel proteasome inhibitors to overcome bortezomib resistance. J Natl Cancer Inst *103*, 1007-1017.

Russell, S.J., and Peng, K.W. (2008). The utility of cells as vehicles for oncolytic virus therapies. Curr Opin Mol Ther *10*, 380-386.

Salmon, S.E., and Hersh, E.M. (1994). Sensitivity of multiple myeloma to imexon in the human tumor cloning assay. J Natl Cancer Inst *86*, 228-230.

Samulitis, B.K., Landowski, T.H., and Dorr, R.T. (2006). Correlates of imexon sensitivity in human multiple myeloma cell lines. Leuk Lymphoma *47*, 97-109.

Santucci, R., Mackley, P.A., Sebti, S., and Alsina, M. (2003). Farnesyltransferase inhibitors and their role in the treatment of multiple myeloma. Cancer Control *10*, 384-387.

Schimke, R.T., Beverley, S., Brown, P., Cassin, R., Federspiel, N., Gasser, C., Hill, A., Johnston, R., Mariani, B., Mosse, E., *et al.* (1984). Gene amplification and drug resistance in cultured animal cells. Cancer Treat Rev *11 Suppl A*, 9-17.

Scott, J., Dorr, R.T., Samulitis, B., and Landowski, T.H. (2007). Imexon-based combination chemotherapy in A375 human melanoma and RPMI 8226 human myeloma cell lines. Cancer Chemother Pharmacol *59*, 749-757.

Sears, K.T., Daino, H., and Carey, G.B. (2008). Reactive oxygen species-dependent destruction of MEK and Akt in Manumycin stimulated death of lymphoid tumor and myeloma cell lines. Int J Cancer *122*, 1496-1505.

Sekhar, K.R., Spitz, D.R., Harris, S., Nguyen, T.T., Meredith, M.J., Holt, J.T., Gius, D., Marnett, L.J., Summar, M.L., and Freeman, M.L. (2002). Redox-sensitive interaction between KIAA0132 and Nrf2 mediates indomethacin-induced expression of gamma-glutamylcysteine synthetase. Free Radic Biol Med *32*, 650-662.

Seo, S.K., Lee, H.C., Woo, S.H., Jin, H.O., Yoo, D.H., Lee, S.J., An, S., Choe, T.B., Park, M.J., Hong, S.I., *et al.* (2007). Sulindac-derived reactive oxygen species induce apoptosis of human multiple myeloma cells via p38 mitogen activated protein kinase-induced mitochondrial dysfunction. Apoptosis *12*, 195-209.

Serafini, A.N. (2000). Samarium Sm-153 lexidronam for the palliation of bone pain associated with metastases. Cancer *88*, 2934-2939.

Serafini, A.N. (2001). Systemic metabolic radiotherapy with samarium-153 EDTMP for the treatment of painful bone metastasis. Q J Nucl Med *45*, 91-99.

Shain, K.H., and Dalton, W.S. (2009). Environmental-mediated drug resistance: a target for multiple myeloma therapy. Expert Rev Hematol *2*, 649-662.

Shimizu, T., Nakazato, T., Xian, M.J., Sagawa, M., Ikeda, Y., and Kizaki, M. (2006). Resveratrol induces apoptosis of human malignant B cells by activation of caspase-3 and p38 MAP kinase pathways. Biochem Pharmacol *71*, 742-750.

Sirohi, B., and Powles, R. (2004). Multiple myeloma. Lancet *363*, 875-887.

Skommer, J., Wlodkowic, D., and Pelkonen, J. (2006). Cellular foundation of curcumin-induced apoptosis in follicular lymphoma cell lines. Exp Hematol *34*, 463-474.

Snowden, J.A., Ahmedzai, S.H., Ashcroft, J., D'Sa, S., Littlewood, T., Low, E., Lucraft, H., Maclean, R., Feyler, S., Pratt, G., *et al.* (2011). Guidelines for supportive care in multiple myeloma 2011. Br J Haematol *154*, 76-103.

Soejima, T., Hirota, S., Tsujino, K., Yoden, E., Fujii, O., Ichimiya, Y., and Mizuno, I. (2007). Total body irradiation followed by bone marrow transplantation: comparison of once-daily and twice-daily fractionation regimens. Radiat Med *25*, 402-406.

Somlo, G., Lashkari, A., Bellamy, W., Zimmerman, T.M., Tuscano, J.M., O'Donnell, M.R., Mohrbacher, A.F., Forman, S.J., Frankel, P., Chen, H.X., *et al.* (2011). Phase II randomized trial of bevacizumab versus bevacizumab and thalidomide for relapsed/refractory multiple myeloma: a California Cancer Consortium trial. Br J Haematol *154*, 533-535.

Spitz, D.R., Azzam, E.I., Li, J.J., and Gius, D. (2004). Metabolic oxidation/reduction reactions and cellular responses to ionizing radiation: a unifying concept in stress response biology. Cancer Metastasis Rev *23*, 311-322.

Stief, A.E., and McCart, J.A. (2008). Oncolytic virotherapy for multiple myeloma. Expert Opin Biol Ther *8*, 463-473.

Sturlan, S., Baumgartner, M., Roth, E., and Bachleitner-Hofmann, T. (2003). Docosahexaenoic acid enhances arsenic trioxide-mediated apoptosis in arsenic trioxide-resistant HL-60 cells. Blood *101*, 4990-4997.

Sun, C., Hu, Y., Liu, X., Wu, T., Wang, Y., He, W., and Wei, W. (2006). Resveratrol downregulates the constitutional activation of nuclear factor-kappaB in multiple myeloma cells, leading to suppression of proliferation and invasion, arrest of cell cycle, and induction of apoptosis. Cancer Genet Cytogenet *165*, 9-19.

Supiot, S., Faivre-Chauvet, A., Couturier, O., Heymann, M.F., Robillard, N., Kraeber-Bodere, F., Morandeau, L., Mahe, M.A., and Cherel, M. (2002). Comparison of the biologic effects of MA5 and B-B4 monoclonal antibody labeled with iodine-131 and bismuth-213 on multiple myeloma. Cancer *94*, 1202-1209.

Supiot, S., Gouard, S., Charrier, J., Apostolidis, C., Chatal, J.F., Barbet, J., Davodeau, F., and Cherel, M. (2005). Mechanisms of cell sensitization to alpha radioimmunotherapy by doxorubicin or paclitaxel in multiple myeloma cell lines. Clin Cancer Res *11*, 7047s-7052s.

Suvannasankha, A., Crean, C.D., Shanmugam, R., Farag, S.S., Abonour, R., Boswell, H.S., and Nakshatri, H. (2008). Antimyeloma effects of a sesquiterpene lactone parthenolide. Clin Cancer Res *14*, 1814-1822.

Terpos, E., Roussou, M., and Dimopoulos, M.A. (2008). Bortezomib in multiple myeloma. Expert Opin Drug Metab Toxicol *4*, 639-654.

Tibodeau, J., Benson, L., Isham, C., Owen, W., and Bible, K. (2008). The Anticancer Agent Chaetocin is a Competitive Substrate and Inhibitor of Thioredoxin Reductase. Antioxid Redox Signal.

Troussard, X., Macro, M., Vie, B., Batho, A., Peny, A.M., Reman, O., Tabah, I., and Leporrier, M. (1995). Human recombinant granulocyte-macrophage colony stimulating factor (hrGM-CSF) improves double hemibody irradiation (DHBI) tolerance in patients with stage III multiple myeloma: a pilot study. Br J Haematol *89*, 191-195.

Tsang, R.W., Gospodarowicz, M.K., Pintilie, M., Bezjak, A., Wells, W., Hodgson, D.C., and Stewart, A.K. (2001). Solitary plasmacytoma treated with radiotherapy: impact of tumor size on outcome. Int J Radiat Oncol Biol Phys *50*, 113-120.

Turesson, I., Velez, R., Kristinsson, S.Y., and Landgren, O. (2010). Patterns of multiple myeloma during the past 5 decades: stable incidence rates for all age groups in the population but rapidly changing age distribution in the clinic. Mayo Clin Proc *85*, 225-230.

Turner, J.H., Claringbold, P.G., Manning, L.S., O'Donoghue, H.L., Berger, J.D., and Glancy, R.J. (1993). Radiopharmaceutical therapy of 5T33 murine myeloma by sequential treatment with samarium-153 ethylenediaminetetramethylene phosphonate, melphalan, and bone marrow transplantation. J Natl Cancer Inst *85*, 1508-1513.

Umemura, K., Itoh, T., Hamada, N., Fujita, Y., Akao, Y., Nozawa, Y., Matsuura, N., Iinuma, M., and Ito, M. (2008). Preconditioning by sesquiterpene lactone enhances H2O2-induced Nrf2/ARE activation. Biochem Biophys Res Commun *368*, 948-954.

Ural, A.U., and Avcu, F. (2007). Therapeutic role of bisphosphonate and radiation combination in the management of myeloma bone disease. Clin Cancer Res *13*, 3432.

van de Donk, N.W., Kamps, S., Mutis, T., and Lokhorst, H.M. (2011). Monoclonal antibody-based therapy as a new treatment strategy in multiple myeloma. Leukemia.

van der Veer, M.S., de Weers, M., van Kessel, B., Bakker, J.M., Wittebol, S., Parren, P.W., Lokhorst, H.M., and Mutis, T. (2011). Towards effective immunotherapy of myeloma: enhanced elimination of myeloma cells by combination of lenalidomide with the human CD38 monoclonal antibody daratumumab. Haematologica *96*, 284-290.

Wang, M., Han, X.H., Zhang, L., Yang, J., Qian, J.F., Shi, Y.K., Kwak, L.W., Romaguera, J., and Yi, Q. (2008). Bortezomib is synergistic with rituximab and cyclophosphamide in inducing apoptosis of mantle cell lymphoma cells in vitro and in vivo. Leukemia *22*, 179-185.

Wang, W., Adachi, M., Kawamura, R., Sakamoto, H., Hayashi, T., Ishida, T., Imai, K., and Shinomura, Y. (2006). Parthenolide-induced apoptosis in multiple myeloma cells involves reactive oxygen species generation and cell sensitivity depends on catalase activity. Apoptosis *11*, 2225-2235.

Willmon, C., Harrington, K., Kottke, T., Prestwich, R., Melcher, A., and Vile, R. (2009). Cell carriers for oncolytic viruses: Fed Ex for cancer therapy. Mol Ther *17*, 1667-1676.

Wong, J.Y., Liu, A., Schultheiss, T., Popplewell, L., Stein, A., Rosenthal, J., Essensten, M., Forman, S., and Somlo, G. (2006). Targeted total marrow irradiation using three-dimensional image-guided tomographic intensity-modulated radiation therapy: an alternative to standard total body irradiation. Biol Blood Marrow Transplant *12*, 306-315.

Wong, J.Y., Rosenthal, J., Liu, A., Schultheiss, T., Forman, S., and Somlo, G. (2009). Image-guided total-marrow irradiation using helical tomotherapy in patients with multiple myeloma and acute leukemia undergoing hematopoietic cell transplantation. Int J Radiat Oncol Biol Phys *73*, 273-279.

Wun, T., Kwon, D.S., and Tuscano, J.M. (2001). Radioimmunotherapy: potential as a therapeutic strategy in non-Hodgkin's lymphoma. BioDrugs *15*, 151-162.

Yasui, H., Hideshima, T., Ikeda, H., Ocio, E.M., Kiziltepe, T., Vallet, S., Okawa, Y., Neri, P., Sukhdeo, K., Podar, K., *et al.* (2007). Novel etodolac analog SDX-308 (CEP-18082) induces cytotoxicity in multiple myeloma cells associated with inhibition of beta-catenin/TCF pathway. Leukemia *21*, 535-540.

Yeginer, M., Roeske, J.C., Radosevich, J.A., and Aydogan, B. (2011). Linear accelerator-based intensity-modulated total marrow irradiation technique for treatment of hematologic malignancies: a dosimetric feasibility study. Int J Radiat Oncol Biol Phys *79*, 1256-1265.

Yeh, H.S., and Berenson, J.R. (2006). Treatment for myeloma bone disease. Clin Cancer Res *12*, 6279s-6284s.

Yu, C., Rahmani, M., Dent, P., and Grant, S. (2004). The hierarchical relationship between MAPK signaling and ROS generation in human leukemia cells undergoing apoptosis in response to the proteasome inhibitor Bortezomib. Exp Cell Res *295*, 555-566.

Zhou, L., Hou, J., Fu, W., Wang, D., Yuan, Z., and Jiang, H. (2008). Arsenic trioxide and 2-methoxyestradiol reduce beta-catenin accumulation after proteasome inhibition and enhance the sensitivity of myeloma cells to Bortezomib. Leuk Res *32*, 1674-1683.

Zhou, P., Kalakonda, N., and Comenzo, R.L. (2005). Changes in gene expression profiles of multiple myeloma cells induced by arsenic trioxide (ATO): possible mechanisms to explain ATO resistance in vivo. Br J Haematol *128*, 636-644.

Zhu, Q., Chen, G., and Huang, Y. (2000). [The relationship between sensitivity to arsenic trioxide and antioxidative capacity of malignant hematopoietic cells]. Zhonghua Zhong Liu Za Zhi *22*, 359-361.

12

Radiation Therapy and Skin Cancer

Jonathan D. Tward, Christopher J. Anker,
David K. Gaffney and Glen M. Bowen
Huntsman Cancer Institute and the University of Utah School of Medicine, USA

1. Introduction

Although radiation therapy (RT) can be very beneficial in a number of cutaneous malignancies, it tends to be a relatively underutilized therapy for a number of reasons. The hesitancy on the part of dermatologists to recommend RT is driven by a number of issues. First and foremost is perception inherent in dermatology training programs. All dermatologists in training encounter patients harmed by the historic overutilization of RT in the past, both by dermatologists themselves and radiation oncologists for non-malignant inflammatory diseases such as eczema, acne, and even fungal infections. Many patients treated in their youth with Grenz rays by dermatologists have gone on to develop multiple cutaneous malignancies later in life. As the role of ultraviolet light in carcinogenesis has come into focus, it can be intellectually awkward for dermatologists to discourage patients from excessive exposure to ultraviolet light and then recommend radiation therapy, especially in the younger population due to perceptions of the consequential increased risk for the latent development of further skin cancers. Dermatologists may focus too much on the carcinogenic consequences of RT (which are statistically quite modest) at the expense of the therapeutic benefits of RT. Another perceptual issue is radiated sites can be characterized by atrophy and telangiectasia which are cosmetically undesirable. Finally, familiarity with standard treatments within the bailiwick of dermatology training (excisional and Mohs micrographic surgery, laser surgery, cryosurgery, curettage and electrodessication, and medical therapies such as topical 5-fluorouracil and imiquimod cream to name a few), are naturally going to be considered first by a dermatologist who is well-versed in these therapies but is relatively unfamiliar with state of the art approaches currently utilized by radiation oncologists.

For these and other reasons, RT is a mostly underutilized treatment modality for many varieties of skin cancer and can have both high cure rates with excellent cosmetic outcomes. This chapter will address the most commonly encountered cutaneous malignancies where RT could be considered as either a first line or adjuvant treatment with acceptable side effect profiles. It remains the obligation of dermatology and radiation oncology training programs to better communicate and discuss the issues surrounding the indications for and selection of RT for skin cancers.

2. Basal cell carcinoma

Basal cell carcinoma (BCC) is, by far, the most common of all cancer types in Caucasians with increased rates directly tied to latitudes approaching the equator (Fears & Scotto, 1983).

Although metastases are very rare with BCC, the tumors are often in cosmetically sensitive areas of the face, scalp, and neck, and represent a major cause of morbidity and take out a major chunk of the health care expenditure pie. The annual cost of treating BCC and squamous cell carcinoma (SCC) in the Medicare population in the United States is estimated to be above $400 million (Liu & Chen, 2000; Mudigonda et al., 2010).

2.1 Epidemiology

There were an estimated 2 million cases of non-melanoma skin cancer diagnosed in the United States in 2010 (Edwards, 2010) and of those, 75-80% are BCC (Lang & Maize, 2005). The incidence of BCC continues to rise worldwide in the Caucasian population and is thought to be mainly a consequence of increased exposure to ultraviolet light.

2.2 Etiology

Ultraviolet light is the causative factor in the majority of BCC and risk is inversely related to inherent melanin content in the skin which serves a protective function. It is uncommon for blacks to get BCC, but if blacks suffer from albinism where the melanocytes fail to manufacture melanin, the risk for BCC increases substantially (Alexander & Henschke, 1981). Multiple BCCs can be seen in a few scenarios such as patients undergoing RT for acne or for other inflammatory conditions in their youth. Doses as low as 450 rads have been associated with the formation of BCC, and in some patients, very high numbers of tumors (Everall & Dowd, 1978). Ultraviolet light therapy given by dermatologists for inflammatory conditions such as psoriasis has also long been known to increase BCC risk (Halprin, 1980; Roenigk et al., 1981; Stern et al., 1979).

2.3 Clinical presentation

The three most commonly encountered subtypes of BCC are the nodular, superficial, and morpheaform variants. The nodular BCC most commonly presents as a nodule that resembles a drop of paraffin wax riddled with telangiectasia. Sometimes the center will begin to necrose and a central ulcer/eschar develops with a surrounding rolled edge (See Figure 1a). The superficial BCC is completely different and presents as a red scaling macule

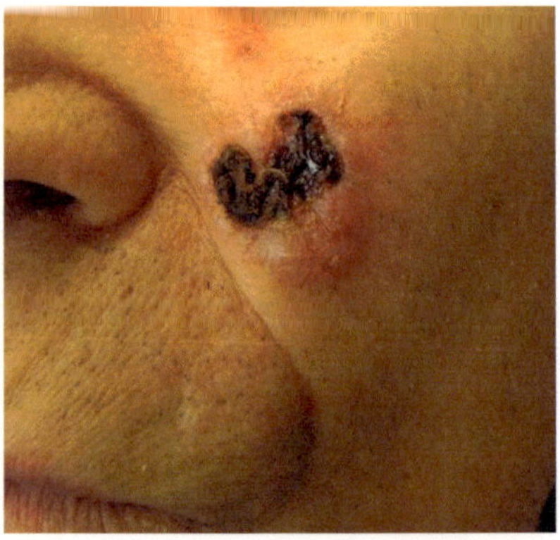

Fig. 1a. Basal cell carcinoma with paraffin wax like opalescence, telangiectasia, and ulceration

that often features notching at the border. Clinically the superficial BCC can be confused with actinic keratosis, nummular eczema, and even tinea (ring-worm). The morpheaform BCC is different still and often is scar-like and depressed compared to the surrounding skin, similar to a compressed snowshoe print. The surface is often shiny as the pores are frequently obliterated by the tumor.

2.4 Histology

The nodular BCC will have a pattern of palisading basophilic basal cells that extend off of the dermoepidermal junction and expand into the dermis. At the leading edge of the expansile tumor mass there is often a thin layer of degradation of the matrix leaving a retraction artifact between the tumor and the surrounding dermis. The tumor slowly dissolves the collagen as it glacially makes its way through the dermis (See Figure 1b). The superficial BCC is similar but has only small buds of the palisading basal cells extending off of the epidermis and lacks the deeper dermal infiltration. The morpheaform tumors arise in a scarred and fibrotic matrix and the basal cells are grouped into numerous thin and filamentous strands surrounded by sclerosis, giving an overall pattern similar to paisley. Since such tumors are much more filamentous and embedded in a sclerotic stroma, they are much less amenable to removal with a curette.

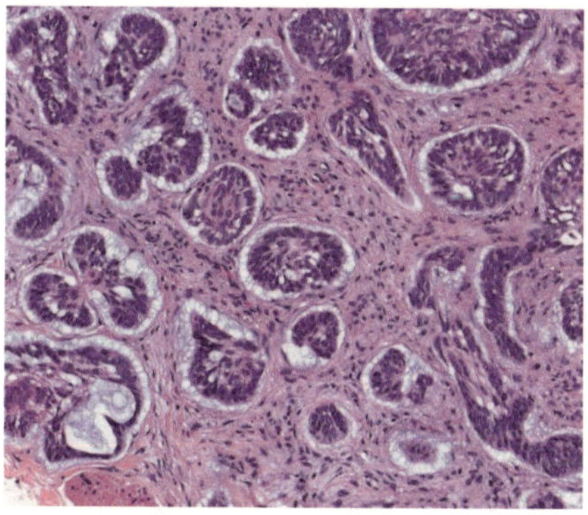

Fig. 1b. Histology of basal cell carcinoma

2.5 Treatment

The majority of BCCs are treated with curettage and electrodessication (CE). CE is not considered for morpheaform tumors, recurrent tumors, large tumors (> 2 cm) or tumors in thickly hair-bearing areas such as the scalp due to lower success rates. Simple excision is also a viable treatment option with cure rates similar to CE (Werlinger et al., 2002). Of all treatment modalities applied to BCC, Mohs micrographic surgery (MMS) has the highest documented cure rates (approximately 99% for most tumors) compared to 90-93 % cure rates for other modalities (Rowe et al., 1989). Pioneered by the late Fredrick Mohs, M.D. of the University of Wisconsin, MMS involves a *staged* excisional approach where the visible tumor is removed and then submitted for fresh tissue frozen sections. Areas of tumor positivity are recorded by the Mohs surgeon who then returns to the patient and removes additional tissue confined to specific tumor roots. This process is repeated until histologically negative margins are verified, and only then is the defect surgically repaired.

The great strength of this procedure is the reduction of the removal of uninvolved tissue to a bare minimum (as opposed to a wide local excision), combined with very high cure rates.

The main drawbacks of MMS are that the surgeries can be very morbid, especially when key anatomic structures are partially or completely removed as part of the surgery in areas on the nose, eyelids, ears, and lips. Although there have been ingenious advances in reconstructive surgery, the Mohs surgeon and facial and plastic reconstructive surgeons are left facing many surgical defects where the reconstructive options are limited. In such cases, radiation therapy can be a very desirable alternative to radical surgical excision.

2.6 Radiation therapy

Radiation therapy (RT) is not used as frequently as in the past, but it remains a very important part of the armamentarium for treating BCC. The chance of success is directly related to the clinician's ability to delineate tumor margins so the lesion is not undertreated. It has been suggested that in tumors with ambiguous borders, scouting biopsies be performed prior to embarking on RT to better define the required field (Lang, 2005). With respect to primary therapy, RT has the great advantage of avoiding reconstructive surgery in critical sites around the lacrimal system and on the nose and ears. (Chahbazian 1984; Leshin et al., 1993; Morrison et al., 1997). Many patients with BCC are poor surgical candidates due to advanced age and co-morbidities, and surgery becomes impractical. Some BCCs are not resectable even in healthy patients and RT is a viable alternative to surgery and, as will be discussed, is curative in most cases (about 95%). In some circumstances, RT may not be curative but palliation is an important consideration for large and complicated tumors. (RD, 1985).

RT has been discouraged in young patients in a wide range of reports (Brady, 1987; Braun-Falco, 1976; Chahbazian, 1980, 1984; Gladstein, 1978; Thissen, 1999), as there are concerns of the latent eruptions of cancer in the irradiated site as well as the tendency for treated sites to develop atrophy and telangiectasias over time that compromise the cosmetic appearance. As predicted, the size and extent of tumor will determine the success rate and long-term cosmetic result. RT has also been discouraged for morpheaform tumors. (Brady et al., 1987; Braun-Falco et al., 1976; Chahbazian, 1980, 1984; Dubin, 1983; Gladstein et al., 1978). There has also been reluctance to treat patients with the basal cell nevus syndrome who harbor a genetic defect in the PTCH tumor suppressor gene where RT may create additional DNA damage, creating higher risk for additional tumor formation. (Gorlin, 1987).

Table 1 is included by permission of the National Comprehensive Cancer Network (NCCN Guidelines®, 2011) with the following clarifications. Because of the wider beam penumbra, wider field margins are necessary when using electron beam than with orthovoltage X-rays. If lead skin collimation is used, tighter field margins can be used with electron beam adjacent to critical structures such as the lacrimal system. When using electron beam, bolus is necessary to achieve adequate surface dose. An electron beam energy should be chosen which achieves adequate surface dose and encompasses the deep margin of the tumor by at least the distal 90% line. Electron beam doses are specified at 90% of the maximal depth dose (Dmax). Orthovoltage X-ray doses are specified at Dmax (skin surface) to account for the approximately 10% higher biologic effectiveness between the two modalities of radiation.

Tumor Diameter	Margins	Dose (Gy)	# of Fractions
< 2 cm	1 - 1.5 cm	64	32 over 6 - 6.4 weeks
		55	20 over 4 weeks
		50	15 over 3 weeks
		35	5 over 5 days
≥ 2 cm	1.5 - 2cm	66	33 over 6 – 6.6 weeks
			55 over 4 weeks
Postoperative adjuvant		50	20 over 4 weeks
		60	30 over 6 weeks

Table 1. Dose Recommendations for Primary BCC and SCC

When the proper patient is selected and RT is expertly executed, response rates are excellent. The five-year complete response rates have been reported to range from 90 to 95%. A study looking at 862 primary BCCs treated with RT alone had a five-year complete response rate of 92.6%. (Silverman et al., 1992). Increasing diameter was associated with recurrence risk with 95.6% cure rates for tumors < 10 mm on the head, but cure rates of 90.5% if the tumors were > 10 mm. (Silverman et al., 1992). Long-term cosmetic outcomes were evaluated in this study and were thought to be satisfactory in 63% of RT-treated patients, 84% for patients who had surgery, and 91% with those treated with curettage and electrodessication. (Silverman et al., 1992). The data on the cosmetic outcome supports the notion that RT ought not to be a first line therapy for young patients if other options are available.

The general consensus in the literature appears to support the notion that tumors < 10 mm can be effectively treated with one dose of radiation. Mendenhall et al. recommend a single dose of 20 Gy, typically reserved only for when late cosmesis is not important and travel for the patient is exceedingly difficult. (Mendenhall et al. 1994). However, fractionation is recommended for tumors > 10 mm due to the improvement in the therapeutic ratio, allowing for less harm to surrounding normal tissue despite a larger biologically effective dose to the tumor.(Lang & Maize, 2005). Fractionation is particularly important for tumors overlying cartilage such as the nose and ears, or over bone. Additional considerations when choosing the extent of fractionation include patient convenience and performance status. For lesions < 3 cm where cosmesis is unimportant, 40 Gy in 8 fractions and 30 Gy in 5 fractions may be considered (Morrison et al., 1993).

The treatment of recurrent tumors with RT is not as effective as it is for primary tumors, which is also true for (MMS). In a study of 221 recurrent tumors that were subsequently treated with RT, the re-recurrence rate was 9.5%. (Silverman et al., 1992). In a second study treating 61 recurrent BCCs with RT the 5-year estimated Kaplan-Meier re-recurrence rate was 4% for tumors $<$ 10 mm and 19% for tumors $>$ 10 mm with a cost that was roughly equivalent to MMS. (Wilder et al., 1991). These data were further stratified by stage where Kaplan-Meier 5-year predicted recurrence rates for stage I/II tumors (tumors of any size but confined to the skin) was 7% compared to a 58% predicted recurrence rate for tumors that were stage III/IV (tumors that invaded underlying bone, muscle, or cartilage)(Wilder et al. 1991). The authors emphasize that *tumor stage* (depth) is much more important than actual tumor size (diameter) in predicting responses to RT.

In summary, RT can be a very effective treatment for BCC and even recurrent BCC, but patient selection is critical as is the cooperation between the Mohs surgeon and the radiation oncologist in making every effort possible in delineating tumor margins in order to inform decisions about field sizes. This may include several scouting biopsies in order to estimate tumor perimeters. RT is particularly well-suited for the elderly or frail patient who is a poor surgical candidate but should be used with great caution in younger patients who may have a poor long-term cosmetic outcome. RT gives the great advantage of avoiding surgical morbidity to critical sites on the nose, eyelids, ears and lips where reconstructive options are limited. Since RT is frequently fractionated, some patients may balk at the need to make several trips to a medical center for treatment, but others jump at the chance to avoid an invasive surgical approach.

3. Squamous cell carcinoma

Squamous cell carcinoma (SCC) is the second most common cutaneous malignancy behind BCC and also occurs in chronically sun-exposed sites, particularly the head and neck. Unlike BCC, SCC does have significant metastatic potential and with the advent of solid organ transplantation, has become a significant risk for morbidity and mortality in that population. SCC is also more frequently associated with underlying states of chronic inflammation such as burn scars, chronic ulcers, chronic thermal injury, and exposure to chemical carcinogens such as arsenic and polycyclic hydrocarbons.

SCC begins with the squamous cells in the epidermis which are situated directly above the basement membrane where the basal cells are located in the stratum basale (See Figure 2). Historically, the layer of the epidermis containing the squamous cells is called the stratum spinosum, which takes its name from the observation under high power light microscopy of desmosomes emanating off of the squamous cells like "spines." Also known as the keratinocytes, the squamous cells undergo an abrupt transition in the stratum granulosum of the epidermis where the intracellular keratin filaments clump, the cell undergoes programmed death, and a keratogenous shell remains with intercellular wax esters acting like "mortar" between the keratinocyte "bricks" forming the watertight barrier of the outer layer of the skin, the stratum corneum. Because of the abundance of keratin protein in squamous cells, in general, SCC is clinically much more scaly and crusted than BCC.

3.1 Incidence and epidemiology

SCC accounts for about 10-20% of all skin cancers (Nguyen, 2005), with an incidence of approximately 700,000 cases per year in the United States with 2500 expected deaths in 2010.(Foundation, 2011). 80% of SCC occurs on sun-exposed sites such as the head and neck, and upper extremities. (Nguyen, 2005). Although BCC outnumbers SCC by 4:1, SCC is

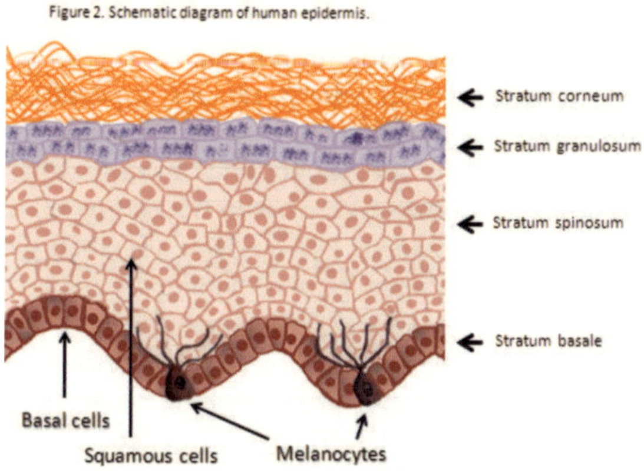

Fig. 2. Schematic diagram of human epidermis

responsible for the vast majority of non-melanoma deaths. In the general population, the rates of SCC have shown a doubling of the incidence over the last 40 years. (Stern, 1978). SCC is becoming a more common problem among solid organ transplant recipients and in patients with lymphoproliferative malignancies, presumably due to decreased immunosurveillance. A renal transplant recipient has an increased risk of developing SCC in the first year of 7%, 45% risk after the first 11 years following transplant, and 70% risk after 20 years. (Bouwes Bavinck et al., 1996). The situation is worse for heart transplant recipients who have a 2-3 times greater risk of SCC than do renal transplant patients.

Risk factors for the development of SCC include fair skinned patients with risk associated with increasing rates of chronic cumulative UV exposure, living in areas of decreasing latitude, increased altitude, prior history of non-melanoma skin cancer, male gender (men are 2-3 times more affected than females), and age > 50. (Nguyen, 2005).

3.2 Etiology

The majority of SCC are caused by mutations induced by ultraviolet light exposure, most notably in the tumor suppressor gene p53. (Sarasin & Giglia-Mari, 2002). A smaller percentage of SCC are not related to UV injury, but rather will occur as a result of sites of chronic skin inflammation which can result from a number of causes such as chronic inflammatory skin diseases, particularly non-healing ulcers. Traumatic insult to the skin such as scar tissue from cuts or thermal or chemical burns has also been associated with SCC. Ionizing radiation as mentioned in the introduction is clearly associated with increased SCC risk as is chronic lymphedema.

Also implicated in the formation of SCC is exposure to chemical carcinogens. Polycyclic aromatic hydrocarbons such as soot, pitch, tar, oil shale, and mineral oil have all been documented associations with SCC. Arsenic has long been appreciated as having a causal link with SCC. (Baca & Dzubow, 1998).

3.3 Clinical presentation

Because squamous cells produce keratin protein which is a key feature in the barrier function of the skin, well-differentiated SCC will be associated with a degree of hyperkeratosis with the clinical correlation of scale (See Figure 3a). As SCC becomes less well-differentiated, they often lose the ability to produce keratin with the clinical correlation of loss of scale and the tumors take on a more erosive appearance not unlike a stewed tomato. The distribution of SCC, as anticipated, favors chronically exposed sites such as the head and neck where 70% of SCC are found and another 15% occur on the upper extremities. (Nguyen, 2005). SCC frequently occurs in the setting of stigmata of chronic sun exposure such as multiple actinic keratosis (AK), which are small patches that are usually red in color and have a slightly roughened surface similar to fine grained sand paper. SCC of the lip has long been recognized as being generally more aggressive, with higher rates of metastases than the other head and neck locations, and is more likely found on the lower lip. (Dinehart & Pollack, 1989). Cigarette and pipe smoking have also been associated with increased rates of lip SCC. (Pentenero et al., 2011).

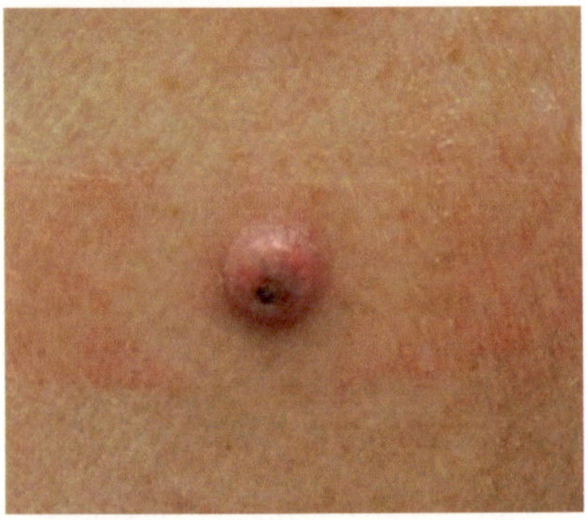

Fig. 3a. Squamous cell carcinoma

Verrucous carcinoma is a variant of SCC that clinically presents as a warty growth that may simulate the surface of a cauliflower. It may occur on the digits, oral mucosa, feet, or the anogenital region where it is known by the eponym "Buschke-Lowenstein tumor." (Schwartz, 1995). Although such tumors may be locally disfiguring, they are less likely to metastasize compared to other SCC subtypes. (Nguyen, 2005).

Bowen's disease is a in situ variant of SCC that presents as an erythematous patch with thin superficial scaling and frequently has notching of the borders which helps to distinguish it from nummular eczema, tinea corporis, or psoriasis. When Bowen's disease occurs on the penis, it has been known in the past as erythroplasia of Querat (Aynaud et al., 1994) although more recently, it has been suggested that the terms "Bowen's disease" and "erythroplasia of Querat" should not be used for lesions in the anogenital region and the preferred terms are vulval intraepithelial neoplasia for females and penile intraepithelial neoplasia for males(Wilkinson, 1992). Although an in situ tumor, Bowen's disease can progress to become an invasive SCC, particularly when it occurs at genital sites.

Although metastases from SCC are statistically much less likely than with melanoma, SCC can metastasize with fatal consequences. Regional metastases have an overall 5-year survival of 25% (Nguyen, 2005) and are a particular problem on the head and neck when SCC invades the perineurium of cranial nerves and gains access to the brain via foramina in the cranium. SCC also spreads via lymphatic channels to involve regional lymph nodes, and from there can gain access to the blood vessels with consequential distant metastases to internal organs. The overall risk for metastases ranges between 2-6% (Nguyen, 2005), but may be associated with factors such as anatomic location, whether the tumor is primary or a recurrence after aggressive therapy, differentiation (well- or poorly-differentiated histologically), the presence of perineural invasion and the size of the diameter of the involved nerve, and the underlying immune status of the patient. Of head and neck sites, the ears and the lips confer higher risk for metastases at 8.8% and 13.7% respectively. (Nguyen, 2005; Ponten, 2003). The disparity in metastatic risk is dramatically increased between *primary* SCC (5.2% risk) to a 30.3% risk for *recurrent* tumors. (Nguyen, 2005). Increasing *size* also confers increasing risk where SCC <2 cm has a metastatic risk of 9.1% compared to 30.3% for tumors > 2 cm. (Nguyen, 2005). Perineural invasion is difficult to quantify but there is some suggestion and it seems logical that increasing diameter of the involved nerve is associated with increased risk. (Jambusaria-Pahlajani, 2010).

3.4 Histology

The histologic appearance of SCC will show keratinocyte hyperplasia and anaplasia, often with areas of dyskeratosis where keratin pearls will appear randomly in islands of squamous cells (See Figure 3b). Poorly-differentiated SCCs will lose the keratin production with a corresponding increase in the anaplastic appearance of the squamous cells. The pathologist may characterize the SCC as "well," "moderately," or "poorly" differentiated but often parses the terminology for intermediate states such as "well to moderately" or "moderate to poorly" differentiated. The poorly-differentiated SCCs warrant a high degree of vigilance since up to 17% of such tumors have been reported to metastasize compared to well-differentiated tumors with an estimated 0.6% risk. (Breuninger et al., 1990).

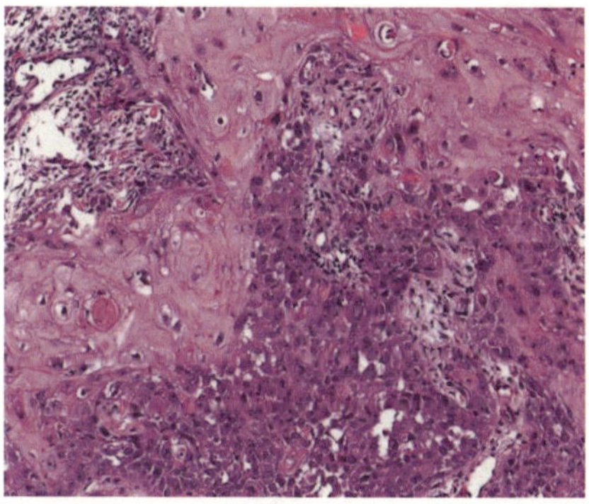

Fig. 3b. Histology of squamous cell carcinoma

3.5 Treatment options

When considering treatment options for SCC, approaches other than RT can be divided into three groups: topical therapies, destructive therapies, and surgical resection with margin control. The selection of a treatment modality will be greatly influenced by a number of variables that contribute to the physician's assessment of the perceived risk of metastases, the cosmetic and functional consequence of the treatment, and the general health status of the patient. RT can be effective as both a primary treatment as well as an adjuvant therapy for high-risk tumors.

When it comes to selecting RT for primary tumors, the first question to address is whether surgical excision with margin control is either feasible or practical since MMS has the highest reported cure rates for SCC compared to all other modalities. However, like all treatments for SCC, MMS cure rates begin to drop with increasing aggressive features such as tumor size, differentiation, discontinuity, perineural invasion, and a history of recurrence(s). Meta-analysis shows that MMS has 5-year recurrence rates for primary cutaneous SCC of 3.1% vs. 7.9% for other treatment options including CE, RT, and surgical excision. (Rowe et al., 1992). When it comes to high-risk tumors, MMS trumps other modalities as well for SCC of the lip, 2.3% (Mohs) *vs.* 10.5% (non-Mohs), on the ear, 5.3% (Mohs) *vs.* 18.7% (non-Mohs), and recurrent SCC 10.0% (Mohs) *vs.* 23.3% (non-Mohs). (Rowe et al., 1992).

In the clinical setting, circumstances arise where either MMS or surgical excision is simply not realistic and RT offers a reasonable alternative. The approximate overall 5-year cure rate for RT for primary SCC is on the order of 90%. (Rowe et al., 1992). There is great hesitancy on the part of many dermatologists to refer patients younger than 60 years of age due to perceptions of long-term cosmetic consequences of RT as previously mentioned, coupled with fears of increased risk of future malignancy in the treatment site. Although RT is associated with a *statistically* increased risk of secondary malignancies, one has to judge the *clinical* significance of the elevated risk before discounting RT in younger patients. Meadows et al. evaluated the risk of secondary neoplasms amongst survivors of childhood cancer from a large, United States, population-based registry. (Meadows et al., 2009). With 30 year follow-up, they demonstrated that the cumulative incidence of non-melanoma skin cancers in children who received RT was 4%, whereas the incidence in non-irradiated patients was 1%. So although the relative risk is high, the absolute excess risk for these non-lethal cancers is arguably low. With regard to other secondary malignancies, the cumulative incidence of secondary cancers over 30 years was 6% in the irradiated children, and 3% in the non-irradiated children.

The discussion of the treatment for SCC in the NCCN website is very candid about diversity of opinions about the role of RT for primary SCC across academic institutions, and more specifically between surgeons and radiation oncologists, stating that the radiation oncologists on the panel wanted to suggest that RT be considered as a first line therapy, whereas the surgeons did not. (NCCNN, 2011). The discussion was based on a large review of published data. (Avila et al., 1977; Collin, 1976; Fischbach et al., 1980; Fitzpatrick, 1995; Johnson et al., 1992; Lovett et al., 1990; Martin et al., 1970; Mazeron et al., 1988; Mendenhall et al., 1987; Petrovich et al., 1987a, 1987b, 1988; Rowe et al., 1989a, 1989b, 1992; Silverman et al; 1991, 1992; Stoll et al, 1964; Traenkle et al., 1962).

3.5.1 RT as primary therapy for SCC

The National Comprehensive Cancer Network published guidelines for RT for primary SCC are reprinted with permission in Table 1. Within these guidelines are specific dosing recommendations for primary SCC and BCC that are stratified by tumors < 2 cm and tumors ≥ 2 cm. In general, control rates for SCC primarily treated with RT are excellent, but slightly less effective than for BCC. In a report by (Solon *et al.*, 1997) the four-year control rate was 96% (426/444) for BCC and 92% (144/156) for SCC with RT as primary therapy. As is the case for BCC, size influences local control rates as demonstrated by (Lovett *et al.*, 1990): control rates for tumors less than 1 cm were 97% (86/89) for BCC *vs* 91% (21/23) for SCC; tumors 1-5 cm in size had control rates of 87% (116/133) for BCC *vs* 76% (39/51) for SCC; and for lesions larger than 5 cm local control was 87% (13/15) for BCC *vs* 56% (9/16) for SCC.

Anatomic site considerations are critical in selecting therapy for primary SCC. MMS has excellent cure rates but is very morbid at specific anatomic sites with the consequence of posing difficult and complicated reconstructive challenges. For instance, SCC of the eyelids can often be treated with RT with high cure rates thus avoiding difficult staged surgical repairs. Also, tumors around the lacrimal system treated with RT can preserve the drainage system (JS, 2005) which is often compromised by reconstructive surgery. It was previously thought that lesions over cartilage, such as the ear, should not be treated with RT due to the risk of chondronecrosis, but it is now known that such sites can be safely treated with fractionated RT. (JS, 2005).

Verrucous carcinoma is particularly radiosensitive and RT can reasonably be considered as a first line therapy. Anogenital SCC is not uncommon, and surgical approaches can be very morbid leading to permanent loss of function and disfigurement. For anal SCC, the standard of care for most T1 and all T2 lesions and above is definitive combined chemoradiotherapy, with surgery reserved only for recurrences. (Bartelink et al., 1997; Flam et al., 1996; UKCCCR, 1996). This therapy results in equivalent cure rates to radical surgery, but offers functional organ preservation.

The medical community has embraced chemoradiotherapy as the gold-standard curative-intent treatment for most advanced squamous cell carcinomas of the head and neck, esophagus, uterine-cervix, and anus, with the goal of functional organ preservation. However, for the more rare entity of penile cancer, surgery remains the most commonly performed treatment. Surgical approaches to SCC of the penis frequently result in distortions of the urethra with consequential disruptions of urine flow direction. Partial or complete penectomies are extremely morbid procedures often with devastating emotional and psychological consequences for the patient. (Sarin et al., 1997). The authors have a series of successfully treated SCC of the penis. As one example, a 49 year old man presented with verrucous carcinoma staged as T1a,N2,M0 (stage IIIB). He presented with conventional condylomas on the base of the penis with a cauliflower morphology, while on the penile shaft there were verrucous plaques with hyperkeratosis and fissuring (See Figure 4). Penectomy and radical inguinal surgery had been recommended, but the patient elected to undergo an organ-sparing approach. He was treated with a regimen that would be considered "standard" for an anal SCC: 5580 cGy to the penis, 4500 cGy to the inguinal and pelvic nodes using an intensity-modulated radiotherapy technique. RT was given with concurrent weekly cisplatinum at 40 mg/m^2.

RT can also be used for SCC that involves the glans penis and encroaches on or involves the urethra where surgery can cause permanent disfiguration. Because penile cancer is a rare entity, no prospectively performed trials with RT exist. However, numerous retrospective studies using interstitial brachytherapy, external beam radiotherapy, or combinations have been reported and are summarized in Table 2. (Sarin et al., 1997) Of note, these studies did not combine chemotherapy with RT and demonstrate local control rates in the 57% to 85% range with RT alone with excellent functional and cosmetic results. It is important to note that the local failures were successfully salvaged with RT, leading to ultimate control rates of 90% to 97%. The most common functional deficit reported after RT is urethral stricture, reported in 9-41% of subjects. (Sarin, 2002).

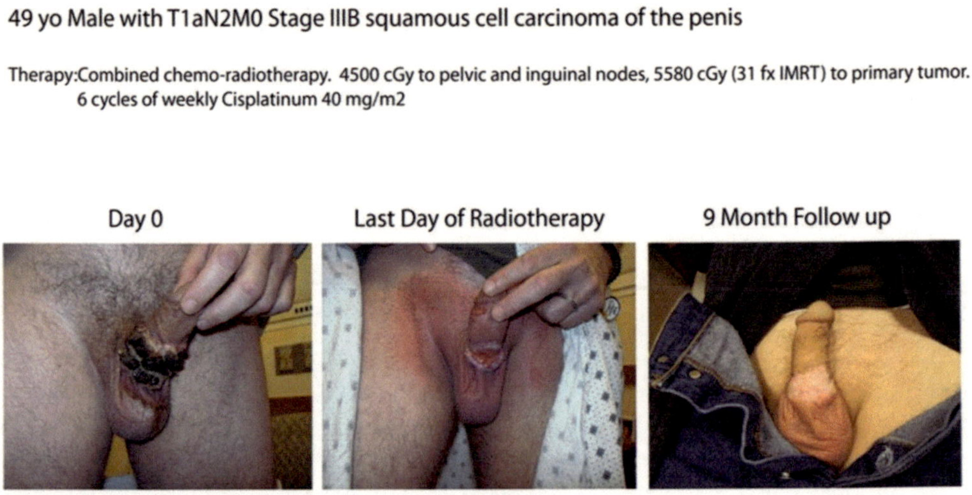

Fig. 4. Verrucous carcinoma of the penis, pre- and post- treatment

Study	Treatment	Local Control		Penectomy for necrosis	Urethral stricture
		Initial	Post salvage		
Rozan *et al* French multi-center 12 years	Implant 63 Gy (185 men) Others (75 men)	85%	94%	7%	30%
Delannes *et al* Toulouse, France 7 years	Implants- 60 Gy (51 patients)	82%	94%	16%	41%
Ravi *et al* Adyar 12 years	EBRT - 50 to 60 Gy (128 patients) Brachy – 60 to 70 Gy (28 patients)	65%	97%	6%	24%
Sarin *et al* Royal Marsden, UK. 5 years	EBRT – 60 Gy (56 patients) Implants – 60 Gy (13 patients)	57%	90%	3%	14%
Chaudhary *et al* TMH 2 years	Ir-192 Implant – 50 Gy (23 patients)	78%	96%	None	9%

Table 2. Radiotherapy for penile cancer (Data compiled in: Treatment Options for Urological Cancer, 2002)

3.5.2 RT as adjuvant therapy for SCC

For patients that have had the SCC removed surgically but are thought to be at high risk for local recurrence due to perineural invasion or aggressive histologic patterns, post-operative adjuvant RT should be considered. This is particularly true where perineural invasion is present despite negative margins on histology, as the patient is known to have a substantial increased risk of local recurrence (Nyguyen, 2005). (See Table 1). RT for SCC showed 5-year actuarial local control of 0.54 for primary SCC, 0.0 for recurrent SCC, and 0.42 for nodal metastases. (Shim & Wilder, 1991).

3.5.3 RT for regional lymph node involvement

Regional disease includes SCC that has spread to regional lymph node basins and is divided into two treatment groups: those patients who have undergone completion lymph node dissection and are deemed at high risk for local recurrence, and those patients who did not. The NCCN guidelines committee suggests that all doses be delivered at 2 Gy per fraction using the shrinking field technique. This means the highest-risk areas should be prescribed the dose specified in Table 3, but lower doses may be prescribed for elective nodal region at less of a risk for relapse. For tumors after lymph node dissection, patients are stratified to those having extracapsular extension (ECE) and those who do not, and dosages are stratified by site: head and neck vs. axilla and groin (See Table 3).

Regional Disease	Dose (Gy)	Duration
After lymph node dissection		
Head and neck; with ECE[1]:	60-66	6- 6.6 weeks
Head and neck; without ECE:	56	5.6 weeks
Axilla, groin; with ECE:	60	6 weeks
Axilla, groin; without ECE:	54	5.4 weeks
No lymph node dissection		
Clinically (-) but at risk for subclinical disease	50	5 weeks
Clinically evident adenopathy: head and neck:	66 - 70	6.6 – 7 weeks
Clinically evident adenopathy: axilla, groin:	66	6.6 weeks
[1]ECE = extracapsular extension		

Table 3. Dose Recommendations for Squamous Cell Carcinoma: Regional Disease (2 Gy per fraction with shrinking field technique). Used by permission of the National Comprehensive Cancer Network

Although the above nodal dose guidelines represent level II consensus guidelines from the NCCN skin panel, it is noteworthy that the doses used to control nodal metastases in anal squamous cell cancers are lower (usually in the 45Gy to 56Gy range) when given in the context of chemotherapy. This could be due to the potent radiosensitizing effect of chemotherapy or an intrinsic radiosensitivity of anal squamous cell primaries relative to other locations.

4. Melanoma

Melanoma is the third most common skin cancer but is responsible for the majority of skin cancer deaths in the United States. The majority of melanomas present either as an unusual appearing nevus relative to the other nevi on the body, or occur as a change in an existing nevus. Melanoma begins with the melanocyte, the pigment-producing cell of the epidermis which is derived from the neural crest during embryology (See Figure 2). The virulence of melanoma may be in part related to its origin, where it appears on the stage very early during embryogenesis and may have the cassette of genetic capabilities for migration not as common in basal cells and squamous cells that appear much later as the fetus develops. For whatever reason, melanoma has a high propensity for metastasis which is directly related to the depth of invasion measured in millimeters from the granular layer to the deepest part of the tumor by a calibrated ocular micrometer, and is referred to as the Breslow depth.(Breslow, 1970). For example, a melanoma with a Breslow depth less than 1 millimeter without other adverse features predicts a 5-year survival of about 95%, whereas a melanoma 4 millimeters in depth without other adverse features has an associated survival of about 65%. If ulceration is present in the primary tumor, survival for a 4 millimeter melanoma drops to about 50% at 5 years. (Balch et al., 2009).

4.1 Epidemiology

It is estimated that in 2010 there were 68,130 new cases of invasive melanoma in the U.S., with 8700 deaths (Jemal et al., 2010). The incidence of melanoma continues to rise as demonstrated by the Surveillance, Epidemiology and End Results (SEER) data, which has shown an increase in incidence in the United States from 6.7 cases per 100,000 in 1973 to 20.4 cases per 100,000 in 1996 in males, and from 5.9 cases per 100,000 to 14.3 cases per 100,000 cases in females. (Berwick, 2003; Surveillance, 2002). Although melanoma can occur in black patients, it is much more common in Caucasian patients with the highest rates of incidence in Australia (40.5/100,000), New Zealand (36.7/100,000), Norway (14.1/100,000) and the United State (13.3/100,000) for males. (Berwick, 2003; Ferlay et al., 2001).

4.2 Etiology

Melanoma appears to be a consequence of both genetic factors and environmental factors, in particular exposure to ultraviolet light. Well-known risk factors for melanoma include a personal or family history of melanoma, UV exposure, childhood sunburns, a high number of nevi, fair skin, blue eyes, easy freckling, higher socioeconomic status, and living at equatorial latitudes, to name a few. (Nestle, 2003). The majority of melanoma cases are spontaneous, but it is estimated that 6% to 14% of melanomas occur in family groups. (Ang et al., 1998). The gene most frequently involved in melanoma resides on chromosome 9p21 and is known as CDKN2A, whose gene product (p16) acts a tumor suppressor. (Pollock et al., 2003).

4.3 Clinical presentation

Melanoma can be roughly divided into various categories that have differing biological and genetic characteristics with varying degrees of causation by sun-exposure. For example, most melanomas appear on intermittently-exposed sites such as the torso. Other subtypes occur in chronically-exposed sites such as the head and neck, non-sun-exposed sites such as the palms, soles, genital/mucosal surfaces and ocular melanoma form their own entity. (Whiteman, et al., 2011).

In general, melanoma will present as a pigmented lesion that visibly changes over time, usually in terms of size and/or color (See Figure 5a). A very useful feature termed the "ugly duckling" sign (Grob & Bonerandi, 1998) connotes the disparity in the appearance of the melanoma compared to surrounding nevi. We have found the "ugly duckling" sign to be of particular effectiveness in identifying melanomas in our high-risk patient group in our mole-mapping clinic at our institution, where total body photography is used to track nevi longitudinally for any signs of change that might herald malignant transformation. (Goodson et al., 2010).

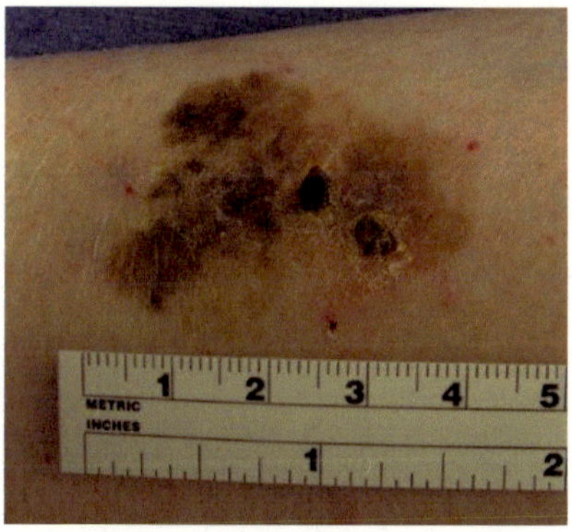

Fig. 5a. Superficial spreading melanoma with border irregularity and color variation.

4.4 Histology

Entire book chapters and even books have been dedicated to the problem of reproducibly making the distinction between melanoma and severely atypical nevi that are not melanoma under the light microscope. For the purposes of this chapter, the hallmark histologic features of melanoma include the proliferation of atypical melanocytes with *lack of maturation* of the melanocytes with depth, (See Figure 5b.) The presence of ulceration, mitotic figures, and angiolymphatic invasion are negative prognostic features. (Nestle, 2003).

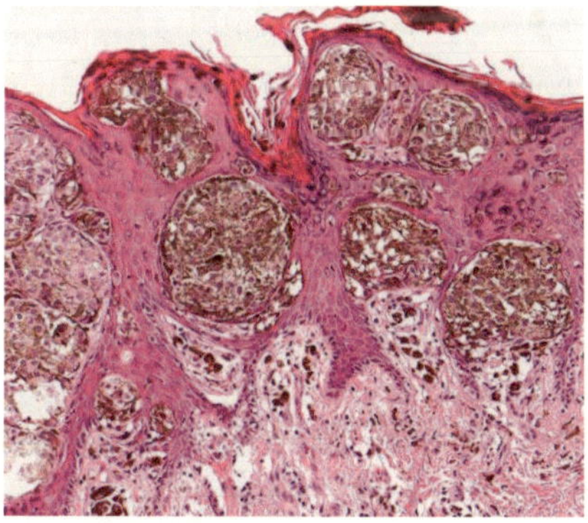

Fig. 5b. Histology of superficial spreading melanoma

4.5 Treatment

Melanoma is, by and large, considered a surgical disease. The role of RT for melanoma can be summarized as the occasional use as primary treatment for lentigo maligna (LM) or unresectable primary tumors and in the adjuvant setting to improve local control or palliation.

4.5.1 Radiation as first-line therapy for melanoma

Lentigo maligna (LM) is a subtype of melanoma that occurs in chronically sun-exposed sites, usually on the head and neck, and is subdivided into lentigo maligna (LM), which is an in situ tumor, and lentigo maligna melanoma (LMM) which connotes histologic invasion. When invasion is present, the recommended treatment is surgical resection. (NCPGION, 2011). However, in situ tumors pose no immediate threat of metastases to the patient and are treated to prevent the future potential for invasion. The actual risk of invasion is unknown but has been predicted to occur at a rate of 2.2%at 45 years and 4.7% at 65 years. (Weinstock & Sober, 1987). In such cases, RT can be considered as a primary therapy for LM, because surgical excisions commonly result in very large defects with significant associated morbidity. (Johnson et al., 1997). When RT is used as primary therapy for LM (Grenz rays or soft X-rays using the Miesher technique), the cure rates range from 86 to 95%. (Farshad et al., 2002; Schmid-Wendtner et al., 2000; Tsang et al., 1994). The largest published study to date included 150 patients with both LM and LMM and a mean follow-up of 8 years with a complete response rate of 93%. (Farshad et al., 2002). For LM, definitive RT achieved a crude local control rate of 95%, with a mean time to recurrence of 45.6 months. Four of the five recurrences were at the edge of the radiation field, so the authors recommended targeting a margin of at least 10 mm around the visible lesion. (Farshad et al., 2002).

4.5.2 RT in the adjuvant setting for local or regional disease

Adjuvant RT is recommended for consideration in the NCCN guidelines under several circumstances: primary disease at risk for local recurrence, regional disease, and metastatic disease. (NCPGiON, 2011). As for primary disease, adjuvant radiation is often considered for the desmoplastic subtype of melanoma (DNM), a spindle cell variant of melanoma that is frequently associated with perineural invasion and is prone to local recurrences. (Vongtama et al., 2003). RT can be considered post-operatively when extensive neurotropism is present, since this will put the patient at an increased risk of a local recurrence (See Table 4). (Chen et al., 2006). RT was also recommended for patients with inadequate margins, which occurred predominately in the head and neck region in this series.

Regional disease includes large primary tumors > 3 centimeters in diameter as well as those cases where there are in transit or satellite metastases around the primary tumor where it is not practical to resect them with the primary tumor (See Table 4). However, even with radiation, patients with satellitosis are at risk for a recurrence in the radiated field. (Chang et al., 2006). Dosing must be tailored to specific sites and sizes of radiation fields. The head and neck area can be treated with orthovoltage or low-energy electrons with bolus in doses ranging from 24 to 36 Gy in four to six fractions. The larger the area, > 50 cm^2, the greater the need for more fractions, especially on the distal lower extremity. It has been recommended that a large surface on the trunk can tolerate 36 Gy in 12 fractions, however, a similar sized field on the lower limbs should be treated with 50 Gy in 25 fractions over 5 weeks. (Ainslie & McKay, 2003).

PRIMARY DISEASE
• Adjuvant treatment for selected patients with desmoplastic melanoma with extensive neurotropism
REGIONAL DISEASE
• Extracapsular extension
• ≥4 involved nodes
• Size ≥ 3 cm
• Cervical **> Axillary > Inguinal Location
• Recurrent disease after prior complete nodal dissection
**For cervical nodes, consider RT if >2 nodes involved or nodes > 2cm in size
METASTATIC DISEASE
• Brain metastases (see NCCN Central Nervous System Cancers Guidelines)
- Definitive or palliative stereotactic radiosurgery and/or whole brain radiation
- Adjuvant radiation following resection of brain metastases
• Other symptomatic or potentially symptomatic soft tissue and/or bone metastases

Table 4. Principles of Radiation Therapy for Melanoma (Reprinted by permission of the National Comprehensive Cancer Network)

For nodal melanoma that has been completely resected, adjuvant RT is considered where a high risk of local recurrence is anticipated (see Table 4). (NCPGiON, 2011). There are no published multi-center trials for RT in this setting, and published data are from single institutions. Most studies demonstrate improved regional control compared to surgery alone. (Ang et al., 1994; Burmeister et al., 1995; Corry et al., 1999; O'Brien et al., 1997; Stevens et al., 2000; Strom & Ross, 1995). The largest retrospective review investigating the role of RT included 615 patients meeting high-risk criteria for relapse. (Agrawal et al., 2009). At a median follow-up of 5 years, regional recurrence occurred in only 10.2% of the radiated patients versus 40.6% of the non-radiated patients. Of note, RT significantly increased treatment-related morbidity (5-year rate of 20% versus 13%, p=0.004), particularly with regards to *lymphedema*. The consensus opinion appears to be that overall survival is not greatly improved with adjuvant RT because four randomized trials involving complete surgical removal of subclinical lymph node involvement with elective lymph node dissections did not significantly improve survival. (Ainslie & McKay, 2003; Balch et al., 2000; Cascinelli et al., 1998; Sim et al., 1986; Veronesi et al., 1982). Retrospective data has not shown one regimen is more efficacious or safe than another, but delivering radiation over a longer time period might decrease the risk of side effects. (Chang et al., 2006). Several published series warn about the concomitant use of RT and interferon-alpha where increased regional toxicity has been observed. (Buckner et al., 2001; Damle et al., 1999; Hazard et al., 2002; Holsti et al., 1987; Stock et al. 1997).

When local, regional, or metastatic melanoma is not completely resected or is deemed unresectable, palliative RT should be considered. Although randomized trials have not shown any superiority of hypofractionated regimens over standard fractionation, a biologically effective dose >39 Gy_{10} (i.e. a regimen stronger than 30 Gy in 10 fractions) is recommended (See Table 4). (Ainslie & McKay, 2003; Olivier et al., 2007; Overgaard et al. 1985; Sause et al., 1991). Radiation effectively improves patient quality of life, as significant symptom relief is achieved for 68% to 84% of patients. (Konefal et al., 1988; Olivier et al., 2007). Reports for clinical complete response (CR) rates range from 17 to 69%, with 49 to 97% achieving either a partial response (PR) or CR. (Overgaard et al., 1985; Sause et al., 1991; Seegenschmiedt et al., 1999).

4.5.3 RT in the setting of metastatic disease

As for metastatic disease, RT is recommended for brain metastases as either definitive stereotactic radiosurgery (SRS) or whole-brain radiotherapy (WBRT) for palliation. RT is also considered post-operatively in cases where a brain metastasis has either been resected or treated with SRS. (Fogarty et al., 2011). For patients with multiple metastatic melanoma lesions in the brain, WBRT is the most frequently utilized form of palliation. In a summary of a large number of series, WBRT provided symptomatic improvement in over two-thirds of patients, however, the medial survival is only 3.5 months. (Morton et al., 2003). A cautionary note is important to consider in the rare cases where patients have prolonged survival and can experience brain atrophy and consequential dementia. Striking functional decline was documented in 30% of women undergoing prophylactic brain irradiation in metastatic breast cancer. (Huang et al., 2009). In another study, brain atrophy developed in 30% of 92 patients undergoing WBRT, although the authors state that this was not necessarily accompanied by decreases in mini-mental status examination scores. (Shibamoto et al., 2008). It should be noted that over 90% of patients with brain metastases have been found to have some level of neurocognitive dysfunction before radiation. (Meyers et al., 2004). Although long-term survival is unlikely following WBRT for melanoma, it is advisable to inform patients and family members of the risk for post-treatment dementia.

For solitary metastatic lesions in the brain, surgical resection (potentially followed by radiation) or SRS should be considered. In two radomized trials for solitary melanoma metastases in the brain, surgical resection followed by WBRT led to better palliation and prolonged disease-free survival than surgery alone. However, overall survival was only increased in one of the two studies (18-month vs. 6-month medial survival; P = 0.002). (Hagen et al., 1990; Skibber et al., 1996).

SRS is becoming an increasingly-used alternative to craniotomy/surgical resection in patients with a small number (generally ≤ 4) of brain metastases, each measuring ≤ 4 cm. (Breneman et al., 1997; Chen et al., 2000; Samlowski et al., 2007). SRS may be a way to quickly address brain metastases to allow for intracranial control, perhaps buying time until more systemic options become available. In general, SRS alone or followed by WBRT yield results similar to surgical resection followed by WBRT. Again, the addition of WBRT after surgery or SRS improves disease-free survival but not necessarily overall survival. (Bafaloukos & Gogas, 2004).

5. Merkel cell carcinoma

Also known as primary neuroendocrine carcinoma of the skin, Merkel cell carcinoma (MCC) has the dubious distinction of being the most deadly of the skin cancers, including melanoma. Despite its virulence, MCC has not achieved the notoriety of melanoma due to its rarity.

5.1 Epidemiology

In the United States, MCC accounts for much less than 1% of all cutaneous malignancies and is estimated to occur at 0.24 tumors per 100,000 person-years compared to 17.0 tumors per 100,000 person-years for melanoma. (Agelli & Clegg, 2003). Overall survival rates are estimated to be 87% (1 year) and 49% (5 years) and a cause-specific survival rate of 62% at five years. (Lewis et al., 2006).

5.2 Etiology

The Merkel cell polyoma virus is detected in 43% to 100% of MCC tissue (Feng et al., 2008; Rollison et al., 2010), but its exact role in the pathogenesis of MCC remains to be sorted out. (Group, 2009). Because MCC tends to occur in a distribution similar to BCC and SCC, ultraviolet light has been suspected as playing a causative role, perhaps in conjuction with polyoma virus infection, but the exact etiology remains elusive.

5.3 Clinical presentation

MCC does not have a distinct and characteristic clinical presentation and is frequently mistaken for both benign and malignant neoplasms, but does have a predilection for sun-exposed sites on the head and neck, sun-exposed extremities and is most commonly seen in elderly Caucasians (See Figure 6a). Clinical signs of regional lymph node involvement is seen in 11-15%, while 50-70% of patients will eventually develop regional nodal metastases. (Hitchcock et al., 1988). Local recurrence rates are very high: 25-30%. Up to 50% of patients develop distant metastases with the most common sites being the liver, bone, brain, lung, skin, and distant lymph nodes. (Akhtar et al., 2000; Gillenwater et al., 2001; Medina-Franco et al., 2001; Taylor & Heilman, 2005).

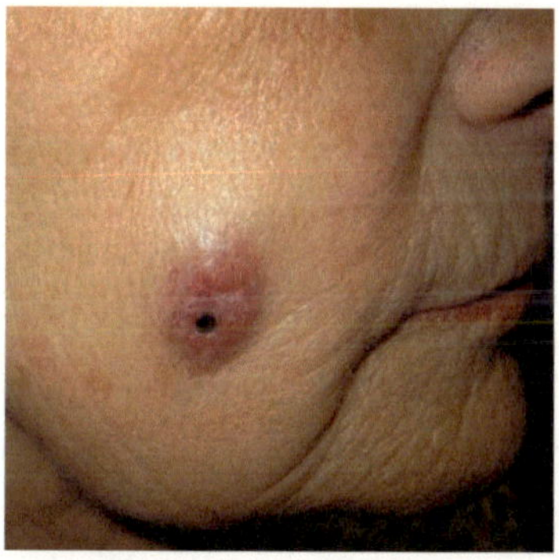

Fig. 6a. Merkel cell carcinoma, with bunch biopsy site in center of tumor

The most important prognostic feature of MCC is the stage followed by tumor size. (Akhtar et al., 2000; Allen et al., 1999; Haag et al., 1995; Kokoska et al., 1997; Medina-Franco et al., 2001; Ott et al., 1999; Ratner et al., 1993; Skelton et al., 1997). The American Joint Committee on Cancer (AJCC) staging system divides the presentation of MCC into local, regional, and disseminated disease. (Edge et al., 2009). 60%-70% of patients present with local disease only (Stage I) and have with an estimated 3-year survival of 55-73%, while up to 40% of patients present with regional node involvement (Stage II) with a 33% 3-year survival estimate. (Yiengpruksawan et al., 1991).

5.4 Histology

The histopathology of MCC may mimic other types of malignancies, particularly basal cell carcinoma. Under light microscopy, the appearance is one of monotonous aggregates of small basophilic cells that are closely packed with a large nuclear:cytoplasmic ratio (See Figure 6b). A high mitotic rate and vesicular nuclei are common. The neoplasm is mostly dermal and is usually separated from the epidermis by a thin layer of uninvolved dermis called the "Grenz zone." Histologically, the closest simulant to MCC is metastatic small cell carcinoma of the lung. The distinction between those two diagnoses can usually be made with immunohistochemistry, since MCC reacts with the immunostain for cytokeratin 20 (CK20) in 89-100% of tumors but does not react with thyroid transcription factor 1(TTF-1) which is expressed in small cell lung cancer but not MCC. (Cheuk et al., 2001; Hanly et al., 2000; Scott & Helm, 1999).

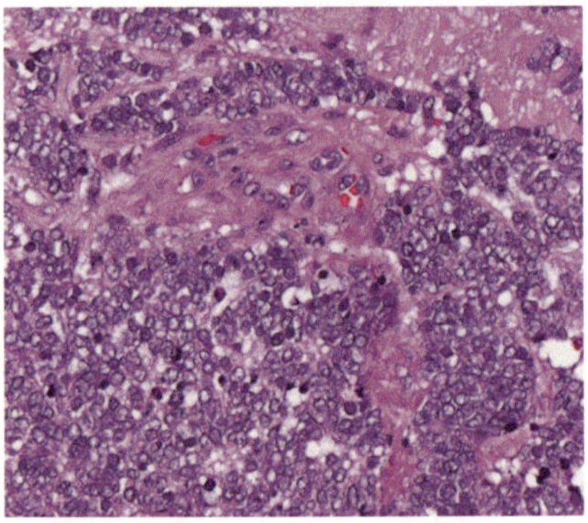

Fig. 6b. Histology of Merkel cell carcinoma

5.5 Treatment

Like melanoma, MCC is a disease best treated surgically. The National Comprehensive Cancer Network™ (NCCN) panel has acknowledged substantial variability in treatment recommendations among physicians representing multiple institutions across the United States, therefore, the treatment recommendations by the NCCN are broad to include disparate attitudes between institutions hampered by the lack of randomized controlled studies to draw upon. (NCPGiON, 2011). In general, a wide local excision is recommended for the primary site with the option of adding a sentinel lymph node biopsy. Postoperative radiation to the primary site is routinely performed at some institutions, whereas others add RT only if the surgical margins are close or positive. The same philosophy applies to regional lymph node basins where some institutions will routinely irradiate regional lymph node basins regardless of sentinel lymph node status, while others reserve postoperative RT for basins with positive nodes, and in some cases, only if the nodes demonstrated extracapsular spread.

The difficulty in making firm treatment recommendations for RT sits firmly in the paucity of prospective randomized data from which to make informed clinical decisions. Despite the fact that the benefits of RT in MCC are mixed in the literature, there is published evidence that postoperative radiation can help to minimize locoregional recurrences but not necessarily improve overall survival. A meta-analysis compared surgery alone compared to surgery followed by RT to the surgical site showing a lower risk of local and regional recurrences but a difference in overall survival did not reach statistical significance. (Lewis et al., 2006). In a series of 82 patients, both relapse-free survival and overall survival were improved with RT to the primary site or the regional lymph node basins (Jabbour et al., 2007), therefore, the NCCN panel includes RT as an option for all stages of MCC. Further support for adjuvant RT comes from a SEER analysis of 1,665 patients, which showed RT was associated with improved survival for all size tumors, especially for those >2 cm. (Mojica et al., 2007).

There are occasions where RT is an acceptable first line therapy for primary tumors, particularly when the tumors are large, have satellite metastases, deemed unresectable, or the patient is a poor surgical candidate. Because MCC tends to occur in an elderly patient cohort, surgical morbidity is often a significant consideration when designing a treatment strategy. In a retrospective series of 43 patients, RT was used as monotherapy and had a reported in-field tumor control rate of 75%. (Veness et al., 2010). Additionally, radiation may be useful for palliation when surgical resection is not feasible for either primary or secondary lesions.

Primary Site	Dose (Gy)
• Negative resection margins	50 - 56
• Microscopic (+) resection margins	56 - 60
• Gross (+) resection margins or unresectable	60- 66
Nodal Bed	
• **No SLNB or LN Dissection**	
➢ Clinically (-) but at risk for subclinical disease	46 - 50
➢ Clinically evident lymphadenopathy	60 - 66
• **After SLNB without LN dissection**	
➢ Negative SLNB: axilla or groin	RT not indicated
➢ Negative SLNB: head and neck, it at risk for false negative biopsy	46 - 50
➢ Microscopic N+ on SLNB: axilla or groin	50 - 54
➢ Microscopic N+ on SLNB: head and neck	50 -60
• **After LN Dissection**	
➢ Lymph node dissection: axilla or groin	50 - 54
➢ Lymph node dissection: head and neck	50 - 60

Table 5. Dose Recommendations for Merkel Cell Carcinoma (2 Gy per fraction with shrinking field technique)

Table 5 with recommended dosages for RT for MCC is reprinted by the permission of the National Comprehensive Cancer Network™ with the following qualifying comments. When MCC on the extremities and torso has been treated with wide local excision (WLE) with negative margins and negative sentinel lymph node biopsy (SLNB), RT is usually reserved for the primary site only (but not given at all institutions). If a SLNB is not performed, RT to the nodal basins can be considered. For head and neck MCC, the risk of falsely negative SLNB is higher and two scenarios may exist: first, if the SLNB is negative, consideration can be given to irradiate the primary site +/- the nodal beds or observe; second, if the WLE is performed without SLNB, consideration should be given to irradiate the primary site +/- nodal beds with the in-transit lymphatics if feasible.

As for the authors, the decisions made regarding the use of RT as either a primary therapy or adjunctive therapy for resected MCC occur after presentation of each individual case in a multidisciplinary tumor conference setting. The multidisciplinary conference has the advantage of including the input of the radiation oncologists, pathologists, surgeons, medical oncologists, and dermatologists. The details of each specific case with regards to staging, tumor histology characteristics, general state of health of the patient, etc. are all factored into tailoring a treatment algorithm appropriate for a given patient and clinical scenario.

6. Cutaneous lymphoma

The cutaneous lymphomas are a subtype of extra-nodal non-Hodgkins lymphomas that are generally divided into three categories: cutaneous T-cell lymphomas, cutaneous B-cell lymphomas, and the natural killer cell (NK cell) lymphomas. As a group, the cutaneous lymphomas are considered as skin-associated lymphoid tissue (SALT) lymphomas. Although there are histologic correlates with the systemic lymphomas, the distinct clinical characteristics and clinical courses of the skin lymphomas requires that the clinician consider them as distinct entities when considering treatment options.

6.1 Cutaneous T-cell lymphoma

When most clinicians hear the term "cutaneous T-cell lymphoma" (CTCL), they are likely to bring images to mind of the two most well-known variants: mycosis fungoides (MF) and Sézary syndrome (SS). There are, in fact, a fascinating myriad of CTCL subtypes including MF, SS, anaplastic large cell lymphoma, subcutaneous panniculitis-like CTCL, granulomatous slack skin, and peripheral T-cell lymphoma to name a few of the most frequently-encountered variants. Both focal electron beam therapy and total body electron beam therapy can greatly reduce morbidity and may provide complete cures in many cases.

6.1.1 History

The term "mycosis fungoides" is an unfortunate appellation that has been a source of confusion since the term was coined by the French physician Alibert, who first described the condition in 1806 in a single patient with tumor stage MF presenting with large fleshy protuberances on the head and neck that indeed resembled mushrooms, hence the name, *"mycosis fungoides."* (JLM, 1806). The report occurred long before fungus was known to cause disease in the skin and was rather a descriptive term that continues to confuse both patients and physicians alike, who naturally associate the diagnosis with an underlying fungal infection. MF is an extra-nodal cutaneous lymphoma predominantly comprised of CD4+ T-cells.

The second most well-known variant is Sézary syndrome (SS), which can be thought of as the leukemic variant of MF where there are a high number of circulating enlarged malignant lymphocytes called "Sézary cells," as described by Sézary in 1936 in a patient with what came to be known as the classic triad of erythroderma (globally inflamed red skin), lymphadenopathy, and more than 10% of the circulating white blood cells have a Sézary cell morphology under light microscopy. (Sezary, 1938). The principles of radiotherapy will be similar for MF, SS, and the various subgroups, therefore, MF and SS will be used as representative diseases for modeling therapy in this chapter.

6.1.2 Epidemiology

In the United States, CTCL is considered a rare disorder. A review data from 13 population-based cancer registries of the Surveillance, Epidemiology, and End Results (SEER) Program of the National Cancer Institute was reviewed from 1973 to 2002 and an incidence of 6.4 cases per million was observed. (Criscione & Weinstock, 2007). An increasing incidence was observed over that observation period but the reason for the increased incidence is unknown. CTCL is most certainly underreported. The majority of cases of MF are low-grade and are treated locally by dermatologists and not referred to cancer treatment centers and consequently, not entered into tumor registries. Also, many cases of early stage MF are not actually diagnosed accurately since MF can easily simulate either psoriasis or eczema and confirming biopsies are never performed, so the diagnosis is never entered into tumor registries.

6.1.3 Clinical presentation

MF has a distinct clinical presentation that classically begins as erythematous patches that can eventually progress to thicker plaques, and then to raised tumors on the skin (See Figure 7a). This progressive nature of the skin involvement is captured in the staging schema with T1 being patches and plaques covering less than 10% of the total skin surface, T2 for > 10% of the surface, T3 for tumors, and T4 for erythroderma or red inflamed skin globally distributed. The distribution of MF lesions is, in general, very reproducible with the patches predominantly located in sun-protected areas in a "bathing-suit" distribution. This implies involvement of the axilla, mid-axillary line, torso, buttocks, and groins. The fact that ultraviolet light is frequently very effective in treating MF may explain why the disease is much less frequently observed in sun-exposed sites. For patch-stage patients, the lesions are curiously positioned in the middle ground between psoriasis plaques and patches of atopic dermatitis (eczema), where the former are very well-demarcated while the later have much less defined borders.

SS presents with an entirely different clinical picture where the patients present with widespread erythema and scaling, and frequently have difficulty maintaining a comfortable body temperature due to the massive loss of heat through the dilated capillaries diffusely in the skin. Patients with SS frequently have severe pruritus accompanied by fissures and erosions that often get infected, and indeed the main cause of death for both MF and SS is sepsis from a skin source. SS patients classically present with palpable lymphadenopathy.

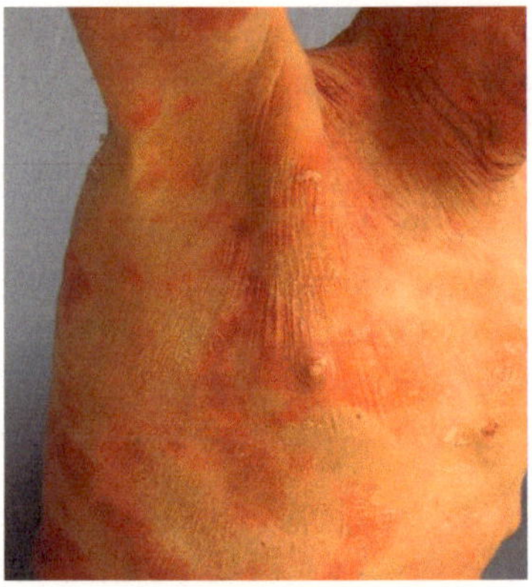

Fig. 7a. Cutaneous T-cell lymphoma (mycosis fungoides)

6.1.4 Histology

The biopsy of classic early stage MF has the distinct feature of swarms of atypical lymphocytes in the upper dermis that home to the epidermis like moths to a street lamp (epidermotropism). The epidermis is peppered with atypical lymphocytes occurring singly and in clusters of T-cell aggregates eponymously called "Pautrier microabscess" (See Figure 7b). (LM, 1937). Problematically, MF is a slowly evolving disease and it may take many years before a skin biopsy displays distinctive diagnostic features. In such ambiguous cases, skin specimens are submitted for clonality testing by looking for overexpression of a singular T-cell receptor on the surface of the CD4+ cells. As the disease progresses from patches and plaques to tumors, the epidermotropism is lost and densely packed aggregates of atypical lymphocytes are seen occupying the superficial and deep dermis without epidermal involvement.

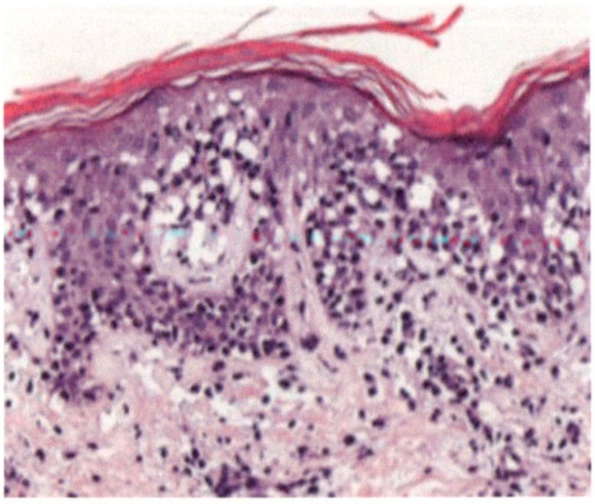

Fig. 7b. Histology of cutaneous T-cell lymphoma, mycosis fungoides subtype

The histology of SS is even more problematic since the classic features of MF are only seen about half of the time, whereas the remaining cases may be impossible for the dermatopathologist to distinguish from generic "dermatitis" where the influx of lymphocytes is so extreme that the distinguishing features of the Pautrier microabscesses are lost. In such cases, flow cytometry of the peripheral blood can be very helpful and cell sorting easily detects an aberrant overexpression of a T-cell clone.

6.1.5 Staging

In general terms, T1 disease includes patients with less than 10% of the skin involved and the five-year survival is not statistically different than that for the control population. (Kim et al., 1996). Patients with T2 disease (> 10% of the skin surface involved) do have an increased mortality with 15% eventually dying from their disease. (Kim et al., 1999). Approximately half of all patients with T3 disease (skin tumors) or T4 disease (erythroderma) die of their disease. (Kim et al., 2003).

6.1.6 Treatment

The majority of patients presenting with MF are treated with skin-based therapies, most frequently ultraviolet light in the form of narrow band UVB or PUVA which is UVA combined with the orally ingested photosensitizer "psoralens" (psoralens + UVA = PUVA). Topical agents include topical chemotherapeutic agents such as nitrogen mustard and carmustine, and a topical retinoid, bexarotene. When these skin-based topical therapies are ineffective, systemic therapies that can be considered include interferon –alpha 2b given as subcutaneous injections (Olsen & Bunn, 1995) and is often combined with a skin-based therapy such as ultraviolet light. Another alternative is oral bexarotene (Olsen & Bunn, 1995), which can also be combined with light therapy. Newer agents include the histone deacetylase inhibitors vorinostat (oral) (Kavanaugh et al., 2010) or romidepsin (intravenous). (Bertino & Otterson, 2011). Denileukin difititox is a fusion protein of interleukin-2 with a diphtheria toxin moiety that acts as a Trojan horse of sorts. The fusion protein is bound to the surface of the T-cell by its IL-2 receptor and the drug is internalized and the diphtheria toxin provokes apoptosis. (Duvic & Cather, 2000).

6.1.7 Radiation therapy for CTCL

Radiation therapy plays a critical role in the treatment of MF and is considered in three scenarios: curative treatment for uni- or paucy lesional MF, intent to achieve a complete response with total skin electron beam therapy (TSEBT), or local palliation with external beam radiation therapy (EBRT).

MF is radiosensitive compared to other skin cancers and doses as low as 2000-3000 cGy can yield high rates of durable remissions. For localized cases of stage I MF, electron beam is the most common radiotherapeutic modality, however, kilovoltage X-rays also have rapid dose fall off and hence work well in this application. Various different electron beam energies can be selected to appropriately treat the lesions. Due to the skin-sparing aspects of electron beam, the use of tissue equivalent bolus material to achieve full dose at the skin surface is standard. Micaily *et al.* reported on 18 patients with unilesional MF treated with electron beam at 30.6 Gy. (Micaily et al., 1998). The complete response rate was 100% and 2 patients failed out of field. The relapse free survival and overall survival rates were 86% and 100%, respectively. Wilson and colleagues reported on 21 patients treated with a median dose of 20 Gy for stage IA MF.(Wilson et al., 1998). Local control was 75% at 10 years, and 85% for patients with unilesional disease. For patients that received >20 Gy, disease free survival was 91%. Dose recommendations are provided in Table 6.

TSEBT is often considered for plaque (T2) and tumor (T3) cases due to the more guarded prognosis than T1 patients (10% of the skin involved). Because the response rates are so good for skin-directed therapies, TSEBT is not commonly used for T1 cases. TSEBT is a specialized technique for radiotherapy departments and is not widely available. The typical dose is 36 Gy in 36 fractions given 4 times weekly for 9 weeks. There are 6 body positions with two gantry positions for each body position. Selective electron boosts are administered to areas that do not receive adequate dose: the scalp, perineum and soles of feet. Also selective blocking is performed to prevent excessive toxicity to the eyes, hands, feet and nails.

In a Stanford study, a total of 180 patients with MF were treated with total skin electron-beam therapy (TSEBT) with or without topical nitrogen mustard. (Navi et al., 2011). The overall response rate was 100% while 60% of patients achieved a complete response (75% with T2 and 47% with T3) and the overall survival was 59% (T2) and 40% (T3). Those patients with incomplete responses or relapse had a 100% overall response rate with a second round of TSEBT. The response durations in this study were 29 months for T2 disease and 9 months for T3 disease (P = 0.006). Attempts have been made to prolong the response rates after TSEBT; however, combined PUVA (Quiros et al., 1997), nitrogen mustard (Navi et al., 2011), and interferon-alpha 2b (Roberge et al., 2007) did not statistically prolong relapse time.

Toxicities associated with TSEBT can be severe. Acute toxicities (defined as those during treatment and within 3 months of completion) include erythema, moist and dry desquamation, and pain. Given the severity of erythema and the resultant pain and discomfort, it is common to give a treatment break for a week or more during TSEBT. Late toxicities include generalized erythema (which can make it difficult for the clinician to separate from recurrence), telangiectasia, xerosis, secondary malignancies, and temporary and sometimes permanent loss of hair follicles, sweat glands, and nails. (KB, 2005). In our experience, hair loss is the main complaint from the patient, and is particularly distressing for females.

EBRT is very useful as a palliative treatment for recurrent MF when isolated tumors are present. MF is relatively radiosensitive compared to other skin cancers and doses as low as 400 cGy in 2 fractions can be very effective. Tumoritis has been known to occur with low doses of RT. This is secondary to prompt apoptosis resulting in rapid desquamation of the tumor site itself with relative sparing of the normal surrounding skin. Advanced MF lesions frequently ulcerate with RT and local wound care is essential to reduce the risk of infection. It is important to be prudent and patient and treat deep lesions with low, palliative doses (2 Gy x 2 or 3 Gy x 2).

6.2 Cutaneous B-Cell Lymphoma (CBCL)

The cutaneous B-cell lymphomas (CBCL) are less common than are the cutaneous T-cell lymphomas and represent between 18.8% and 26.4% of all cutaneous lymphomas. (Fink-Puches et al., 2002; Willemze et al., 1997). As opposed to most CTCL, CBCL presents as either solitary or paucilesional nodules on the skin with a red to violaceous color without any surface change such as scaling. The vast majority of CBCL are very low-grade malignancies with an excellent prognosis. It is important to make a distinction between primary cutaneous B-cell lymphomas which appear to arise in and stay in the skin as opposed to secondary cutaneous B-cell lymphoma where the skin is a secondary organ of involvement from a systemic lymphoma. Clinically the primary and secondary cutaneous B-cell lymphomas can be indistinguishable, therefore, a search for a systemic lymphoma is undertaken before the diagnosis of primary CBCL is confidently made. Radiotherapy is the mainstay of treatment for the majority of CBCL cases.

6.2.1 Epidemiology

Due to the rarity of CBCL, the actual incidence is difficult to determine, although there is a curious range in incidence across geographical areas. CBCL has been reported to comprise 4.5% of all cutaneous lymphomas in the United States (Zackheim et al., 2000) to 26.4% of cases in Austria. (Fink-Puches et al., 2002). The varying regional incidence of CBCL is likely due to do differences in how CBCL is classified and entered into tumor registries. However, there is also good evidence that etiologic factors may well contribute to regional disparities in incidence. For example, it has been reported for many years now that infection with *Borrelia burgdorferi* has been linked with CBCL in Europe but a similar link has not been demonstrated in the U.S. (Cerroni et al., 1997; Goodlad et al., 2000).

6.2.2 Etiology

The hypothesis linking an underlying infection to CBCL is the notion that chronic antigenic stimulation may eventually lead to the selection of B-cell subclones over time. A similar hypothesis has been proposed for the development of small bowel lymphoma in patients with chronic bowel inflammation with celiac sprue or the development of MALT (mucosa associated lymphoid tissue) lymphomas arising in the gastric mucosa of patients with chronic *Helicobactor pylori* infection. (Fischbach et al., 2005). The reality is that the majority of cases of CBCL arise spontaneously and multiple attempts to pinpoint a specific cause have been unsuccessful. Putative etiologic agents include *Borrelia spp.,* Epstein-Barr virus, or human herpes viruses. (Nagore et al., 2000; Zochling et al., 1998). Interestingly, there are cases of presumed benign reactive lymphoid hyperplasia (pseudolymphoma) simulating CBCL that have occurred secondary to drugs, vaccinations, arthropod bites, or chronic folliculitis. (Bergman, 2010). The question as to whether pseudolymphomas can progress to true CBCL has not been definitively answered, although it seems logical that with chronic antigenic stimulation, the ultimate selection of a B-cell clone remains a possibility.

6.2.3 Clinical presentation

There are many variants of CBCL, but the three entities discussed in this chapter are the two most common forms with indolent behavior (cutaneous follicle center cell lymphoma and cutaneous marginal zone lymphoma) and a more aggressive variant, large B-cell lymphoma of the leg as described in the European Organization for the Treatment of Cancer (EORTC) classification scheme. (Willemze et al., 1997).

Primary cutaneous follicle center cell lymphoma (FCCL) and primary cutaneous marginal zone lymphoma (MZL) are frequently indistinguishable clinically and can only be separated histologically. Both usually present as either solitary or grouped red to plum-colored nodules that have a smooth surface without the scaling common to CTCL (See Figure 8a.). Compared to CTCL, the color tends to shift slightly on the spectrum from red to purple. The nodules are generally non-tender and are usually distributed on the head and neck or the torso. It is more uncommon to see lesions on the extremities. The prognosis for both entities is excellent.

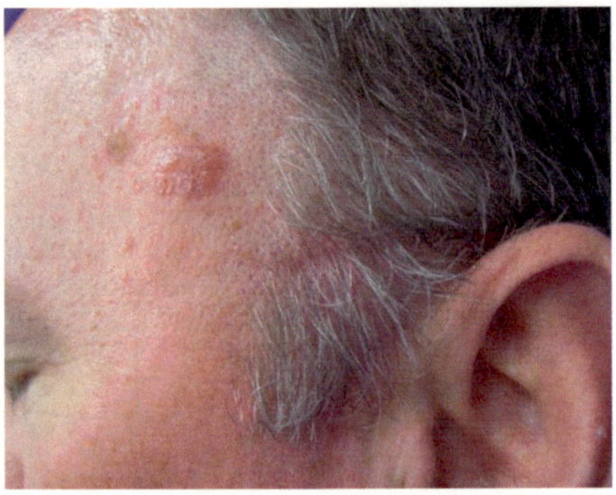

Fig. 8a. Cutaneous B-cell lymphoma

Large B-cell lymphoma of the leg (LBCLL) is one of the dermatologic diseases with a striking predilection for a specific anatomic site. The patients are almost uniformly elderly with a female to male predominance and the presentation is one of plum-shaped and plum-colored nodules usually confined to one leg, most commonly below the knee. (Kerl & Cerroni, 1996; Vermeer et al., 1996). As the name implies, the B-cells show a striking large cell morphology compared to the relatively small-sized B-cells in FCCL and MZL. Unlike the other cutaneous B-cell lymphomas, it is common for nodules in LBCLL to ulcerate and lesions are usually grouped and multiple as opposed to solitary. The prognosis is much more guarded in LBCLL, with an estimated 5-year survival at about 50%. (Fink-Puches et al., 2002; Willemze, 1997). There are other systemic large B-cell lymphomas that can secondarily involve the skin of the leg, therefore, a search for an underlying systemic lymphoma is essential in these patients. (Cerroni, 2003).

6.2.4 Pathology

Follicle Center Cell Lymphoma (FCCL):

Skin biopsies of FCCL have a fascinating appearance under light microscopy: a dermal infiltrate is comprised of lymphocytes simulating the germinal centers/follicles seen in lymph nodes (See Figure 8b). In about 25% of cases, the follicular architecture resembling the lymph node is classic, while the majority of cases will have less well-formed follicular

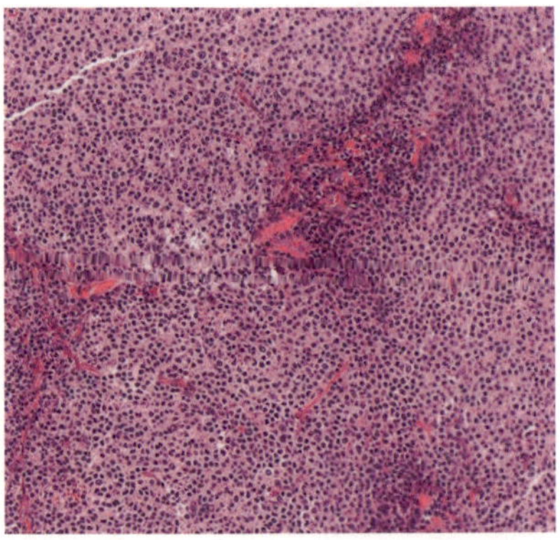

Fig. 8b. Histology of cutaneous B-cell lymphoma, follicle center cell type.

architecture. (Cerroni et al., 2000). There is often an admixture of other inflammatory cells, including T-cells, histiocytes, eosinophils, and plasma cells. The B-cells express the B-cell markers CD-20 and CD-79a, but unlike nodal lymphomas, do not express the Bcl-2 protein. (Cerroni et al., 1994; Cerroni & Kerl, 1994). The tumors stain positively for kappa or lambda light chains but may or may not have an identifiable restriction of one light chain favoring another. Not present in FCCL is the 14;18 translocation normally found in nodal follicular lymphomas, and when present, should alert the clinician to search for an underlying systemic lymphoma. (Cerroni et al., 1994; Delia et al., 1989).

Marginal Zone Lymphoma (MZL):

The infiltrate in MZL is also dermal-based with sparing of the epidermis, but the follicular architecture is much less well-defined. There is, however, an infiltrate surrounding the neoplasm with pale-staining cells with abundant pale cytoplasm referred to by hematopathologists as "marginal zone cells." These are centrocyte-like cells that are CD-20 and -CD-79a positive, and unlike FCCL, Bcl-2 protein is usually positive. The majority of cases of MZL with demonstrate light chain restriction with a predominance of one of the light chains over the other.

Large B-cell Lymphoma of the Leg (LBCLL):

In contrast to FCCL and MZL, the infiltrates of LBCLL will abut the epidermis and can often percolate into the epidermis and simulate the epidermotropism of CTCL. (Cerroni et al, 2009). The B-cells are larger and consist of immunoblasts, centroblasts, and large centrocytes. Cells express CD-20 and CD79a as expected but staining for the Bcl-2 protein is variable. Another pitfall may be the presence of CD30 positive large cells, which may simulate anaplastic large cell lymphoma derived from T-cells. In such cases, the CD30 does not have prognostic import as it does in anaplastic large cell lymphoma where its presence connotes a more favorable prognosis.

6.2.5 Treatment

Because most FCCL and MZL follow an indolent course, observation can be an acceptable practice with follow-up visits scheduled every six months or if new lesions arise. It has been our experience that many solitary lesions will actually involute with an incisional biopsy, which is to say that even if a small portion of the tumor is biopsied, the lesion may undergo spontaneous regression. In these cases it has been our practice to observe the lesion over time and treat only in cases of a local recurrence. In most cases, local RT is the first line treatment and provides high response rates coupled with excellent cosmesis and minimal morbidity.

Radiotherapy was evaluated in a meta-analysis study of 460 patients with FCCL and the complete response rate was 99% (457/460), with doses ranging from 20-54 Gy but in general averaging above 30 Gy. The relapse rate was 30%, but not defined as to whether these were in-field vs. extra-field recurrences. That same study included a meta-analysis of 132 patients with MZL treated with radiotherapy, and again showed a 99% complete response rate (130/132), with average dosages between 30-45 Gy and margins ranging from 1 to 5 cm. (Senff et al., 2008). The recurrence rate was 46%, but again not specified if these were in-field recurrences or other site recurrences. We have examined our own experience for patients with cutaneous B cell lymphoma. We identified 38 cases treated. The overall survival for the entire study population at 5 and 10 years was 97% and 78%, respectively, whereas disease-specific overall survival has not shown any deaths in the FCCL and MZL subgroups. Among patients that relapsed, in-field control was achieved in 82% of patients. Patients treated with other local therapies had a significant higher rate of failure (p=0.01). (Tward et al., 2009).

Other treatment modalities included surgical excision for small solitary lesions, intralesional interferon-alpha, or intralesional or systemic rituximab therapy which is a monoclonal antibody to CD20. Due to ease of use and response rates, RT is the most commonly used first line treatment with doses recommended of 30 Gy with margins of 1 to 1.5 cm of uninvolved surrounding skin (See Table 6). (Senff et al., 2008).

Disease	Dose (Gy)	# of Fractions
MF (unilesional)	24-36	10-18
MF (TSEBT)	30-36	30-36
MF (palliative)	4-6	2-3
FCCL	30	15
MZL	30	15
FCCL or MZL (palliative)	4	2
LBCLL	30-40	15-20

Table 6. Dose Recommendations for Cutaneous Lymphomas

For LBCLL where the prognosis is much more guarded, the EORTC recommends that these cases be treated with multi-agent chemotherapy similar to treatment for diffuse large B-cell lymphoma. (Senff et al., 2008). Recommended is R-CHOP (rituximab plus cyclophosphamide, doxorubicin (Adriamycin), oncovin (Vincristine), and prednisone with or without adjuvant radiotherapy. Unfortunately, there are no controlled trials available to help make an informed decision in such cases. At our institution, the decision as to whether or not to add radiotherapy to R-CHOP is made in the setting of the Multidisciplinary Lymphoma Conference, and is based on the extent of the cutaneous disease.

7. Conclusion

Radiation therapy is an excellent treatment choice for a variety of skin cancers in a wide number of anatomic locations, which frequently offers high rates of cure with excellent cosmesis and provides patients an alternative to aggressive and often disfiguring surgery. RT also fills an important role as an adjunctive therapy when combined with other systemic agents with or without surgery, and should be included in the armamentarium of any multidisciplinary approach to skin cancer. Improved communication between academic training programs in dermatology and radiation oncology are needed to overcome historical biases that contribute to the underutilization of this very useful treatment modality.

8. References

Agelli, M and L.X. Clegg, *Epidemiology of primary Merkel cell carcinoma in the United States.* J Am Acad Dermatol, 2003. 49(5): p. 832-41.

Agrawal, S., et al., *The benefits of adjuvant radiation therapy after therapeutic lymphadenectomy for clinically advanced, high-risk, lymph node-metastatic melanoma.* Cancer, 2009. 115(24): p. 5836-44.

Jill Ainslie, L.J.P., Michael J. McKay, *Radiotherapy for primary and regional melanoma,* in *Cutaneous Melanoma*

Akhtar, S., K.K. Oza, and J. Wright, *Merkel cell carcinoma: report of 10 cases and review of the literature.* J Am Acad Dermatol, 2000. 43(5 Pt 1): p. 755-67.

Alexander, G.A. and U.K. Henschke, *Advanced skin cancer in Tanzanian albinos: preliminary observations.* J Natl Med Assoc, 1981. 73(11): p. 1047-54

Allen, P.J., Z.F. Zhang, and D.G. Coit, *Surgical management of Merkel cell carcinoma.* Ann Surg, 1999. 229(1): p. 97-105.

Ang, C.G., et al., *Characteristics of familial and non-familial melanoma in Australia.* Melanoma Res, 1998. 8(5): p. 459-64.

Ang, K.K., et al., *Postoperative radiotherapy for cutaneous melanoma of the head and neck region.* Int J Radiat Oncol Biol Phys, 1994. 30(4): p. 795-8.

A.N.H. Charles M. Balch, Arthur J. Sober, Seng-jaw Soong, Editor 2003, Quality Medical Publishing, Inc.: St. Louis, Missouri. p. 449-471.

Avila, J., et al., *Carcinoma of the pinna.* Cancer, 1977. 40(6): p. 2891-5.

Aynaud, O., M. Ionesco, and R. Barrasso, *Penile intraepithelial neoplasia. Specific clinical features correlate with histologic and virologic findings.* Cancer, 1994. 74(6): p. 1762-7.

Baca HJ, B.S., Dzubow L, *Risk Factors,* in *Cutaneous Oncology: pathophysiology, diagnosis, and management,* M.M. Miller SJ, Editor 1998, Blackwell Science: Malden, MA. p. 382-390.

Bafaloukos, D. and H. Gogas, *The treatment of brain metastases in melanoma patients.* Cancer Treat Rev, 2004. 30(6): p. 515-20.

Balch, C.M., et al., *Long-term results of a multi-institutional randomized trial comparing prognostic factors and surgical results for intermediate thickness melanomas (1.0 to 4.0 mm). Intergroup Melanoma Surgical Trial.* Ann Surg Oncol, 2000. 7(2): p. 87-97.

Balch, C.M., et al., *Final version of 2009 AJCC melanoma staging and classification.* J Clin Oncol, 2009. 27(36): p. 6199-206.

Bartelink, H., et al., *Concomitant radiotherapy and chemotherapy is superior to radiotherapy alone in the treatment of locally advanced anal cancer: results of a phase III randomized trial of the European Organization for Research and Treatment of Cancer Radiotherapy and Gastrointestinal Cooperative Groups.* J Clin Oncol, 1997. 15(5): p. 2040-9.

Bergman, R., *Pseudolymphoma and cutaneous lymphoma: facts and controversies.* Clin Dermatol, 2010. 28(5): p. 568-74.

Bertino, E.M. and G.A. Otterson, *Romidepsin: a novel histone deacetylase inhibitor for cancer.* Expert Opin Investig Drugs, 2011. 20(8): p. 1151-8.

Berwick M, W.M., *Epidemiology: Current Trends,* in *Cutaneous Melanoma,* B.C.e. al, Editor 2003, Quality Medical Publishing, Inc: St. Louis, MO. p. 16-17.

Bouwes Bavinck, J.N., et al., *The risk of skin cancer in renal transplant recipients in Queensland, Australia. A follow-up study.* Transplantation, 1996. 61(5): p. 715-21.

Brady LW, B.S., Fitzpatrick PJ, *Skin Cancer*, in *Principles and Practice of Radiation Oncology*, B.L. Perez CA, Editor 1987, Lippincott: Philadelphia, PA. p. 377.

Braun-Falco O, L.S., Goldschmidt H, in *Dermatologic Radiotherapy*1976, Springer: New York, NY. p. 69.

Breneman, J.C., et al., *Stereotactic radiosurgery for the treatment of brain metastases. Results of a single institution series.* Cancer, 1997. 79(3): p. 551-7.

Breslow, A., *Thickness, cross-sectional areas and depth of invasion in the prognosis of cutaneous melanoma.* Ann Surg, 1970. 172(5): p. 902-8.

Breuninger, H., B. Black, and G. Rassner, *Microstaging of squamous cell carcinomas.* Am J Clin Pathol, 1990. 94(5): p. 624-7.

Buckner, J.C., et al., *A phase III study of radiation therapy plus carmustine with or without recombinant interferon-alpha in the treatment of patients with newly diagnosed high-grade glioma.* Cancer, 2001. 92(2): p. 420-33.

Burmeister, B.H., et al., *Radiation therapy for nodal disease in malignant melanoma.* World J Surg, 1995. 19(3): p. 369-71.

Cascinelli, N., et al., *Immediate or delayed dissection of regional nodes in patients with melanoma of the trunk: a randomised trial. WHO Melanoma Programme.* Lancet, 1998. 351(9105): p. 793-6.

Cerroni, L. and H. Kerl, *Aberrant bcl-2 protein expression provides a possible mechanism of neoplastic cell growth in cutaneous basal-cell carcinoma.* J Cutan Pathol, 1994. 21(5): p. 398-403.

Cerroni, L., et al., *bcl-2 protein expression and correlation with the interchromosomal 14;18 translocation in cutaneous lymphomas and pseudolymphomas.* J Invest Dermatol, 1994. 102(2): p. 231-5.

Cerroni, L., et al., *Infection by Borrelia burgdorferi and cutaneous B-cell lymphoma.* J Cutan Pathol, 1997. 24(8): p. 457-61.

Cerroni, L., et al., *Primary cutaneous follicle center cell lymphoma with follicular growth pattern.* Blood, 2000. 95(12): p. 3922-8.

Cerroni L, K.H., *B-cell lymphomas of the skin*, in *Dermatology*, J.J. Bolognia JL, Rapini RP, et al, Editor 2003, Mosby: London. p. 1907-1919.

Cerroni L, G.K., Kerl H, *Skin Lymphoma: The Illustrated Guide*, 2009, Wiley-Blackwell Oxford. p. 156-163.

Chahbazian CM, B.G., *Radiation therapy for carcinoma of the skin of the face and neck. Special considerations.* JAMA, 1980. 244: p. 1135-1137.

Chahbazian CM, B.G., *Skin cancer*, in *Modern Radiation Oncology Classic Literature and Current Management*, e. Gilbert HA, Editor 1984, Harper & Row: Philadelphia, PA. p. 158.

Chang, D.T., et al., *Adjuvant radiotherapy for cutaneous melanoma: comparing hypofractionation to conventional fractionation.* Int J Radiat Oncol Biol Phys, 2006. 66(4): p. 1051-5.

Chen, J.C., et al., *Stereotactic radiosurgery in the treatment of metastatic disease to the brain.* Neurosurgery, 2000. 47(2): p. 268-79; discussion 279-81.

Chen, J.Y., et al., *Desmoplastic neurotropic melanoma: a clinicopathologic analysis of 128 cases.* Cancer, 2008. 113(10): p. 2770-8.

Cheuk, W., et al., *Immunostaining for thyroid transcription factor 1 and cytokeratin 20 aids the distinction of small cell carcinoma from Merkel cell carcinoma, but not pulmonary from extrapulmonary small cell carcinomas.* Arch Pathol Lab Med, 2001. 125(2): p. 228-31.

Collin, J.R., *Basal cell carcinoma in the eyelid region.* Br J Ophthalmol, 1976. 60(12): p. 806-9.

Corry, J., et al., *Nodal radiation therapy for metastatic melanoma.* Int J Radiat Oncol Biol Phys, 1999. 44(5): p. 1065-9.

Criscione, V.D. and M.A. Weinstock, *Incidence of cutaneous T-cell lymphoma in the United States, 1973-2002.* Arch Dermatol, 2007. 143(7): p. 854-9.

Damle, S., et al., *Phase I-II study of 5-fluorouracil, recombinant interferon alpha2a, and cisplatin in combination with external beam radiation therapy followed by surgery in patients with locally advanced carcinoma of the esophagus.* Am J Clin Oncol, 1999. 22(4): p. 391-5.

Delia, D., et al., *Clonal immunoglobulin gene rearrangements and normal T-cell receptor, bcl-2, and c-myc genes in primary cutaneous B-cell lymphomas.* Cancer Res, 1989. 49(17): p. 4901-5.

Dinehart, S.M. and S.V. Pollack, *Metastases from squamous cell carcinoma of the skin and lip. An analysis of twenty-seven cases.* J Am Acad Dermatol, 1989. 21(2 Pt 1): p. 241-8.

Dubin N, K.A., *Multivariate risk score for recurrent cutaneous basal cell carcinomas.* Arch Dermatol, 1983. 119: p. 373-377.

Duvic, M. and J.C. Cather, *Emerging new therapies for cutaneous T-cell lymphoma.* Dermatol Clin, 2000. 18(1): p. 147-56.

Edge SB, C.M., Byrd DR, *Merkel Cell Carcinoma*, in *AJCC Cancer Staging Manual*, C.M. Edge SB, Byrd DR, Editor 2009, Springer-Verlag New York, LLC: New York.

Edwards, B.K., et al., *Annual report to the nation on the status of cancer, 1975-2006, featuring colorectal cancer trends and impact of interventions (risk factors, screening, and treatment) to reduce future rates.* Cancer, 2010. 116(3): p. 544-73.

Everall, J.D. and P.M. Dowd, *Influence of environmental factors excluding ultra violet radiation on the incidence of skin cancer.* Bull Cancer, 1978. 65(3): p. 241-7.

Farshad, A., et al., *A retrospective study of 150 patients with lentigo maligna and lentigo maligna melanoma and the efficacy of radiotherapy using Grenz or soft X-rays.* Br J Dermatol, 2002. 146(6): p. 1042-6.

Fears, T.R. and J. Scotto, *Estimating increases in skin cancer morbidity due to increases in ultraviolet radiation exposure.* Cancer Invest, 1983. 1(2): p. 119-26.

Feng, H., et al., *Clonal integration of a polyomavirus in human Merkel cell carcinoma.* Science, 2008. 319(5866): p. 1096-100.

Ferlay J, B.F., Pisani P, Parkin DM. *GLOBOCAN 2000. Cancer Incidence, Mortality and Prevalence Worldwide, version 1.0. IARC.* 2001; Available from: Limited version accessed at http://www.-dep.iarc.fr/globocan/globocan.html.

Fink-Puches, R., et al., *Primary cutaneous lymphomas: applicability of current classification schemes (European Organization for Research and Treatment of Cancer, World Health Organization) based on clinicopathologic features observed in a large group of patients.* Blood, 2002. 99(3): p. 800-5.

Fischbach, W., A.O. Chan, and B.C. Wong, *Helicobacter pylori and Gastric Malignancy.* Helicobacter, 2005. 10 Suppl 1: p. 34-9.

Fitzpatrick, P.J., *Organ and functional preservation in the management of cancers of the eye and eyelid.* Cancer Invest, 1995. 13(1): p. 66-74.

Flam, M., et al., *Role of mitomycin in combination with fluorouracil and radiotherapy, and of salvage chemoradiation in the definitive nonsurgical treatment of epidermoid carcinoma of the anal canal: results of a phase III randomized intergroup study.* J Clin Oncol, 1996. 14(9): p. 2527-39.

Fogarty, G., et al., *Whole brain radiotherapy after local treatment of brain metastases in melanoma patients--a randomised phase III trial.* BMC Cancer, 2011. 11: p. 142.

Foundation, S.C. *Skin Cancer Information, Squamous cell carcinoma.* 2011; Available from: http://www.skincancer.org/scc-treatment-options.html.

Gillenwater, A.M., et al., *Merkel cell carcinoma of the head and neck: effect of surgical excision and radiation on recurrence and survival.* Arch Otolaryngol Head Neck Surg, 2001. 127(2): p. 149-54.

Gladstein AH, K.A., Bart RS, *Radiotherapy of cutaneous malignancies*, in *Physical Modalities in Dermatologic Therapy. Radiotherapy, Electrosurgery, Phototherapy, Cryosurgery*, G. H, Editor 1978, Springer-Verlag: New York, NY. p. 95.

Goodlad, J.R., et al., *Primary cutaneous B-cell lymphoma and Borrelia burgdorferi infection in patients from the Highlands of Scotland.* Am J Surg Pathol, 2000. 24(9): p. 1279-85.

Goodson, A.G., et al., *Comparative analysis of total body and dermatoscopic photographic monitoring of nevi in similar patient populations at risk for cutaneous melanoma.* Dermatol Surg, 2010. 36(7): p. 1087-98.

Gorlin, R.J., *Nevoid basal-cell carcinoma syndrome.* Medicine (Baltimore), 1987. 66(2): p. 98-113.

Grob, J.J. and J.J. Bonerandi, *The 'ugly duckling' sign: identification of the common characteristics of nevi in an individual as a basis for melanoma screening.* Arch Dermatol, 1998. 134(1): p. 103-4.

Group, R.M.C.C., *Merkel cell carcinoma: recent progress and current priorities on etiology, pathogenesis, and clinical management.* J Clin Oncol, 2009. 27: p. 4021-4026.

Guidelines™), N.C.P.G.i.O.N. *Melanoma. Version 1.2012.* 2011; Available from: http://www.nccn.org/professionals/physician_gls/pdf/melanoma.pdf.

Guidelines™), N.C.P.G.i.O.N. *Merkel Cell Carcinoma, Version 1.2012.* 2011; Available from: http://www.nccn.org/professionals/physician_gls/pdf/mcc.pdf.

Guidelines, N.C.C.N.N. *Squamous Cell Skin Cancers.* 2011 2011]; Version 2.2011:[Available from: http://www.nccn.org/professionals/physician_gls/pdf/nmsc.pdf.

Haag, M.L., L.F. Glass, and N.A. Fenske, *Merkel cell carcinoma. Diagnosis and treatment.* Dermatol Surg, 1995. 21(8): p. 669-83.

Hagen, N.A., et al., *The role of radiation therapy following resection of single brain metastasis from melanoma.* Neurology, 1990. 40(1): p. 158-60.

Halprin, K.M., *Psoriasis, skin cancer, and PUVA.* J Am Acad Dermatol, 1980. 2(5): p. 432-4.

Hanly, A.J., et al., *Analysis of thyroid transcription factor-1 and cytokeratin 20 separates merkel cell carcinoma from small cell carcinoma of lung.* J Cutan Pathol, 2000. 27(3): p. 118-20.

Hazard, L.J., W.T. Sause, and R.D. Noyes, *Combined adjuvant radiation and interferon-alpha 2B therapy in high-risk melanoma patients: the potential for increased radiation toxicity.* Int J Radiat Oncol Biol Phys, 2002. 52(3): p. 796-800.

Hitchcock, C.L., et al., *Neuroendocrine (Merkel cell) carcinoma of the skin. Its natural history, diagnosis, and treatment.* Ann Surg, 1988. 207(2): p. 201-7.

Holsti, L.R., et al., *Enhancement of radiation effects by alpha interferon in the treatment of small cell carcinoma of the lung.* Int J Radiat Oncol Biol Phys, 1987. 13(8): p. 1161-6.

Huang, F., et al., *Prophylactic cranial irradiation in advanced breast cancer: a case for caution.* Int J Radiat Oncol Biol Phys, 2009. 73(3): p. 752-8.

Jabbour, J., et al., *Merkel cell carcinoma: assessing the effect of wide local excision, lymph node dissection, and radiotherapy on recurrence and survival in early-stage disease--results from a review of 82 consecutive cases diagnosed between 1992 and 2004.* Ann Surg Oncol, 2007. 14(6): p. 1943-52.

Jambusaria-Pahlajani, A., et al., *Uncertainty in the perioperative management of high-risk cutaneous squamous cell carcinoma among Mohs surgeons.* Arch Dermatol, 2010. 146(11): p. 1225-31.

Jemal, A., et al., *Cancer statistics, 2010.* CA Cancer J Clin, 2010. 60(5): p. 277-300.

JLM, A. *Tableau de plan fongoide: Description des maladies de la peau observe a l'hopital St. Louis et exposition des meilleures methods suivies pour leur traitment.* in *Barior l'Aine et Files.* 1806. Paris, France.

Johnson, T.M., et al., *Squamous cell carcinoma of the skin (excluding lip and oral mucosa).* J Am Acad Dermatol, 1992. 26(3 Pt 2): p. 467-84.

Johnson, T.M., et al., *Usefulness of the staged excision for lentigo maligna and lentigo maligna melanoma: the "square" procedure.* J Am Acad Dermatol, 1997. 37(5 Pt 1): p. 758-64.

Kavanaugh, S.M., L.A. White, and J.M. Kolesar, *Vorinostat: A novel therapy for the treatment of cutaneous T-cell lymphoma.* Am J Health Syst Pharm, 2010. 67(10): p. 793-7.

KB, H., *Cutaneous T-cell lymphoma: mycosis fungoides and Sézary syndrome,* in *Cancer of the Skin,* R.D.e. al., Editor 2005, Elseiver. Inc.: Philadelphia, PA. p. 349-362.

Kerl, H. and L. Cerroni, *The morphologic spectrum of cutaneous B-cell lymphomas.* Arch Dermatol, 1996. 132(11): p. 1376-7.

Kim, Y.H., et al., *Clinical stage IA (limited patch and plaque) mycosis fungoides. A long-term outcome analysis.* Arch Dermatol, 1996. 132(11): p. 1309-13.

Kim, Y.H., et al., *Clinical characteristics and long-term outcome of patients with generalized patch and/or plaque (T2) mycosis fungoides.* Arch Dermatol, 1999. 135(1): p. 26-32.

Kim, Y.H., et al., *Long-term outcome of 525 patients with mycosis fungoides and Sezary syndrome: clinical prognostic factors and risk for disease progression.* Arch Dermatol, 2003. 139(7): p. 857-66.

Kokoska, E.R., et al., *Early aggressive treatment for Merkel cell carcinoma improves outcome.* Am J Surg, 1997. 174(6): p. 688-93.

Konefal, J.B., B. Emami, and M.V. Pilepich, *Analysis of dose fractionation in the palliation of metastases from malignant melanoma.* Cancer, 1988. 61(2): p. 243-6.

Lang, P.G., Maize, J.C., *Basal cell carcinoma,* in *Cancer of the Skin,* D.L. Rigel DS, Reitgen DS, Bystryn JC, Marks R, Editor 2005, Elsevier-Saunders: Philadelphia, PA. p. 101-132.

Leshin, B., et al., *Management of periocular basal cell carcinoma: Mohs' micrographic surgery versus radiotherapy.* Surv Ophthalmol, 1993. 38(2): p. 193-212.

Lewis, K.G., et al., *Adjuvant local irradiation for Merkel cell carcinoma.* Arch Dermatol, 2006. 142(6): p. 693-700.

Liu, Z. and J. Chen, *[The research advance of brain derived neurotrophic factor].* Sheng Wu Yi Xue Gong Cheng Xue Za Zhi, 2000. 17(4): p. 454-6, 460.

LM, P., *Erythrodermie quasi-généralisée, mais respectant des ilôts de peau saine, avec petites tumeurs à formule histologique de mycosis fongoide.* Bull Soc Derm Syphil, 1937. 44: p. 1307-1319.

Lovett, R.D., et al., *External irradiation of epithelial skin cancer.* Int J Radiat Oncol Biol Phys, 1990. 19(2): p. 235-42.

Martin H, S.E., Spiro RH, *Radiation-induced cancer of the head and neck.* Cancer, 1970(25): p. 61-71.

Mazeron, J.J., et al., *Radiation therapy of carcinomas of the skin of nose and nasal vestibule: a report of 1676 cases by the Groupe Europeen de Curietherapie.* Radiother Oncol, 1988. 13(3): p. 165-73.

Meadows, A.T., et al., *Second neoplasms in survivors of childhood cancer: findings from the Childhood Cancer Survivor Study cohort.* J Clin Oncol, 2009. 27(14): p. 2356-62.

Medina-Franco, H., et al., *Multimodality treatment of Merkel cell carcinoma: case series and literature review of 1024 cases.* Ann Surg Oncol, 2001. 8(3): p. 204-8.

Mendenhall, W.M., et al., *T2-T4 carcinoma of the skin of the head and neck treated with radical irradiation.* Int J Radiat Oncol Biol Phys, 1987. 13(7): p. 975-81.

Mendenhall, W.M., et al., *Carcinoma of the Skin*, in *Management of Head and Neck Cancer: A Multidisciplinary Approach*, R.R. Million and N.J. Cassisi, Editors. 1994, J.B. Lippincott Company: Philadelphia. p. 643-691.

Meyers, C.A., et al., *Neurocognitive function and progression in patients with brain metastases treated with whole-brain radiation and motexafin gadolinium: results of a randomized phase III trial.* J Clin Oncol, 2004. 22(1): p. 157-65.

Micaily, B., et al., *Radiotherapy for unilesional mycosis fungoides.* Int J Radiat Oncol Biol Phys, 1998. 42(2): p. 361-4.

Mojica, P., D. Smith, and J.D. Ellenhorn, *Adjuvant radiation therapy is associated with improved survival in Merkel cell carcinoma of the skin.* J Clin Oncol, 2007. 25(9): p. 1043-7.

Morrison, W., P. Wong, and L. Peters, *Radiotherapy for Basal and Squamous Cell Skin Cancinomas.* Cancer Bulletin, 1993. 45: p. 256-260.

Morrison, W.H., A.S. Garden, and K.K. Ang, *Radiation therapy for nonmelanoma skin carcinomas.* Clin Plast Surg, 1997. 24(4): p. 719-29.

Morton DL, E.R., Balch CM, *Surgical excision of distant metastases*, in *Cutaneous Melanoma*, B.C.e. al., Editor 2003, Quality Medical Publishing, Inc.: St. Louis, MO. p. 547-572.

Mudigonda, T., et al., *The economic impact of non-melanoma skin cancer: a review.* J Natl Compr Canc Netw, 2010. 8(8): p. 888-96.

Nagore, E., et al., *Detection of Epstein-Barr virus and human herpesvirus 7 and 8 genomes in primary cutaneous T- and B-cell lymphomas.* Br J Dermatol, 2000. 143(2): p. 320-3.

Navi, D., et al., *The Stanford University experience with conventional-dose, total skin electron-beam therapy in the treatment of generalized patch or plaque (T2) and tumor (T3) mycosis fungoides.* Arch Dermatol, 2011. 147(5): p. 561-7.

Nestle FO, K.H.I.e.D., *Melanoma*, in *Dermatology*, J.J. Bolognia J, Rapini RP, et al, Editor 2003, Mosby: London. p. 1789-1815.

Nguyen TH, Y.J., *Squamous Cell Carcinoma*, in *Cancer of the Skin*, D.L. Rigel DS, Reitgen DS, Bystryn JC, Marks R, Editor 2005, Elsevier Saunders: Philadelphia. p. 133-150.

O'Brien, C.J., et al., *Adjuvant radiotherapy following neck dissection and parotidectomy for metastatic malignant melanoma.* Head Neck, 1997. 19(7): p. 589-94.

Olivier, K.R., et al., *A higher radiotherapy dose is associated with more durable palliation and longer survival in patients with metastatic melanoma.* Cancer, 2007. 110(8): p. 1791-5.

Olsen, E.A. and P.A. Bunn, *Interferon in the treatment of cutaneous T-cell lymphoma.* Hematol Oncol Clin North Am, 1995. 9(5): p. 1089-107.

Ott, M.J., et al., *Multimodality management of Merkel cell carcinoma.* Arch Surg, 1999. 134(4): p. 388-92; discussion 392-3.

Overgaard, J., H. von der Maase, and M. Overgaard, *A randomized study comparing two high-dose per fraction radiation schedules in recurrent or metastatic malignant melanoma.* Int J Radiat Oncol Biol Phys, 1985. 11(10): p. 1837-9.

Pentenero, M., et al., *Evidence for a possible anatomical subsite-mediated effect of tobacco in oral potentially malignant disorders and carcinoma.* J Oral Pathol Med, 2011. 40(3): p. 214-7.

Petrovich, Z., et al., *Treatment results and patterns of failure in 646 patients with carcinoma of the eyelids, pinna, and nose.* Am J Surg, 1987. 154(4): p. 447-50.

Petrovich, Z., et al., *Carcinoma of the lip and selected sites of head and neck skin. A clinical study of 896 patients.* Radiother Oncol, 1987. 8(1): p. 11-7.

Petrovich, Z., et al., *Treatment of carcinoma of the skin with bone and/or cartilage involvement.* Am J Clin Oncol, 1988. 11(2): p. 110-3.

Pollock PM, W.A., Trent JM, *Genetics and molecular staging,* in *Cutaneous Melanoma,* B.C.e. al., Editor 2003, Quality Medical Publishing, Inc.: St. Louis, MO. p. 687-712.

Ponten F, L.J., *Principles of tumor biology and pathogenesis of BCCs and SCCs,* in *Dermatology,* J.J. Bolognia JL, Rapini RP et al, Editor 2003, Mosby: London. p. 1675.

Quiros, P.A., et al., *Total skin electron beam therapy followed by adjuvant psoralen/ultraviolet-A light in the management of patients with T1 and T2 cutaneous T-cell lymphoma (mycosis fungoides).* Int J Radiat Oncol Biol Phys, 1997. 38(5): p. 1027-35.

Ratner, D., et al., *Merkel cell carcinoma.* J Am Acad Dermatol, 1993. 29(2 Pt 1): p. 143-56.

RD, H., *Skin,* in *The Radiotherapy of Malignant Disease,* P.R. Eassen EC, Editor 1985, Springer-Verlag: New York, NY. p. 135.

Roberge, D., et al., *Does adjuvant alpha-interferon improve outcome when combined with total skin irradiation for mycosis fungoides?* Br J Dermatol, 2007. 156(1): p. 57-61.

Roenigk, H.H., Jr. and W.A. Caro, *Skin cancer in the PUVA-48 cooperative study.* J Am Acad Dermatol, 1981. 4(3): p. 319-24.

Rollison, D.E., A.R. Giuliano, and J.C. Becker, *New virus associated with merkel cell carcinoma development.* J Natl Compr Canc Netw, 2010. 8(8): p. 874-80.

Rowe, D.E., R.J. Carroll, and C.L. Day, Jr., *Long-term recurrence rates in previously untreated (primary) basal cell carcinoma: implications for patient follow-up.* J Dermatol Surg Oncol, 1989. 15(3): p. 315-28.

Rowe, D.E., R.J. Carroll, and C.L. Day, Jr., *Mohs surgery is the treatment of choice for recurrent (previously treated) basal cell carcinoma.* J Dermatol Surg Oncol, 1989. 15(4): p. 424-31.

Rowe, D.E., R.J. Carroll, and C.L. Day, Jr., *Prognostic factors for local recurrence, metastasis, and survival rates in squamous cell carcinoma of the skin, ear, and lip. Implications for treatment modality selection.* J Am Acad Dermatol, 1992. 26(6): p. 976-90

Samlowski, W.E., et al., *Multimodality treatment of melanoma brain metastases incorporating stereotactic radiosurgery (SRS).* Cancer, 2007. 109(9): p. 1855-62.

Sarasin, A. and G. Giglia-Mari, *p53 gene mutations in human skin cancers.* Exp Dermatol, 2002. 11 Suppl 1: p. 44-7.

Sarin, R., et al., *Treatment results and prognostic factors in 101 men treated for squamous carcinoma of the penis.* Int J Radiat Oncol Biol Phys, 1997. 38(4): p. 713-22.

Sarin R, T.H., *Current Concepts in the Management of Penile Cancer*, in *Treatment Options in Urological Cancer*, W. J, Editor 2002, Blackwell Science, Ltd.: Oxford, UK. p. 366-381.

Sause, W.T., et al., *Fraction size in external beam radiation therapy in the treatment of melanoma.* Int J Radiat Oncol Biol Phys, 1991. 20(3): p. 429-32.

Schmid-Wendtner, M.H., et al., *Fractionated radiotherapy of lentigo maligna and lentigo maligna melanoma in 64 patients.* J Am Acad Dermatol, 2000. 43(3): p. 477-82.

Schwartz, R.A., *Verrucous carcinoma of the skin and mucosa.* J Am Acad Dermatol, 1995. 32(1): p. 1-21; quiz 22-4.

Scott, M.P. and K.F. Helm, *Cytokeratin 20: a marker for diagnosing Merkel cell carcinoma.* Am J Dermatopathol, 1999. 21(1): p. 16-20.

Seegenschmiedt, M.H., et al., *Palliative radiotherapy for recurrent and metastatic malignant melanoma: prognostic factors for tumor response and long-term outcome: a 20-year experience.* Int J Radiat Oncol Biol Phys, 1999. 44(3): p. 607-18.

Senff, N.J., et al., *European Organization for Research and Treatment of Cancer and International Society for Cutaneous Lymphoma consensus recommendations for the management of cutaneous B-cell lymphomas.* Blood, 2008. 112(5): p. 1600-9.

Sézary A, B.Y., *Erythrodermie avec présence de cellules monstrueuses dans le derme et le sang circulant.* Bull Soc Derm Syphil, 1938. 45: p. 254-260.

Shibamoto, Y., et al., *Incidence of brain atrophy and decline in mini-mental state examination score after whole-brain radiotherapy in patients with brain metastases: a prospective study.* Int J Radiat Oncol Biol Phys, 2008. 72(4): p. 1168-73.

Shimm, D.S. and R.B. Wilder, *Radiation therapy for squamous cell carcinoma of the skin.* Am J Clin Oncol, 1991. 14(5): p. 383-6.

Silverman, M.K., et al., *Recurrence rates of treated basal cell carcinomas. Part 4: X-ray therapy.* J Dermatol Surg Oncol, 1992. 18(7): p. 549-54.

Silverman, M.K., et al., *Recurrence rates of treated basal cell carcinomas. Part 1: Overview.* J Dermatol Surg Oncol, 1991. 17(9): p. 713-8.

Sim, F.H., et al., *Lymphadenectomy in the management of stage I malignant melanoma: a prospective randomized study.* Mayo Clin Proc, 1986. 61(9): p. 697-705.

Skelton, H.G., et al., *Merkel cell carcinoma: analysis of clinical, histologic, and immunohistologic features of 132 cases with relation to survival.* J Am Acad Dermatol, 1997. 37(5 Pt 1): p. 734-9.

Skibber, J.M., et al., *Cranial irradiation after surgical excision of brain metastases in melanoma patients.* Ann Surg Oncol, 1996. 3(2): p. 118-23.

Solan MJ, B.L., Binnick SA, *Skin*, in *Principles and Practice of Radiation Oncology*, B.L. Perez CA, Editor 1997, Lippincott-Raven: Philadelphia, PA. p. 723-744.

Stern, R.S., et al., *Risk of cutaneous carcinoma in patients treated with oral methoxsalen photochemotherapy for psoriasis.* N Engl J Med, 1979. 300(15): p. 809-13.

Stevens, G., et al., *Locally advanced melanoma: results of postoperative hypofractionated radiation therapy.* Cancer, 2000. 88(1): p. 88-94.

Stock, R.G., et al., *Enhanced radiosensitization with interferon-alpha-2b and cisplatin in the treatment of locally advanced cervical carcinoma.* Gynecol Oncol, 1997. 67(3): p. 309-15.

Stoll HL Jr, M.H., Traenkle HL, *Results of Roentgen therapy of carcinoma of the nose.* Arch Dermatol, 1964. 90: p. 577-580.

Strom, E.A. and M.I. Ross, *Adjuvant radiation therapy after axillary lymphadenectomy for metastatic melanoma: toxicity and local control.* Ann Surg Oncol, 1995. 2(5): p. 445-9.

Surveillance, E., and End Result (SEER) Program, Public Use Data (1973-1999). 2002; Available from: http://www.seer.cancer.gov.

Taylor G, M.D., Heilman ER, *Merkel cell carcinoma*, in *Cancer of the Skin*, R.D.e. al., Editor 2005, Elsevier-Saunders: Philadelphia, PA. p. 323327.

Thissen, M.R., M.H. Neumann, and L.J. Schouten, *A systematic review of treatment modalities for primary basal cell carcinomas.* Arch Dermatol, 1999. 135(10): p. 1177-83.

Traenkle, H.L., H.L. Stoll, Jr., and A. Lonkar, *Results of roentgen therapy of carcinoma of the lip.* Arch Dermatol, 1962. 85: p. 488-9.

Tsang, R.W., et al., *Lentigo maligna of the head and neck. Results of treatment by radiotherapy.* Arch Dermatol, 1994. 130(8): p. 1008-12.

Tward, J.D., et al., *Survival and recurrence in nonmycosis fungoides primary cutaneous lymphoma.* Cancer J, 2009. 15(1): p. 87-92.

UKCCCR Anal Cancer Trial Working Party. Epidermoid anal cancer: results from the UKCCCR randomised trial of radiotherapy alone versus radiotherapy, 5-fluorouracil, and mitomycin. UK Co-ordinating Committee on Cancer Research. Lancet, 1996. 348(9034): p. 1049-54.

Veness, M., et al., *The role of radiotherapy alone in patients with merkel cell carcinoma: reporting the Australian experience of 43 patients.* Int J Radiat Oncol Biol Phys, 2010. 78(3): p. 703-9.

Vermeer, M.H., et al., *Primary cutaneous large B-cell lymphomas of the legs. A distinct type of cutaneous B-cell lymphoma with an intermediate prognosis. Dutch Cutaneous Lymphoma Working Group.* Arch Dermatol, 1996. 132(11): p. 1304-8.

Veronesi, U., et al., *Delayed regional lymph node dissection in stage I melanoma of the skin of the lower extremities.* Cancer, 1982. 49(11): p. 2420-30.

Vongtama, R., et al., *Efficacy of radiation therapy in the local control of desmoplastic malignant melanoma.* Head Neck, 2003. 25(6): p. 423-8.

Weinstock, M.A. and A.J. Sober, *The risk of progression of lentigo maligna to lentigo maligna melanoma.* Br J Dermatol, 1987. 116(3): p. 303-10.

Werlinger, K.D., G. Upton, and A.Y. Moore, *Recurrence rates of primary nonmelanoma skin cancers treated by surgical excision compared to electrodesiccation-curettage in a private dermatological practice.* Dermatol Surg, 2002. 28(12): p. 1138-42; discussion 1142.

Whiteman, D.C., W.J. Pavan, and B.C. Bastian, *The melanomas: a synthesis of epidemiological, clinical, histopathological, genetic, and biological aspects, supporting distinct subtypes, causal pathways, and cells of origin.* Pigment Cell Melanoma Res, 2011.

Wilder, R.B., et al., *Recurrent basal cell carcinoma treated with radiation therapy.* Arch Dermatol, 1991. 127(11): p. 1668-72.

Willemze, R., et al., *EORTC classification for primary cutaneous lymphomas: a proposal from the Cutaneous Lymphoma Study Group of the European Organization for Research and Treatment of Cancer.* Blood, 1997. 90(1): p. 354-71.

Wilkinson, E.J., *Normal histology and nomenclature of the vulva, and malignant neoplasms, including VIN.* Dermatol Clin, 1992. 10(2): p. 283-96.

Wilson, L.D., B.M. Kacinski, and G.W. Jones, *Local superficial radiotherapy in the management of minimal stage IA cutaneous T-cell lymphoma (Mycosis Fungoides).* Int J Radiat Oncol Biol Phys, 1998. 40(1): p. 109-15.

Yiengpruksawan, A., et al., *Merkel cell carcinoma. Prognosis and management.* Arch Surg, 1991. 126(12): p. 1514-9.

Zackheim, H.S., et al., *Relative frequency of various forms of primary cutaneous lymphomas.* J Am Acad Dermatol, 2000. 43(5 Pt 1): p. 793-6.

Zochling, N., et al., *Human herpesvirus 8-specific DNA sequences in primary cutaneous B-cell lymphomas.* Arch Dermatol, 1998. 134(2): p.246-7.

Part 5

Radiation Induced Effects and Overcoming Strategies

13

Critical Normal Tissue and Radiation Injury: The Stomach

Mineur Laurent, Jaegle Enric,
Pourel Nicolas and Garcia Robin
Gastrointestinal and Liver Oncology-Radiotherapy Unit
Institut Sainte Catherine Avignon,
France

1. Introduction

There is limited detailed information regarding morbidity and toxicity following radiation treatment in gastric cancer. The curative effectivity of external radiotherapy necessitates exposure of normal tissues with significant radiation doses, and hence must be associated with an accepted rate of side effects. Based on the time of first treatment, complications can range from the acute side effects of external beam radiation therapy which resolve shortly after treatment to those that develop 6 months to 2 years after completion radiotherapy.

2. An outline of anatomy and normal histology of the stomach

2.1 Anatomy

Anatomically stomach is divided into 4 regions: the cardiac region is surrounded by esophageal sphincter. The fundus lies against the diaphragm.Body (corpus) ensue from fundus and pylorus (pyloric antrum): ends at the pyloric sphincter which is a thickening of the muscle walls. (figure 1 et 2)

2.2 Types of cells present in the stomach

Mucous secreting cells (goblet cells) line the luminal surface of the stomach and gastric pits and gastric glands which produce mucus and bicarbonate. Mucous neck cells are present in the neck of the gland andproduce mucin. Parietal cells (oxyntic cells) are distributed throughout the length of the gland, but numerous in the middle portion. Large, rounded cells with eosinophilic cytoplasm and centrally located nucleus. Chief cells (peptic or zymogenic cells) produce gastric acid. They are clustered at the base of the gland and identified by basally located nuclei and strongly basophilic granular cytoplasm. They produce pepsinogen, which digests protein.

2.3 Normal histological features

The gastric mucosa consists of surface epithelium, gastric pits and gastric glands. The gastric glands extend from the muscular mucosae to extend into the stomach lumen via gastric pits.The foveolar cells lining the surface and gastric pits are identical throughout the stomach. Glands differ in different regions of the stomach. Gastric pits occupy approximately 25% of the mucosa. Pits lie parallel to one another. These are separated by the lamina propria. There is more lamina propria separating the pits than between the glands. In normal gastric biopsy, degree of pit and glandular separation should be same throughout the biopsy. Cardia is a small area of predominantly mucus secreting glands surrounding the entrance of the esophagus. Glands are less coiled than in the antralglands.The pits are shorter than the antropyloric pits.Fundus and body are a major histological region consisting of straight, tubular glands. Strands of muscularis mucosae extend between the glands from the base. The glands secrete gastric juices as well as protective mucus. In pylorusbranched glands open into deep irregular shaped pits and are composed of mucus secreting cells. Mucus secreted by pyloric glands lubricate and protect entrance to the duodenum.Scattered 'G' cells (endocrine cells), secrete gastrin. Gastric mucosa forms a barrier todiffusion of gastric acid from the gastric lumen.The fundus is best seen by inverting the gastroscope to see the dome-shaped upper portion of the stomach.When the patient is lying on the left side, the gastric juice is seen to the left in the fundus, called "the mucous lake".

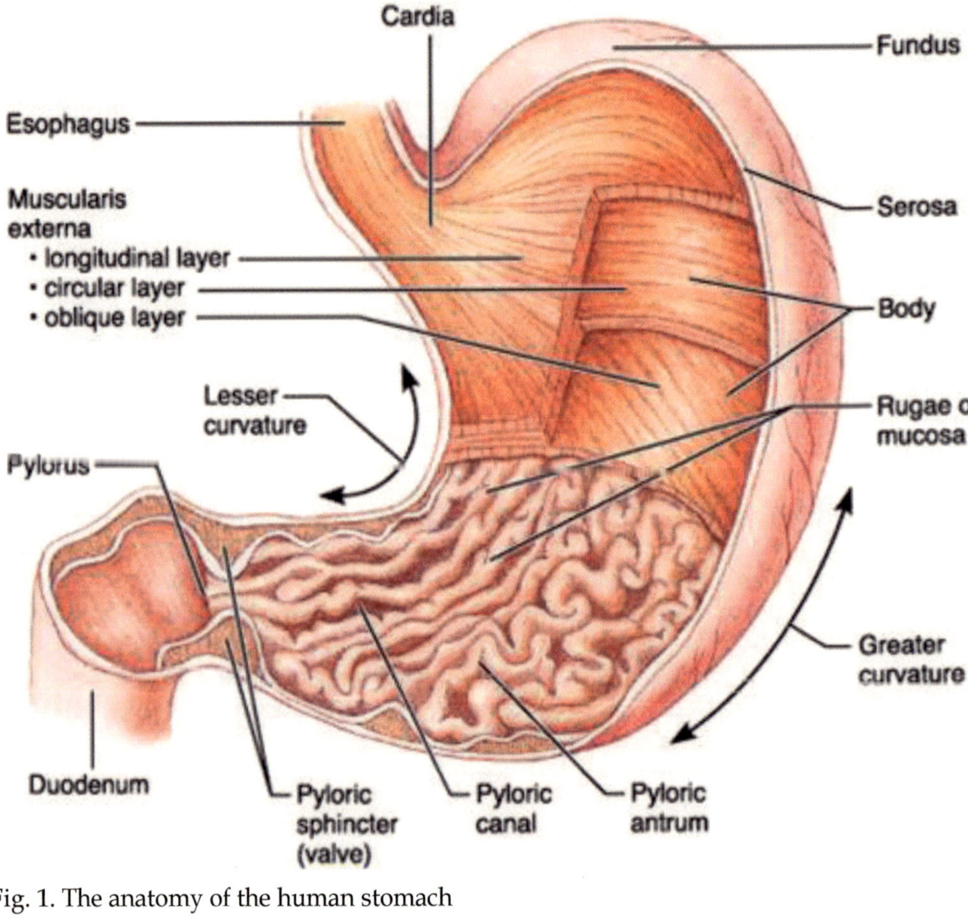

Fig. 1. The anatomy of the human stomach

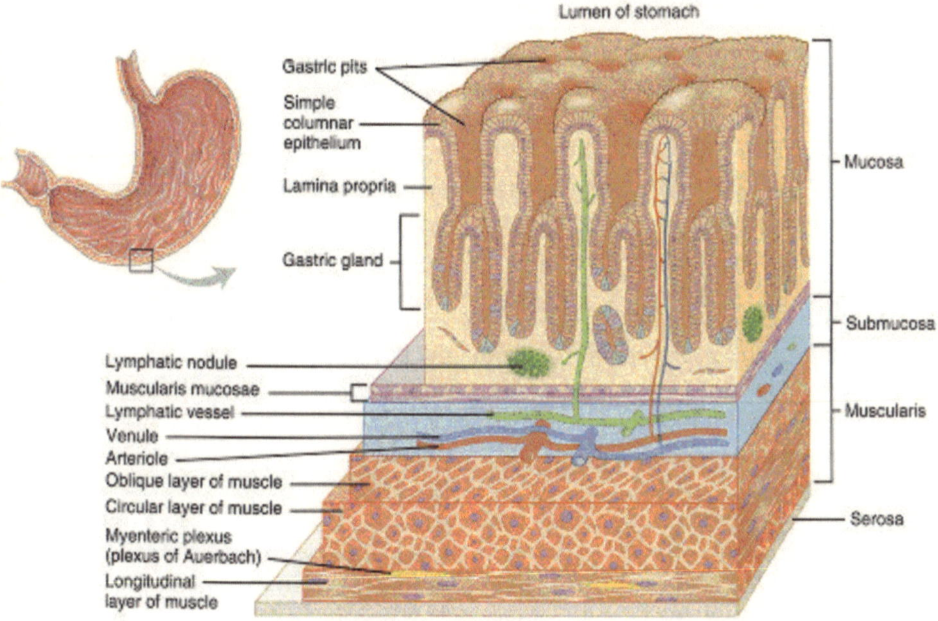

Fig. 2. Histology of the stomach. a three-dimensional view of layers of the stomach

3. Critical normal tissue

Early reactions per definition occur within 90 days after onset of the radiation exposure. They are based on impairment of cell production in turnover tissues, which in face of ongoing cell loss results in hypoplasia and eventually a complete loss of functional cells. The latent time is largely independent of dose and is defined by tissue biology (turn-over time). Usually, complete healing of early reactions is observed. Late radiation effects can occur after symptom-free latent times of months to many years, with an inverse dependance of latency on dose. Late normal tissue changes are progressive and usually irreversible. They are based on a complex interaction of damage to various cell populations (organ parenchyma, connective tissue, capillaries), with a contribution from macrophages. Late effects are sensitive for a reduction in dose rate (recovery effects). A number of biologically based strategies for protection of normal tissues or for amelioration of radiation effects was and still is tested in experimental systems, yet, only a small fraction of these approaches has so far been introduced into clinical studies. Published data suggest that the risk of moderately severe (>or=Grade 3) radiation-induced acute small-bowel, pulmonary or others organs toxicity can be predicted with a threshold model whereby for a given dose level, D, if the volume receiving that dose or greater (VD) exceeds a threshold quantity, the risk of toxicity escalates. One advantage of most of the methods is that they may be effective even if the treatment starts way after the end of radiation exposure. For a clinical exploitation, availability of early indicators for the progression of subclinical damage in the individual patient would be desirable. Moreover, there is need to further investigate the molecular pathogenesis of normal tissue effects in more detail, in order to optimise biology based preventive strategies, as well as to identify the precise mechanisms of already tested approaches (e.g. stem cells).

3.1 Radiation tolerance (radiation gastritis)

Any discussion about the use of radiation therapy in the treatment of any cancer with critical normal tissue than stomach must prefaced by a description of the radiotolerance of stomach. It is commonly held that there is a low risk of gastritis or ulcer if the whole stomach receives less than 45Gy with a classic fractionation (1.8Gy or 2 Gy once a day 5 times per week).In the rat after irradiation with single doses, three distinct gastric disorders were observed which occurred at different latency times. Acute death 2-3 weeks after irradiation was caused by ulcerative gastritis and occurred in all animals given 28.5 Gy without diet, in 17% of the animals given 28.5 Gy plus diet, and in 13% of the animals given 23 Gy. Subacute to chronic fatal disorders 4 weeks to 7 months after irradiation were seen as stomach dilatation and gastroparesis, associated with the replacement of the normal gastric mucosa by a hyperkeratinized multilayered squamous epithelium(Breiter et al1989). These disorders occurred in 40–100% of the animals after doses between 16 Gy and 28.5 Gy (+diet). Late gastric obstruction exceeding 7 months after irradiation was seen in the rats because of profound changes in the gastric wall in 13–18% of the animals after doses between 23 Gy and 14 Gy. In animals surviving these three periods, an atrophic mucosa and intestinal metaplasia developed. From functional and morphohistological studies, it can be concluded that there are differences in the pathogenesis of the fatal radiation damage for each of these periods after irradiation. The most sensitive cells in fundus would be the parietal cells and chief cells. They are the first two to necrose. Glands are colonized by mucous neck cells and are an inflamatorylymphocitic reaction seat. Mucous gland metaplasia, the reversible replacement of differentiated cells, occurs in the setting of severe damage of the gastric glands, which then waste away (atrophic gastritis) and are progressively replaced by mucous glands.Acute gastric ulcers may develop and are rather linked to a mucosa desquamation. Vascular necrosis occurs second time and is responsible for late perforation(golgraber et al 1975, roswit,et al 1972))

Gastric acid is produced by cells lining the stomach, which are coupled to radiotherapy, radiation ionizing decrease earlier acid production but only with low dose radiotherapy. Stemcells of the gastric mucosa seems to be insensitive to radiotherapy. In one hour ,they repair 60% of ADN lesions and the stems cells doubling time is fast (about 43hours) (chen et al 1972). However, these in vitro data are not always correlated with clinical data on late effects of stomach radiotherapy. Clinically nauseas occur in the first hours after irradiation. Vomiting and anorexia are reported in clinical trial MALT lymphoma, lomboaortic radiotherapy or in preoperative gastric radiation (ajani et al 2004). Moderate (2–3 Gy of radiation) exposure is associated with nausea and vomiting beginning within 12–24 hours after exposure.

3.2 Acute and chronic side effects

3.2.1 Ulcer and gastritis

The late effects of radiation gastric therapy can include ulcer and gastritis (bush 1993, otsuka 2008). Ulcer is usually alone and in pyloric antrum. Microscopic appearance is telangectasia, fibrin deposit and numerous fibroblasts. Microscopic chronic gastritis appears with lympocytic cells mass,gastric gland are infrequent, not typical architecture, perivascular lymphocytic cells cluster, submucosal fibrosis dividing muscularis fibers, Intestinal metaplasia typically begins in response to chronic mucosal injury in the stomach and may extend to the body. Gastric mucosa cells change to look like intestinal mucosa. (coia et al 1995).

Grigsby reported a series of 30 patients received 48Gy(1.2 Gy twice a day). One patient declared a toxicity grade 3. In a similar review of the literature of patients treated with 26Gy and a lomboaortic field (normal fractionation) in seminoma stage I, any patient presented late gastro intestinal toxicity (Grisgby et al 2001).

3.2.2 Previous gastric radiation and incidence of gastric cancer

In previous radiation therapy for benign gastric disease there are limited data to suggest that radiation delivered may be a risk factor for gastric cancer. Gastric radiation was used in the ulcer gastro-duodenal treatment. Doses ranging from 15Gy to 20 Gy in 10 fractions were used before prescribing the inhibitors of pump proton. In small series of patients Peters et al described patients with partial gastrectomy and radiotherapy had an increased incidence of gastric cancer. Grien and coll reported 1831 patients with ulcers disease who were treated with radiation therapy and compared them to a similar group of medically managed patient over an average of 22 years. Radiotherapy was linked to an increased relative risk for cancers of the stomach (rr=2.77 and 95% as was partial gastrectomy(rr=2.6). So if surgery was combined with radiation therapy the risk increased 10 fold (Griem et al 1994).

3.3 Gastric tolerance dose

3.3.1 Conventional fractionation

The most of clinical information come from trials about hodgkin lymphoma, MALT (mucosa associated lymphoid tissue) or postoperative radiation of the paraaortic and ipsilateral pelvic lymph nods in testiculars tumors, or uterine cervix carcinoma. Two trials are relevant and expose acute and late toxicity in radiation gastric adenocarcinoma treatment. Forty three patients received concurrent radiation 45Gy(1.8Gy/fraction) and chemotherapy (infusional fluorouracil and weekly paclitaxel). Resection was attempted 5 to 6 weeks after chemoradiotherapy was completed. Any ulcer but gastritis grade 3 in one patient were noticed (ajani et al 2004). Cosset et al reported the toxicity results of 2 trials and 516 patients were entered in two consecutive EORTC trials for supra-diaphragmatic Hodgkin's lymphoma received an sub-diaphragmatic irradiation and, 36 (7%) developed late radiation injuries of the gastrointestinal tract. Twenty-five patients presented with ulcers (stomach or duodenum), two with severe gastritis. Although there was a significant improvement of complication in previous laparotomy the complication rate was 2.7% without any previous abdominal surgery, it was 11.5% after laparotomy ($p < 0.001$). Fractionation was also found to be of importance in the occurrence of complications: three different weekly schedules were used -5 x 2 Gy, 4 x 2.5 Gy and 3 x 3.3 Gy; the GIT complication rates were 4, 9 and 22%, respectively (p less than 0.001). When combining laparotomy and fractionation, we found that the patients who were treated using 5 weekly fractions of 2 Gy without any prior laparotomy had a very low rate of late digestive complications (1%), whereas the patients who received 3 weekly fractions of 3.3 Gy after laparotomy presented a 39% complication rate. The other subgroups of patients were at an intermediate risk (from 5 to 13%) of late digestive injuries. The median time occurring gastric ulcer was 18 months(3-92 months). Authors illustrate the effect of the difference with linear quadratic equation often used to model biological response to radiation. For the acute effect if the ratio α/β =10, 40Gy(2 Gy per fraction) is an equivalent to 41.7 Gy(2.5Gy per fraction) and 44.3Gy(3.3Gyper fraction). So a ratio α/β =2.5 40Gy(2 Gy per fraction) is an equivalent to 44.4 Gy(2.5Gy per fraction) and 51.8Gy (3.3Gy per fraction)(Cosset al 1988). Mohiuddin studied 81 patients with localized, unresectable carcinoma of the pancreas with a combination of intraoperative Iodine-125 implantation, external beam radiation 50-55Gy with standard fractionation, and peri-operative systemic chemotherapy and 22% of the patients developed gastric bleeding mainly antral bleeding(mohiuddin et al 1992). Grisby reported a series of 30 patients received 48Gy(1.2 Gy twice a day). one patient declared a toxicity grade 3(Grigsby et al 2001). In a similar review of the literature of patients treated with 26Gy and a lomboaortic field (normal fractionation) in seminoma stage I, any patient presented late gastro intestinal toxicity(classen et al 2004).Yang et al reported radiation toxicity in 153 Patients treated for an hepatocarcinoma. The median dose received was 38Gy into15 fractions and a median dose of 3Gy. Twenty percent of the patients received an oesogastric endoscopy. Six percent of the patients presented gastro-duodenal ulcer and 5% gastritis. Fifty percent of the gastric ulcer and 75% of duodenal ulcer were in the irradiated target volume. Gastric bleeding was significantly commonly for femal, but not in cirhosis, total dose, dose per fraction and this study doesn't identify an high risk group(yang et al 2005). In the paraaortic lymph node treatment, 121 patients received 50Gy normal dose per fraction, 5% presented gastro duodenal ulcer and 2 patients gastrectomy (Goldstein et al 1975).

3.3.2 Hypofractionation of radiation

In general, rates of gastric-specific toxicity are inconsistently reported in the literature, often in combination with other GI toxicities. Additionally, other side effects such as anorexia, nausea, and vomiting are often included in gastrointestinal toxicity profiles. Other authors have also reported GI toxicity with hypofractionated regimens. Hoyer, et al conducted, Phase-II study on stereotactic radiotherapyof locally advanced pancreatic carcinoma, conducted a Phase II trial in which 22 patients with locally advanced pancreatic adenocarcinoma received 45 Gy split into three fractions. These investigators reported significantly greater acute GI toxicity than in our previous pancreatic cancer SBRT studies. The greater toxicity observed in the Hoyer study may be attributed in part to the significantly larger treatment volumes reported in this cohort of patients. Five patients (23%) experienced severe mucositis of the stomach or duodenum, including one who developed a gastric perforation. In each of these patients, part of the stomach or duodenum received at least 67% of the prescribed dose (30Gy over three fractions)(Hoyer et al 2005).Guidelines and different trials phase I- II propose to respect Dose volume histograms (DVHs) and to analyze V0.5 (volume of gastric in cm^3 receiving 30 Gy), and the maximum dose to $0.5cm^3$ of stomach ($D_{max} \leq 30Gy$)(Murphy et al 2009, lee et al 2009)

Streitparth reported gastric mucosa acute tolerance to high dose rate brachytherapy. Study treated 33 patients and liver segment tumor II and III(near stomach). In all patients a minimum dose applied to 1 ml of the gastric wall ($D_{1\,ml}$) ranged from 6.3 to 34.2 Gy; median, 14.3 Gy. Toxicity was present in 18 patients (55%). Nausea was present in 16 patients (69%), emesis in 9 (27%), cramping in 13 (39%), weight loss in 12 (36%), gastritis in 4 (12%), and ulceration in 5 patients (15%). they found a threshold dose $D_{1\,ml}$ of 11 Gy for general gastric toxicity and 15.5 Gy for gastric ulceration verified by an univariate analysis (p = 0.01).Authors conclude for a single fraction, small volume irradiation we found in the upper abdomen a threshold dose $D_{1\,ml}$ of 15.5 Gy for the clinical endpoint ulceration of the gastric mucosa the tolerance dose of gastric mucosa after EBT applied as a single fraction dose can be estimated at 15.7 Gy, converted by using the linear-quadratic model with an assumed α/β of 12 Gy for gastric tissue for a TD50/5 of 70 Gy (probability of 50% during the next 5 years after irradiation to one third of the total gastric volume) (Streitparth et al 2006) these data are in adequacy with Emami who assessed perforation gastric risk for 60Gy and one third volume of gastric treated, 5% rate severe complication at 5 years was 60Gy.(Emami et al 1991.)

3.4 Guidelines to minimizing toxicity

There are several means to reduce normal tissue toxicity when combining radiotherapy and chemotherapy. Generally radiotherapy fields should cover clinically evident disease only and not attempt to cover areas where subclinical carcinoma is likely to present. But in gastric cancer and postoperative radiation or in trial with perioperative radiotherapy fields and clinical target volume cover all stomach organ. In the current state of the medical knowledge, there remains an uncertainty about dose and stomach tolerance and especially the ratio dose/volume and dose/time. Scientific Studies evaluating the dose and critical stomach organ tolerance are difficult to compare. It seems not necessary to delineate stomach and it is important to verify any hot point more than 35Gy. The ICRU "hot spot" (i.e. the dose outside of the PTV with a volume of at least 1.8 cc) should not exceed the prescription dose by more than 7%.(matzinger et al 2009)

Thirty five Gy is probably the threshold beyond the ulcer risk increase. Fifty four Gy maximal dose may be administred in a small volume with a classical fractionation.In SBRT Guidelines propose to respect Dose volume histograms (DVHs) and to analyze V0.5 (volume of gastric in cm^3 receiving 30 Gy), and the maximum dose to $0.5cm^3$ of stomach ($D_{max} \leq 30Gy$)(Murphy et al 2009).

4. Conclusion

Stomach is a critical normal tissue that need to be considered in the radiation treatment but toxicities are rare and it seems not necessary to delineate stomach. But in the current state of knowledge there is still uncertainty about the stomach dose tolerance and especially the ratio dose volume and dose time. Acute toxicities, vomiting, nausea occur precociously. Stomach tolerance dose if the totality is irradiated is 45Gy and classical fractionation35Gy is the threshold beyond which the ulcer risk appear and it is possible to prescribe 54 Gy in a reduce volume.

5. References

Breiter N, Trott KR, Sassy T. Effect of X-irradiation on the stomach of the rat. Int J Radiat Oncol Biol Phys 1989;17:779-84.

Goldgraber MB, Rubin CE, Palmer WL, Dobson RL, Massey BW. The early gastric response to irradiation; a serial biopsy study. Gastroenterology 1954;27:1-20.

Classen J, Schmidberger H, Meisner C, et al. Para-aortic irradiation for stage I testicular seminoma: results of a prospective study in 675 patients. A trial of the German testicular cancer study group (GTCSG). Br J Cancer 2004;90:2305-11.

RoswitB, Malsky SJ, Reid CB. Severe radiation injuries of the stomach, small intestine, colon and rectum. Am J Roentgenol Radium Ther Nucl Med 1972; 114:460-75.

Chen KY, Withers HR. Survival characteristics of stem cells of gastric mucosa in C 3 H mice subjected to localized gamma irradiation. Int J Radiat Biol Relat Stud Phys Chem Med 1972;21:521-34.

Ajani JA, Winter K, Okawara GS, et al. Phase II trial of preoperative chemoradiation in patients with localized gastric adenocarcinoma (RTOG 9904): quality of combined modality therapy and pathologic response. J Clin Oncol 2006;24:3953-8.

Busch DB. Radiation and chemotherapy injury: pathophysiology, diagnosis, and treatment. Crit Rev Oncol Hematol 1993;15:49-89.

Otsuka T, Noda T, Yokoo M, Ibaraki K. Recurrent gastric perforation as a late complication of radiotherapy for mucosa-associated lymphoid tissue lymphoma of the stomach. Intern Med 2008;47:1407-9.

Coia LR, Myerson RJ, Tepper JE. Late effects of radiation therapy on the gastrointestinal tract. Int J Radiat Oncol Biol Phys 1995;31:1213-36.

Grigsby PW, Heydon K, Mutch DG, Kim RY, Eifel P. Long-term follow-up of RTOG 92-10: cervical cancer with positive para-aortic lymph nodes. Int J Radiat Oncol Biol Phys 2001;51:982-7.

Griem MI, Kleinerman RA, Boice JD, Stowall M, Shefner D, Lubin JH. Cancer following radiotherapy for peptic ulcer. J Natl Cancer Inst 1994 ;86 :842-9.

Cosset JM, Henry-Amar M, Burgers JM, et al. Late radiation injuries of the gastrointestinal tract in the H2 and H5 EORTC Hodgkin's disease trials: emphasis on the role of exploratory laparotomy and fractionation. Radiother Oncol 1988;13:61-8.

Mohiuddin M, Rosato F, Barbot D, Schricht A, Biermann W, Cantor R. Long term results of combined modality treatment with I-125implantation for carcinoma of the pancreas. Int J RadiatOncol Biol Phys 1992;23:305_11.

Yang MH, Lee JH, Choi MS, et al. Gastrointestinal complications after radiation therapy in patients with hepatocellular carcinoma. Hepatogastroenterology 2005;52:1759-63.

Goldstein H, Rogers L, Fletcher G, Dodd G. Radiological manifestations of radiation-induced injury to the normal upper gastrointestinal tract. Radiology 1975;117:135-40.

Hoyer M, Roed H, Sengelov L, et al. Phase-II study on stereotactic radiotherapy of locally advanced pancreatic carcinoma. Radiotherapy and oncology : journal of the European Society for Therapeutic Radiology and Oncology 2005;76:48-53.

Murphy JD, Christman-Skieller C, Kim j, Dieterich s, Chang DT. A Dosimetric Model of Duodenal Toxicity After Stereotactic Body Radiotherapy for Pancreatic Cancer. Int J Radiat Oncol Biol Phys, 2009;78:1420-1426a

Lee MT, Kim JJ, Dinniwell R, etal.Phase I study of individualized stereotactic body radiotherapy of liver metastases. J Clin Oncol 2009; 27:1585–1591.

Streitparth F, Pech M, Böhmig M, et al. In vivo assessment of the gastric mucosal tolerance dose after single fraction, small volume irradiation of liver malignancies by computed tomography-guided, high-dose-rate brachytherapy. Int J Radiat Oncol Biol Phys 2006;65:1479-86.

Emami B, Lyman J, Brown A, et al. Tolerance of normal tissue to therapeutic irradiation. Int J Radiat Oncol Biol Phys 1991;21:109-22.

Matzinger O, Gerber E, Bernstein Z, Maingon P, Haustermans K, Bosset JF, Gulyban A, Poortmans P, Collette L, Kuten A.EORTC-ROG expert opinion: radiotherapy volume and treatment guidelines for neoadjuvant radiation of adenocarcinomas of the gastroesophageal junction and the stomach. Radiother Oncol. 2009 Aug;92(2):164-75. Epub 2009 Apr 15.

14

The Cytoprotective Effect of Amifostine Against Radiation Induced Toxicity

Vassilis E. Kouloulias[1] and John R. Kouvaris[2]
[1]*Attikon Hospital University of Athens, Medical School,*
[2]*Aretaieion Hospital University of Athens, Medical School, Athens,*
Greece

1. Introduction

Radiation treatment is an important therapeutic option for a number of malignancies (American Cancer Society), but its use is frequently limited by adverse effects on normal tissues (Stone et al., 2003). Thus, the goal of most oncology treatments is to maximize the antineoplastic effect while minimizing deleterious outcomes for the patient. WR-2721 was developed by the U.S. Army Anti-Radiation Drug Development Program for its potential to protect against damage caused by ionizing radiation (Yuhas &Stoner, 1969). Today, WR-2721 is known as amifostine (Ethyol®; MedImmune Oncology, Inc., Gaithersburg, MD). Initial preclinical studies demonstrated that amifostine could protect treated mice from lethal doses of radiation, and this protection did not extend to transplanted mammary tumor cells (Yuhas &Stoner, 1969).

Amifostine, a thiol that protects cells from damage by scavenging oxygen-derived free radicals, was later evaluated for a potential role in reducing the toxicities from radiation and chemotherapeutic agents, such as alkylating agents and platinum agents. In contrast to organ-specific protectants, amifostine is considered a broad-spectrum cytoprotective agent (Hensley et al., 1999). Preclinical studies demonstrated that amifostine can selectively protect almost all normal tissues from the cytotoxic effects of some chemotherapeutic agents and radiation therapy. Neoplastic tissues do not benefit from amifostine's protection (Koukourakis, 2003; Sasse et al., 2006; Yuhas et al., 1980). Amifostine is an inactive prodrug that cannot protect until dephosphorylated to the active metabolite, WR-1065, by alkaline phosphatase in the plasma membrane (Calabro-Jones et al., 1985). The selective protection of normal tissue is the result of a greater accumulation of WR-1065 in normal tissues than in tumor cells. Tumors are relatively hypovascular, thus resulting in comparative hypoxia and a low interstitial pH. Furthermore, alkaline phosphatase expression is reduced in malignant tissues. Taken together, the combination of hypovascularity, low pH, and reduced enzyme levels results in low accumulation of active drug in tumor tissues. Thus, normal tissues may be able to maintain as much as a 100-fold greater concentration of the free thiol than tumor tissue (Yuhas, 1980).

Once inside the cell, WR-1065 scavenges free radicals, protecting cellular membranes and DNA from damage. However, other studies have suggested that additional mechanisms may also play important roles in the action of amifostine. In vitro studies have shown that oxidation of WR-1065 to its polyamine-like disulfide metabolite (WR-33278) is followed by a rapid consumption of oxygen in culture medium, suggesting that induction of cellular anoxia may be a mechanism for radioprotection (Purdie et al., 1983). This was supported by a study by Glover et al. that showed a rapid increase in the oxygen saturation of the venous blood after i.v. administration of amifostine without affecting the oxygen dissociation curves of hemoglobin, again suggesting that a decrease in oxygen consumption by normal tissues may be involved in amifostine-related radioprotection (Glover et al., 1984). In another study, high concentrations of WR-33278 condensed DNA, thereby limiting potential target sites for free-radical attack (Savoye et al., 1997). This activity would clearly account for a decrease in the number of double-strand breaks after radiotherapy, in turn leading to a reduction of the transient block at the G2 phase of cell division induced by radiation (Rubin et al., 1996). The enhanced cellular proliferation that results from a reduction in damage to DNA may be an important pathway to accelerated recovery of endothelial tissues that are affected soon after radiation exposure (Rubin et al., 1996) and seems to be important for the recovery of irradiated mucosa (Koukourakis et al., 1999). In addition, amifostine, indirectly through hypoxia, may upregulate the expression of a variety of proteins involved with DNA repair and inhibition of apoptosis, such as Bcl-2 and hypoxia-inducible factor-1 (Carmeliet et al., 1998; Kajstura et al., 1996; Shimizu et al., 1996).

Early phase I trials with amifostine were not able to demonstrate a maximum-tolerated dose but did establish a tolerable dose range of 740–910 mg/m2 for use in phase II studies (Blumberg et al., 1982). Amifostine is generally well tolerated, although transient adverse events may be dose related and include hypotension, nausea, vomiting, sneezing, somnolence, a metallic taste during infusion, and occasional allergic reactions that may include rash, fever, and anaphylactic shock (Blumberg et al., 1982). Although hypotension is the most clinically significant adverse event, treatment interruptions caused by a significant decline in blood pressure are rare, occurring in <5% of patients receiving amifostine. Emesis can be reduced with judicious use of an antiemetic regimen before amifostine administration. Transient hypocalcemia caused by inhibition of parathyroid hormone secretion has also been reported (Glover et al., 1983). The incidence and severity of amifostine-related adverse events have been shown to vary based on the route of administration. A recent meta-analysis of randomized studies using amifostine reported a significantly greater risk for grade 3 or 4 hypotension when amifostine was administered as a slow i.v. infusion (Sasse et al., 2006). Studies examining the s.c. administration of amifostine have demonstrated a lower incidence of hypotension and nausea/vomiting than with i.v. administration (Koukourakis et al., 2000; Anne & Curran, 2002; Anne et al., 2007). However, s.c. administration of amifostine has been reported to be associated with a higher incidence of fever and cutaneous reactions than with i.v. administration in these studies (Sasse et al., 2006; Koukourakis et al., 2000; Anne & Curran, 2002; Anne et al., 2007).

Pharmacokinetic studies in patients have demonstrated that amifostine is rapidly cleared from the plasma compartment, with a half-life of <1 minute, and >90% cleared within 6 minutes (Shaw et al., 1986). However, very little amifostine, or the metabolites WR-1065 and WR-33278, is excreted in urine 1 hour after injection. These data show that once amifostine enters the plasma, it is rapidly metabolized and distributed in the tissues, whereas the excretion of the metabolic products is very slow. Timely administration of amifostine relative to radiation or chemotherapeutic treatment is necessary. One study by Buentzel et al., in which amifostine was administered 30 minutes before chemoradiotherapy, demonstrated no significant difference in the incidence of grade 2 acute or chronic xerostomia or grade 3 oral mucositis between patients receiving i.v. amifostine and those receiving placebo (Buentzel et al., 2006). On days when combined radiochemotherapy was administered, timing between amifostine and radiotherapy may have exceeded 60 minutes. The authors suggest that timing of the amifostine doses relative to the beginning of radiotherapy may have influenced efficacy because of inadequate exposure to amifostine. In addition, the observed rates of grade 2 acute xerostomia and grade 3 oral mucositis in the placebo group were unexpectedly low, reducing the ability of the study to show significant benefit with amifostine. In contrast, studies in which amifostine was administered within 30 minutes of radiotherapy have shown promise with regard to protection from acute and chronic xerostomia (Andonadou et al., 2002; Brizel et al., 2000; Vacha et al., 2003). Taken together, it appears that administration of amifostine within 30 minutes of radiotherapy or chemoradiotherapy may provide optimal benefit for cytoprotection of normal tissues.

Of primary concern with the use of any substance or technique that is intended to spare normal tissues from treatment-related toxicities is the unintended and undesirable protection of tumor cells. Clearly, procedures that protect tumors are not clinically useful. A recent meta-analysis of the available clinical data concluded that, in addition to reducing the toxicities associated with radiation therapy, amifostine does not affect the efficacy of radiotherapy (Sasse et al., 2006). To the contrary, patients receiving amifostine with radiotherapy achieved higher rates of complete response, presumably the result of fewer treatment interruptions because of reduced acute toxicity of the treatment.

1.1 Xerostomia and oral mucositis

Xerostomia and mucositis are significant and potentially debilitating toxicities associated with radiation therapy. The risk for these complications depends on the area receiving radiation, the dose and schedule of therapy, whether radiation therapy is combined with chemotherapy, and other factors (Sonis & Fey, 2002). Although rarely life threatening, the acute and long-term consequences can be significant, causing discomfort, reduced nutrition, and a diminished quality of life. Xerostomia is the most common toxicity associated with standard fractionated radiation therapy to the head and neck. Whereas acute xerostomia from radiation is the result of an inflammatory reaction, late xerostomia, observed 1 year after radiation, is usually a permanent result of fibrosis of the salivary gland. The dry mouth of xerostomia affects the patient's ability to eat and speak. The decreased salivary output in patients with xerostomia can be responsible for an increased risk for dental caries, oral infections, and osteonecrosis.

The results of numerous randomized controlled studies suggest that amifostine may protect against radiation- and chemoradiation-induced toxicity in patients with head and neck cancer (Table 1) (Sasse et al., 2006). In one study by Buntzel et al., 28 patients received radiation therapy in conjunction with carboplatin (Buntzel et al., 1998). Amifostine was administered to 14 patients on the day of carboplatin at a fixed dose of 500 mg (equivalent to 250–340 mg/m2). Acute grade 3 or 4 mucositis was experienced by 12 of 14 patients (86%) treated with radiochemotherapy alone compared with none of the amifostine-treated

patients ($p < .001$). Additionally, at a 12-month follow-up, 17% of patients who received amifostine experienced late grade 2 xerostomia, compared with 55% of the patients treated without amifostine ($p = .05$). An international phase III trial of radiation therapy with and without amifostine was conducted in 315 patients with squamous cell carcinoma of the head

Study	Number of patients	Treatment	Main conclusions
RT			
McDonald et at., 1994	9	RT + i.v, amifostine, 100 mg/m^2	Flow rates of unstimulated whole saliva recovered to 20% of baseline at 12 months post-treatment
Wagner etal., 1998	14	RT 4 · i.v. amifostine. 200 mg/m^2	i.v. amifostine treatment led to significant reduction in oral symptoms and duration of mucositis
Bourhis et al., 2000	26	RT + i.v. amifostine, 150 mg/m^2, versus RT alone	i.v. amifostine treatment led to significant reduction in duration of acute mucositis and duration of feeding tube use compared with RT treatment alone
Koukourakis et al., 2000	40	RT + s.c. amifostine, 500 mg, versus RT alone	s.c. amifostine led to significant reduction in severity of oral mucositis compared with RT treatment alone
Brizel et al., 2000	315	RT + i.v. amifostine, 200 mg/m^2, versus RT alone	i.v. amifostine led to significant reduction inacute and chronic xerostomia versus RT alone and increased saliva production versus RT alone; no significant reduction in grade ≥3 mucositis versus RT
Wasserman et al., 2005	315	2-yr follow-up of Brizel etal. (2000) [26]	i.v. amifostine led to significant decrease in severity and duration of xerostomia at 2 yrs post-treatment without compromising tumor control
Anne el al., 2002, 2007	54	RT + s.c. amifostine, 500 mg	Incidence of acute grade ≥2 xerostomia, 56%; 1-yr rates of locoregional tumor control, progression-free survival, and overall survival, 78%, 75%, and 85%, respectively
RCT			
Buntzel et al., 1998	39	RT + carboplatin + i.v. amifostine. 500 mg, versus RT + carboplatin (control)	i.v. amifostine treatment led to significant reductions in acute xerostomia, grade ≥3 mucositis, and grade ≥3 thrombocytopenia compared with control treatment
Peters et al., 1999	28	RT + carboplatin + i.v. amifostine, 500 mg, versus RT + carboplatin (control)	i.v. amifostine treatment had no significant effect on xerostomia or mucositis compared with control treatment
Atttonadou et al., 2002	50	RT + carboplatin + i.v. amifostine, 300 mg/m^2, versus RT + carboplatin (control)	i.v. amifostine treatment led to significant reduction in acute and late grade ≥2 xerostomia and grade ≥3 mucositis compared with control treatment

Study	Number of patients	Treatment	Main conclusions
Vacha et al., 2003	52	RT + carboplatin + i.v. amifostine. 250 mg, versus RT + carboplatin (control)	i.v. amifostine treatment led to significant reduction in xerostomia compared with control treatment: reduction in mucositis was not significant between treatment groups
Bucntzel et al., 2006	132	RT + carboplatin + i.v, amifostine, 200-300 mg/ m², versus RT + carboplatin (control)	No difference between i.v, amifostine treatment and control treatment with regard to incidence of grade ≥2 xerostomia or grade ≥3 mucositis: low incidence of grade ≥2 xerostomia and grade ≥3 mucositis in control patients; no evidence of tumor protection was Observed with either treatment
Abbreviations: CRT, radiochemotherapy: RT, radiotherapy.			

Table 1. Clinical trials of amifostine therapy during radiation therapy or radiochemotherapy for head and neck cancer

and neck in which at least 75% of each parotid gland was present in the radiation fields (Brizel et al., 2000). The amifostine dose was 200 mg/m2 daily, 15–30 minutes before each fraction of radiation therapy (1.8–2.0 Gy/day, 5 days per week for 5–7 weeks, to a total dose of 50–70 Gy). Amifostine significantly reduced acute and late xerostomia and associated symptoms. Using Radiation Therapy Oncology Group (RTOG) grading criteria, patients receiving amifostine had a lower incidence of grade 2 or higher acute xerostomia (51% versus 78%; $p < .001$) and a lower incidence of grade 2 or higher late xerostomia (34% versus 57%; $p = .002$). The proportion of patients with meaningful saliva production after 1 year was significantly higher with amifostine (72% versus 49%; $p = .003$). Despite a trend toward lower severity of mucositis with amifostine ($p = .14$), the difference in the incidence of grade 3 or higher mucositis was not statistically significant ($p = .48$). Importantly, at 1 year, with a median follow-up of 20 months, the locoregional tumor control rates did not differ, and disease-free and overall survival times were comparable. Two-year follow-up data from this study demonstrate the continued benefits of amifostine treatment on the incidence of grade 2 xerostomia ($p = .002$ versus patients who did not receive amifostine) (Wasserman et al., 2005). Furthermore, no significant differences in locoregional tumor control rate, progression-free survival time, or overall survival rates were observed 2 years post-treatment between the amifostine group and the control group (Wasserman et al., 2005).

In another study, 50 patients with head and neck cancer were randomized to receive radiotherapy plus carboplatin with or without amifostine (Antonadou et al., 2002). Treatment interruptions were more frequent in the control group. Consequently, patients receiving amifostine experienced significantly shorter treatment durations ($p = .013$). Patients treated with amifostine experienced less severe acute mucositis and dysphagia; all patients who did not receive amifostine in the control group experienced grade 2 mucositis by week 3. In contrast, only 9% of patients treated with amifostine experienced grade 2 mucositis ($p < .001$). By the fifth week, grade 4 mucositis was experienced by 52% and 4.5%

of the patients in the respective groups ($p < .001$). Dysphagia was similarly reduced among patients given amifostine. After 3 months of follow-up, grade 2 xerostomia was reported in 27% and 74% of patients treated with and without amifostine, respectively ($p < .001$).

Another consideration in the treatment of head and neck cancer is the tolerance dose of the parotid glands and the potential for raising this threshold with amifostine. Eisbruch et al. [31] reported a threshold of 26 Gy, using conformal or intensity-modulated radiotherapy, as a mean dose to spare parotid gland function (Eisbruch et al., 1999). With the use of amifostine, the threshold radiation dose for chronic xerostomia may be increased, allowing for greater dose coverage (Munter et al., 2007).

1.2 Esophagitis and pneumonitis

Damage from radiation treatment is also a major complication in the treatment of thoracic cancers, with higher rates of acute and late toxicity associated with concurrent chemoradiotherapy. Several studies have investigated the cytoprotective efficacy of amifostine against radiation-induced esophagitis and pneumonitis. Antonadou et al., in a multicenter trial of patients with advanced lung cancer, investigated whether daily pretreatment with amifostine could reduce the incidence of acute and late lung toxicity and esophagitis without affecting antitumor efficacy of radiation treatment (Antonadou et al., 2001). One hundred forty-six patients received radiotherapy in daily fractions of 2 Gy, 5 days per week, to a total of 55–60 Gy with or without daily amifostine, 340 mg/m2 15 minutes before irradiation. There was a significantly lower incidence of grade 2 or higher pneumonitis among patients receiving amifostine (9% versus 43%; $p < .001$). At 6 months post-treatment, fibrosis was present in 53% of patients not receiving amifostine compared with 28% of patients receiving amifostine ($p < .05$). The incidence of grade 2 or higher esophagitis during the fourth week of treatment was 4% among patients receiving amifostine, compared with 42% of patients receiving radiotherapy alone ($p < .001$). No evidence of tumor protection by amifostine was noted: complete or partial responses were observed in 75% and 76% of patients receiving amifostine or radiotherapy alone, respectively.

Komaki et al. evaluated the cytoprotective role of amifostine for esophagitis and hematologic and pulmonary toxicities in a randomized study of patients with stage II or III non-small cell lung cancer receiving concurrent chemoradiotherapy. Patients in the study group received amifostine, 500 mg i.v., twice weekly before chemoradiation, and patients in the control group received chemoradiation without Amifostine (Komaki et al., 2004). The median survival time was longer, but not significantly so, for patients receiving amifostine (26 months versus 15 months). Significantly fewer patients who received amifostine also received morphine to relieve severe esophagitis (7.4%) than patients who received chemoradiotherapy alone (31%; $p = .03$). Amifostine treatment was also associated with a significantly lower incidence of acute pneumonitis (3.7% versus 23%; $p = .037$). Although not statistically significant, 26% of patients receiving amifostine had a complete response, compared with 8% of patients who did not receive amifostine ($p = .07$).

Despite a limited number of studies, a recent meta-analysis reported that amifostine treatment was observed to reduce the incidence of pneumonitis and esophagitis for patients undergoing radiotherapy for lung cancer [6]. However, results from the largest multicenter study conducted to date were unable to show a reduction in the incidence of esophagitis with amifostine treatment (Movsas et al., 2005). A phase III study conducted by the RTOG (trial 9801) treated 243 patients with favorable-prognosis inoperable stage II–IIIA/B non-small cell lung cancer with concurrent hyperfractionated radiotherapy plus paclitaxel (50 mg/m2) and carboplatin (dosed to achieve area under the concentration–time curve of 2) (Movsas et al., 2005). Half of the patients also received i.v. amifostine (500 mg) before the afternoon radiation treatment. During the course of the study, esophagitis was measured via National Cancer Institute Common Toxicity Criteria maximum esophagitis grade, physician dysphagia log, and patient daily self-assessment of swallowing ability. No significant differences in esophagitis were observed for patients receiving amifostine compared with those who did not receive amifostine, with the exception of the patient-reported lower rate of swallowing dysfunction observed in amifostine-treated patients (z test, $p = .025$). The authors attributed the lack of significant reduction in esophagitis with amifostine to several factors, including the timing of amifostine administration (Movsas et al., 2005), given that preclinical studies suggest that a single morning dose of amifostine provides superior radioprotection than with a single afternoon dose (Bachy et al., 2003; Fanzenbaker et al., 2003). The randomized trials involved in cytoprotection for lung irradiation are shown in table 2.

1.3 Lower gastrointestinal mucositis

Lower gastrointestinal mucositis frequently results from pelvic irradiation. Several clinical trials have demonstrated that amifostine pretreatment before radiotherapy or chemoradiotherapy can reduce the incidence and severity of gastrointestinal toxicities that commonly occur following these treatments (Table 2) (Koukourakis et al., 2000; Antonadou et al., 2004; Athanasiou et al., 2003; Ben-Josef et al., 2002; Dunst et al., 2000; Kouloulias et al., 2004; Kouloulias et al., 2005; Kouvaris et al., 2003; Liu et al., 1992; Simone et al., 2005; Singh et al., 2006). Guidelines published by the Mucositis Study Group of the Multinational Association of Supportive Care in Cancer/International Society for Oral Oncology recommend the use of amifostine (340 mg/m^2) to prevent proctitis in patients receiving standard-dose radiotherapy (Bensadoun et al., 2006). Furthermore, these studies demonstrate that various routes of administration of amifostine (i.v., s.c., and intrarectal) are effective at reducing radiation- and chemoradiation-induced gastrointestinal toxicities in patients with pelvic malignancies. One study, conducted in 53 patients with prostate or gynecologic cancer, directly compared intrarectal amifostine administration with s.c. administration and found that intrarectal administration was more effective at reducing radiotherapy-induced rectal toxicities, whereas s.c. administration was more effective at reducing radiotherapy-induced urinary toxicities (Table 3) (Kouloulias et al., 2005). These results suggest that optimal cytoprotection may be achieved by combining routes of amifostine administration during treatment.

1.4 Dermatitis

Protection by amifostine against radiation-induced dermatitis was assessed in a retrospective analysis in which 100 patients with pelvic tumors treated with radiotherapy and amifostine were compared with 120 historical controls who was not administered

Reference	Radiation dose	Chemotherapy	Amifostine dose	Comment
Movsas et al., 2005 (n = 242)	69.6 Gy at 1.2 Gy (hyperfractionation)	Induction PC	500 mg i.v. 4doses/wk between RT fractions	No difference by NCI-CTC esophagitis, Swallowing diaries (p < 0.03) and weight loss (p < 0.05) favour amifostine (median survival, 15.6 and 15.8 mo)
Leong et al. , 2003 (n = 60)	60–66 Gy at 2.0 Gy	Induction PC	740 mg/m² with each chemo (Days 1, 22, 43, 50, 57,64, 71, 78)	Esophagitis Grade 2–3: 43% in amifostine, 70% in control (not significant) (median survival, 12.5 and 14.5 mo)
Senzer et al., 2002 (n = 63)	64.8 Gy at 1.8 Gy	Concurrent PC, gemcitabine and cisplatin X 3 after chemoradiation	500 mg i.v. before weekly chemo; 200 mg i.v. daily before RT	No difference in toxicity, no survival data (ongoing trial)
Antonadou et al., 2001 (n = 146)	55–60 Gy at 2.0 Gy.	None	340 mg/m²/d before RT	↓Pneumonitis ↓Esophagitis (no survival data)
Antonadou et al., 2003 (n = 73)	55–60 Gy at 2.0Gy.	Concurrent weekly Por C	300 mg/m²/d before chemoradiation and RT	↓esophagitis (p <0.001) ↓pneumonitis (p = 0.009) (no survival data)
Komaki et al., 2004 (n = 62)	69.6 Gy at 1.2 Gy (hyperfractionation)	Concurrent i.v. cisplatin Days 1, 8,29, 36; Oral etoposide Days1-5 8-12, 29-33,36-40	500 mg i.v. 1st, 2nd day each wk before chemo and 1st RT fraction	↓Degree of esophagitis, ↓Pneumonitis, ↓Neutropenic fever (median survival,19 and 20 mo)

Abbreviations: P = paclitaxel; C = carboplatin; RT = radiotherapy; NCI-CTC = National Cancer Institute-Common Toxicity Criteria.

Table 2. Randomized trials with amifostine in lung cancer

amifostine (Kouvaris et al., 2002). There was a 77% lower risk for radiation-induced dermatitis with amifostine use. The severity of dermatitis was also significantly lower among patients receiving amifostine compared with historical controls: the mean gross dermatitis scores were 0.18 ± 0.09 versus 1.0 ± 0.11 (p < .001). In another study of 40 patients receiving radiation treatment for pelvic tumors, grade 2 or 3 dermatitis of the perineal/vulvar area was observed in all patients with gynecologic and rectal cancer who did not receive amifostine (500 mg s.c.) (Koukourakis et al., 20000). Among patients who received amifostine, only grade 1 dermatitis was noted.

Author	N	Rectal Toxicity (Control vs Amifostine)	P Value	Remarks
Liu et al., 1992	100	14% vs 0% ; moderate or severe late toxicities	0.03	Randomized (intravenous)
Dunst et al., 2000	30	1.07±1.03 vs 0.40±0.63; maximum diarrhea score	0.044	Nonrandomized (intravenous)
Antonadou et al., 2004	124	5.6% for amifostine vs 22.2% for control	<0.011	Randomized
Kligerman et al., 1992	100	5% vs 0% moderate or severe late toxicity	<0.01	Randomized (intravenous)
Kouvaris et al., 2002	220	Grade I/II toxicity, 70% vs 42%	<0.001	Nonrandomized (retrospective, intravenous)
Ben-Josef et al., 2002	29	Grade I/II toxicity, 50% (500-1000 mg amifostine) vs 15% (1500-2500 mg amifostine)	0.0325	Nonrandomized (intrarectal)
Koukourakis et al., 2000	40	Grade III/IV observed in 15% for control vs 0% for amifosine group	<0.05	Randomized
Kouvaris et al., 2003	36	Grade I/II toxicity, 88% vs 11%	<0.001	Randomized (intravenous)
Muller et al., 2004	6	Leukocytes and lymphocytes irradiated were radioprotected (comet assay measurements)	<0.05	Nonrandomized
Athanasiou et al., 2003	205	Grade II/III acute toxicity, 22.1% vs 5.5% (3rd wk of radiation)	0.001	Randomized (intravenous)
Kouloulias et al., 2004	67	Grade I/II acute toxicity, 44% vs 15% for IR	0.026	Randomized (intrarectal)
Singh et al., 2006; Simone et al., 2005	30	33% in 1gr IR vs 0% in 2gr IR	0.06	Randomized
Kouloulias et al., 2005	53	Grade I/II acute toxicity, 42% for s.c. Vs 11% for IR	0.04	Randomized (subcutaneous vs intrarectal)

IR=intrarectal; s.c.=subcutneous

Table 3. Clinical Trials of Amifostine With Radiotherapy in Pelvic Tumors

2. Different ways of administration

2.1 i.v. Amifostine

The American Society of Clinical Oncology guidelines for cytoprotective agents recommend amifostine at a dose of 200 mg/m^2 daily, given as a slow i.v. push over 3 minutes, 15-30 minutes before each fraction of radiation therapy (Hensley et al., 1999). Adverse events are reduced at this lower dose. Nonetheless, administration of amifostine requires close patient monitoring. Many patients require antiemetics. Hypotension associated with amifostine at this dose is less frequent but still requires close monitoring. Blood pressure should be measured before and immediately after the 3-minute amifostine infusion.

2.2 s.c. Amifostine

The s.c. administration of amifostine has been proposed to reduce treatment-related and dose-limiting adverse events (Koukourakis et al., 20000). In a pharmacokinetic study, the plasma concentration of WR-1605 after s.c. injection of 500 mg of amifostine was 67% of that after a 200 mg/m2 i.v. dose (Shaw et al., 1997). Lower plasma levels of amifostine after s.c. injection do not necessarily translate to lower tissue concentrations. Because the amount of amifostine that is absorbed and converted to the active metabolites is not dependent on plasma pharmacokinetics, i.v. or s.c. administration may not have a significant impact on whether therapeutic levels are achieved in the tissues. Precise determination of the protective efficacy of different routes of administration will require more comprehensive studies that measure intracellular levels of the metabolites or assess radiation-induced DNA double-strand breaks in tissues after i.v. or s.c. administration of amifostine. Nonetheless, the efficacy of s.c. amifostine administration is best addressed in the context of a clinical trial.

A phase II randomized trial with 140 patients assessed the feasibility, tolerance, and activity of the s.c. route (Koukourakis et al., 20000). A dose of amifostine of 500 mg s.c. was administered 20 minutes before each fraction of radiotherapy. The s.c. administration of amifostine was well tolerated by 85% of patients. In approximately 15% of patients, amifostine therapy was interrupted because of cumulative asthenia or a fever/rash reaction. Mild nausea was frequent (29%), and the incidence of hypotension was negligible (3%). Significantly less pharyngeal, esophageal, and rectal mucositis was observed among patients receiving amifostine ($p < .04$). Treatment delays because of grade 3 mucositis were significantly longer in patients treated with radiotherapy alone ($p < .04$).

2.3 Endorectal

Initial attempts with rectal administration of amifostine admixed in a foam did not demonstrate protection in patients receiving large pelvic fields of radiation (Montana et al., 1992). However, after successful topical application of amifostine in the rectum of male rats (Ben-Josef et al., 1995), subsequent significant clinical benefit of endorectal administration of amifostine was demonstrated in a phase I study (Ben-Josef et al., 2002).

A randomized trial of 67 patients undergoing radiotherapy for prostate cancer further assessed intrarectal administration of amifostine (Kouloulias et al., 2002). Patients were treated with or without amifostine at a dose of 1,500 mg intrarectally 20–30 minutes before each radiotherapy session. All patients receiving amifostine completed therapy without amifostine-related toxicities, suggesting that intrarectal amifostine was feasible and well tolerated. According to RTOG grading criteria, amifostine was superior to no treatment, with a significantly lower incidence of rectal mucositis (15% versus 44%; $p < .04$). The mean rectal mucositis index of patients who received amifostine was 0.3 ± 0.1 compared with 2.2 ± 0.4 in patients without cytoprotection ($p < .001$). The severity of rectal mucositis was significantly lower in patients who received amifostine ($p < .001$). Urinary toxicity was comparable between the two groups ($p = .76$). A more recent study suggests that the efficacy of intrarectal amifostine may be dose dependent. Although not statistically significant, the incidence of acute grade 2 rectal mucositis was lower in patients receiving a 2-g suspension of amifostine ($n = 12$) than in those receiving 1 g ($n = 18$; $p = .06$) (Sigh et al., 2006). No breaks in treatment for radiation-induced toxicities were required in that study. A combination of intrarectal and s.c. amifostine administration might be optimal for cytoprotection with pelvic irradiation.

3. New grading scale for acute pelvic radiation induced toxicity

The past years all the trials with amifostine and pelvic radiotherapy used as endpoints the WHO and RTOG/EORTC scales as shown in table 4.

	Grade 0	Grade 1	Grade 2	Grade 3	Grade 4
WHO Toxicity Grade	None	Increase of 2–3 stools per d over pretreatment	Increase of 4– 6 stools per d, or nocturnal stools, or moderate cramping	Increase of 7–9 stools per d, or incontinence, or severe cramping	Increase of >10 stools per d or grossly bloody diarrhea, or need for parenteral support
EORTC-RTOG scale for lower gastro-intestinal	None	Increased frequency or change in quality of bowel habits not requiring medication, rectal discomfort not requiring analgesics	Diarrhea requiring parasympatholytic drugs, mucous discharge not necessitating sanitary pads, rectal or abdominal pain requiring analgesics	Diarrhea requiring parenteral support, severe mucous or blood discharge necessitating sanitary pads/abdominal distension (flat plate radiograph demonstrates distended bowel loops)	Acute or subacute obstruction, fistula or perforation; gastrointestinal bleeding requiring transfusion; abdominal pain or tenesmus requiring tube decompression or bowel diversion

EORTC-RTOG= European Organization for Research and Treatment of Cancer/Radiation Therapy Oncology Group; WHO=World Health Organization.

Table 4. WHO Toxicity Criteria and RTOG Acute Radiation Morbidity Scoring Criteria

A specific analytical for subjective and objective measurements was introduced. Endoscopy offers accurate endpoints for the evaluation of tissue damage, whereas the criteria of rectosigmoidoscopy findings are still not well defined in the literature. The literature deals mainly with symptomatic patients presenting with rectal bleeding, pain, increased stool frequency, urgency and incontinence, whereas systematic endoscopic analysis including asymptomatic patients rarely exists. A valid scoring system is essential for adequate description of acute rectal toxicity. For the benefit of sharing and comparing data collected from endoscopy after RT, we have introduced a graduation system based on rectosigmoidoscopic criteria focused on acute effects and standardized terminology published by the European Society for Gastrointestinal Endoscopy. The scale is shown in table 5.

	Grade 1	Grade 2	Grade 3	Grade 4
Subjective				
Tenesmus	Occasional urgency	Intermittent urgency	Persistent urgency	Refractory
Mucosal loss	Occasional	Intermittent	Persistent	Refractory
Sphincter control	Occasional	Intermittent	Persistent	Refractory
Stool frequency	2–4 per d	4–8 per d	>8 per d	Uncontrolled diarrhoea
Pain	Occasional & minimal	Intermittent & tolerable	Persistent & intense	Refractory & excruciating
Objective				
Bleeding	Occult	Occasionally >2 per wk	Persistent, daily	Gross hemorrhage
Mucosa surface	Localized spotted, congested mucosa	Punctate, congested mucosa	Diffused, congested mucosa	Bleeding mucosa
Ulceration	Superficial ≤1 cm^2	Superficial >1 cm^2	Deep ulcer	Surgical intervention

*Modification to Subjective Objective Management Analytic scale to fit radiation-induced acute toxicity to the rectum. Subjective and objective items were used for evaluation of acute radiation-induced rectal mucositis. The second and third items of the objective scale were based on findings from flexible rectosigmoidoscopy and were in accordance with the endoscopic terminology of the World Organization for Digestive Endoscopy. The final score was the sum of scores of the 8 items (score=0 in the absence of toxicity).

Table 5. Rectal Toxicity Grade*

Our experience has shown that the terminology is practicable and provides a definition of terms usable by radiation oncologist and endoscopists. The S-RS scale showed a satisfied clinical validity and reliability (Kouvaris etal., 2003). In a previous publication (Kouloulias et al., 2004) the rectosigmoidoscopic findings for amifostine versus no amifostine showed significant differences, as shown in figure 1.

4. Conclusion

Normal tissues vary in the extent that they are protected from radiation damage by amifostine. Because amifostine does not cross the blood-brain barrier, the central nervous system, often the dose-limiting organ in radiotherapy, is not protected (Millar et al., 1982; Washburn et al., 1976). Protection factors for other tissues range from three in the hematopoietic system and salivary glands to approximately one in the lung, kidney, and bladder (Rojas et al., 1984; Rojas et al., 1986; Travis 1984). Within the same tissues, a range of protection factors has been reported (Rojas et al., 1984; Mori et al., 1984). Discrepancies in WR-1065 concentrations in tissues within 15–30 minutes of administration (Utley et al., 1984) and the normal interval between administration of amifostine and radiotherapy may explain these differences. Nonhomogenous distribution of amifostine and its metabolites within a tissue, even at the level of the DNA (savoye et al., 1997), may also contribute to this heterogeneity.

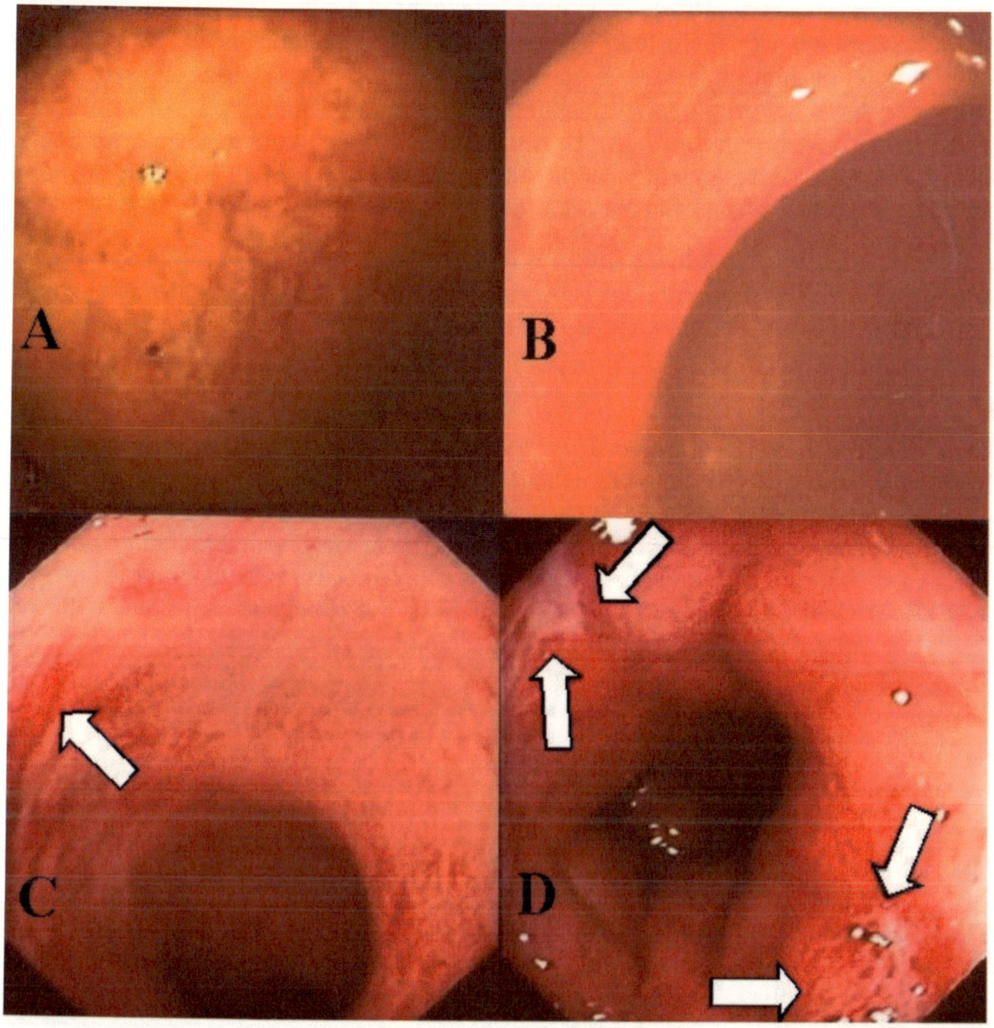

Fig. 1. Rectosigmoidoscopic findings. Panels A and B illustrate a regular rectal mucosa in patients after intrarectal administration of amifostine. Panels C and D are from patients who did not receive amifostine and illustrate congested mucosa with superficial ulceration >1 cm^2 (indicated by the arrows).

The U.S. Food and Drug Administration has approved the i.v. use of amifostine to reduce the cumulative renal toxicity associated with repeated administration of cisplatin in patients with advanced ovarian cancer and to reduce the incidence of moderate to severe xerostomia in patients undergoing postoperative radiation treatment for head and neck cancer, where the radiation port includes a substantial portion of the parotid glands. Amifostine has been proven to be a useful addition to the arsenal of the radiation oncologist, helping improve patients' quality of life and in some cases allowing more aggressive radio- and chemotherapeutic regimens. Currently, s.c. administration of amifostine is a standard practice for patients with head and neck cancer as well as for patients with recurrent ovarian carcinoma (Hensley et al., 2009), while there is evidence now that the s.c is not superior to i.v. administration of amifostine (Bardet et al., 2011). Refinements in doses and administration of amifostine lead to constant improvement in the adverse event profile, resulting in fewer interruptions in treatment and ultimately improving patient outcomes (Jellena et al., 2006; Bourhis et al., 2011). At last but not least, recent meta-analysis from 16 randomized trials (1,554 patients) confirmed the lack of any tumor protection in routine radiotherapy practice when amifostine is administered (Bourhis et al., 2011).

It has to be mentioned that the graduation system designed by our group is user-friendly and more than this, it is an interface between radiation-oncologists and gastroenterologists by means of common terminology between specializations for radiation induced rectal toxicity.

5. References

American Cancer Society. Cancer Facts and Figures 2005. Atlanta, GA: American Cancer Society, 2005:60.

Anne PR, Curran WJ Jr. A phase II trial of subcutaneous amifostine and radiation therapy in patients with head and neck cancer. Semin Radiat Oncol 2002;12(suppl 1):18–19.

Anne PR, Machtay M, Rosenthal DI et al. A Phase II trial of subcutaneous amifostine and radiation therapy in patients with head-and-neck cancer. Int J Radiat Oncol Biol Phys 2007;67:445–452.

Antonadou D, Athanassiou H, Sarris N et al. Final results of a randomized phase III trial of chemoradiation treatment + amifostine in patients with colorectal cancer: Clinical Radiation Oncology Hellenic Group. Presented at the 46th Annual Meeting of the American Society for Therapeutic Radiology and Oncology; October 3–7; Atlanta, GA, 2004.

Antonadou D, Coliarakis N, Synodinou M et al. Randomized phase III trial of radiation treatment +/- amifostine in patients with advanced-stage lung cancer. Int J Radiat Oncol Biol Phys 2001;51:915–922.

Antonadou D, Pepelassi M, Synodinou M et al. Prophylactic use of amifostine to prevent radiochemotherapy-induced mucositis and xerostomia in head-and-neck cancer. Int J Radiat Oncol Biol Phys 2002;52:739–747.

Antonadou D, Throuvalas N, Petridis A, et al. Effect of amifostine on toxicities associated with radiochemotherapy in patients with locally advanced non-small-cell lung cancer. Int J Radiat Oncol Biol Phys 2003;57:402–408.

Athanassiou H, Antonadou D, Coliarakis N et al. Protective effect of amifostine during fractionated radiotherapy in patients with pelvic carcinomas: Results of a randomized trial. Int J Radiat Oncol Biol Phys 2003;56:1154–1160.

Bachy CM, Fazenbaker CA, Kifle G et al. Daily dosing with amifostine is necessary for full protection against oral mucositis caused by fractionated radiation in rats: Protection and pharmacokinetics. Proc Am Soc Clin Oncol 2003;22:518.

Bardet E, Martin L, Calais G, et al. Subcutaneous compared with intravenous administration of amifostine in patients with head and neck cancer receiving radiotherapy: final results of the GORTEC2000-02 phase III randomized trial.J Clin Oncol. 2011 Jan 10;29(2):127-33.

Ben-Josef E, Han S, Tobi M et al. A pilot study of topical intrarectal application of amifostine for prevention of late radiation rectal injury. Int J Radiat Oncol Biol Phys 2002;53:1160–1164.

Ben-Josef E, Mesina J, Shaw LM et al. Topical application of WR-2721 achieves high concentrations in the rectal wall. Radiat Res 1995;143:107–110.

Bensadoun RJ, Schubert MM, Lalla RV et al. Amifostine in the management of radiation-induced and chemo-induced mucositis. Support Care Cancer 2006;14:566–572.

Blumberg AL, Nelson DF, Gramkowski M et al. Clinical trials of WR-2721 with radiation therapy. Int J Radiat Oncol Biol Phys 1982;8:561–563.

Bourhis J, Blanchard P, Maillard E, et al. Effect of Amifostine on Survival Among Patients Treated With Radiotherapy: A Meta-Analysis of Individual Patient Data. J Clin Oncol. 2011 May 16.

Bourhis J, De Crevoisier R, Abdulkarim B et al. A randomized study of very accelerated radiotherapy with and without amifostine in head and neck squamous cell carcinoma. Int J Radiat Oncol Biol Phys 2000;46:1105–1108.

Brizel DM, Wasserman TH, Henke M et al. Phase III randomized trial of amifostine as a radioprotector in head and neck cancer. J Clin Oncol 2000;18:3339–3345.

Buentzel J, Micke O, Adamietz IA et al. Intravenous amifostine during chemoradiotherapy for head-and-neck cancer: A randomized placebo-controlled phase III study. Int J Radiat Oncol Biol Phys 2006;64:684–691.

Buntzel J, Glatzel M, Frohlich D et al. Intensification of radiochemotherapy with Ethyol in head and neck cancer. Proc Am Soc Clin Oncol 1998;17:403A.

Buntzel J, Kuttner K, Frohlich D et al. Selective cytoprotection with amifostine in concurrent radiochemotherapy for head and neck cancer. Ann Oncol 1998;9:505–509.

Calabro-Jones PM, Fahey RC, Smoluk GD et al. Alkaline phosphatase promotes radioprotection and accumulation of WR-1065 in V79–171 cells incubated in medium containing WR-2721. Int J Radiat Biol Relat Stud Phys Chem Med 1985;47:23–27.

Carmeliet P, Dor Y, Herbert JM et al. Role of HIF-1alpha in hypoxia-mediated apoptosis, cell proliferation and tumour angiogenesis. Nature 1998;394:485–490.

Dunst J, Semlin S, Pigorsch S et al. Intermittent use of amifostine during postoperative radiochemotherapy and acute toxicity in rectal cancer patients. Strahlenther Onkol 2000;176:416–421.

Eisbruch A, Ten Haken RK, Kim HM et al. Dose, volume, and function relationships in parotid salivary glands following conformal and intensity-modulated irradiation of head and neck cancer. Int J Radiat Oncol Biol Phys 1999;45:577–587.

Fazenbaker CA, Bachy CM, Kifle G et al. Dose and schedule dependency of amifostine protection against hyperfractionated radiotherapy in a rat model. Proc Am Soc Clin Oncol 2003;22:518.

Glover D, Negendank W, Delivoria-Papadopoulos M et al. Alterations in oxygen transport following WR-2721. Int J Radiat Oncol Biol Phys 1984;10:1565–1568.

Glover D, Riley L, Carmichael K et al. Hypocalcemia and inhibition of parathyroid hormone secretion after administration of WR-2721 (a radioprotective and chemoprotective agent). N Engl J Med 1983;309:1137–1141.

Hensley ML, Hagerty KL, Kewalramani T, et al. American Society of Clinical Oncology 2008 clinical practice guideline update: use of chemotherapy and radiation therapy protectants. J Clin Oncol. 2009 Jan 1;27(1):127-45.

Hensley ML, Schuchter LM, Lindley C et al. American Society of Clinical Oncology clinical practice guidelines for the use of chemotherapy and radiotherapy protectants. J Clin Oncol 1999;17:3333–3355.

Jellema AP, Slotman BJ, Muller MJ, et al. Radiotherapy alone, versus radiotherapy with amifostine 3 times weekly, versus radiotherapy with amifostine 5 times weekly: A prospective randomized study in squamous cell head and neck cancer. Cancer. 2006 Aug 1;107(3):544-53.

Kajstura J, Cheng W, Reiss K et al. Apoptotic and necrotic myocyte cell deaths are independent contributing variables of infarct size in rats. Lab Invest 1996;74:86–107.

Kligerman MM, Liu T, Liu Y, Scheffler B, He S, Zhang Z. Interim analysis of a randomized trial of radiation therapy of rectal cancer with/without WR-2721. Int J Radiat Oncol Biol Phys. 1992;22(4):799-802

Komaki R, Lee JS, Milas L, et al. Effects of amifostine on acute toxicity from concurrent chemotherapy and radiotherapy for inoperable non-small-cell lung cancer: report of a randomized comparative trial. Int J Radiat Oncol Biol Phys. 2004;58(5):1369-77

Koukourakis MI, Flordellis CS, Giatromanolaki A et al. Oral administration of recombinant human granulocyte macrophage colony-stimulating factor in the management of radiotherapy-induced esophagitis. Clin Cancer Res 1999;5:3970–3976.

Koukourakis MI, Kyrias G, Kakolyris S et al. Subcutaneous administration of amifostine during fractionated radiotherapy: A randomized phase II study. J Clin Oncol 2000;18:2226–2233.

Koukourakis MI. Amifostine: Is there evidence of tumor protection? Semin Oncol 2003;30(suppl 18):18–30.

Kouloulias VE, Kouvaris JR, Pissakas G et al. A phase II randomized study of topical intrarectal administration of amifostine for the prevention of acute radiation-induced rectal toxicity. Strahlenther Onkol 2004;180:557–562.

Kouloulias VE, Kouvaris JR, Pissakas G et al. Phase II multicenter randomized study of amifostine for prevention of acute radiation rectal toxicity: Topical intrarectal versus subcutaneous application. Int J Radiat Oncol Biol Phys 2005;62:486–493.

Kouvaris J, Kouloulias V, Kokakis J et al. The cytoprotective effect of amifostine in acute radiation dermatitis: A retrospective analysis. Eur J Dermatol 2002;12:458–462.

Kouvaris J, Kouloulias V, Kokakis J, Matsopoulos G, Balafouta M, Miliadou A, Vlahos L. Cytoprotective effect of amifostine in radiation-induced acute mucositis - a retrospective analysis. Onkologie. 2002 Aug;25(4):364-9.

Kouvaris J, Kouloulias V, Malas E et al. Amifostine as radioprotective agent for the rectal mucosa during irradiation of pelvic tumors. A phase II randomized study using various toxicity scales and rectosigmoidoscopy. Strahlenther Onkol 2003;179:167–174.

Leong SS, Tan EH, Fong K, et al. Randomized double-blind trial of combined modality treatment with or without amifostine in unresectable stage III non-small-cell lung cancer. J Clin Oncol 2003;21:1767–1774.

Liu T, Liu Y, He S et al. Use of radiation with or without WR-2721 in advanced rectal cancer. Cancer 1992;69:2820–2825.

McDonald S, Meyerowitz C, Smudzin T et al. Preliminary results of a pilot study using WR-2721 before fractionated irradiation of the head and neck to reduce salivary gland dysfunction. Int J Radiat Oncol Biol Phys 1994;29:747–754.

Millar JL, McElwain TJ, Clutterbuck RD et al. The modification of melphalan toxicity in tumor bearing mice by s-2-(3-aminopropylamino)- ethylphosphorothioic acid (WR 2721). Am J Clin Oncol 1982;5:321–328.

Montana GS, Anscher MS, Mansbach CM 2nd et al. Topical application of WR-2721 to prevent radiation-induced proctosigmoiditis. A phase I/II trial. Cancer 1992;69:2826–2830.

Mori T, Nikaido O, Sugahara T. Dephosphorylation of WR-2721 with mouse tissue homogenates. Int J Radiat Oncol Biol Phys 1984;10:1529–1531.

Movsas B, Scott C, Langer C et al. Randomized trial of amifostine in locally advanced non-small-cell lung cancer patients receiving chemotherapy and hyperfractionated radiation: Radiation Therapy Oncology Group trial 98-01. J Clin Oncol 2005;23:2145–2154.

Müller AC, Pigorsch S, Beyer C, et al. Radioprotective effects of amifostine in vitro and in vivo measured with the comet assay. Strahlenther Onkol. 2004 Aug;180(8):517-25.

Munter MW, Hoffner S, Hof H et al. Changes in salivary gland function after radiotherapy of head and neck tumors measured by quantitative pertechnetate scintigraphy: Comparison of intensity-modulated radiotherapy and conventional radiation therapy with and without amifostine. Int J Radiat Oncol Biol Phys 2007;67:651–659.

Peters K, Mucke R, Hamann D et al. Supportive use of amifostine in patients with head and neck tumors undergoing radio-chemotherapy. Is it possible to limit the duration of the application of amifostine? Strahlenther Onkol 1999;175(suppl 4):23–26.

Purdie JW, Inhaber ER, Schneider H et al. Interaction of cultured mammalian cells with WR-2721 and its thiol, WR-1065: Implications for mechanisms of radioprotection. Int J Radiat Biol Relat Stud Phys Chem Med 1983;43:517–527.

Rojas A, Denekamp J. The influence of X ray dose levels on normal tissue radioprotection by WR-2721. Int J Radiat Oncol Biol Phys 1984;10:2351–2356.

Rojas A, Stewart FA, Soranson JA et al. Fractionation studies with WR-2721: Normal tissues and tumour. Radiother Oncol 1986;6:51–60.

Rubin DB, Drab EA, Kang HJ et al. WR-1065 and radioprotection of vascular endothelial cells. I. Cell proliferation, DNA synthesis and damage Radiat Res, 1996;145:210–216.

Sasse AD, Clark LG, Sasse EC et al. Amifostine reduces side effects and improves complete response rate during radiotherapy: Results of a meta-analysis. Int J Radiat Oncol Biol Phys 2006;64:784–791.

Savoye C, Swenberg C, Hugot S et al. Thiol WR-1065 and disulphide WR-33278, two metabolites of the drug ethyol (WR-2721), protect DNA against fast neutron-induced strand breakage. Int J Radiat Biol 1997;71:193–202.

Senzer N. A phase III randomized evaluation of amifostine in stage IIIA/IIIB non-small cell lung cancer patients receiving concurrent carboplatin, paclitaxel, and radiation therapy followed by gemcitabine and cisplatin intensification: Preliminary findings. Semin Oncol 2002;29:38–41.

Shaw L, Brown W, Schein P et al. A phase I study comparing the relative bioavailability of intravenous (i.v.) and subcutaneous (s.c.) administration of amifostine (Ethyol). Proc Am Soc Clin Oncol 1997;16:250a.

Shaw LM, Turrisi AT, Glover DJ et al. Human pharmacokinetics of WR-2721. Int J Radiat Oncol Biol Phys 1986;12:1501–1504.

Shimizu S, Eguchi Y, Kamiike W et al. Induction of apoptosis as well as necrosis by hypoxia and predominant prevention of apoptosis by Bcl-2 and Bcl-XL. Cancer Res 1996;56:2161–2166.

Simone N, Menard C, Singh A. Intrarectal amifostine suspension may protect against acute proctitis during prostate radiation therapy: A pilot study. Presented at the Radiological Society of North America Annual Meeting; November 27–December 2; Chicago, IL, 2005.

Singh AK, Menard C, Guion P et al. Intrarectal amifostine suspension may protect against acute proctitis during radiation therapy for prostate cancer: A pilot study. Int J Radiat Oncol Biol Phys 2006;65:1008–1013.

Sonis ST, Fey EG. Oral complications of cancer therapy. Oncology (Williston Park) 2002;16:680–686, 691–692, 695; discussion 686.

Stone HB, Coleman CN, Anscher MS et al. Effects of radiation on normal tissue: Consequences and mechanisms. Lancet Oncol 2003;4:529–536.

Travis EL. The oxygen dependence of protection by aminothiols: Implications for normal tissues and solid tumors. Int J Radiat Oncol Biol Phys 1984;10:1495–1501.

Utley JF, Seaver N, Newton GL et al. Pharmacokinetics of WR-1065 in mouse tissue following treatment with WR-2721. Int J Radiat Oncol Biol Phys 1984;10:1525–1528.

Vacha P, Fehlauer F, Mahlmann B et al. Randomized phase III trial of postoperative radiochemotherapy +/- amifostine in head and neck cancer. Is there evidence for radioprotection? Strahlenther Onkol 2003;179:385–389.

Wagner W, Prott FJ, Schonekas KG. Amifostine: A radioprotector in locally advanced head and neck tumors. Oncol Rep 1998;5:1255–1257.

Washburn LC, Rafter JJ, Hayes RL. Prediction of the effective radioprotective dose of WR-2721 in humans through an interspecies tissue distribution study. Radiat Res 1976;66:100–105.

Wasserman TH, Brizel DM, Henke M et al. Influence of intravenous amifostine on xerostomia, tumor control, and survival after radiotherapy for head-and-neck cancer: 2-year follow-up of a prospective, randomized, phase III trial. Int J Radiat Oncol Biol Phys 2005;63:985–990.

Yuhas JM, Spellman JM, Culo F. The role of WR-2721 in radiotherapy and/or chemotherapy. Cancer Clin Trials 1980;3:211–216.

Yuhas JM, Storer JB. Differential chemoprotection of normal and malignant tissues. J Natl Cancer Inst 1969;42:331–335.

Yuhas JM. Active versus passive absorption kinetics as the basis for selective protection of normal tissues by S-2-(3-aminopropylamino)-ethylphosphorothioic acid. Cancer Res 1980;40:1519–1524.

15

Abscopal Effect of Radiation Therapy: Current Concepts and Future Applications

Kenshiro Shiraishi
Department of Radiology,
The University of Tokyo Hospital,
Japan

1. Introduction

Radiation therapy is one of the most important treatment tools in cancer therapy. It has a wide variety of indications for many malignant tumors, mostly for local control, whether a curative or palliative outcome is the intent, or as pre- or post-operative treatment as either neoadjuvant or adjuvant therapy. Radiation therapy is commonly used along with hormone therapy or chemotherapy. The full scope of the capabilities of radiation therapy is achieved particularly in combination settings with various anti-tumor modalities, the so-called multidisciplinary approach. To enhance the therapeutic efficacy of radiation sufficiently, one may choose radiation therapy in combination with cytotoxic chemotherapeutic agents or with warming devices used for hyperthermia treatment or utilize newly developing physical approaches as typified by intensity modulated radiation therapy, stereotactic radiation therapy, brachytherapy and image-guided radiation therapy. Moreover, an immunoenhancing agent might be selected in combination with radiation therapy from the standpoint of immunobiology in the treatment of cancer. Some promising data have been shown on the basis of immunological activation with ionizing radiation, demonstrating cytotoxic T lymphocyte (CTL) amplification and dendritic cell (DC) stimulation or maturation (Demaria, et al., 2004,Ganss, et al., 2002,Nikitina and Gabrilovich, 2001,Schuler, et al., 2003).

Radiation therapy plays a crucial role in enhancing tumor immunogenicity by promoting cross-priming and eliciting anti-tumor T-cell responses, and generates inflammatory signals via induction of tumor cell death (Hong, et al., 1999,Quarmby, et al., 1999,Watters, 1999). Thus, ionizing radiation can achieve not only direct cancer cell death but also has an indirect and systemic anti-tumor mechanism outside of the irradiated field, which has been reported in some clinical settings (Antoniades, et al., 1977,Ehlers and Fridman, 1973,Kingsley, 1975,Nobler, 1969,Perego and Faravelli, 2000,Rees, 1981,Rees and Ross, 1983,Sham, 1995). Local irradiation resulted in an anti-tumor effect at a non-irradiated location in a patient with hepatocellular carcinoma that regressed after palliative local radiotherapy for pain control of bone metastases (Ohba, et al., 1998). This rare phenomenon is known as the abscopal effect and is defined as a reaction following irradiation but occurring outside the zone of actual radiation absorption (Mole, 1953). The word "abscopal" is derived from the Latin prefix "ab," meaning "away from," and the Greek word "scopos," meaning "target."

The abscopal mechanism of action remains to be clarified, although a variety of underlying biological events can be hypothesized, mainly those induced by the immune system (Macklis, et al., 1992,Uchida, et al., 1989). Thus far, immunological activation with local irradiation has been explained by multiple possible mechanisms (Awwad and North, 1990,Cameron, et al., 1990,Chiang, et al., 1997,Dybal, et al., 1992,Younes, et al., 1995,Younes, et al., 1995).

This chapter gives an overview of theoretical mechanisms of the abscopal effect being progressively elucidated in the development of multidisciplinary approaches for cancer therapy.

2. Speculation on the mechanism of the abscopal effect

2.1 Possible cytokine contribution

Historically, the abscopal effect has been described in various tumors with possible underlying mechanisms explaining each observed case. A 76-year-old patient with hepatocellular carcinoma was irradiated to control his bone metastases as palliative, not curative, therapy. Yet following this palliative radiotherapy the primary liver tumor regressed (Ohba, et al., 1998). Ohba *et al.* also found in this patient an increase in blood levels of tumor necrosis factor alpha (TNF-α), which has known anti-tumor activity. They suggested that the primary tumor regression might have been caused by an immune response spearheaded by TNF-α. TNF-α has a paradoxical role in cancer by promoting growth, invasion, and metastasis in some tumors, while having a reverse effect in other cancers through destruction of blood vessels and cell-mediated killing. One wonderful review of the relation between TNF-α and cancer is found in the Lancet Oncology (Szlosarek and Balkwill, 2003).

2.2 Hyperthermia-related abscopal effect

Abscopal effects are usually associated with radiation therapy, however, one could sometimes see after other treatments as well, such as surgery or even hyperthermia. For example, in an experiment conducted in India, administering hyperthermia to the hind leg of a mouse for 40 min before transplanting a fibrosarcoma reduced the growth of the tumor in the heated leg (Vartak, et al., 1993). More surprisingly, it inhibited the growth of a tumor transplanted to the unheated leg as well. Actually, two to three weeks after hyperthermic treatment, tumor growth retardation had ceased in the leg that had been heated, but was still noticeable in the leg that had not been heated. Although the mechanism for this effect had not been investigated, the abscopal effect from hyperthermia turned out to be greater than its direct effect on the local target tumor. The authors concluded that local hyperthermia induced both direct and abscopal anti-tumor effects that might probably be the result of a systemic effect of hyperthermia in the host animal.

2.3 Radiation-related abscopal effect

In the clinical setting, Konoeda *et al.* conducted a practical study to investigate the mechanism of the abscopal effect in patients with breast cancer (Konoeda, 1990). Study subjects were 62 women with advanced breast cancer who received radiation therapy before surgery and then underwent mastectomy or tumor resection. Physical examination, including palpation, indicated an abscopal effect on metastatic lymph nodes in 15 out of 42 cases (35.7%). Pathologic findings revealed an even greater tendency for regression, with an abscopal effect demonstrated in tissue samples from 22 of 42 cases (52.4%). Thus, more than half of these patients with advanced breast cancer exhibited some sort of abscopal effect following irradiation and surgery. The incidence of the abscopal effect was significantly higher in patients under 55 years old and was most frequent in patients who had "infiltrating lymphocytes around the degenerated cancer cells in the irradiated primary tumor nests." In other words, under the favorable condition of a vigorous immune reaction to the tumor as indicated by the presence of abundant lymphocytes, the host was more likely to attack the tumor and bring about an abscopal response as a result. Among the types of lymphocytes, the authors claimed that the most prevalent cells had been identified as primarily CD8 and CD4 lymphocytes, which play a role in cellular defense against pathogens, malignant cells, and other foreign substances. According to the authors, their findings suggested that the abscopal effect was caused by activated cellular immunity in the hosts. Although the study was not large enough for data to yield statistically significant results, the survival rate among patients who exhibited the abscopal effect was higher than among those patients who showed no such reaction.

The logical inference from this research is that the abscopal effect is a desirable and common systemic reaction to localized cancer treatment. Since the abscopal effect is dependent on a healthy immune system, one might infer that immune-damaging treatments should be kept to a minimum. In terms of this point, the trend in most parts of the world is in the undesirable direction, and immunosuppressive chemotherapy is given at every opportunity. The recruitment of leukocytes may have been inhibited by the antitumor chemotherapeutic agents, which would support the assumption that some types of recruited leukocytes play a role in the enhancement of the efficacy of radiation and the abscopal effect.

2.4 Surgery-related abscopal effect

Blay *et al.* reported that higher pretreatment interleukin (IL)-6 and C-reactive protein (CRP) levels in renal cell carcinoma were associated with a diminished response to cytokine therapy and poorer survival. Survival appeared to be better in those patients that had elevated CRP values that decreased to normal levels after nephrectomy compared to those whose CRP did not decrease to normal. For those whose pre-treatment CRP was within normal limits, there was no difference in survival between those who did or did not undergo nephrectomy (Blay, et al., 1992). Fujikawa *et al.* proposed that an IL-6-induced inflammatory response might inhibit the immune anti-tumor response. They suggested the following mechanism: in the setting of metastatic renal cell carcinoma and a primary tumor predominantly expressing IL-6, an associated drop in CRP following nephrectomy appears to curb the inflammatory response while simultaneously inducing immune activation (Fujikawa, et al., 2000).

3. Basic research for induction of radiation-related abscopal effect

3.1 Basic research on the basis of immunological mechanisms

Fms-like tyrosine kinase receptor 3 ligand (Flt3-L) is a growth factor that stimulates production of DCs and has been shown to induce antitumor immunity to several mouse tumors, although its effects as a single agent are limited to early and more immunogenic tumors (Maraskovsky, et al., 1996,Maraskovsky, et al., 1997). The first study to test the combination of Flt3-L with local irradiation used the Lewis lung model of metastatic carcinoma (Chakravarty, et al., 1999). When Flt3-L was administered after the ablation of the primary tumor by high-dose local irradiation with 60 Gy, lung metastasis formation was inhibited and disease-free survival was enhanced compared with that of mice treated with irradiation or Flt3-L alone. Importantly, the anti-metastatic effect required T cells because this effect was not observed in nude (T cell-deficient) mice. These results provide preliminary evidence in support of the hypothesis that radiation-induced tumor cell death can release antigens for DCs to present to T cells. The high single dose of radiation used in this study limits its clinical applicability in addition to the fact that the intrinsic tumor immunogenicity could explain these responses. Nevertheless, these studies provided initial proof of the principle and stimulated some groups to further investigate whether more clinically relevant radiation doses could be used to elicit systemic antitumor immunity in combination with Flt3-L.

Demaria *et al.* used mouse mammary carcinoma 67NR, a moderately immunogenic syngeneic tumor. A radiation dose sufficient to cause growth delay of the irradiated tumor, in this case 2 Gy, was able to induce a systemic antitumor effect only in combination with Flt3-L administration. Inhibition of tumor growth outside of the irradiated field was specific and required T cells, confirming that it was immune-mediated (Demaria, et al., 2004).

Other groups have used a slightly different approach based on the same hypothesis, that radiation can free tumor-derived antigens for DC uptake and presentation. Nikitina *et al.* used *in vitro* bone marrow-derived DCs that were injected i.v. and s.c. around the tumor after local irradiation (Nikitina and Gabrilovich, 2001) whereas Teitz-Tennenbaum *et al.* used intratumoral injection of DCs (Teitz-Tennenbaum, et al., 2003). In both cases, the administration of DCs after radiation therapy was able to induce a potent antitumor immune response. Yasuda *et al.* reported intratumoral IL-2 injection after irradiation to colon adenocarcinoma enhances antitumor local control and abscopal metastatic inhibition via CD4 positive lymphocytes (Yasuda, et al., 2011). In another study, p53 appeared to mediate a radiation-induced abscopal effect in mice that was dose dependent (Camphausen, et al., 2003). Table 1 summarizes the possible underlying mechanisms for the abscopal effects observed preclinically or clinically.

Table 1. Type of malignancies in relation to abscopal effect reported and possible underlying mechanism.

Author	Tumor type	Treated sites (treatment)	Observed abscopal effect	Putative intrinsic mediator that induces abscopal effect
Preclinical				
Vartak *et al*.	fibrosarcoma	hind leg (HT)	tumor growth inhibition of unheated leg	unknown
Chakravarty *et al*.	LLC	primary tumor (RT)	lung metastasis regression	DC
Demaria *et al*.	mammary carcinoma 67NR	primary tumor (RT)	distant tumor growth inhibition	DC
Teitz-Tennenbaum *et al*.	melanoma/sarcoma	primary tumor (RT)	lung metastasis regression	DC
Camphausen *et al*.	LLC/fibrosarcoma	hind leg (RT)	distant tumor growth inhibition	p53
Shiraishi *et al*.	colon adenocarcinoma/LLC/fibrosarcoma	primary tumor (RT)	distant tumor growth inhibition/longer survival	CD8 and CD4 lymphocytes/NK
Iida *et al*.	hepatocellular carcinoma	primary tumor (RFA)	distant tumor growth inhibition	DC
Yasuda *et al*.	colon adenocarcinoma	primary tumor (RT)	liver metastasis inhibition	CD4 lymphocytes
Clinical				
Ohba *et al*.	hepatocellular carcinoma	bone metastasis (RT)	primary tumor regression	TNF-a
Konoeda *et al*.	breast cancer	breast (RT)	metastatic lymph node regression	CD8 and CD4 lymphocytes
Blay *et al*.	renal cell cacinoma	nephrectomy (surgery)	longer survival	IL-6 and CRP
Fujikawa *et al*.	metastatic renal cell carcinoma	nephrectomy (surgery)	longer survival	IL-6

Abbreviations: RT, radiation therapy; HT, hyperthermia; RFA, radiofrequency ablation; LLC, Lewis lung carcinoma; DC ,dendritic cells; NK, natural killer; TNF, tumor necrosis factor; IL, interleukin; CRP, C-reactive prot

Table 1. Possible mechanisms for the abscopal effect

Important factor is that radiation therapy appears to cause less immunosuppression compared to surgery or other invasive treatment modalities. Therefore, radiation therapy potentially should have the more favorable biological activity for inducing an abscopal effect than surgical procedures if the major underlying mechanism is based upon immune activation.

The abscopal effect apparently operates through mechanisms that are paralleled in gene therapy, local immunotherapy, hyperthermia, and post-surgical distant bystander effects. Recently, some investigators have suggested that the definition of the abscopal effect should have been broadened to include other forms of local therapy that have systemic effects, *i.e.*, a distant bystander effect (Perego and Faravelli, 2000,Vartak, et al., 1993). Whether or not the definition should be extended to include local therapies other than radiation therapy that have a distant effect is a matter of debate. However, to unravel the abscopal effect of radiation, it seems prudent to evaluate other directed therapies that are associated with systemic effects (Kaminski, et al., 2005). Since the literal meaning is the same for abscopal and distant bystander, the terms could be used interchangeably to refer to any local therapy with a distant impact.

3.2 Possible mechanisms via DC activation

In recent years, the crucial role played by innate immunity, and in particular by DCs in enhancing T cell activation, has been widely clarified. DCs are lineage-negative, bone marrow-derived mononuclear cells found in peripheral blood or in many organs (O'Neill, et al., 2004). DCs can be broadly divided into myeloid or plasmacytoid DCs (MDCs and PDCs, respectively) on the basis of phenotypic, morphologic, and functional differences. Antigens acquired both endogenously (*i.e.*, synthesized within the DC cytosol) or exogenously (acquired from the extracellular environment) are processed into peptides, which are loaded onto major histocompatibility complex class I and II (MHC I and II) molecules and transported to the cell surface for recognition by antigen-specific T cells. DCs most efficiently capture antigens when they are in the immature phase. The terminal process of differentiation termed as *maturation* transforms DCs with weak immunostimulatory properties for antigen capture into cells specialized for T cell stimulation in the lymphoid organ. This process is accompanied by cytoskeletal reorganization, loss of adhesiveness, acquisition of cellular motility with development of characteristic cytoplasmic extensions, migration to lymphoid tissues, reduced phagocytic uptake, and T cell activation potential (O'Neill, et al., 2004). Natural killer (NK) cells are activated by type I interferon (IFN) produced from tumor tissues as a "danger signal" to attack tumor cells. Immature DCs uptake tumor tissue-derived products such as apoptotic bodies and necrotic bodies with tumor-associated antigens (Moretta, 2002). Mature DCs can secrete chemokines and cytokines that attract other immune cells and activate resting T cells. DCs can prime resting NK cells via proinflammatory cytokines such as IL-12 or IL-15 and NK-inducing chemokines such as IL-8 or macrophage inflammatory protein 1-alpha (MIP-1α), and enhance their own maturation by attachment with activated NK cells. However, NK cells negatively regulate the function of DCs also by killing immature DCs in peripheral tissues. Moreover, a subset of NK cells, after migration to secondary lymphoid tissues, might have a role in the editing of mature DCs based on the selective killing of mature DCs that do not express optimal surface densities of MHC class I molecules. Maturation of DCs can be induced by a growing number of exogenous and endogenous molecular signals, generally referred to as "danger signals" (Matzinger, 1994). Danger signals include host-derived proinflammatory cytokines, such as TNF, IL-1, IL-6, and type I IFN, and a variety of molecules released not only by microbes but also by damaged host tissues, including tumor involvement (Skoberne, et al., 2004). These noncytokine molecules signal primarily through transmembrane receptors related to Drosophila Toll protein, known as Toll-like receptors (TLR) (Kopp and Medzhitov, 2003), which are expressed by DCs.

The major concern is whether ionizing radiation-induced apoptosis can increase tumor immunogenicity. The immunostimulatory activity associated with cell lysates (endogenous adjuvant activity) was shown to be elevated once the cells were stressed by ultraviolet radiation, indicating that injury can modulate this effect (Gallucci, et al., 1999,Shi, et al., 2000). Some examples exist in which apoptotic cells show immunostimulatory features (Rock, et al., 2005). Immunization with apoptotic cells or *in situ* induction of tumor cell apoptosis induced T cell responses *in vivo* as exemplified in some reports (Kotera, et al., 2001,Nowak, et al., 2003,Ronchetti, et al., 1999). Injection of immature DCs into tumor tissue after irradiation-induced tumor cell apoptosis can stimulate strong antitumor immunity (Kim, et al., 2004). These studies suggest that under some favorable conditions for an immunocompetent host, radiation-induced tumor cell death might be associated with the production of ideal maturation signals for DCs (Demaria, et al., 2005).

The possible contribution of radiation-induced apoptosis vs. necrosis to immunostimulation has not been fully elucidated, and no significant difference was seen in capabilities of both kinds of cell death for antigen presentation *in vitro* (Larsson, et al., 2001). Endogenous factors released from necrotic cells might be responsible for the ability of the necrotic body to activate DCs (Skoberne, et al., 2004). Examples of these are immunostimulatory self-DNA that binds TLR9, self-single-strand RNA that stimulates TLR7 and TLR8, secondary structures of messenger RNA that activate TLR3, and heat shock proteins that stimulate TLR4 (Demaria, et al., 2005). The induction of necrosis *in vivo* could be accompanied by the release not only of self-antigens but also inflammatory factors that may cause DC maturation and the whole immune response. Candidates for cell-associated antigens being cross-presented from dying cells could include heat shock protein-associated proteins, native proteins (Shen and Rock, 2004), peptides (Neijssen, et al., 2005), or other constituents. In general, it is considered that DC maturation signals are essential to convert cross-tolerance to cross-priming (Steinman and Nussenzweig, 2002).

Opinion is divided as to the ability of ionizing radiation to generate the signals required for DC maturation; however, the combined approach of inducing cell death by irradiation in combination with the administration of a chemotactic agent that activates DCs can lead to the priming or enhancement of antitumor responses (Shiraishi, et al., 2008).

3.3 Attempts to consistently induce the abscopal effect

Based on the theory of immunological activation with ionizing radiation, Shiraishi *et al.* have chosen MIP-1α in combination with radiotherapy and investigated whether MIP-1α could cause a broad-spectrum enhancement of the efficacy of radiotherapy in tumor-bearing mice. Although there are many reports concerning anti-cancer (Crittenden, et al., 2003,Nakashima, et al., 1996,Taub, et al., 1995,Zibert, et al., 2004) and anti-metastasis effects of MIP-1α (van

Deventer, et al., 2002), enhancement of radiation efficacy had not been investigated sufficiently. Radiation treatment at tumor bearing sites is known to induce strong inflammation in the irradiated field and to recruit tumor-specific T lymphocytes and DCs, which seem to play an important role in the remission of tumors (Friedman, 2002,Garnett, et al., 2004,Teitz-Tennenbaum, et al., 2003). MIP-1α or CCL3, is a chemokine known to be secreted from various leukocytes including T lymphocytes and activated macrophages, and to recruit CCR1- and/or CCR5-expressing leukocytes such as monocytes, DCs, NK cells and T lymphocytes (Rollins, 1997). It was also reported that MIP-1α could enhance survival of DCs (Sumida, et al., 2004) and primed T lymphocytes to generate IFN-γ (Lillard, et al., 2003).

An active variant of human MIP-1α with improved pharmaceutical properties that carries a single amino acid substitution of the 26Asp to Ala was reported (Hunter, et al., 1995), which has a reduced tendency to form large aggregates at physiological pH and ionic strength. Myelosuppressive effect of the active variant (Arango, et al., 1999,Arango, et al., 2001,Gilmore, et al., 1999,Lord, et al., 1995) was investigated in several clinical trials of patients receiving chemotherapy (Bernstein, et al., 1997,Broxmeyer, et al., 1998,Clemons, et al., 1998,Marshall, et al., 1998). We previously showed that the recombinant MIP-1α variant, now called ECI301, strikingly enhanced the antitumor efficacy of subcutaneous tumor irradiation and induced an abscopal effect (Shiraishi, et al., 2008). Our study resulted in tumor-free mice with long-term survival without significant toxicity and complete rejection by surviving mice to a re-challenge with the same tumor cells. In accordance with our findings, no significant side effects of a compound with the same structure (BB-10010) had been reported previously when administered to human patients. Moreover, we observed a tumor-type- and mouse-strain-independent abscopal effect, indicating that the antitumor effect of ECI301 may be exerted via systemic inflammation and immune response. Marked infiltration of CD4+ and CD8+ cells was observed both in irradiated and non-irradiated sites. It was reported that DC precursors were mobilized into the circulation by administration of MIP-1α (Zhang, et al., 2004) and radiofrequency ablation-treated hepatocellular carcinomas (Iida, et al., 2010); however, we did not observe an increase in $CD11c^{+}$ cell infiltration into the tumor tissue in this model. Depletion of CD8+ T cells by antibodies diminished the effect of combination treatment at the irradiated site, indicating that CD8+ T cells are involved in the antitumor effect. Furthermore, rejection of the same tumor type in the cured mice may have been mediated by the presence of these types of lymphocytes. An increased number of splenocytes with tumor-specific IFN-γ-generating ability with the combination treatment also supports this assumption (Shiraishi, et al., 2010). Depletion of CD4+ T lymphocytes or NK1.1 cells by antibodies diminished the abscopal effect, indicating that these cells are involved in the remission either directly or indirectly. CD4+ T cells may play a role in generating cytokines such as IFN-γ, which may also activate other leukocytes (Dorner, et al., 2002,Pender, et al., 2005,Shiraishi, et al., 2008).

Further studies using C3H/HeN, C3H/HeJ and athymic mice will show whether the high mobility group box 1 (HMGB1) RNA level, an important mediator of local and systemic inflammation, is up-regulated at each tumor-bearing site (unpublished data). Results might clarify the underlying HMGB1-dependent mechanism for the abscopal effect via TLR4-mediated inflammation (Fig. 1).

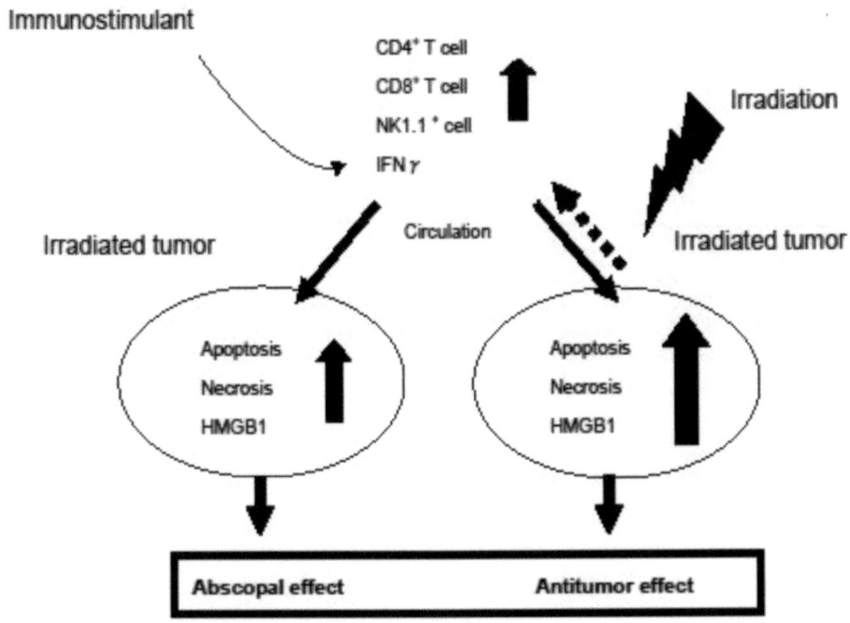

Fig. 1. Possible mechanism for radiation-induced abscopal effect.

Ionizing radiation induces tumor cell death in the irradiated tumor, causes inflammation and activates the immune system via chemokines with HMGB1. Length of arrows means relative strength of the effects.

HMGB1 = high mobility group box 13.4 Implications for future therapies

For future development, further insights into the mechanisms underpinning abscopal signaling are required. Theoretical elucidation of the relevance of abscopal responses in radiation-induced carcinogenesis is also required, including molecular pathways and targets outside of directly exposed fields.

A balance between angiogenic and anti-angiogenic molecules seems to be one of the key factors behind tumor growth. For example, several experimental animal models indeed suggest that the growth of a primary tumor can inhibit the production of distant metastases, probably due to inhibition of angiogenesis (Gorelik, 1983,Prehn, 1991). In contrast, the angiogenic inhibitors, angiostatin and endostatin, are known to function in tumor inhibition (O'Reilly, et al., 1997,O'Reilly, et al., 1994). Hartford *et al.* reported that the effect of irradiation of a primary tumor on angiogenesis at a distal site may differ from the effect of surgical removal of the primary tumor with respect to angiostatin production (Hartford, et al., 2000). They clearly demonstrated that, unlike surgery, irradiation of a tumor might enhance angiogenic suppression at a distal site. The involvement of angiogenic regulation in a radiation-induced abscopal effect should be emphasized as a clinical advantage in contrast to other invasive procedures, which may reduce possible angiogenic inhibition.

4. Conclusion

In conclusion, data that possibly support an intriguing concept as an abscopal effect are reviewed. These data will encourage future therapeutic gain of immunostimulants utilization in the treatment of advanced or metastatic cancer. The development of safer, reasonable, and targeted therapies will be facilitated as we clarify the mechanisms for the abscopal effects. Future therapies will need to be optimized with tumor-type tailoring in consideration of various intra- or inter-tissue signals if these are to affect treatment outcome.

Hopefully, a more aggressive effort for investigating and developing a potentially novel application of ionizing radiation in combination with immunotherapy will be needed. When the effectiveness of "immunoradiotherapy" in a clinical setting is established in a desirable manner, it could lead to a new era of cancer treatment, with common availability of established modalities, without significant adverse events.

5. List of abbreviations and expansions in the order corresponding to apperances

CTL, cytotoxic T lymphocyte
DC, dendritic cell
TNF-α, tumor necrosis factor alpha
IL , interleukin
CRP, C-reactive protein
Flt3-L, fms-like tyrosine kinase receptor 3
MHC, major histocompatibility complex
NK, natural killer
IFN, interferon
MIP-1α, macrophage inflammatory protein 1-alpha
TLR, Toll-like receptors
HMGB1, high mobility group box 1

6. References

Antoniades *et al.* (1977). Lymphangiographic demonstration of the abscopal effect in patients with malignant lymphomas. *Int J Radiat Oncol Biol Phys*, Vol.2, No.1-2, pp. 141-147, ISSN 0360-3016 (Print)

Arango *et al.* (1999). BB-10010, an analogue of macrophage inflammatory protein-1 alpha, reduces proliferation in murine small-intestinal crypts. *Scand J Gastroenterol*, Vol.34, No.1, pp. 68-72, ISSN 0036-5521 (Print)

Arango *et al.* (2001). BB-10010, an analog of macrophage inflammatory protein-1alpha, protects murine small intestine against radiation. *Dig Dis Sci*, Vol.46, No.12, pp. 2608-2614, ISSN 0163-2116 (Print)

Awwad & North. (1990). Radiosensitive barrier to T-cell-mediated adoptive immunotherapy of established tumors. *Cancer Res*, Vol.50, No.8, pp. 2228-2233, ISSN 0008-5472 (Print)

Bernstein *et al.* (1997). A randomized phase II study of BB-10010: a variant of human macrophage inflammatory protein-1alpha for patients receiving high-dose etoposide and cyclophosphamide for malignant lymphoma and breast cancer. *Br J Haematol*, Vol.99, No.4, pp. 888-895, ISSN 0007-1048 (Print)

Blay *et al.* (1992). Serum level of interleukin 6 as a prognosis factor in metastatic renal cell carcinoma. *Cancer Res*, Vol.52, No.12, pp. 3317-3322, ISSN 0008-5472 (Print) 0008-5472 (Linking)

Broxmeyer *et al.* (1998). Myeloid progenitor cell proliferation and mobilization effects of BB10010, a genetically engineered variant of human macrophage inflammatory protein-1alpha, in a phase I clinical trial in patients with relapsed/refractory breast cancer. *Blood Cells Mol Dis*, Vol.24, No.1, pp. 14-30, ISSN 1079-9796 (Print)

Cameron *et al.* (1990). Synergistic antitumor activity of tumor-infiltrating lymphocytes, interleukin 2, and local tumor irradiation. Studies on the mechanism of action. *J Exp Med*, Vol.171, No.1, pp. 249-263, ISSN 0022-1007 (Print)

Camphausen *et al.* (2003). Radiation abscopal antitumor effect is mediated through p53. *Cancer Res*, Vol.63, No.8, pp. 1990-1993, ISSN 0008-5472 (Print)

Chakravarty *et al.* (1999). Flt3-ligand administration after radiation therapy prolongs survival in a murine model of metastatic lung cancer. *Cancer Res*, Vol.59, No.24, pp. 6028-6032, ISSN 0008-5472 (Print) 0008-5472 (Linking)

Chiang *et al.* (1997). Effects of IL-3 gene expression on tumor response to irradiation in vitro and in vivo. *Cancer Res*, Vol.57, No.18, pp. 3899-3903, ISSN 0008-5472 (Print)

Clemons *et al.* (1998). A randomized phase-II study of BB-10010 (macrophage inflammatory protein- 1alpha) in patients with advanced breast cancer receiving 5-fluorouracil, adriamycin, and cyclophosphamide chemotherapy. *Blood*, Vol.92, No.5, pp. 1532-1540, ISSN 0006-4971 (Print)

Crittenden *et al.* (2003). Expression of inflammatory chemokines combined with local tumor destruction enhances tumor regression and long-term immunity. *Cancer Res*, Vol.63, No.17, pp. 5505-5512, ISSN 0008-5472 (Print)

Demaria *et al.* (2005). Combining radiotherapy and immunotherapy: a revived partnership. *Int J Radiat Oncol Biol Phys*, Vol.63, No.3, pp. 655-666, ISSN 0360-3016 (Print) 0360-3016 (Linking)

Demaria *et al.* (2004). Ionizing radiation inhibition of distant untreated tumors (abscopal effect) is immune mediated. *Int J Radiat Oncol Biol Phys*, Vol.58, No.3, pp. 862-870, ISSN 0360-3016 (Print)

Dorner *et al.* (2002). MIP-1alpha, MIP-1beta, RANTES, and ATAC/lymphotactin function together with IFN-gamma as type 1 cytokines. *Proc Natl Acad Sci U S A*, Vol.99, No.9, pp. 6181-6186, ISSN 0027-8424 (Print)

Dybal *et al.* (1992). Synergy of radiation therapy and immunotherapy in murine renal cell carcinoma. *J Urol*, Vol.148, No.4, pp. 1331-1337, ISSN 0022-5347 (Print)

Ehlers & Fridman. (1973). Abscopal effect of radiation in papillary adenocarcinoma. *Br J Radiol*, Vol.46, No.543, pp. 220-222, ISSN 0007-1285 (Print)

Friedman. (2002). Immune modulation by ionizing radiation and its implications for cancer immunotherapy. *Curr Pharm Des*, Vol.8, No.19, pp. 1765-1780, ISSN 1381-6128 (Print) 1381-6128 (Linking)

Fujikawa *et al.* (2000). Serum immunosuppressive acidic protein and natural killer cell activity in patients with metastatic renal cell carcinoma before and after nephrectomy. *J Urol*, Vol.164, No.3 Pt 1, pp. 673-675, ISSN 0022-5347 (Print) 0022-5347 (Linking)

Gallucci *et al.* (1999). Natural adjuvants: endogenous activators of dendritic cells. *Nat Med*, Vol.5, No.11, pp. 1249-1255, ISSN 1078-8956 (Print) 1078-8956 (Linking)

Ganss *et al.* (2002). Combination of T-cell therapy and trigger of inflammation induces remodeling of the vasculature and tumor eradication. *Cancer Res*, Vol.62, No.5, pp. 1462-1470, ISSN 0008-5472 (Print)

Garnett *et al.* (2004). Sublethal irradiation of human tumor cells modulates phenotype resulting in enhanced killing by cytotoxic T lymphocytes. *Cancer Res*, Vol.64, No.21, pp. 7985-7994, ISSN 0008-5472 (Print)

Gilmore *et al.* (1999). Protective effects of BB-10010 treatment on chemotherapy-induced neutropenia in mice. *Exp Hematol*, Vol.27, No.2, pp. 195-202, ISSN 0301-472X (Print)

Gorelik. (1983). Concomitant tumor immunity and the resistance to a second tumor challenge. *Adv Cancer Res*, Vol.39, pp. 71-120, ISSN 0065-230X (Print) 0065-230X (Linking)

Hartford *et al.* (2000). Irradiation of a Primary Tumor, Unlike Surgical Removal, Enhances Angiogenesis Suppression at a Distal Site: Potential Role of Host-Tumor Interaction. *Cancer Res*, Vol.60, No.8, pp. 2128-2131, ISSN

Hong *et al.* (1999). Rapid induction of cytokine gene expression in the lung after single and fractionated doses of radiation. *Int J Radiat Biol*, Vol.75, No.11, pp. 1421-1427, ISSN 0955-3002 (Print)

Hunter *et al.* (1995). BB-10010: an active variant of human macrophage inflammatory protein-1 alpha with improved pharmaceutical properties. *Blood*, Vol.86, No.12, pp. 4400-4408, ISSN 0006-4971 (Print)

Iida *et al.* (2010). Antitumor effect after radiofrequency ablation of murine hepatoma is augmented by an active variant of CC Chemokine ligand 3/macrophage inflammatory protein-1alpha. *Cancer Res*, Vol.70, No.16, pp. 6556-6565, ISSN 1538-7445 (Electronic) 0008-5472 (Linking)

Kaminski *et al.* (2005). The controversial abscopal effect. *Cancer Treat Rev*, Vol.31, No.3, pp. 159-172, ISSN 0305-7372 (Print)

Kim *et al.* (2004). Direct injection of immature dendritic cells into irradiated tumor induces efficient antitumor immunity. *Int J Cancer*, Vol.109, No.5, pp. 685-690, ISSN 0020-7136 (Print) 0020-7136 (Linking)

Kingsley. (1975). An interesting case of possible abscopal effect in malignant melanoma. *Br J Radiol*, Vol.48, No.574, pp. 863-866, ISSN 0007-1285 (Print)

Konoeda. (1990). [Therapeutic efficacy of pre-operative radiotherapy on breast carcinoma: in special reference to its abscopal effect on metastatic lymph-nodes]. *Nippon Gan Chiryo Gakkai Shi*, Vol.25, No.6, pp. 1204-1214, ISSN 0021-4671 (Print)

Kopp & Medzhitov. (2003). Recognition of microbial infection by Toll-like receptors. *Curr Opin Immunol*, Vol.15, No.4, pp. 396-401, ISSN 0952-7915 (Print) 0952-7915 (Linking)

Kotera *et al.* (2001). Comparative analysis of necrotic and apoptotic tumor cells as a source of antigen(s) in dendritic cell-based immunization. *Cancer Res*, Vol.61, No.22, pp. 8105-8109, ISSN 0008-5472 (Print) 0008-5472 (Linking)

Larsson *et al.* (2001). Efficiency of cross presentation of vaccinia virus-derived antigens by human dendritic cells. *Eur J Immunol*, Vol.31, No.12, pp. 3432-3442, ISSN 0014-2980 (Print) 0014-2980 (Linking)

Lillard *et al.* (2003). MIP-1alpha and MIP-1beta differentially mediate mucosal and systemic adaptive immunity. *Blood*, Vol.101, No.3, pp. 807-814, ISSN 0006-4971 (Print)

Lord *et al.* (1995). Mobilization of early hematopoietic progenitor cells with BB-10010: a genetically engineered variant of human macrophage inflammatory protein-1 alpha. *Blood,* Vol.85, No.12, pp. 3412-3415, ISSN 0006-4971 (Print)

Macklis *et al.* (1992). Lymphoid irradiation results in long-term increases in natural killer cells in patients treated for Hodgkin's disease. *Cancer,* Vol.69, No.3, pp. 778-783, ISSN 0008-543X (Print)

Maraskovsky *et al.* (1996). Dramatic increase in the numbers of functionally mature dendritic cells in Flt3 ligand-treated mice: multiple dendritic cell subpopulations identified. *J Exp Med,* Vol.184, No.5, pp. 1953-1962, ISSN 0022-1007 (Print) 0022-1007 (Linking)

Maraskovsky *et al.* (1997). Dramatic numerical increase of functionally mature dendritic cells in FLT3 ligand-treated mice. *Adv Exp Med Biol,* Vol.417, pp. 33-40, ISSN 0065-2598 (Print) 0065-2598 (Linking)

Marshall *et al.* (1998). Clinical effects of human macrophage inflammatory protein-1 alpha MIP-1 alpha (LD78) administration to humans: a phase I study in cancer patients and normal healthy volunteers with the genetically engineered variant, BB-10010. *Eur J Cancer,* Vol.34, No.7, pp. 1023-1029, ISSN 0959-8049 (Print)

Matzinger. (1994). Tolerance, danger, and the extended family. *Annu Rev Immunol,* Vol.12, pp. 991-1045, ISSN 0732-0582 (Print) 0732-0582 (Linking)

Mole. (1953). Whole body irradiation; radiobiology or medicine? *Br J Radiol,* Vol.26, No.305, pp. 234-241, ISSN 0007-1285 (Print)

Moretta. (2002). Natural killer cells and dendritic cells: rendezvous in abused tissues. *Nat Rev Immunol,* Vol.2, No.12, pp. 957-964, ISSN 1474-1733 (Print) 1474-1733 (Linking)

Nakashima *et al.* (1996). A candidate for cancer gene therapy: MIP-1 alpha gene transfer to an adenocarcinoma cell line reduced tumorigenicity and induced protective immunity in immunocompetent mice. *Pharm Res,* Vol.13, No.12, pp. 1896-1901, ISSN 0724-8741 (Print)

Neijssen *et al.* (2005). Cross-presentation by intercellular peptide transfer through gap junctions. *Nature,* Vol.434, No.7029, pp. 83-88, ISSN 1476-4687 (Electronic) 1476-4687 (Linking)

Nikitina & Gabrilovich. (2001). Combination of gamma-irradiation and dendritic cell administration induces a potent antitumor response in tumor-bearing mice: approach to treatment of advanced stage cancer. *Int J Cancer,* Vol.94, No.6, pp. 825-833, ISSN 0020-7136 (Print)

Nobler. (1969). The abscopal effect in malignant lymphoma and its relationship to lymphocyte circulation. *Radiology,* Vol.93, No.2, pp. 410-412, ISSN 0033-8419 (Print)

Nowak *et al.* (2003). Induction of tumor cell apoptosis in vivo increases tumor antigen cross-presentation, cross-priming rather than cross-tolerizing host tumor-specific CD8 T cells. *J Immunol,* Vol.170, No.10, pp. 4905-4913, ISSN 0022-1767 (Print) 0022-1767 (Linking)

O'Neill *et al.* (2004). Manipulating dendritic cell biology for the active immunotherapy of cancer. *Blood,* Vol.104, No.8, pp. 2235-2246, ISSN 0006-4971 (Print) 0006-4971 (Linking)

O'Reilly *et al.* (1997). Endostatin: an endogenous inhibitor of angiogenesis and tumor growth. *Cell,* Vol.88, No.2, pp. 277-285, ISSN 0092-8674 (Print) 0092-8674 (Linking)

O'Reilly *et al.* (1994). Angiostatin: a novel angiogenesis inhibitor that mediates the suppression of metastases by a Lewis lung carcinoma. *Cell*, Vol.79, No.2, pp. 315-328, ISSN 0092-8674 (Print) 0092-8674 (Linking)

Ohba *et al.* (1998). Abscopal regression of hepatocellular carcinoma after radiotherapy for bone metastasis. *Gut*, Vol.43, No.4, pp. 575-577, ISSN 0017-5749 (Print)

Pender *et al.* (2005). Systemic administration of the chemokine macrophage inflammatory protein 1alpha exacerbates inflammatory bowel disease in a mouse model. *Gut*, Vol.54, No.8, pp. 1114-1120, ISSN 0017-5749 (Print)

Perego & Faravelli. (2000). Unexpected consequence of splenectomy in composite lymphoma. The abscopal effect. *Haematologica*, Vol.85, No.2, pp. 211, ISSN 0390-6078 (Print)

Prehn. (1991). The Inhibition of Tumor Growth by Tumor Mass. *Cancer Res*, Vol.51, No.1, pp. 2-4, ISSN

Quarmby *et al.* (1999). Radiation-induced normal tissue injury: role of adhesion molecules in leukocyte-endothelial cell interactions. *Int J Cancer*, Vol.82, No.3, pp. 385-395, ISSN 0020-7136 (Print)

Rees. (1981). Abscopal regression in lymphoma: a mechanism in common with total body irradiation? *Clin Radiol*, Vol.32, No.4, pp. 475-480, ISSN 0009-9260 (Print)

Rees & Ross. (1983). Abscopal regression following radiotherapy for adenocarcinoma. *Br J Radiol*, Vol.56, No.661, pp. 63-66, ISSN 0007-1285 (Print)

Rock *et al.* (2005). Natural endogenous adjuvants. *Springer Semin Immunopathol*, Vol.26, No.3, pp. 231-246, ISSN 0344-4325 (Print) 0344-4325 (Linking)

Rollins. (1997). Chemokines. *Blood*, Vol.90, No.3, pp. 909-928, ISSN 0006-4971 (Print)

Ronchetti *et al.* (1999). Immunogenicity of apoptotic cells in vivo: role of antigen load, antigen-presenting cells, and cytokines. *J Immunol*, Vol.163, No.1, pp. 130-136, ISSN 0022-1767 (Print) 0022-1767 (Linking)

Schuler *et al.* (2003). The use of dendritic cells in cancer immunotherapy. *Curr Opin Immunol*, Vol.15, No.2, pp. 138-147, ISSN 0952-7915 (Print)

Sham. (1995). The abscopal effect and chronic lymphocytic leukemia. *Am J Med*, Vol.98, No.3, pp. 307-308, ISSN 0002-9343 (Print)

Shen & Rock. (2004). Cellular protein is the source of cross-priming antigen in vivo. *Proc Natl Acad Sci U S A*, Vol.101, No.9, pp. 3035-3040, ISSN 0027-8424 (Print) 0027-8424 (Linking)

Shi *et al.* (2000). Cell injury releases endogenous adjuvants that stimulate cytotoxic T cell responses. *Proc Natl Acad Sci U S A*, Vol.97, No.26, pp. 14590-14595, ISSN 0027-8424 (Print) 0027-8424 (Linking)

Shiraishi *et al.* (2008). Enhancement of antitumor radiation efficacy and consistent induction of the abscopal effect in mice by ECI301, an active variant of macrophage inflammatory protein-1alpha. *Clin Cancer Res*, Vol.14, No.4, pp. 1159-1166, ISSN 1078-0432 (Print)

Shiraishi *et al.* (2010). Enhancement of antitumor radiation efficacy and the abscopal effect by ECI301 mediated TLR4 dependent innate immunity in mice, *Proceedings of American Association for Cancer Research 101st Annual Meeting 2010.* p. 5617

Skoberne *et al.* (2004). Danger signals: a time and space continuum. *Trends Mol Med*, Vol.10, No.6, pp. 251-257, ISSN 1471-4914 (Print) 1471-4914 (Linking)

Steinman & Nussenzweig. (2002). Avoiding horror autotoxicus: the importance of dendritic cells in peripheral T cell tolerance. *Proc Natl Acad Sci U S A*, Vol.99, No.1, pp. 351-358, ISSN 0027-8424 (Print) 0027-8424 (Linking)

Sumida *et al.* (2004). Recruitment and expansion of dendritic cells in vivo potentiate the immunogenicity of plasmid DNA vaccines. *J Clin Invest*, Vol.114, No.9, pp. 1334-1342, ISSN 0021-9738 (Print)

Szlosarek & Balkwill. (2003). Tumour necrosis factor alpha: a potential target for the therapy of solid tumours. *Lancet Oncol*, Vol.4, No.9, pp. 565-573, ISSN 1470-2045 (Print) 1470-2045 (Linking)

Taub *et al.* (1995). Alpha and beta chemokines induce NK cell migration and enhance NK-mediated cytolysis. *J Immunol*, Vol.155, No.8, pp. 3877-3888, ISSN 0022-1767 (Print)

Teitz-Tennenbaum *et al.* (2003). Radiotherapy potentiates the therapeutic efficacy of intratumoral dendritic cell administration. *Cancer Res*, Vol.63, No.23, pp. 8466-8475, ISSN 0008-5472 (Print)

Uchida *et al.* (1989). Effects of X-ray irradiation on natural killer (NK) cell system. II. Increased sensitivity to natural killer cytotoxic factor (NKCF). *Immunopharmacol Immunotoxicol*, Vol.11, No.2-3, pp. 521-534, ISSN 0892-3973 (Print)

van Deventer *et al.* (2002). Transfection of macrophage inflammatory protein 1 alpha into B16 F10 melanoma cells inhibits growth of pulmonary metastases but not subcutaneous tumors. *J Immunol*, Vol.169, No.3, pp. 1634-1639, ISSN 0022-1767 (Print)

Vartak *et al.* (1993). Antitumor effects of local hyperthermia on a mouse fibrosarcoma. *Anticancer Res*, Vol.13, No.3, pp. 727-729, ISSN 0250-7005 (Print)

Watters. (1999). Molecular mechanisms of ionizing radiation-induced apoptosis. *Immunol Cell Biol*, Vol.77, No.3, pp. 263-271, ISSN 0818-9641 (Print)

Yasuda *et al.* (2011). Intratumoral injection of interleukin-2 augments the local and abscopal effects of radiotherapy in murine rectal cancer. *Cancer Sci*, Vol.102, No.7, pp. 1257-1263, ISSN 1349-7006 (Electronic) 1347-9032 (Linking)

Younes *et al.* (1995). Local tumor irradiation augments the response to IL-2 therapy in a murine renal adenocarcinoma. *Cell Immunol*, Vol.165, No.2, pp. 243-251, ISSN 0008-8749 (Print)

Younes *et al.* (1995). Radiation-induced effects on murine kidney tumor cells: role in the interaction of local irradiation and immunotherapy. *J Urol*, Vol.153, No.6, pp. 2029-2033, ISSN 0022-5347 (Print)

Zhang *et al.* (2004). Mobilization of dendritic cell precursors into the circulation by administration of MIP-1alpha in mice. *J Natl Cancer Inst*, Vol.96, No.3, pp. 201-209, ISSN 1460-2105 (Electronic)

Zibert *et al.* (2004). CCL3/MIP-1alpha is a potent immunostimulator when coexpressed with interleukin-2 or granulocyte-macrophage colony-stimulating factor in a leukemia/lymphoma vaccine. *Hum Gene Ther*, Vol.15, No.1, pp. 21-34, ISSN 1043-0342 (Print)

Part 6

Emerging Dosimeters and New QA Practices

16

Quality Assurance (QA) for Kilovoltage Cone Beam Computed Tomography (CBCT)

Joerg Lehmann[1,2] and Stanley Skubic[2]
[1] *University of California Davis, Davis, CA,*
[2]*Radiological Associates of Sacramento, Sacramento, CA, USA*

1. Introduction

The use of an imaging modality for verification of patient and target location immediately prior to application of radiation therapy has spread widely in the last years. It is referred to as Image Guided Radiation Therapy (IGRT).

Precise targeting of the tumor is very important and imaging can be immensely helpful in this process. Hence the quality assurance of the imaging systems used to guide the beam is also essential. One could argue that it is as critical as quality assurance for the radiation dose output, since the right amount of radiation dose delivered to the wrong location can be as detrimental as delivering the wrong amount of radiation dose to the right location.

Kilovoltage (kV) Cone Beam Computed Tomography (CBCT) is a premiere modality for image guidance, since it produces three-dimensional images of the region of interest of the patient in a relatively short time (1-3 min) in a sufficient quality for the purpose of patient alignment. Popular systems are the Varian OBI (Varian Medical Systems, Inc., Palo Alto, CA, USA) and the Elekta XVI system (Elekta AB, Stockholm, Sweden), shown in figure 1.

This chapter introduces the different tests needed for a comprehensive QA of kV CBCT. The tests are related to (1) System safety, (2) Geometric accuracy, that is agreement of the CBCT imaging isocenter with the isocenter of the MV treatment beam, (3) Registration and correction accuracy, that is how accurately can the system position a patient, (4) Image quality, and (5) x-ray tube and generator performance (dose to the patient). It further suggests a frequency with which each of the tests should be performed, organizing them in being part of the daily, monthly and annual QA procedures. In addition, tests which should be performed following system repair or upgrades are described.

Responsibility for all tests should be with a Qualified Medical Physicist. Generally the physicist will delegate some of the tests, likely the daily tests to other members of the Radiation Therapy team, for instance the Radiation Therapists. (Klein et al., 2009).

For information about in-room kV X ray imaging beyond CBCT QA see the report of AAPM Task Group 104 (Yin et al., 2009) and (Jaffrey et al., 1999).

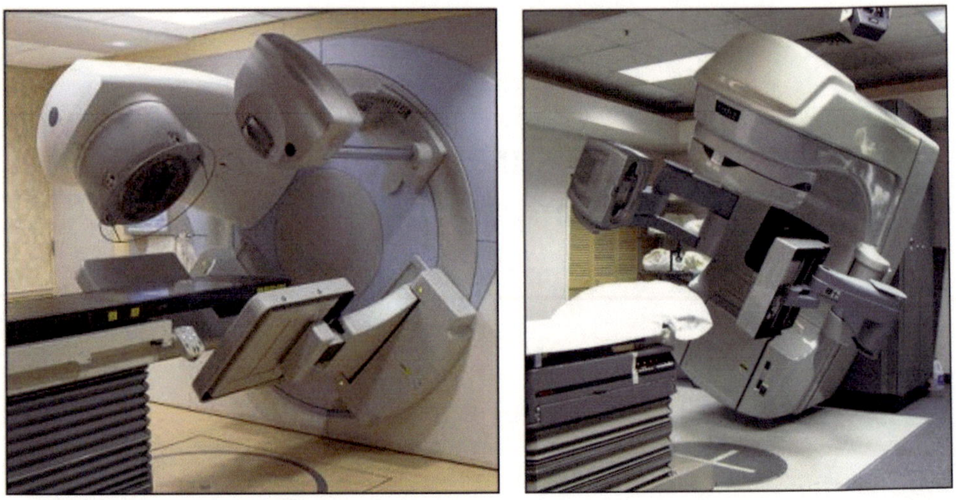

Fig. 1. Linear accelerators equipped with kV imaging systems capable of CBCT, left: Elekta Synergy® with XVI, right: Varian 21EX with OBI (all images by the author JL)

2. CBCT test descriptions

2.1 System safety

The system safety checks are concerned with gross safety issues, such as measures to avoid injury to the patient through collision with a remote controlled part of the system and access control to the room. Some of the tests will be very system dependent and can be modeled on the procedures in the acceptance document provided by the manufacturer. Often, the tests can be taken directly from the acceptance document and performed in reasonable intervals, which should depend on the severity of the issue checked and on the complexity of the test.

All kV CBCT systems provide safety interlocks that are triggered by switches on the kV tube housing and/or imaging panel as well as the (robotic) arms holding these components. Depending on the system, once triggered an interlock will inhibit all motion of the entire linear accelerator system until a reset button is pressed or just until the interlock switch is released. The functionality of each interlock switch, also referred to as touch guard or collision detector, needs to be tested in regular intervals. Testing should involve attempts to move the couch or gantry while the interlock is engaged and verifying that all motion is stopped.

System interlocks that prevent operation of the kV imaging system under undesirable conditions (door open, kV source arm not fully extended) should also be tested by creating these conditions and verifying that the operation ceases immediately or not begin. The same applies to any "Terminate" or "Emergency off" key. Such key is tested by pressing it while X-rays are being emitted.

2.2 Geometric accuracy

This group of tests assesses how well the isocenter of the CBCT imaging system agrees with the isocenter of the MV treatment beam of the machine. Correct alignment of the patient depends directly on this agreement, and therefore these tests are very important and need to be performed daily. Accuracy levels are described in the literature (Klein et al., 2009). However, precise tests (~1 mm) of the agreement of kV CBCT and MV treatment isocenters are very comprehensive and time consuming. Therefore multiple levels of tests are generally used: faster and less precise ones for daily checks by Radiation Therapists and more precise tests for monthly checks and checks following system repair or upgrade by the Medical Physicist.

All tests have in common that a structure, often a sphere of some type, is imaged with the CBCT system and alignment with the MV treatment beam is checked. In the less involved tests, which will be called Level 1 in the following, the cross hair of the machine or even the room lasers are used to describe isocenter of the MV treatment beam. In the precise tests (Level 2) the treatment beam is engaged from multiple angles to describe its isocenter. The methodology of these tests goes back to Winston and Lutz (Lutz et al., 1988).

Current CBCT systems offer multiple imaging geometries (modes) with varying position of imaging panel and kV aperture to accommodate different patient sizes and treatment locations. If each of these modes has a separate calibration file within the kV imaging system, as is the case for the current Varian OBI and Elekta XVI systems, then it needs to be checked separately for isocenter accuracy, provided the mode is used clinically. This also applies to the direction of the gantry rotation during imaging. For a current Varian OBI system (including True Beam), accuracy tests would need to be performed for four scan types: (1) body scan, clockwise, (2) body scan, counter clockwise, (3) head scan, clockwise, and (4) head scan, counter clockwise. Similarly for the Elekta XVI system with the small, medium and large field of view, six accuracy tests would be needed, if all modes are used clinically. To keep the workload reasonable, these tests can be performed on a rotating schedule. Specific suggestions are made later in this chapter.

2.2.1 Level 1 geometric accuracy test

For daily checks the kV CBCT isocenter should be compared to the cross hair of the machine as a representation of the MV beam isocenter. It is also possible to use the room lasers instead of the cross hair with the understanding that this introduces an additional uncertainty, which depends on the accuracy of the lasers.

The test requires a phantom that contains a structure, which can be well imaged with kV CBCT and also aligned visually with the cross hair. In a simple approach, one or more patient markers (BB) could be fixed on plate (Lehmann et al., 2007), as shown in figure 2. However, plastic markers should be used instead of the metal ones seen in the image, since they will lead to an improved CBCT image quality. Metal markers often cause streak artifacts in the CBCT image due to the metal - air and metal - plastic interfaces, which make alignment challenging.

A well suitable phantom is the Penta Guide phantom (Modus Medical system, London, ON, Canada). The phantom is shown in figure 3.

It is made from a plastic material and contains a spherical air-filled cavity in its center. It also contains additional spherical air-filled cavities. The location of the center air cavity is aligned with cross hair markers on the outside of the cube. The cross hair markers feature

Fig. 2. Simple Level 1 geometry accuracy test phantom. Consisting of three small steel balls (BBs) mounted half-sunken in a Lucite plate on a Styrofoam base, this phantom is used verify agreement of the isocenters of the kilovoltage and megavoltage beams.

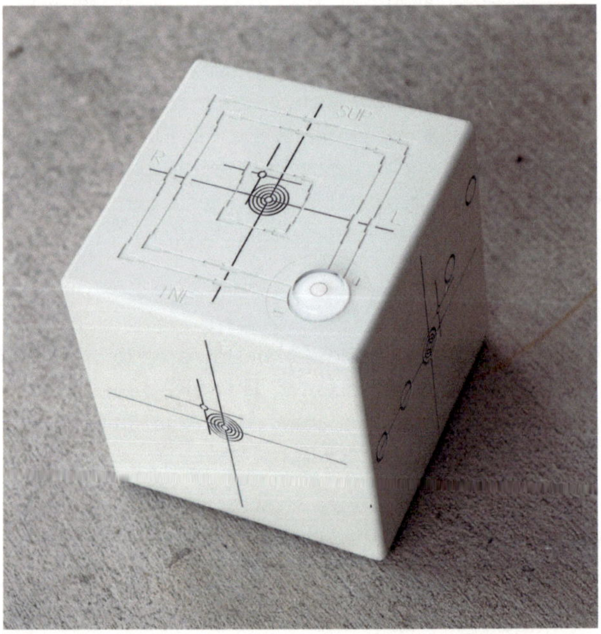

Fig. 3. Penta Guide phantom (Modus Medical, London, ON, Canada)

concentric circles, which allow for easy quantification of deviation. The phantom also has a second set of cross hair markers, which are correspond to a location 1 cm right, 1.2 cm up, 1.4 cm towards gantry with respect to the center of the cube. These can be used for a test of the position correction as described in section 2.3.

Another phantom is the "ISO Cube" (CIRS Inc, Norfolk, VA, USA). It is made from a plastic material and contains a unique center point fiducial as well as an offset blind target. The phantom measures 12 cm x 12 cm x 12 cm, which offers advantages for systems with a shorter length of scanned volume.

Fig. 4. ISO Cube (CIRS Inc, Norfolk, VA, USA)

To perform a Level 1 Geometric Accuracy Test for CBCT, the phantom is first lined up manually with the MV isocenter using the field cross hair at two gantry angles: 0° and 90° (or 270°) IEC scale. Then a CBCT imaging procedure is performed. Agreement of the center of the phantom in the CBCT software is checked against the center of the image, as marked in the system. In the simplest case any reference image data set can be used and agreement of the location of the center of the center sphere is checked against the marked isocenter of the system. For a routine procedure it is recommended to use a CT image of the phantom as reference data set. In order to create such reference data set, the phantom is scanned with a conventional CT scanner using a thin slice thickness of 1 or 1.5 mm. The CT data set is then sent to the treatment planning system, where the isocenter is carefully located and marked. A treatment beam is assigned to the isocenter. For some systems it is also necessary to create at least one structure. If applicable, a test patient is created in the record and verify system and the reference image set including structure and beam data is sent there, just like a the information of a patient would be.

With these preparatory steps the Level 1 Geometric Accuracy Test can now be performed using a reference image of the phantom and aligning the CBCT to it. For obvious reasons it is crucial to have the isocenter marked correctly in the reference data set.

One could also perform the test in reverse order, which is to first align the phantom using CBCT to the isocenter of the CBCT system and then check the alignment against the MV system using cross hair (or laser). While conceptually more aligned with the clinical process of patient setup, a drawback of this approach is the limited resolution (and accuracy) of the couch movement of many current systems (1 mm), which limits the accuracy with which the phantom can be aligned with the kV CBCT .

Still, the latter order is often used, as it allows for easy combination with the check of the registration accuracy as described in section 2.3.

2.2.2 Level 2 geometric accuracy test

For a more precise check of the agreement of the kV CBCT isocenter with the MV beam isocenter, the MV beam isocenter is described with MV imaging at multiple gantry and possibly collimator angles.

This test requires a phantom that can be imaged well with kV CBCT and also with MV imaging. The Penta Guide (Figure 3) offers limited image quality for MV imaging. However, the authors have used it successfully over several years. The ISO Cube phantom (Figure 4) offers better MV imaging, but has less favorable kV CBCT image quality. A phantom that offers very good kV CBCT and MV image quality has been adopted by the authors from Greg Gibbs and Jerry White of the Colorado Associates in Medical Physics. The phantom consists of a platform with a small tube upright mounted on it. The tube can hold one of two identically sized balls (diameter ~8 mm). For kV CBCT a plastic (Lucite) ball is placed on the tube and for MV imaging a Tungsten ball. The phantom, which has been called Two Ball Phantom (TBP), is shown in figures 5 and 6.

The procedure for the level 2 test is analogous to the one for level 1. The phantom is setup to one modality, in this case the kV CBCT, as it allows for better adjustments of the position. The before mentioned problem of the limited precision of the table movement can be eased by recording the final deviation from which no more moves are possible (due to precision limit) in each dimension and comparing it later with the deviation seen with the MV imaging.

If the precision of the on-screen display is also only 1 mm, which is common and reasonable for clinical applications, the deviation can be estimated by first aligning the images as good as possible on the screen and then making discrete steps (using the keyboard) in each direction until the next digit turns. This needs to be done both ways to determine how many "clicks" equal 1 mm.

After best alignment with kV CBCT has been achieved and additional deviations are noted as described, the phantom is imaged with the MV beam at gantry angles 0°, 90°, 180°, and 270°. This can be done using film or the portal imager. Film measurements will require precise alignment of the film perpendicular to the beam and subsequent scanning for best results. The use of portal imaging is recommended and will be described in the following. However, the method can also be applied to film.

To reduce errors in the position of the portal imager, a fixed jaw or MLC geometry is used and the position of the image of the phantom is evaluated relative to the so described radiation field. To reduce errors due to small inaccuracies in jaw or MLC position the images are taken at two different collimator angles (preferably 90° and 270°) for each gantry angle.

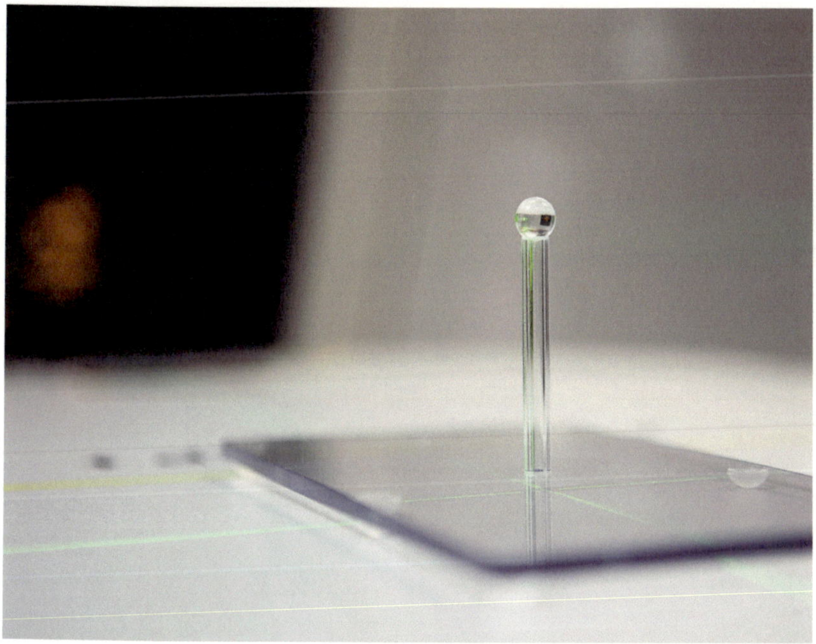

Fig. 5. Two Ball Phantom: A platform with a small tube as a riser holds one of two same-size balls for imaging, a plastic ball shown here for kV CBCT and a Tungsten ball for MV imaging as shown in figure 6.

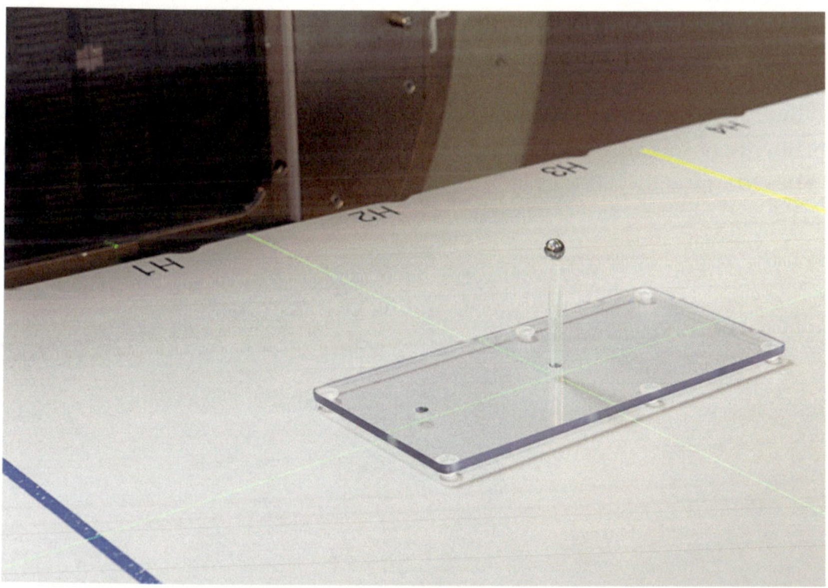

Fig. 6. Two Ball Phantom setup with Tungsten ball for MV imaging and small MLC shaped field.

On each image the position of the ball relative to the field edges is analyzed. This can be done visually for a first and approximate confirmation. Image analysis tools of the MV portal imaging software or another software can be used to assist in manual analysis. Figure 7 shows an example of the center of the Penta Guide phantom being imaged.

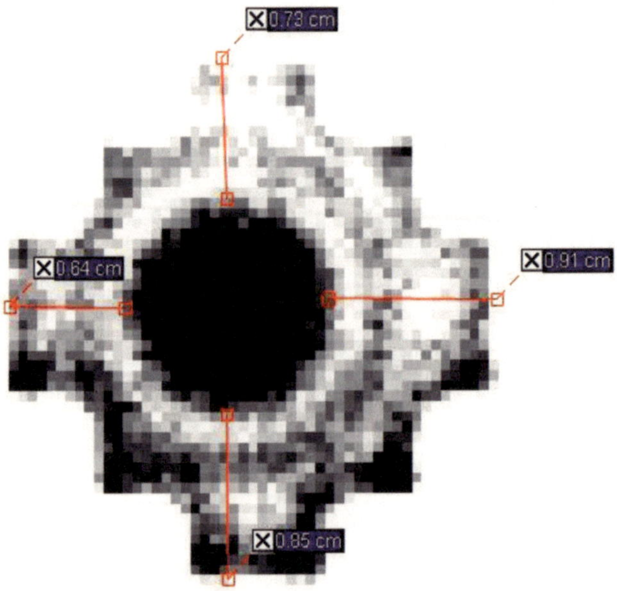

Fig. 7. MV image of the central sphere in the Penta Guide phantom with a small MLC shaped field. The measurement tool in the OBI software has been used to estimate the position of the image of the sphere relative to the field. Note: The poor image quality is (mainly) the result of the underlying physics, not of the reproduction.

Figures 8 shows kV and MV images of the plastic and Tungsten ball of the TBP. Clearly the MV image quality of the TBP is superior to the one of the Penta Guide. Both MV images also show the lines from the measurement tool that was used to assess the distance of the edge of the cavity (Penta Guide) or ball (TBP) from the edge of the field.

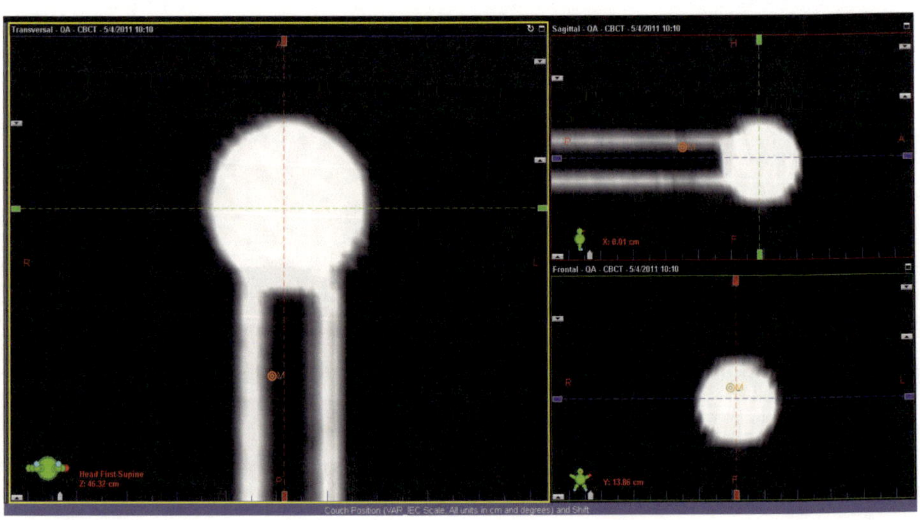

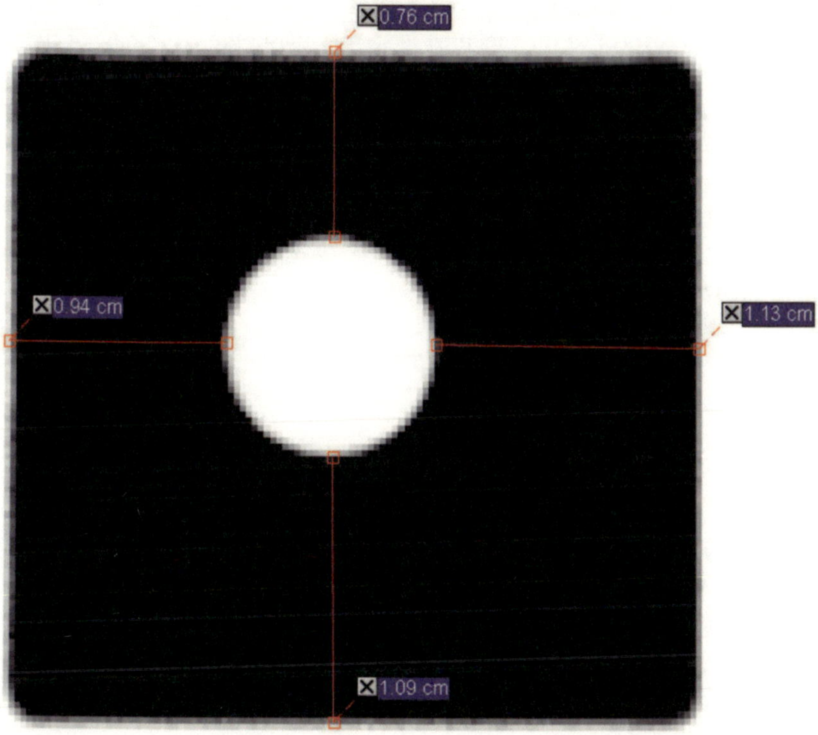

Fig. 8. Two Ball Phantom, top: kV CBCT image of the plastic ball in three views, bottom: MV port film image of the Tungsten ball (Varian OBI).

2.3 Registration and correction accuracy

In addition to checking the accuracy of the isocenter of the imaging system, it needs to be verified that the system calculates shifts of a slightly misaligned patient's planned isocenter to the machine isocenter correctly. It should also be verified that these shift are executed correctly (couch movement), if this feature is used clinically. If this process could not be trusted, a confirmation CBCT of the patient in the final position would need to be taken after each shift to verify that patient treatment and machine isocenter agree following the shift. The magnitude of the shifts to be checked should be as large as the largest magnitude expected in clinical use.

This test uses the same phantom as the isocenter accuracy test and is best done with a prepared reference CT scan of the phantom as described in section 2.2 above. The phantom is setup to an off center location. It is then CBCT imaged and the shifts are determined with the CBCT software. Some phantoms, such as the Penta Guide offer a second setup point with known shifts from the center. If that is used the values of the determined shifts can be compared to the expected values. Otherwise, or in general for a more thorough QA (monthly), the shifts are applied (couch moved), a second CBCT is performed and the alignment of the phantom is verified in the CBCT software. If the system functions properly, only very small shifts within the accuracy of the couch movement should now be needed for alignment of CBCT with the reference image.

Following the alignment of the images in the software, the physical position of the phantom can be verified with respect to the machine isocenter using the cross hair of the accelerator. This effectively combines this registration and correction accuracy check with the isocenter accuracy check described in section 2.2. However, the combined test is potentially limited by the accuracy of the couch motion.

2.4 Image quality

Image quality needs to be measured in regular intervals to ensure stability of the imaging chain and to guarantee the lowest imaging dose to the patient. A suitable tool is the Catphan (The Phantom Lab, Salem, NY), as shown in figure 9. The phantom consists of a housing containing several modules with geometrical structures to test variousimage quality parameters.

Currently, both manufacturers of kV CBCT systems provide a version of the phantom with the CBCT system. The Catphan phantom is setup on the treatment couch of the linac system. It hangs from its wooden box over the end of the couch, so that no part of the couch is between phantom and image source or panel. For consistency it is important to keep the phantom leveled. (Figure 10)

The phantom should be setup with the machine isocenter at approximately its center (vertical laser between second and third module). With this approach all modules of the phantom can be captured in one CBCT image set.

There is another approach, which is currently used in the acceptance testing of the Elekta XVI system. Here separate scans are performed for the separate modules of the phantom. A smaller field size is used in longitudinal direction. The phantom is setup to each module individually with the machine isocenter placed at the center of the particular module. This will potentially lead to better results, as in higher measured parameters. However, since the objective of the ongoing QA is to monitor changes in performance, imaging the entire phantom at once seems more suitable with regard to workflow.

Following the scan(s) of the Catphan, each module is analyzed. This is done best according to the procedure laid out in the acceptance document of the machine or using an independent software.

It can be argued that the spatial resolution test is most important, since it can detect changes to the focal spot of the x-ray tube. The test uses line pairs that are visually analyzed, as can be seen in figure 11. Visual analysis is subjective and can vary with the environment. However, it is widely used. The phantom section with the line pairs offers also a small element that can be used for automatic analysis. The image of a thin wire (small white dot towards the center of the image in figure 11) can be analyzed with suitable software to determine to modulation transfer function of the system.

Another very important set of tests are the geometric tests, which test parameters of the imaging geometry: Using measurement tools of the software or a third party application, it is determined how accurate known distances in the phantom are represented in the image (Figure 12). This is important since shifts of the patient following the CBCT imaging will be determined based on distances in the image.

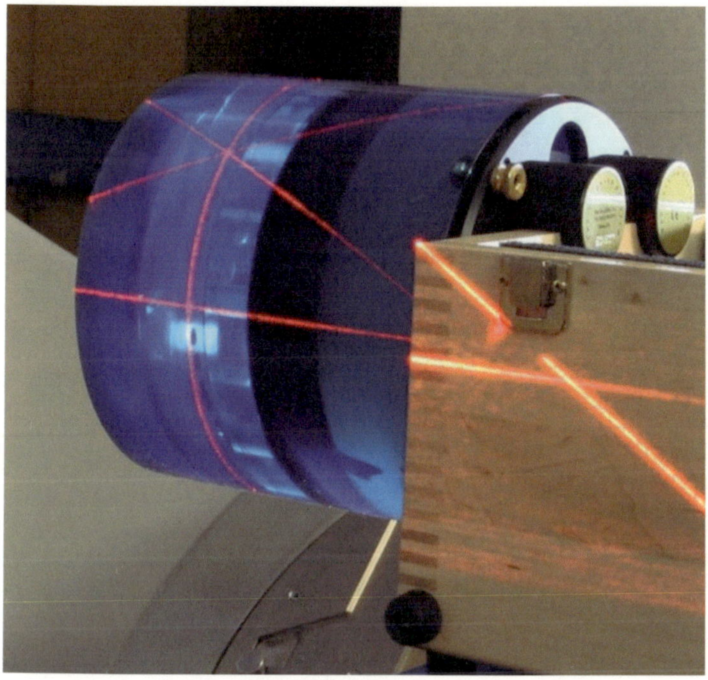

Fig. 9. Catphan: Phantom for image quality tests.

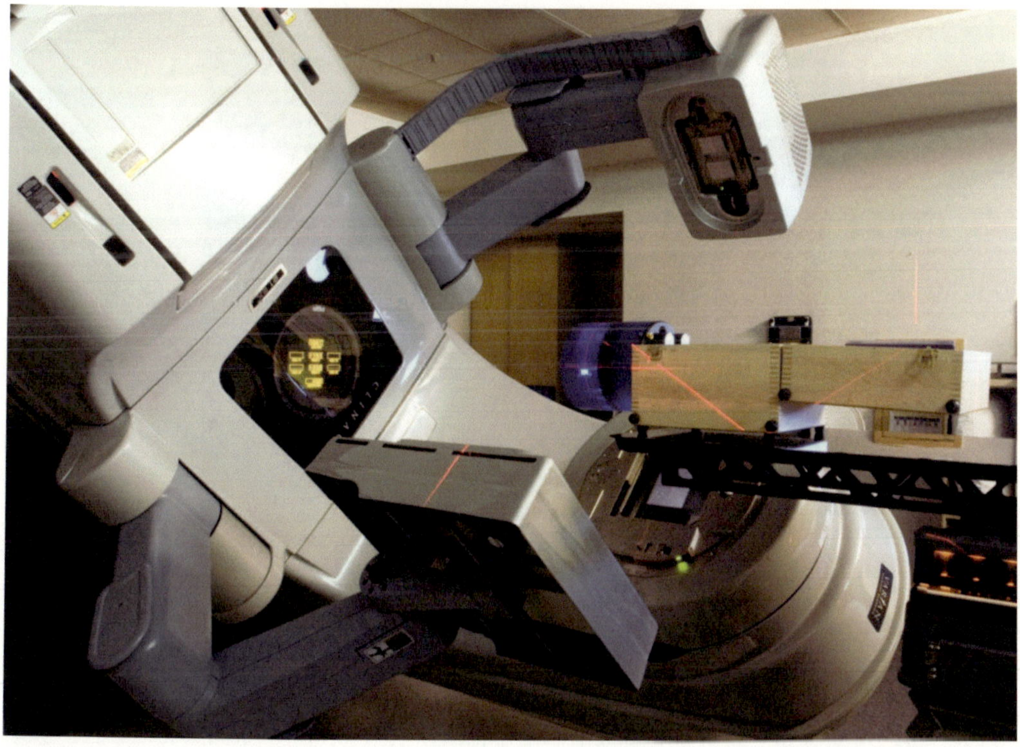

Fig. 10. Setup of Catphan at machine Isocenter to be imaged for image quality tests.

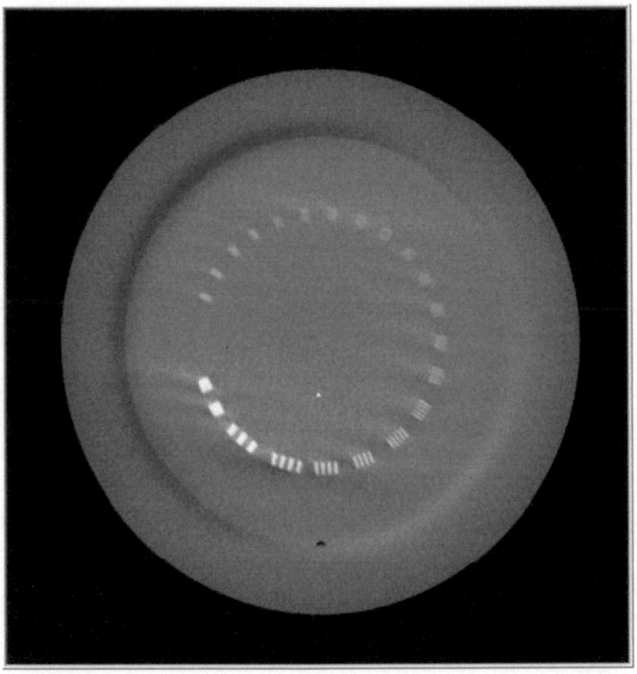

Fig. 11. CBCT image of a slice of the Catphan phantom used to determine the spatial resolution of the imaging system.

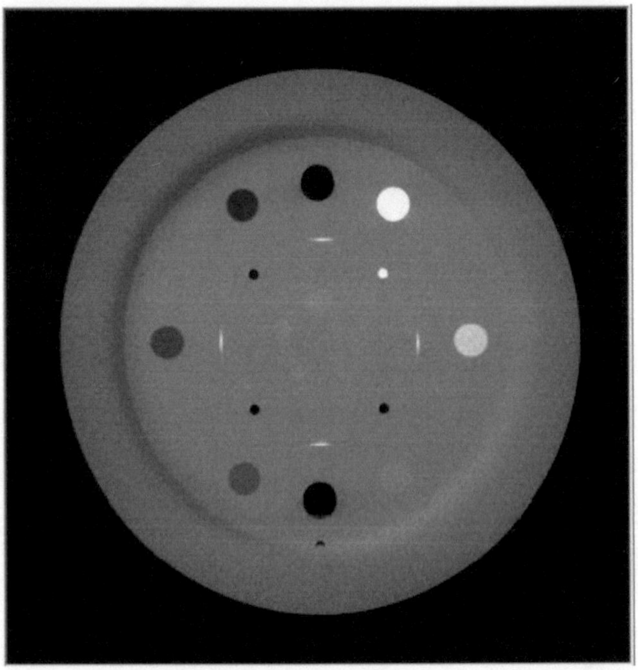

Fig. 12. CBCT image of a slice of the Catphan phantom with samples of various materials of known d density (outer ring of circles) as well as with precise geometrical markers.

CT number linearity can also be checked with the Catphan. The module that is used for the geometry tests also provides samples of various materials with known densities (Figure 12). The Acceptance Test of the Elekta XVI system uses these CT numbers to calculate low contrast visibility. Markers in this module can be used to calculate slice thickness, which is, however, less relevant for CBCT.

Low contrast visibility can also be determined with another module (Figure 13) and is currently done so for the Varian Acceptance testing. This test depends on the user being able to see circles of different diameters in the image.

Fig. 13. CBCT image of a slice of the Catphan phantom used to determine Low Contrast Visibility.

Lastly, there is a uniformity check, which compares the CT number in different parts of the image of a homogenous phantom. This test can be done easily by hand with the tools in the imaging software.

For all image quality checks consistency of the check execution is more important than how exactly the test is done. The image quality tests should be done with a parameter set that produces a high image quality. The tests do not have to be repeated with all clinically used imaging parameters, as any significant changes in the system would be visible in test with the high quality.

All tests can be done with the tools available on the imaging software. However, user independent tests using software to analyze the images give more objective measures of the stability of the system. Several vendors have products that allow user independent analysis and also varying degree of automation of the QA. Companies with such products are

Standard Imaging (Middleton, Wisconsin, USA) and Radiological Imaging Technology (Colorado Springs, Colorado, USA)

2.5 X-ray tube and generator tests

The CBCT imaging chain consists of the x-ray generator, x-ray tube, and the digital imaging device. It is important to assure that the generator and x-ray tube are performing properly in a known manner so that imaging technique parameters can be confidently adjusted by the user. For example, if the user wishes a less noisy image, he should be confident that increasing the mA setting by 50% results in 50% more radiation being incident on the patient.

A fairly comprehensive set of measurements should be performed at time of acceptance. A more streamlined set should be performed periodically thereafter and following major component changes. The desired data are easily taken with a number of commercially available diagnostic imaging products, e.g. by Unfors (Billdal, Sweden) and RTI Electronics AB (Mölndal, Sweden) that provide μGy, exposure time, and HVL in a single measurement. These tests do require that the CBCT be operated in a radiographic, as opposed to CT, mode. The way to operate in this manner varies with vendor.

The tests to be performed and parameters to be measured are the same for the baseline acceptance tests and the periodic tests. The latter is a subset of the former and concentrates on those settings used for clinical CBCT protocols. The tests should be performed in a readily reproduced geometry that allows the measurement device to obtain accurate readings. Since most of the devices require a certain minimum amount of radiation for a reading, it may be necessary to set the detector some 60 to 80 cm from the x-ray source. This is easily accomplished through of simple platforms made of readily available polystyrene blocks.

The required tests fall into three general categories:

1. kVp accuracy and HVL: Measure the kVp and compare to the value chosen on the generator. Also record the μGy at these setting so that the μGy/mAs performance can be recorded for future reference. Perform these measurements for the full range kVp settings and for both large and small focal spots. Record the HVL at a clinically used kVp such as 120 kVp. Note that the measured HVL for a CT tube is substantially higher than a conventional diagnostic x-ray tube.
2. Timer accuracy and linearity: At a fixed, clinically used kVp, measure the exposure time and μGy for a full range of allowed timer setting. Timer values should agree with the set time readings, while the μGy/mAs should remain constant.
3. mA linearity: It is difficult to measure tube mA directly in a non-invasive fashion. The true mA is less important than the output, (μGy), tracking in a linear fashion with the set mA. Measure the μGy for the full range of allowed mA settings at fixed, clinically used kVp, for both small and large focal spots. Compute the μGy/mAs by dividing the μGy by set mAs. This value should be approximately constant over the full range of mA settings and between focal spots.

A comprehensive baseline should be obtained at time of acceptance. Acceptable performance occurs when the measured value is within 10% of the set value (kVp, mA) or measured mean value (μGy/mAs). Any unacceptable results should be resolved with the vendor. Ongoing checks should be performed on an annual basis or following a change in major system component, e.g. the x-ray tube. The annual test can be limited to those kVp settings used for clinical protocols and to a limited number of mA and time settings that encompass the range used in the clinical protocols.

2.6 End-to-end test

As for all systems that help position the patient prior to the application of radiation therapy, the ultimate verification is an end-to-end test, in which a phantom is put through the entire chain of procedures that the patient goes through and where the received radiation dose in the phantom can be recorded and visualized.

Radiochromic film, which is relatively insensitive to visible light but changes its color proportional to the amount of ionizing radiation received, is an elegant tool for end-to-tests. This also applies to kV CBCT systems since the kV radiation involved in CBCT imaging does not have a visible effect on the film. For straightforward, visual verification an end-to-end test is planned with an arc field as the treatment field to be delivered to the phantom. An arc, which covers close to 360 degrees, produces a circular darkening of the film around the isocenter when the film is placed perpendicular to the couch (Figure 14). Using a precisely machined hole, which meets the film at center of the phantom, a needle can be used to mark the center of the phantom on the film.

Fig. 14. Phantom for and end-to-end test with radiochromic film after an arc of therapeutic radiation has been delivered to it.

Using CBCT, the phantom is setup to align its center with the isocenter of the beam. After the arc field has been delivered, the relationship between the needle poke and the circular darkening can be evaluated to determine the agreement of the alignment in anterior-posterior and left-right direction. Visual evaluation can be supplemented with software-based analysis of the film.

To assess the isocenter agreement in superior-inferior direction, the film needs to be rotated by 90° around the anterior-posterior axis. The shown phantom is build to accommodate precise rotation of the block with the film (Figure 15) by 90°, while maintaining isocenter position. Alternatively a separate test can be performed with the entire phantom rotated.

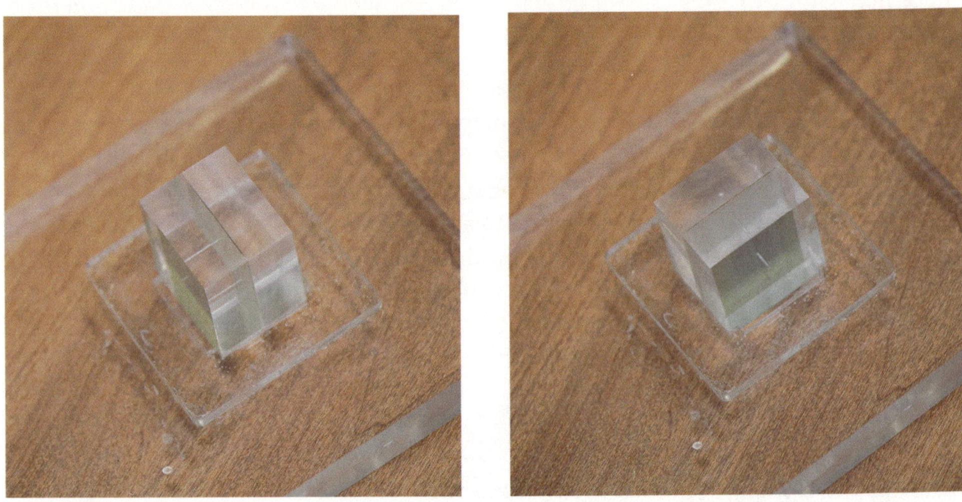

Fig. 15. Phantom for and end-to-end test using a plastic cube as holder for radiochromic film, which can be rotated precisely by 90° while maintaining isocenter.

3. Frequency of tests

3.1 Daily QA

Safety checks: Depending on the specifics of the system the available interlocks should be either all tested, or tested on a rotational basis throughout the week.

Level 1 Geometric Accuracy Test: For daily checks the kV CBCT isocenter should be compared to the cross hair of the machine.

Registration and correction accuracy: A phantom should be manually setup to an off-center location. The phantom should be CBCT imaged, aligned and moved to isocenter. This test can be combined with the geometric accuracy test, if the final location of the phantom is compared to the cross hair of the beam or to the in-room laser system, provided that has been checked.

3.2 Monthly QA

Safety checks: All available interlocks should be tested monthly

Level 2 Geometric Accuracy Test: The isocenter of the kV CBCT system should be compared to the isocenter of the MV system as represented by a portal image with the treatment beam.

Registration and correction accuracy: The daily QA of the Registration and correction accuracy should be repeated by the physicist performing the monthly QA.

Image quality check: The described image quality checks with the Catphan should be performed. Preferably all tests should be done each month.

3.3 Annual QA

X-ray Tube and Generator Tests should be performed annually, as described above. An end-to-end test should also be performed.

3.4 QA following system repairs or upgrades

Following a system repair or upgrade that did not involve the x-ray tube and/or generator the tests of the monthly QA should be performed.

If x-ray generator and/or generator were involved in the intervention, an annual QA should be done in addition to the tests of the monthly QA.

Depending on the severity of the repair or upgrade, an end-to-end test should also be performed. If the repair or upgrade on affected the CBCT system but not the first part of the chain (CT scanner, treatment planning system, record and verify system), only the final step of the end-to-end test needs to be done.

4. Summary

Image guided therapy in general and kV Cone Beam Computed Tomography in particular are powerful tools in a modern Radiation Oncology facility. Assuring the correct function of all systems is of utmost importance in the interest of accurate patient treatment.

The described tests, which should be performed or supervised by a Qualified Medical Physicist, help assure a high level of patient care. For additional information refer to current literature, such as (Bissonnette 2007) and (Bissonnette et al., 2008), and to upcoming publications from professional organizations, such as the AAPM Task Group on CBCT QA.

5. List of abbreviation and acronyms

AAPM – American Association of Physicists in Medicine
CBCT – Cone Beam Computed Tomography
CT – Computed Tomography
Gy – Gray – Unit for the absorbed radiation dose in medium (1 Gray = 1 Joule / 1 Kilogram)
HVL – Half Value Layer
IGRT – Image Guided Radiation Therapy
kV – Kilo voltage
kVp – Kilo voltage peak value (max kV in a beam with photons of a range of kV)
mA - Milliampere

mAs - Milliampere Seconds
MLC - Multi Leaf Collimator (Beam shaping device of the linear accelerator)
MV - Mega voltage
OBI - On Board Imaging (product of Varian Medical Systems, Inc.)
QA - Quality Assurance
QMP - Qualified Medical Physicist
TBP - Two Ball Phantom
TG - Task Group (here: Task Group of the AAPM)
XVI - X-ray Volume Imaging (product of Elekta AB)

6. References

Bissonnette, J. P. "Quality assurance of image-guidance technologies." Semin Radiat Oncol 17(4): 278–286, 2007.

Bissonnette, J. P., D. Moseley et al. "Quality assurance for the geometric accuracy of cone beam CT guidance in radiation therapy." Int J Radiat Oncol Biol Phys 71(1 Suppl):S57–S61., 2008

Jaffray, D. A., D. G. Drake et al. "A radiographic and tomographic imaging system integrated into a medical linear accelerator for localization of bone and soft-tissue targets." Int J Radiat Oncol Biol Phys 45(3): 773–789. 1999

Klein, Eric E., Joseph Hanley, John Bayouth et al "Task Group 142 report: Quality assurance of medical accelerators" Med. Phys. 36 (9), 2009

Lehmann Joerg, Julian Perks, Sheldon Semon, Rick Harse, and James A. Purdy, "Commissioning experience with cone-beam computed tomography for image-guided radiation therapy" JACMP 8(3), 2007

Lutz W, Winston KR, Maleki N. "A system for stereotactic radiosurgery with a linear accelerator" IJROBP 14(2):373-81. 1988

Yin, Fang-Fang, John Wong, et al "The Role of In-Room kV X-Ray Imaging for Patient Setup and Target Localization" Report of AAPM Task Group 104 ISBN: 978-1-888340-89-1, 2009

17

Polymer Gel Dosimetry for Radiation Therapy

Senthil Kumar Dhiviyaraj Kalaiselven and
James Jebaseelan Samuel Emmanvel Rajan
*Photonics Division, School of Advanced Sciences,
VIT University, Vellore, Tamil Nadu
India*

1. Introduction

Although conventional dosimeters such as ionization chambers measure absolute dose to a high precision at single points, their finite size makes resolving areas of high dose gradient difficult. Improved spatial resolution can be achieved with thermo luminescent dosimeters (TLDs), silicon diodes and diamond detectors. Film also offers high spatial resolution in a single, two dimensional planes and provides excellent relative dose information and absolute dose measurements when appropriately calibrated. More complete three-dimensional (3D) dose measurements can be produced by positioning film in multiple planes, although accurate positioning of film in several layers can be a difficult and time-consuming process. The search for a dosimetry technique that allows full 3D imaging of a radiation dose distribution has led to the development of radiation-sensitive gels. These gels have a long history, with gels containing Folin's phenol, which change colour upon irradiation, first investigated by (Day & Stein, 1950). Later, measurements of photon and electron depth doses were made by (Andrews et al., 1957) using agar gels. Several studies subsequent to this have used Fricke solutions and gels by (Day, 1990). An alternative technique, based on radiation-induced polymerization in solutions of monomers and polymers, was studied by (Hoecker & Watkins, 1958), with dosimetry investigations carried out by (Audet & Schreiner, 1991).

The interest in the gel dosimetry technique follows on from a type of dosimetry gel proposed by (Maryanski et al., 1993). Here, acrylic molecules embedded within a gel matrix polymerize upon irradiation, with the degree of polymerization being strongly related to the absorbed dose received by the gel. The spatially localized polymerization can then be imaged by MRI or optical scanning methods. When imaged in an MRI scanner the relaxation rate of the polymerized region was reported to be linearly proportional to absorbed dose in a dose range applicable to the study of clinical radiation therapy (0-15 Gy). The first international workshop dedicated to gel dosimetry (Schreiner, 1999) took place, demonstrating the growing interest in development and application of the gel dosimetry technique. (Gore et al., 1996; Maryanski et al., 1996) demonstrated the potential of Optical CT as an alternative imaging technique to MRI for PAG-type polymer gel dosimeters. This technique was further investigated by (Oldham et al., 2001, 2003) and (Oldham & Kim,

2004). In 2000 Hilts et al demonstrated the use of x-ray CT to image PAG-type gels and subsequently used x-ray CT to investigate stereotactic dose distributions. (Mather et al., 2002b) demonstrated the use of ultrasound to image polymer gel dosimeters. (Rintoul et al., 2003) demonstrated the use of Raman imaging to evaluate an electron depth dose in an irradiated PAG dosimeter. Although polymer-type dosimeters did not have the diffusion limitations of Fricke-type gel dosimeters, there was another significant limitation to their use. Due to the nature of their free radical chemistry, polymer gel dosimeters were susceptible to atmospheric oxygen inhibiting the polymerization processes. As a result, these gel dosimeters had to be manufactured in an oxygen-free environment, for example in a glove box flushed with inert gas such nitrogen or argon (Baldock et al., 1998a; De Deene et al., 1998a). A significant development in the field of gel dosimetry was reported by (Fong et al., 2001). This development was a new type of polymer gel dosimeter, known as MAGIC, in which atmospheric oxygen was bound in a metallo-organic complex thus removing the problem of oxygen inhibition and enabling polymer gels to be manufactured on the benchtop in the laboratory. These types of polymer gel dosimeters became known as the new class of normoxic gel dosimeters. The existing PAG dosimeters subsequently became known as hypoxic or anoxic gel dosimeters. The MAGIC polymer gel formulation consisted of methacrylic acid, ascorbic acid, gelatin and copper. The principle behind removing the problem of oxygen in the MAGIC gel is in the use of ascorbic acid, commonly known as vitamin C. Ascorbic acid binds free oxygen contained within the aqueous gelatin matrix into metallo-organic complexes in a process initiated by copper sulfate (De Deene et al., 2002b). It was subsequently shown that other antioxidants could also be used in the manufacture of normoxic gels including tetrakis (hydroxymethyl) phosphonium chloride (THPC) (De Deene et al., 2002a; Baldock, 2006). Numerous authors subsequently published results of work investigating different compositions and formulations of normoxic polymer gel dosimeters which have been summarized by (Senden et al., 2006). With the introduction of normoxic gel dosimeters, MRI studies were undertaken to investigate their usefulness for IMRT (Gustavsson et al., 2003), and radionuclide therapy (Courbon et al., 2006; Gear et al., 2006; Braun et al., 2007, 2009). There have been a limited number of previous reviews on polymer gel dosimetry (McJury et al., 2000). Further additional information on gel dosimetry has been published in the proceedings of the DOSGEL conferences (DOSGEL, 1999, 2001, 2004, 2006, 2008). The fundamental science underpinning the dosimetry technique is reviewed along with the various evaluation techniques and associated issues for the purposes of clinical dosimetry applications.

2. Principles

Polymerized gel, prepared with monomers such as Acrylamide, is widely used in biochemistry as a medium for electrophoresis for protein and nucleic acid separation (Hoecker et al., 1958; Gordon et al., 1973; Chambers et al., 1967; Alexander et al., 1954; Hsu et al., 1984). Polymer gels are typically composed of Acrylamide monomer and a cross-linking agent such as N, N′-Methylene- Bis Acrylamide (BIS). Polymerization of gels used for electrophoresis is normally initiated and controlled chemically using a free radical initiator such as ammonium/sulphate. However, polymerization can also be induced by radiation via free radical production during water radiolysis. The radiation-induced polymerization has been explained by (Charlesby, 1987). The first polymer gels designed for radiation dosimetry were made by adding the monomer Acrylamide and the cross-linking agent BIS to a simple Agarose-based gel solution (Maryanski et al., 1993, 1994). These were given the acronym "BANANA", from Bis Acrylamide Nitrous oxide and Agarose. Greater sensitivity was later achieved with gelatin-based gel solutions, which led to the current generation of BANG (Bis Acrylamide Nitrogen and Gelatin) polymer gels (Maryanski et al., 1994). The gases nitrous oxide and nitrogen included in the acronyms are required to displace oxygen from the gel during manufacture. After irradiation, a gel will contain regions that are polymerized and cross-linked. This is the origin of the gel's spatial dose-response characteristics, as the degree of polymerization depends on the initial quantity of free radicals generated by the incident radiation and, therefore, on the absorbed dose. Within the polymerized regions, certain populations of water molecules alter their state of binding to, and exchange protons with, the polymer network. This can be investigated using nuclear magnetic resonance (NMR) relaxation time measurements. The relaxation characteristics of the trapped water protons are determined by both the polymer structure (e.g. polymer porosity, as the size of network spaces will determine the efficiency of trapping water molecules) and its concentration (which depends on the absorbed dose). The water R_2 relaxation rate (reciprocal of the transverse relaxation time T2) measured by MRI of BANG gels is linearly related to dose in the range 1-10 Gy (Maryanski et al., 1993, 1994, 1996; Back et al., 1998; Oldham et al., 1998). This range encompasses the dose used in a standard external beam radiotherapy treatment fraction (2 Gy). Previously, there are two types of BANG gel in 1996, although many variations on the basic formulation are possible. BANG-1™ and BANG-2 ™ gels are current models of BANG ™ gels. The former is made using acrylamide in powder form, while the latter replaces acrylamide with acrylic acid and NaOH to buffer the pH. Gel response is improved with acrylic acid compared to acrylamide, allowing larger relaxation-rate changes per unit dose. Acrylamide is a neurotoxin that can lead to nervous system disorders. Safe handling of acrylamide is essential and appropriate care should be taken when disposing of the gel. (McJury, 2000). BANG-3TM gels have been developed after 2000. This type of gel has strong optical and MR responses (Oldham, 2001). With the BANG-3TM gel, acrylic acid is replaced with methacrylic acid. Table 2.1 summarizes the gel types and compositions. Of the BANGTM polymer gels, BANG-3TM polymer gel reportedly has the highest MR sensitivity upon photon irradiation (Ramm, 2000).

3. Gel dosimeter manufacture

Deionized, distilled water is used, which is degassed with an inert gas such as pure N_2 to remove any dissolved oxygen. The gas is humidified to minimize water loss during the degassing process. Oxygen is a free radical scavenger that inhibits gel polymerization and must therefore be removed. To avoid exposure to oxygen during preparation, the gel is usually manufactured either in sealed reaction flasks (Oldham et al., 1998; Baldock et al., 1998; De Deene et al., 1998) or in a sealed glove box with a nitrogen environment (De Deene et al., 1998). A BANG-1 type gel is made by adding typically 5% (all percentages by weight) gelatin (type A, approx. 300 bloom) to the degassed water at room temperature (Maryanski et al., 1994; Baldock et al., 1998). This is allowed to dissolve, and is heated to approximately 50 °C to prevent gel setting. 3% Acrylamide is added to the solution, followed by 3% BIS. When the monomers have completely dissolved in the solution, the gel can be transferred to airtight containers and allowed to cool and set. Oxygen impermeable vessels, such as glass or BAREX (BP Chemicals Inc., London, UK), are usually used because plastic containers (e.g.

Perspex, polystyrene etc.) allow transfer of oxygen through the vessel wall (Maryanski et al., 1994). This may have a minimal effect on applications where the irradiated regions are distant from the container walls. It is advantageous to irradiate the gel dosimeter soon after manufacture owing to potential oxygen contamination and exposure to light. Once irradiated, the dosimeters are usually imaged, or "read", within a few days. The optimum time between irradiation and imaging is approximately 3-4 days (McJury et al., 1999).

4. Characteristics of the gel dosimeter

4.1 Light and oxygen contamination

Once the monomer and cross-linker are added to the gelatin solution, the mixture should be shielded from light as photo polymerization may begin and consequently degrade the overall sensitivity of the gel (Maryanski et al., 1993, 1994; Richmond et al., 1987).

The gel polymerization process is initiated by free radical reactions, and molecular oxygen is an efficient "scavenger" of free radicals. The effect of dissolved oxygen upon the sensitivity of Polyacrylamide gels has been reported in (Maryanski et al., 1993, 1994; Baldock et al., 1998; Richmond et al., 1987; Hepworth et al., 1999), where it was observed that even trace amounts led to a complete failure of the polymerization reaction. A typical result of oxygen contamination is shown in Figure 1.

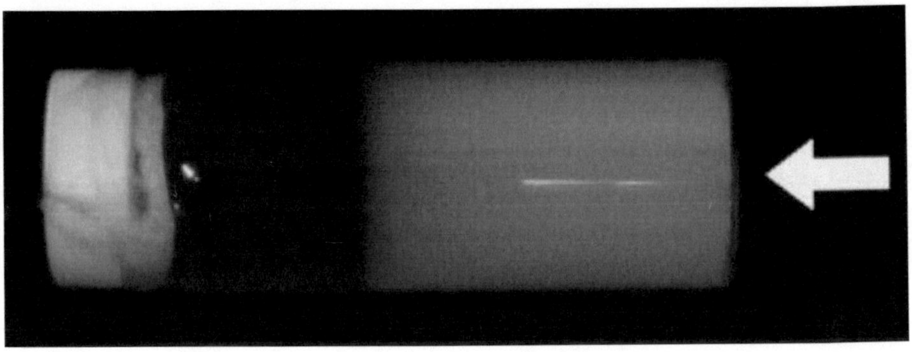

Fig. 1. 35 ml gel sample in a glass vial that has been irradiated with 6 MV photons as indicated by the arrow. Owing to oxygen diffusion through the gel, polymerization in a large component of the sample is prevented.

Here, a calibration vessel irradiated with a single 6 MV photon beam (direction indicated) had a faulty seal at the top of the flask. A strong gradient in gel polymerization is clearly seen, which renders the gel useless as a dosimeter.

4.2 Temperature

The temperature of the gel at irradiation is reported to have little effect on final R_2 measurements (Maryanski et al., 1994). However, at imaging the sensitivity of the gel increases with decreasing temperature (Maryanski et al., 1997, 1995; De Deene et al., 1998, 1999). Figure 2 shows the dose response of a BANG-1 gel imaged at a number of temperatures between 5 ^{o}C and 40 ^{o}C, as demonstrated by (Maryanski et al., 1997).

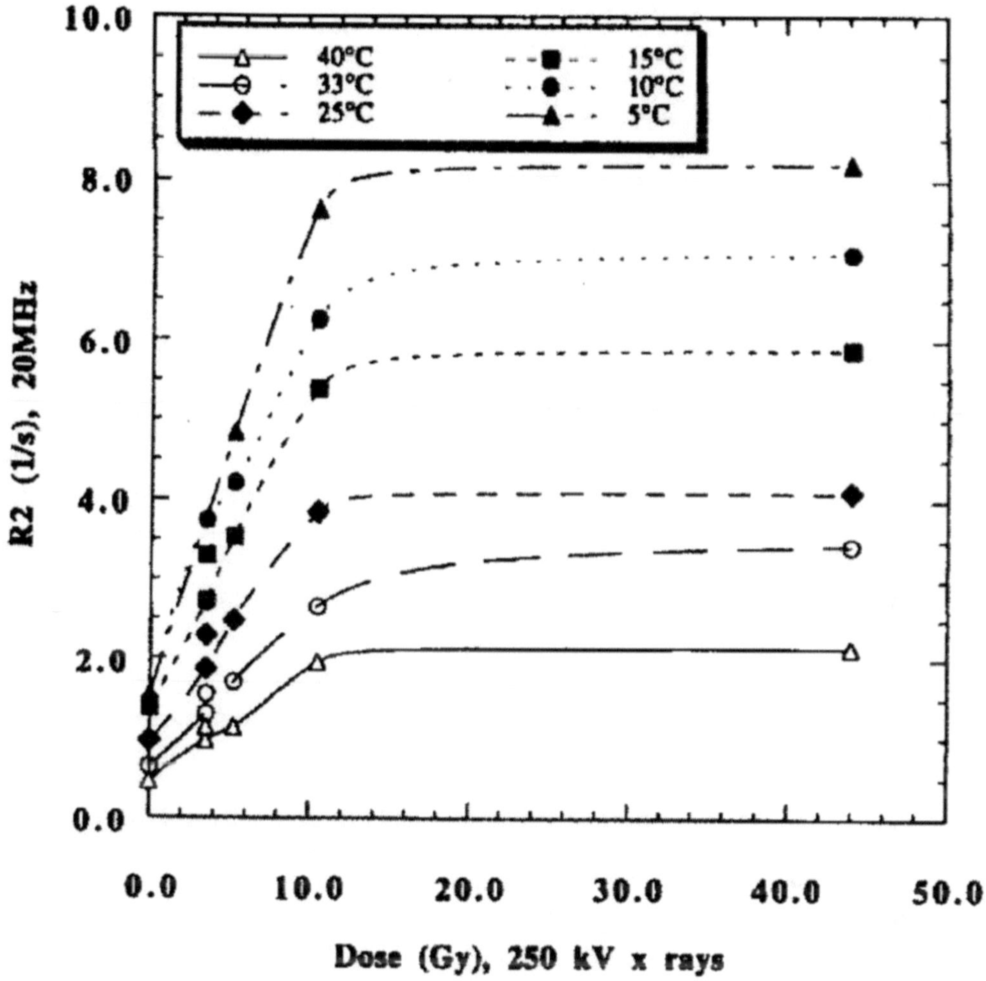

Fig. 2. The effect of temperature during imaging on the water R_2 relaxation rate dose-response curve. (Phys Med Biol 1997; 42: 303-11 [Maryanski et al., 1997])

The effect is explained by the change in proton correlation times and proton exchange rates in the gel with temperature. These increase as the motions of the polymer chains become slower with decreasing temperature. The relaxation rate of gelatin increases with decreasing temperature (Vachier et al., 1996). It is therefore important that a gel is allowed to equilibrate to a uniform temperature prior to MRI and that the experimental gel and the calibration gels be at the same temperature.

4.3 Concentration of polymer and cross-linker

Both the absolute and relative weight fractions of monomer and cross-linker agents in the gel may be varied (Oldham et al., 1998; Maryanski et al., 1995, 1997; Audet et al., 1995; Baldock et al., 1996). Increasing the total monomer content of the gel, for example from 3% to 6%, increases sensitivity (Maryanski et al., 1997; Baldock et al., 1996; Farajollahi et al.,

1997) and extends the dose saturation point. The increase of dose saturation point with increasing cross-linker density can be explained by two competing mechanisms: (i) the reactivity of monomers, and (ii) the relaxation properties of the gel, which depend on polymer composition and concentration. It has been reported (Maryanski et al., 1997; Kennan et al., 1996) that as the fraction of cross-linker increases, the conversion of monomers to polymer per unit dose decreases due to the lower reactivity of the BIS cross-linker relative to Acrylamide. Therefore, the sensitivity of the gel to radiation decreases. As the cross-linker density increases, the polymerized gel becomes more rigid, with the BIS molecules cross-linking tightly with each other as well as with the residual Acrylamide.

4.4 Gel matrix

Gelatin produces clearer gels than Agarose, allowing greater visibility of the polymerized region and facilitating imaging of the gel by optical means. Other gelling agents can also be used, such as Sephadex-200 and Sumikagel N-100 (Zebrowska et al., 1995). Increasing the strength of the gelatin increases the melting point of the gel. Generally, 300-bloom gelatin is used, which has a comparatively high melting point (30-35 ^{o}C (Vachier et al., 1996)). Increasing gelatin concentration decreases R_2 and overall gel sensitivity (Maryanski et al., 1994; Audet et al., 1995), thus a gel consisting of 5% gelatin by weight is usually used.

4.5 Gel pH

For gelatin, R_2 increases with pH at all temperatures and concentrations (Vachier et al., 1996). Maryanski et al noted that a more reproducible dose-response was achieved with neutral gel pH following initial studies with acidic gels (Maryanski et al., 1993). In simple gels constructed only from solutions of Acrylamide and BIS, and polymerized by chemical means, an increase in pH (greater than pH 8) was also associated with an increase in relaxation rate (Kennan et al., 1996). This is consistent with a chemical-exchange mediated interaction between water protons and the polymer. Further work by (Gochberg et al., 1998) indicated that the dependence of relaxation rate on pH was not entirely consistent with a simple acid/base catalyzed chemical exchange.

4.6 Gel fogging

Bubbling N_2 gas through a gel during mixing can cause fogging of the gel, leading to an increased background R_2 value (Maryanski et al., 1994). This spontaneous polymerization is thought to be due to free radical impurities in some of the gel manufacturing materials (Maryanski et al., 1993; Richmond et al., 1987). It is therefore important to use high-grade chemicals when manufacturing these gels.

4.7 High dose edge effects

An increased response at the edges of irradiated regions in BANG gels has been reported (Maryanski et al., 1994). This occurs for high dose irradiations, beyond the linear region where R_2 is proportional to absorbed dose. The effect is not completely understood but is thought to be due to monomers slowly diffusing from low dose to high dose regions and interacting with long-lived macro-radicals.

4.8 Gel ageing

It is not clear whether the saturation doses employed in NMR studies lead to complete polymerization. It may take from a few hours to perhaps several weeks for the polymerization reactions to be complete after irradiation, with the reaction slowing down quasi-exponentially (McJury et al., 1999). Imaging within a few hours of irradiation, when the rate of polymerization is greatest, could lead to errors, particularly if there is a time delay between imaging a set of calibration gels and the target phantom.

4.9 Toxicity

Acrylamide is a neurotoxin. Repeated skin contact or ingestion can lead to nervous system disorders (Richmond et al., 1987; Sittig et al., 1985). Guidelines for handling Acrylamide include wearing gloves at all times and working with the powder chemicals within a fume cupboard. Care should also be taken when disposing of gels after use. Some of these concerns have been reduced by replacing Acrylamide with acrylic acid, as used in the BANG-2 gel formulation.

5. Imaging of the dosimeter

5.1 MR relaxation time imaging

MRI allows the measurement of the longitudinal and transverse relaxation rates (R_1 and R_2) of the dosimeter gels, from which dose maps can be calculated. Conventionally, the corresponding relaxation times (T_1 and T_2) are measured, from which the rates can be computed. Relaxation times are measured by applying radiofrequency (RF) pulses to excite the magnetization of the spin system, and then sampling during the return to equilibrium. The transverse relaxation time T_2 (equals to $1/R_2$) is measured by fitting data collected from at least two points on the transverse relaxation curve following excitation. Two points are the minimum needed to obtain a T_2 value, but errors will be reduced if a larger number of points on the relaxation curve are collected. Two main approaches to data collection are used, each technique having its own intrinsic advantages and disadvantages (Baustert et al., 2000). (a) Single echo or Hahn spin echo sequence method (Hahn et al., 1950). Each acquisition collects only a single echo, but can be repeated with different echo times. (b) Multiple spin echo method (Carr et al., 1956; Meiboom et al., 1958). A sequence that acquires a train of spin echoes, each corresponding to a different echo time. Concern about errors in measured R_2 values owing to imperfect refocusing with 180° pulses, leading to standing wave effects or RF attenuation in large aqueous samples, has led to a preference for single echo sequence methods by some authors (Maryanski et al., 1994, 1996).

5.2 X-Ray CT scanning

Hilts et al showed that x-ray CT could be used to investigate changes in irradiated PAG polymer gels and demonstrated that in cases of high dose gradients this evaluation technique had potential (Hilts et., 1999, 2000). Trapp et al further evaluated the technique for investigating different compositions of PAG polymer gels (Trapp et al., 2001). It was shown that the CT-dose sensitivity increased with the concentration of co monomers used, and by varying the gelling agent from gelatin to agarose. Hill et al further investigated the change in CT number with dose for normoxic polymer gels (Hill et al., 2003). Trapp et al showed that the observed contrast obtained in CT images of irradiated PAG gels was due to density changes in the gels (Trapp et al., 2002) and was further demonstrated for normoxic MAGIC gels by (Brindha et al., 2004). Hilts et al suggested that polymer gels for x-ray CT evaluation should have a more optimal sensitivity for an extrinsic density change and maximum polymer yield in the gel dosimeter (Hilts et al., 2004). A significant limitation of x-ray CT evaluation is the relatively small change of CT numbers or Hounsfield units in the irradiated polymer gel with absorbed dose along with image noise (Trapp et al., 2001). Hilts and Duzenli showed that the issue of image noise could potentially be tackled through image processing techniques (Hilts & Duzenli, 2004). The phantom wall material used has the potential to induce artifacts in the resulting CT image and is particularly significant for glass (Trapp et al., 2001). To overcome these artifacts, alternative plastic wall materials have been investigated (Hill et al., 2003). Audet et al undertook a clinical dosimetry study of CT gel dosimetry and showed that high dose regions produced by stereotactic irradiations could be accurately localized (Audet et al., 2002). Further clinical dosimetry studies are required to fully evaluate the potential of the x-ray CT technique. However, as a result of the easy access to x-ray CT scanners in clinical departments, CT does potentially offer a relatively simple, convenient and inexpensive method of implementing clinical polymer gel dosimetry (Hilts et al., 2004). The main advantage of this evaluation technique is that limited access to MRI facilities is not an issue or is the need to build or purchase specialized optical scanning equipment. It should be noted that where, for instance, MRI has evaluation problems such as in homogeneities or temperature dependence (De Deene et al., 2000a, 2000b; De Deene & Wagter, 2001), x-ray CT evaluation has its own limitations. However, as is the case with MRI evaluation of polymer gel dosimeters (De Deene, 2001), it has been shown that x-ray CT evaluation techniques require the CT scanner be commissioned specifically for gel dosimetry before use (Hill et al., 2005).

5.3 Ultrasound scanning

Mather et al showed that ultrasound could be used to investigate changes in irradiated PAG polymer gels (Mather et al., 2001, 2002a). In these studies, acoustic speed of propagation, attenuation and transmitted signal intensity showed a strong variation with absorbed dose indicating the potential of this technique. Comparative studies of PAG and MAGIC polymer gels indicated that differences in acoustic properties with absorbed dose were due to differences in the elastic modulus of the materials (Mather et al., 2002b). Further acoustic studies (Mather et al., 2003a) showed that the overall acoustic attenuation, dose sensitivity and dynamic range were dependant on dosimeter formulation. These studies, along with those of acoustic attenuation coefficients highlighted the complex nature of the acoustic properties of polymer gel dosimeters and indicated that further studies are required to fully understand the properties (Mather et al., 2003b). Mather and Baldock undertook a preliminary study to evaluate the potential of acoustic imaging of

irradiated PAG polymer gel dosimetry phantoms (Mather & Baldock, 2003c). An ultrasound computer tomography (UCT) system was developed and evaluated. Results from the UCT system indicated that transmission-generated images produced greater contrast than time-of-flight-generated images. However, time-of-flight-generated images produced superior geometrical accuracy of the dose distribution than transmission-generated images. The motivation to develop UCT of polymer gels was to produce a more economically viable evaluation methodology.

5.4 Optical scanning

As the gel polymerizes it becomes opaque. It is therefore possible to use optical scanning to generate two-dimensional dose distributions. A prototype optical tomography system has been designed previously for reading the optical density of irradiated BANG gels (Gore et al., 1996). The technique depends on light scattering from the polymer micro particles within an irradiated gel. A detailed study of light attenuation within a BANG-1 gel (Maryanski et al., 1996) indicated that light was not absorbed by the polymer particles, only scattered. Light attenuation can therefore be related to polymer density and, consequently, to absorbed dose. Using a gel sample irradiated with a series of uniform dose regions the attenuation coefficient for 500 nm wavelengths light increased by approximately 0.7 mm^{-1} when the dose increased from 0 to 5 Gy. The attenuation was directly related to absorbed dose over the range 0-10 Gy. The shape of the dose-response curve was also found to depend upon the fraction of the cross linking monomer in the initial mixture and on the wavelength of light used. The main conclusion of the preliminary optical scanning studies was that the technique could replace, or at least complement, the NMR imaging method of dose measurement.

5.5 Calibration of the dosimeter

The accuracy and sensitivity of an individual gel batch is dependent upon the exact conditions of manufacture and the purity of chemicals used. It is therefore recommended that each gel batch is calibrated separately at the time of use (Baldock et al., 1998). Several different methods of dosimeter calibration have been reported. In all cases, a quantity of the gel batch to be used experimentally is transferred to a calibration phantom (or phantoms) and irradiated with a range of known doses. MRI of the calibration phantom produces a T_2 relaxation map and a plot of known dose against relaxation rate R_2 ($1/T_2$). This can then be used to calibrate the rest of the experimental data (Maryanski et al., 1994). The quality of a calibration method may be judged by the errors in fitting the calibration data and, to a lesser extent, on the quantity of gel needed for calibration. There are three main calibration methods reported in the literature.

5.5.1 Multibeam method

A large (approx. 1-1.5 litres) flask filled with gel is irradiated with several small beams (either using stereotactic cones or small square fields) to a number of doses over the sensitivity range of the gel. Figure 3a shows the T_2 map of a calibration phantom using the "Multibeam" method.

The phantom had a diameter of 20 cm and a thickness of 6 cm, and was filled with 1.5 litres of BANG-1 gel. It was irradiated with six 2x2 cm^2 fields to doses of 2, 4, 6, 8, 9 and 10 Gy (Oldham et al., 1998). The resulting calibration plot is shown in Figure 3b, with errors on the slope and intercept of 1.5% and 5.0%, respectively.

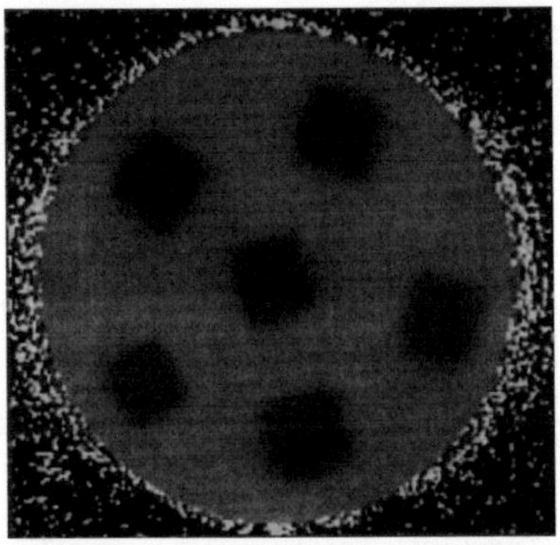

Fig. 3(a). T_2 image through a calibration gel using the "Multibeam" method. The gel sample received 2 x 2 cm^2 irradiations of 2, 4, 6, 8, 9 and 10 Gy. (Phys Med Biol 1998; 43:1113-2 [Oldham et al., 1998]).

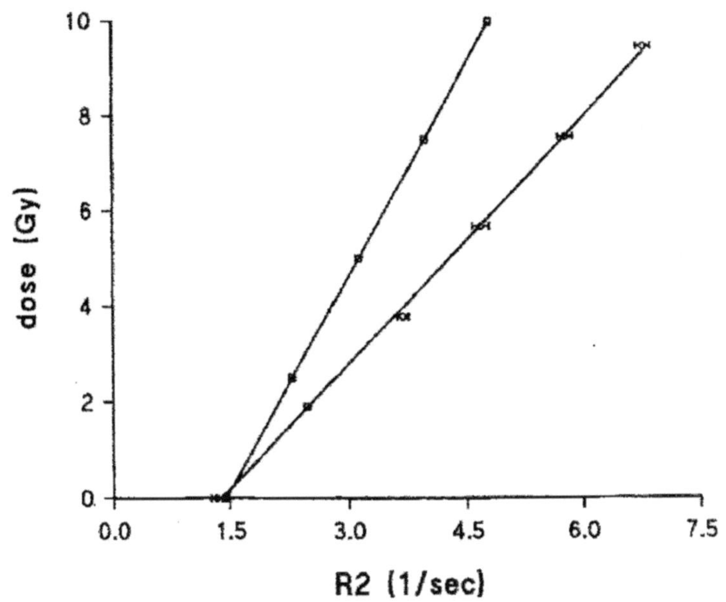

Fig. 3(b). Calibration plot of water relaxation rate R_2 against dose obtained using the "Multibeam" method for gel with a 3% N, N'-methylene-bis-acrylamide (BIS) and Acrylamide formulation and a gel with 6% BIS and Acrylamide. The plot demonstrates the enhanced sensitivity obtained when using higher concentrations of monomers. (Phys Med Biol 1998; 43: 1113-2 [Oldham et al., 1998]).

Alternatively, a "spoke" irradiation can be carried out where a phantom is irradiated with an odd number of coplanar, circular beams, arranged equidistantly around the midline and converging on a centrally located target point (Maryanski et al., 1996). The dose is calibrated at a depth d_{max} for each beam. The main disadvantages of this technique are that a considerable amount of gel must be used and only a small number of data points are obtained for a calibration plot.

5.5.2 Multi flask method

Here, a number of small flasks are filled with gel and each flask is irradiated to a known dose with a parallel-opposed dual beam arrangement to produce a uniform dose distribution throughout the entire gel sample (Back et al., 1996; Baldock et al., 1998; Ibbott et al., 1997). Depending on the number of gel samples used, similar errors in a calibration fit may be found as with the Multibeam method, but less gel will be required. A potential drawback to this calibration technique is the risks of inter flask variability. Care must also be taken to ensure full backscatter conditions are met so that the gels receive a uniform irradiation dose, i.e. flasks should be thin walled, be completely filled with gel leaving no air cavities and be surrounded by a tissue/water equivalent phantom that is thick enough to ensure charged particle equilibrium.

5.5.3 Depth-dose method

A long test tube of gel is positioned vertically in a water tank and irradiated from the closed end with a single radiation beam. This results in the gel recording a characteristic R_2 "depth-dose" distribution. These data may be plotted against a known depth-dose distribution for the beam size and energy, or against ion chamber measurements in an identical geometry. To adequately cover the dose range 0-10 Gy, a number of short test tubes of gel irradiated to different doses should be used in preference to a single, long test tube owing to the potential limitations of RF coil homogeneity. At each depth, several adjacent points may be averaged together to increase the signal-to-noise ratio.

Two 80 ml flasks irradiated to 8.5 Gy and 5 Gy were used in this example. The gel shows good sensitivity and linearity up to 7 Gy. Errors on the slope and intercept are 0.4% and 0.6%, respectively. Overlaid onto this figure are data points obtained using the "multi flask" method with gel from the same batch. The five points correspond to flasks irradiated with 10x10 cm^2 fields to doses of 2, 4, 6, 8 and 10 Gy. Errors on the slope and intercept for the five-point fit are 2.5% and 3.7%, respectively. Other calibration methods have also been reported. Rather than generating a T_2 map of the calibration phantom, two T_2 weighted images with widely differing echo times may be used to generate a look-up table of transverse relaxation rates for known absorbed doses (Maryanski et al., 1994; Ibbott et al., 1997).

6. Applications of gel dosimetry

Polyacrylamide gel dosimetry has been used to verify a number of increasingly complex treatment techniques using a variety of radiation modalities. Attention was first focused on validating the response of the gels to simple radiation fields to assess their response characteristics (Maryanski et al., 1993, 1994, 1996; De Deene et al., 1998). Following this, more complex treatment deliveries have been investigated, exploring more thoroughly the versatility of the 3D imaging sequences.

6.1 Brachytherapy

Clinical brachytherapy systems are capable of delivering very high doses with high dose gradients. It is important therefore to be able to verify accurately the doses calculated by brachytherapy treatment planning. Conventional dosimetry has limited resolution and may give rise to large errors through the use of TLDs or mini ionization chambers that measure the dose only at a single point. BANG gel dosimetry has been used to investigate the dosimetry of clinical ^{192}Ir (Maryanski et al., 1996; McJury et al., 1999) and ^{137}Cs (Maryanski et al., 1994; Audet et al., 1996) sources with considerable success. Good spatial and dosimetric agreement with treatment planning calculations has been demonstrated. Figure 4 illustrates a T_2 weighted image obtained using a clinical ^{192}Ir high dose rate source irradiating a spherical phantom filled with BANG-1 gel (McJury et al., 1999).

The catheter was positioned within the phantom so that the iridium source would irradiate from a single point at the centre of the phantom. The source was positioned using remote after loading equipment and the absolute dose was found to be within 7% of that at the dose prescription point. Dose maps have also been derived from gels exposed to multiple positions of a ^{192}Ir source (Maryanski et al., 1996). To resolve sharp dose gradients in close proximity to a brachytherapy source, high field strength (5 T+), and small bore MR scanners can be used (Hasson et al., 1998). With such equipment, resolution down to 0.1 mm x 0.1 mm x 0.1 mm is achievable.

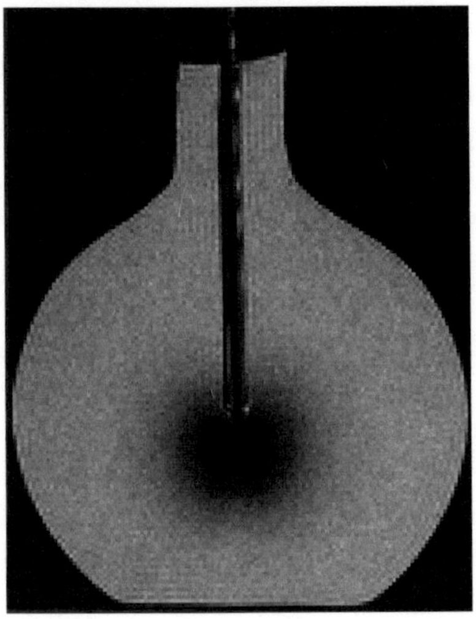

Fig. 4. T_2 weighted image of a brachytherapy phantom. Signal loss close to the source indicates a high degree of polymerization of the gel. (Phys Med Biol 1999; 44: 2431-44 [McJury et al., 1999])

6.2 Conformal and intensity modulated radiotherapy

Advances in conformal and intensity modulated radiotherapy techniques allow highly complex field shapes with high dose gradients to be planned and delivered. The capacity for imaging dose distributions fully in three dimensions is graphically illustrated by the photograph of a gel phantom in Figure 5.

This specially designed phantom was used to verify a stereotactically guided conformal treatment plan of a small brain lesion. The measurement is useful quantitatively for visualizing the irradiation geometry, while NMR scans of the phantom will provide dose maps to compare with the treatment plan. BANG-1 gels were used to verify the planning and delivery of intensity modulated radiotherapy using the NOMOS MIMiC tomotherapy system (NOMOS Corp., Sewickley, PA, USA) (Oldham et al., 1998). The MIMiC is a multiple

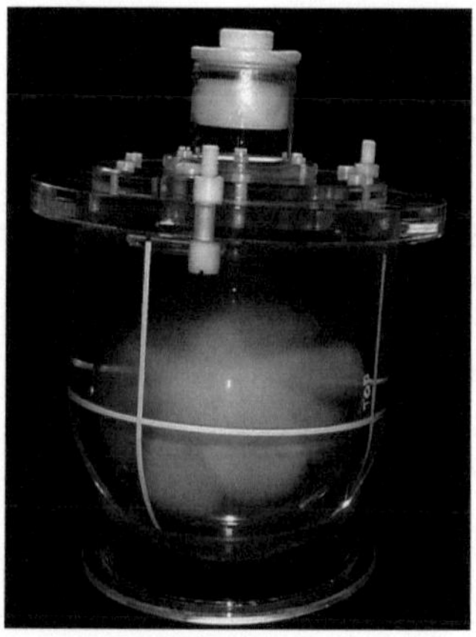

Fig. 5. Phantom used to verify a stereotactically guided conformal treatment plan. The phantom consists of an inner glass bulb containing the Acrylamide gel, which is surrounded by water in a Perspex container.)

vane, slit shaped collimator with two banks of 20 moveable leaves along the slit (Carol et al., 1992). As the MIMiC is rotated around a target, the leaves are programmed to open and close, generating intensity-modulated beam. Dose distributions measured using the BANG-1 gel was compared with those produced by the Peacock planning system and by an in-house optimized version of the planning system.

6.3 Stereotactic radiosurgery

Stereotactic radiosurgery dosimetry must be performed with high spatial resolution as the irradiated target volumes are both small and contain regions of steep dose gradient. Ibbott and co-workers have performed a series of experiments, from the placement of simple, single target points (Maryanski et al., 1996; Ibbott et al., 1996) to complicated dose distributions planned with multiple target points using a Gamma knife system (Ibbott et al., 1997). In the single target experiments there was good agreement with the calculated dose in the isocentric region, with differences increasing to 10% at the 20-30% isodose level. Although the measured isodose distribution agreed well with planning data (within 2-3 mm), doses measured by the gel were consistently lower than those predicted by the Gamma knife planning system.

7. Conclusion

Polymer gel dosimetry offers a method of acquiring 3D maps of complex radiotherapy dose distributions with a spatial resolution of the order of 1 mm, depending upon the scanning and imaging specifications. If care is taken during manufacture, particularly ensuring that there is no oxygen contamination, gels with good reproducibility can be obtained. As an alternative, it is possible to purchase ready-made gels (MGS, Guildford, New Haven, CT, USA). However, continued care is required between gel storage, irradiation and imaging to prevent any alteration or degradation of response. The dosimetry technique has been applied to a number of different radiation modalities and applications where measurements with conventional dosimeters are either difficult or of limited scope. One particular advantage of the gel is that it can be used in phantoms of any shape so that realistic irradiation geometry can be reproduced (De Deene et al., 1998). Polymer gels are still in a period of rapid development. There are many choices for each component of the gel, such as solvent, gelling agent, monomer and cross-linker. More sensitive, accurate and easier to manufacture polymer gel formulation is expected. The requirement for MRI makes the use of the gels prohibitively expensive for many radiotherapy departments. The possibility of using optical techniques to measure polymer density distributions could extend the use of gel dosimeters beyond research groups with easy access to MRI scanners. Further research has to be undergone to characterize the optical properties of the polymer gels and to produce a precise and effective optical scanner.

8. Acknowledgments

We are grateful to the management of VIT University for their continuous encouragement and the Atomic Energy Regulatory Board, Defence Research and Development Organization for their financial support. We are thankful to the United States-India Educational Foundation and Institute of International Education for their support through Fulbright program. We also acknowledge the valuable suggestions provided by Dr. B. Paul Ravindran, Dr. L. Keshava, Dr. Yoichi Watanabe, Dr. P. Sellakumar and Dr. N. Gopishankar.

9. References

Alexander A, Charlesby A, Ross M. The degradation of solid poly (methylmethacrylate) by ionising radiation. Proc R Soc Lond B Biol Sci 1954; A223: 392-404.

Andrews HL, Murphy RE, and LeBrun EJ. Gel dosimeter for depth dose measurements. Rev Sci Instrum 1957; 28:329-32.

Acrylamide polymerization - a practical approach, Bio-Rad Bulletin #1156. Richmond, CA: Bio-Rad, 1987.

Audet C, Schreiner LJ. Radiation dosimetry by NMR relaxation time measurement of irradiated polymer solutions. In: Proc Soc Magn Reson Med 10th Annual Scientific Meeting; 1991 August; San Francisco, CA. California: Society for Magnetic Resonance in Medicine: 705.

Audet C, Maryanski MJ, Gore JC. Dose response of the BANG polymer gel dosimeter: dependence on composition. Med Phys 1995; 22:951

Audet C, Duzenli C, Kwa W, Tsang V, Mackay A. An example of MRI polymer dosimetry applied to 3D conformal radiotherapy. Med Phys 1996; 23:803.

Audet C, Hilts M, Jirasek A and Duzenli C 2002 CT gel dosimetry technique: Comparison of a planned and measured 3D stereotactic dose volume J. Appl. Clin. Med. Phys. 3 110–8

Baldock C, Burford RP, Billingham NC, Cohen D, Keevin SF. Polymer gel composition in MRI dosimetry. Med Phys 1996; 23:1070.

Baldock C, Burford RP, Billingham NC, Wagner GS, Patval S, Badawi R, et al. Experimental procedure for the manufacture and calibration of Polyacrylamide gel (PAG) for magnetic resonance imaging (MRI) radiation dosimetry. Phys Med Biol 1998; 43:695-702.

Baldock C 2006 Historical overview of the development of gel dosimetry J. Phys.: Conf. Ser. 56 14–22

Back SA. Implementation of MRI gel dosimetry in radiation therapy. PhD thesis, Department of Radiation Physics, Malmo Lund University, Sweden, 1998.

Baustert IC, Oldham M, Smith TAD, Hayes C, Webb S, Leach MO. Optimized MR imaging for Polyacrylamide gel dosimetry. Phys Med Biol 2000; 45:847-58.

Braun K, Brown S, Hill B, Bailey D and Baldock C 2007 Use of polymer gels for radionuclide dosimetry Australas. Phys. Eng. Sci. Med. 30 63

Braun K, Bailey D, Hill B and Baldock C 2009 Preliminary investigation of PAGAT polymer gel radionuclide dosimetry of Tc-99m J. Phys.: Conf. Ser. 164 012050

Brindha S, Venning AJ, Hill B and Baldock C 2004 Experimental study of attenuation properties of normoxic polymer gel dosimeters Phys Med Biol 2004; 49:N353-361.

Carr H, Purcell E. Effects of diffusion on free precession in nuclear magnetic resonance experiments. Phys Rev 1956; 94:630-8.

Chambers KW, Collinson E, Dainton FS, Seddon WA. Pulse radiolysis: adducts of vinyl compounds and simple free radicals. Trans Faraday Soc 1967; 63:1699-1711.

Charlesby A. Radiation chemistry of polymers. In: Farhataziz, Rodgers MAJ, editors. Radiation chemistry. Weinheim, Germany: Wiley-VCH, 1987.

Carol M. An automatic 3D treatment planning and implementation system for optimized conformal therapy by the NOMOS Corporation. Int J Radiat Oncol Biol Phys 1992; 24(Suppl. 1): 158.

Courbon F, Love P, Chittenden S, Flux G, Ravel P and Cook G 2006 Preparation and use of I-131 MAGIC gel as a dosimeter for targeted radionuclide therapy Cancer Biother. Radiopharm. 21 427–36

Day MJ, Stein G. Chemical effects of ionising radiation in some gels. Nature 1950; 166:141-7.

Day MJ. Radiation dosimetry using nuclear magnetic resonance: an introductory review. Phys Med Biol 1990; 35:1605-9.

De Deene Y, De Wagter C, Van Duyse B, Derycke S, De Neve W, Achten E. Three-dimensional dosimetry using polymer gel and magnetic resonance imaging applied to the verification of conformal radiation therapy in head-and-neck cancer. Radiother Oncol 1998; 48:283-91.

De Deene, De Wagter, De Neve, Achten E. 3D polymer gel dosimetry in conformal radiotherapy: evaluation artifacts and future perspectives. Radiat, Oncol 1998; Suppl. 48:550.

De Deene Y, De Wagter C, De Neve W and Achten E 2000a Artefacts in multi-echo T2 imaging for high-precision gel dosimetry. I. Analysis and compensation of eddy currents Phys. Med. Biol. 45 1807–23

De Deene Y, De Wagter C, De Neve W and Achten E 2000b Artefacts in multi-echo T2 imaging for high-precision gel dosimetry. II. Analysis of B1 field inhomogeneity Phys. Med. Biol. 45 1825–39

De Deene Y and De Wagter C 2001 Artefacts in multi-echo T2 imaging for high-precision gel dosimetry. III. Effects of temperature drift during scanning Phys. Med. Biol. 46 2697–711

De Deene Y 2001 Fundamentals of MRI measurements for gel dosimetry Proc. 2nd Int. Conf. on Radiotherapy Gel Dosimetry (Brisbane, Australia) ed C Baldock and Y De Deene

De Deene Y, Hurley C, Venning A, Vergote K, Mather M, Healy B J and Baldock C 2002a A basic study of some normoxic polymer gel dosimeters Phys. Med. Biol. 47 3441–63

De Deene Y, Venning A, Hurley C, Healy B J and Baldock C 2002b Dose–response stability and integrity of the dose distribution of various polymer gel dosimeters Phys. Med. Biol. 47 2459–70

DOSGEL 1999 Proc. 1st Int. Workshop on Radiation Therapy Gel Dosimetry (Canadian Organization of Medical Physicists, Lexington, KY) ed L J Schreiner and C Audet

DOSGEL 2001 Proc. 2nd Int. Conf. on Radiation Therapy Gel Dosimetry (Queensland University of Technology, Brisbane, Australia) ed C Baldock and Y De Deene

DOSGEL 2004 Proc. 3rd Int. Conf. on Radiation Therapy Gel Dosimetry (Ghent University, Ghent, Belgium) ed Y De Deene and C Baldock

DOSGEL 2006 Proc. 4th Int. Conf. on Radiation Therapy Gel Dosimetry (Universit´e de Sherbrooke, Sherbrooke, Canada) ed M Lepage, A Jirasek and L J Schreiner

DOSGEL 2008 Proc. 5th Int. Conf. on Radiation Therapy Gel Dosimetry (University of Crete, Greece) ed T G Maris and E Pappas

Farajollahi AR, Bonnett DE, Aukett RJ, Radcliffe AJ. The advantages and limitations ofpolymer gel dosimetry in brachytherapy. J Int Fed Med Biol Eng 1997;35:1012.

Fong P M, Keil D C, Does M D and Gore J C 2001 Polymer gels for magnetic resonance imaging of radiation dose distributions at normal room atmosphere Phys. Med. Biol. 46 3105–13

Gear J I, Flux G D, Charles-Edwards E, Partridge M, Cook G and Ott R J 2006 Theapplication of polymer gel dosimeters to dosimetry for targeted radionuclide therapy Phys. Med. Biol. 51 3503–16

Gordon AH. Electrophoresis of proteins in Polyacrylamide and starch gels. In: Work TS, Work E, editors. Laboratory techniques in biochemistry and molecular biology. New York: American Elsevier, 1973.

Gore JC, Ranade M, Maryanski MJ, Schulz RJ. Radiation dose distributions in 3D from tomographic optical density scanning of polymer gels: I. Development of an optical scanner. Phys Med Biol 1996; 41:2695-704.

Gochberg DF, Kennan RP, Maryanski MJ, Gore JC. The role of specific side groups and pH in magnetization transfer in polymers. J Magn Reson 1998; 131:191-8.

Gustavsson H, Karlsson A, B¨ack S °A J, Olsson L E, Haraldsson P, Engstr¨om P and Nystr¨om H 2003 MAGICtype polymer gel for three-dimensional dosimetry: intensity-modulated radiation therapy verification Med. Phys. 30 1264–71

Hahn E. Spin echoes. Phys Rev 1950; 80:580-94.

Hilts M, Audet C, Duzenli C and Jirasek A 1999 Polymer gel dosimetry using x-ray computer tomography: feasibility and potential application to stereotactic radiosurgery Proc. 1st Int. Workshop on Radiation Therapy Gel Dosimetry (Lexington, USA) ed L J Schreiner and C Audet

Hilts M, Audet C, Duzenli C and Jirasek A 2000 Polymer gel dosimetry using x-ray computer tomography: A feasibility study Phys. Med. Biol. 45 2559–71

Hill B, Venning AJ and Baldock C 2003 Computer tomography evaluation of normoxic gel dosimeters Proc. World Congress on Medical Physics and Biomedical Engineering (Sydney, Australia)

Hill B, Venning AJ and Baldock C 2005 Dose response of normoxic polymer gel dosimeters measured using x-ray computer tomography Brit. J. Radiol. 78:623-30.

Hilts M, Jirasek A and Duzenli C 2004 Effects of gel composition on the radiation induced density change in PAG polymer gel dosimeters: a model and experimental investigations. Phys. Med. Biol. 49 2477–90 139

Hilts M and Duzenli C 2004 Image filtering for improved dose resolution in CT polymer gel dosimetry Med. Phys. 31 39–49

Hoecker FE, Watkins IW. Radiation polymerization dosimetry. Int J Appl Rad Iso 1958; 3:31-5.

Hsu TP, Cohen C. Observations on the structure of a Polyacrylamide gel from electron micrographs. Polymer 1984; 25:1419-23.

Hasson BF. Chemical dosimetry in the near-zone of brachytherapy sources. Med Phys 1998; 25:2076.

Hepworth SJ, Leach MO, Doran SJ. Kinetics of polymerization in Polyacrylamide gel (PAG) dosimeters: (ii) modelling oxygen diffusion. Phys Med Biol 1999;44:1875-84.

Ibbott GS, Bova FJ, Maryanski MJ, Zhang Y, Holcomb S, Avison RG, et al. Use of BANG polymer gel dosimeter to evaluate repeat-fixation stereotactic radiation therapy. Med Phys 1996;23: 1070.

Ibbott GS, Maryanski MJ, Eastman P, Holcomb SD, Zhang Y, Avison R, et al. Three-dimensional visualization and measurement of conformal dose distributions using magnetic resonance imaging of BANG gel dosimeters. Int J Radiat Oncol Biol Phys 1997; 38:1097-103.

Kennan RP, Richardson KA, Zhong J, Maryanski MJ, Gore JC. The effects of cross-link density and chemical exchange on magnetization transfer in Polyacrylamide gels. J Magn Reson 1996; 110:267-77.

Meiboom S, Gill D. Modified spin-echo method for measuring nuclear relaxation times. Rev Sci Instr 1958; 29:688.

Maryanski MJ, Gore JC, Kennan RP, Schulz RJ. NMR relaxation enhancement in gels polymerized and cross-linked by ionizing radiation: a new approach to 3D dosimeters by MRI. Magn Reson Imaging 1993;11:253-8.

Maryanski MJ, Gore JC, Schulz RJ. Three dimensional detection, dosimetry and imaging of an energy field by formation of a polymer in a gel. US Patent #5 321 357, 1994.

Maryanski MJ, Schulz RJ, Ibbott GS, Gatenby JC, Xie J, Horton D, et al. Magnetic resonance imaging of radiation dose distributions using a polymer-gel dosimeter. Phys Med Biol 1994; 39:1437-55.

Maryanski MJ, Audet C, Gore JC. Dose response of a BANG polymer gel: temperature dependence. Med Phys 1995; 22:951.

Maryanski MJ, Audet C, Gore JC. Dose response of the BANG polymer gel: effects of cross-linking density. Med Phys 1995; 22:950.

Maryanski MJ, Ibbott GS, Schulz RJ, Gore JC. Radiation therapy dosimetry using magnetic resonance imaging of polymer gels. Med Phys 1996; 23: 699-705.

Maryanski MJ, Zastavker YZ, Gore JC. Radiation dose distributions in 3D from tomographic optical density scanning of polymer gels: II. Optical properties of the BANG polymer gel. Phys Med Biol 1996; 41:2705-17.

Maryanski MJ, Audet C, Gore JC. Effects of cross-linking and temperature on the dose response of a BANG polymer gel dosimeter. Phys Med Biol 1997; 42:303-11.

Mather M, Whittaker A K and Baldock C 2001 Ultrasound – a new method of evaluation of polymer gel dosimeters Proc. 2nd Int. Conf. on Radiotherapy Gel Dosimetry (Brisbane, Australia) ed C Baldock and Y De Deene

Mather M, Whittaker A K and Baldock C 2002a Ultrasound evaluation of polymer gel dosimeters Phys. Med. Biol. 47 1449–58

Mather M, De Deene Y, Whittaker A K, Simon G, Rutgers R and Baldock C 2002bInvestigation of ultrasonic properties of PAG and magic polymer gel dosimeters Phys. Med. Biol. 47 4397–409

Mather M L, Collings A F, Bajenov N, Whittaker A K and Baldock C 2003a Ultrasonic absorption in polymer gel dosimeters Ultrasonics 41 551–9

Mather M L, Charles P and Baldock C 2003b Measurement of ultrasonic attenuation coefficient in polymer gel dosimeters Phys. Med. Biol. 48 N269–75

Mather M L and Baldock C 2003c Ultrasound tomography imaging of radiation dose distributions in polymer gel dosimeters Med. Phys. 30 2140–8

McJury M, Oldham M, Leach MO, Webb S. Dynamics of polymerization in Polyacrylamide gel (PAG) dosimeters: (i) aging and long-term stability. Phys Med Biol 1999; 44:1863-73.

McJury M, Tapper PD, Cosgrove VP, Murphy PS, Griffin S, Leach M, et al. Experimental 3D dosimetry around a high dose-rate clinical 192Ir source using a Polyacrylamide gel (PAG) dosimeter. Phys Med Biol 1999; 44:2431-44.

McJury M, Oldham M, Cosgrove V P, Murphy P S, Doran S, Leach M O and Webb S 2000 Radiation dosimetry using polymer gels: methods and applications Br. J. Radiol. 73 919–29

Oldham M, Baustert I, Lord C, Smith TAD, McJury M, Leach MO, et al. An investigation into the dosimetry of a nine-field tomotherapy irradiation using BANG-gel dosimetry. Phys Med Biol 1998; 43:1113-2.

Oldham M, McJury M, Baustert I, Webb S, Leach MO. Improving calibration accuracy in geldosimetry. Phy Med Biol 1998; 43:2709-20.

Oldham M, Siewerdsen J H, Shetty A and Jaffray D A 2001 High resolution gel-dosimetry by optical-CT and MR scanning Med. Phys. 28 1436–45

Oldham M, Siewerdsen J H, Kumar S,Wong J and Jaffray D A 2003 Optical-CT gel dosimetry I: basic investigations Med. Phys. 30 623–34

Oldham M and Kim L 2004 Optical-CT-gel dosimetry: II. Optical artifacts and geometric distortion Med. Phys. 31 1093–104

Ramm U, Weber U, Bock M, Kramer M., Bankamp A, Damrau M, Thilmann C, BottcherH D, Schad L R and G Kraft 2000 Three-dimensional BANGTM gel dosimetry in conformal carbon ion radiotherapy, Phys. Med. Biol. 45: N95-N102.

Rintoul L, Lepage M and Baldock C 2003 Radiation dose distribution in polymer gels by Raman spectroscopy Appl. Spectrosc. 57 51–7

Senden R J, De Jean P, McAuley K B and Schreiner L J 2006 Polymer gel dosimeters with reduced toxicity: a preliminary investigation of the NMR and optical dose-response using different monomers Phys. Med. Biol. 51 3301–14

Sittig M. Handbook of toxic and hazardous chemicals and carcinogens (2nd edn). New Jersey, USA: Noyes Publications, 1985:41.

Schreiner LJ, Henri C, Evans MDC, Podgorsak EB. 3-D MRI radiation dosimetry using the magnetization to dose (M&D) calibration technique. Med Phys 1995; 22:951.

Schreiner JL, editor. Proceedings of the 1st International Workshop on Radiation Therapy Gel Dosimetry; 1999 July 21±23; Lexington KY. Edmonton, AB, Canada: Canadian Organization of Medical Physicists.

Trapp J, Bäck S Å J, Lepage M, Michael G and Baldock C 2001 An experimental study of thedose response of polymer gel dosimeters imaged with x-ray computed tomography Phys. Med Biol. 46 2939–51

Trapp JV, Michael G, De Deene Y and Baldock C 2002 Measurements of density and linear attenuation coefficient changes in polymer gel dosimeters Phys. Med. Biol. 47 4247–58

Vachier MC, Rutledge DN. Influence of temperature, pH, water content, gel strength and their interactions on NMR relaxation of gelatins IĐ analysis of the calculated relaxation times. J Magn Reson Analysis 1996;2:311-20.

Zebrowska G, Husson F, Lewa CJ, de Certaintes JD. MRI of radiation dose distribution in a new tissue equivalent gel dosimeter (PIRA). In: Proc Soc Magn Reson 3rd Scientific Meeting; 1995 August; Nice, France. California, USA: Society for Magnetic Resonance in Medicine: 1102.

18

Digital Filtering Techniques to Reduce Image Noise and Improve Dose Resolution in X-Ray CT Based Normoxic Gel Dosimetry

N. Gopishankar[1], S. Vivekanandhan[1], A. Jirasek[2], S. S. Kale[1], G. K. Rath[1]
Sanjay Thulkar[1], V. Subramani[1], S. Senthil Kumaran[1] and R. K. Bisht[1]
[1]*All India Institute of Medical Sciences, New Delhi,*
[2]*Department of Physics and Astronomy, University of Victoria, B.C.*
[1]*India*
[2]*Canada*

1. Introduction

Gel dosimetry is a promising technique for the implementation of 3D dose verification within the radiation therapy clinic, since it is the only methodology which provides comprehensive 3D dose measurement of conformal treatments such as Intensity Modulated Radiation Therapy (IMRT), Stereotactic radiosurgery (SRS) and Stereotactic Radiotherapy (SRT), and volumetric arc therapies (VMAT). In polymer gel dosimetry, monomer molecules polymerize and are fixed in the gel matrix when exposed to radiation. The spatial dose information thus can be obtained by gel imaging. Gel imaging methods which are currently under investigation are MRI, optical computed tomography and x-ray computed tomography. While promising, x-ray CT images of irradiated polymer gel exhibit low contrast due to small change in density that occurs during polymerization. The response of CT contrast to dose is weak and noise reduction is critical in order to achieve adequate dose resolution in gel dosimetry using x-ray computed tomography. Ideally, image noise is minimized using high tube current, long scan times and high number of image averages [Hilts et al 2005]. Digital image filtering is an effective method in reducing image noise while maintaining accurate spatial dose information. Generally it is performed in either frequency domain or spatial domain. In a previous work, several spatial filters were applied to Stereotactic Radiosurgery (SRS) irradiated polymer gel image [Hilts and Duzenli 2004]. In their study it was found that the two highest performing filters were the adaptive mean (Wiener) and smallest univalue segment assimilating nucleus (SUSAN). In another study a new method of signal to noise ratio (SNR) enhancement by 2D two-point maximum entropy regularization method (TPMEM) was investigated [Jirasek 2006]. However, a comprehensive examination of the best performing filters from these two studies has not been undertaken. Here, we test the Weiner and TPMEM digital filters on X-ray CT image of the irradiated PAGAT gel in order to reduce noise and improve dose resolution. This work builds off of previous studies and evaluates and compares the highest-performing filters from past individual works [Hilts and Duzenli 2004, Jirasek et al 2006].

1.1 Concept of dose resolution

Dose resolution or minimal detectable difference (MDD) in dose is one of the most important features of gel dosimeter [Baldock et.al. PMB 2001]. In general, $D^{P}\Delta$ is related to σ_D which is the standard deviation or dose uncertainty in the measured dose.

$$D^{P}\Delta = K_P\sqrt{2}\,\sigma_D \quad (1)$$

Where Kp is the coverage factor for coincidence interval p. σ_D is the standard deviation of the sample. Dose resolution ($D^{95\%}\Delta$) is defined as the minimal separation between two absorbed doses such that they may be distinguished with a given level of confidence, p. For a 95% confidence level the dose resolution becomes $D^{95\%}\Delta = 2.77.\sigma_D$.

In CT gel dosimetry, σ_D is due to uncertainty in measured N_{CT} (σN_{CT}) and in the dose response curve for a linear dose response, σ_D is related to σN_{CT} by the slope of the response.

$$\sigma_D = [\partial D/\partial N_{CT}]\,\sigma N_{CT} \quad (2)$$

Combining Equation (1) an (2), $D^{P}\Delta$ is improved by a decrease in the slope of the dose to N_{CT} response and as investigated here a decrease in σN_{CT}.

1.1.1 Theory

1.1.1.1 CT image noise

Noise in CT images is primarily due to quantum noise inherent in photon detection and electronic noise. Quantum noise results from counting a finite number of random events (photons hitting the detector) and is, theoretically, Poisson distributed. A characteristic of Poisson distributions is that the variance equals the mean. That is for Poisson distributed noise the variance depends on signal strength (number of photons counted). In theory, noise in CT projection data should display this characteristic. However the noise in a final CT image is Gaussian distributed due to the presence of electronic noise. As a result, the noise characteristics of a CT imaged depend on the scan parameters and reconstruction algorithm used [Hilts 2004].

1.1.1.2 Digital image filtering

The gel images used to evaluate filters in this project has been preprocessed for artifact removal using image averaging and background subtraction [Hilts and Jirasek 2008] and its image degradation is predominantly caused by noise. Mask-based filters are all based on a similar principle: a mask (m x n pixels) is centered on each pixel of the image, g(s,t), and a function is applied to the image pixels in the region of the mask (Sxy) so that the center pixel is replaced with a new value, f(x,y). The TPMEM filter is not based on a masking approach but, rather, the minimization of an entropy and fidelity function, as described below.

For this study slope of linear region (0 to 12 Gy) from calibration gel was used to calculate σ_D. The filters are applied to a gel irradiated CT image of a simulated clinical treatment. For all cases, the effects of filter kernel size, number of filter iterations, and TPMEM stopping criterion are investigated.

1.2 Filters used for noise reduction in this study

1.2.1 Adaptive filter

Adaptive filters are a class of filters for which filtering power is adapted based on local image statistics. The adaptive mean or Wiener filter is a spatial filter based on moving kernel (similar to a traditional mean filter). The difference lies in that the adaptive filter weights its filtering power based on the ratio of the local variance (σ^2_L) to the overall image noise (σ^2_N). The filtered value is given by

$$f(x,y) = g(x,y) - ((\sigma^2_L) / (\sigma^2_N))*[g(x,y) - m_L] \tag{3}$$

where g(x,y) is the unfiltered value and m_L is the local mean value of the image. All image filtering was performed in Matlab (The Mathworks, Natick, MA, USA), which provides a built-in function for this adaptive mean filter.

1.2.2 TPMEM

The aim of TPMEM is to minimize the function

$$T = -S + \lambda \chi^2 \tag{4}$$

where S is the entropy function, χ^2 is the chi square statistic which characterizes the degree of agreement between the initial (unsmoothed) and recovered (smoothed) full dataset and λ is a Lagrange multiplier that is to be determined. The balance between recovered data smoothness and fidelity is optimized by Lagrange multiplier λ.

Recently, the traditional TPMEM algorithm has been extended to (i) incorporate an additional parameter (X, see equation 5) to allow for a user defined tuning of the ratio between smoothness and image fidelity, and (ii) allow for the minimization to occur in 2D, as opposed to 1D spectra, as was originally reported. [Jirasek 2006 and Foist 2010].

$$T = -S + X \lambda \chi^2 \tag{5}$$

The underlying theory of TPMEM is presented in detail elsewhere (Greek et al 1995, Jirasek et al 2006).

2. Materials and methods

2.1 Gel preparation

Normoxic gel was prepared under normal atmospheric conditions. The components used for the preparation of the dosimeter were 6% gelatin (300 Bloom from Porcine skin), 3% Acrylamide, 3% N, N′ methylene-bis-acrylamide (BIS), 88% distilled water and 10mM tetrakis(hydroxy methyl) phosphonium chloride (THP). (All from Sigma Aldrich, India). For preparation of PAGAT gel, 6% of gelatin by total weight of the dosimeter was mixed with 88% distilled water and allowed to soak for 45 minutes before heating to 50°C in specially designed water bath with temperature control (thermostat), also crossed checked with a thermometer. A waterbath was designed in-house specifically to prepare 10 to 15 Liters of gel. The container with gelatin was heated to 50°C and simultaneously stirred with overhead stirrer. BIS (3%) was added and stirred thoroughly in the gel solution till it dissolved completely for 30 minutes. 3% Acrylamide was added subsequently and the whole content was stirred for an hour. Once a clear solution was obtained 10mM of THPC was added and stirred and the container was removed from water bath. A final volume of 4.5 Litres of gel was prepared. The gel was immediately poured into the three PET containers (each with 1mm thickness and 1500mL capacity) till screw top and tightly sealed with clingfilm and container lid. One container was used for calibration, one for clinical beam exposure and one for background subtraction purpose. The gel containers were immediately covered with black plastic cover to avoid exposure to daylight that might cause photopolymerization and then were refrigerated at $5^{o}C$.

2.2 Gel calibration and irradiation

Irradiations were done approximately 23h post manufacture on a Varian Clinac2300CD linear accelerator (Varian Assoc., Palo Alto, CA). The PET containers (10cm diameter, 19cm height) for exposure were left in the linac room for 1 hr prior to irradiation. The first PET container (1) was irradiated to doses from 0 Gy to 14 Gy at D_{max} by large flask geometry method (Taylor et al. 2007). The second PET container (2) was irradiated with four intersecting 4 x $4cm^2$ 6MV field doses of 2, 5, 7, and 11Gy at the depth of maximum dose. The third PET container (3) was used for background subtraction. The background subtraction technique in general is successful in removing beam hardening artifacts and ring artifacts due to miscaliberated detector in the scanner. Reference marks were made on the containers for scanning reproducibility. After irradiation the gel containers were exposed to atmosphere one day after irradiation in order to make gels inactive and thus prevent further polymerization. Acquisition parameters for dose response experiments were as follows: tube voltage 140 kV, tube current 200 mA, scan time 1sec, FOV 130 x 130, image thickness 2.5 mm, Reconstruction algorithm B30 medium.

The relationship of N_{CT} to dose was obtained from the image of the calibration gel as shown in Figure 1. The average N_{CT} was measured for each region of uniform dose using MATLAB and a dose response curve was made. This curve was then used to convert the clinical treatment gel image into a dose map for quantitative image analysis. N_{CT} is shown as the value above background increases with dose. The response is well modeled by a linear function up to a dose of 12Gy, ($R^2 = 0.996$) with a slope of 222 cGy/H. Measured noise σN_{CT} (H) from region of interest in the gel exposed with clinical beams and calculated values of σ_D and dose resolution ($D^{95\%}\Delta$) were analyzed for dose regions in gel irradiated for simulated clinical treatment. The response had dose region of slope 222 cGy/H ($R^2 = 0.996$) and $D^{95\%}\Delta$ of 512.06 cGy.

2.3 Image filtering

The processed gel images were filtered using a Wiener filter described in Sec. 1.2.1 for 3x3, 5x5, 7x7, 11x11 pixel mask. All filtering was performed using MATLAB (Mathworks, USA). We further did filtering analysis with the TPMEM filter and compared those results with Wiener filter results. For this study we took several quantitative measures such as (i) profiles through the image (ii) FFT analysis of unfiltered image, Wiener filtered image and TPMEM filtered image (iii) SNR of filtered images (iv) root mean squared error (RMSE) (v) Pearsons correlation coefficient (PEARS). In order to compare RMSE with PEARS all the values quoted in this study are 1 – PEARS and are labeled as PEARSC.

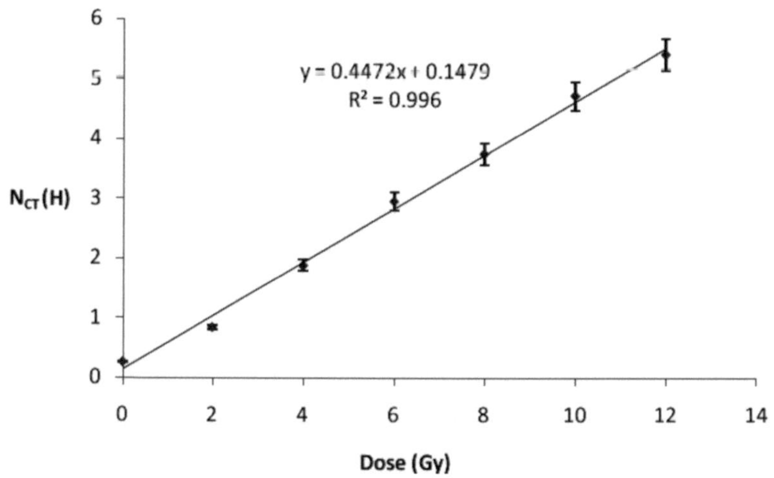

Fig. 1. Dose to N_{CT} response curve obtained from the calibration gel.

3. Results and discussions

3.1 Wiener filter

Figure 2 shows the radiation exposed gel image. The results of filtering the gel image with the filters described in Sec. 1.2.1 are shown in Table 1. In our study excellent dose resolution was obtained by applying adaptive mean filter with 11 x 11 pixel mask at second iteration which showed a $D^{95\%}\Delta$ of 24.65 cGy.

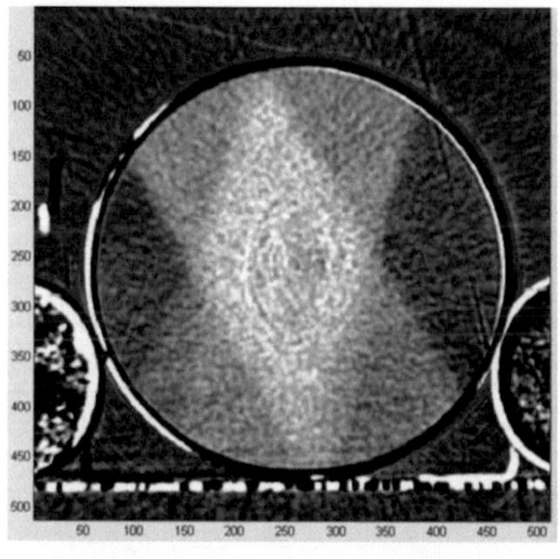

Fig. 2. X-ray CT image of gel in PET container irradiated with four intersecting 4 x 4cm^2 6MV field doses of 2, 5, 7, and 11Gy at the depth of maximum dose. This image was obtained after background subtraction and image averaging.

Measured σN_{CT} and calculated values of σ_D and $D^{95\%}\Delta$ are provided in Table 1 for Wiener filtered image and the unfiltered image. Results show that the improvement increases with the mask size as expected since a large mask provides a greater smoothing area.

Figure 3a represents the unfiltered gel image. Figure 3b and 3c represent the gel image after applying an 11 x 11 Wiener filter at first and second iteration respectively.

Filter	Mask	σN_{CT}(H)	σ_D(cGy)	$D^{95\%}\Delta$ (cGy)	Profile Width
Unfiltered		0.8327	184.86	512.06	35.6174
Wiener filter	3 x 3	0.6270	139.19	385.56	34.3367
	5 x 5	0.5069	112.53	311.71	34.447
	7 x 7	0.3936	87.38	242.04	34.40312
	11 x 11(1st iteration)	0.1457	32.35	89.61	34.7228
	11 x 11(2nd iteration)	0.0401	8.90	24.65	33.881

Table 1. Summary of measured noise (σN_{CT}) and calculated dose uncertainty (σ_D) and dose resolution with 95% confidence ($D^{95\%}\Delta$) for the unfiltered and filtered images.

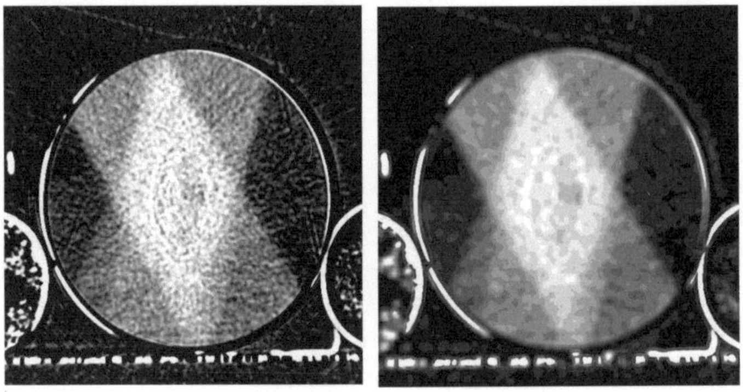

(a) (b)

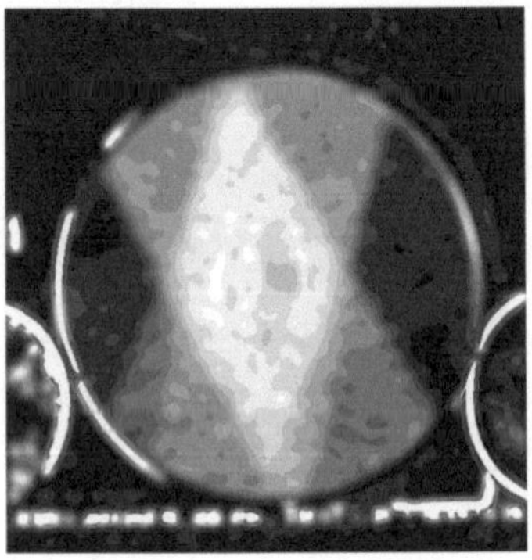

(c)

Fig. 3. (a) Unfiltered Image, (b) Wiener filtered Image (11 x 11 mask, 1st iteration), (c) Wiener filtered Image (11x11mask, 2nd iteration).

Figure 4a and 4b are horizontal and vertical profiles of the unfiltered and Wiener filtered images. The unfiltered gel image is characterized by several spikes throughout the profile. By applying Wiener filter the spikes are significantly reduced and smooth profile is obtained.

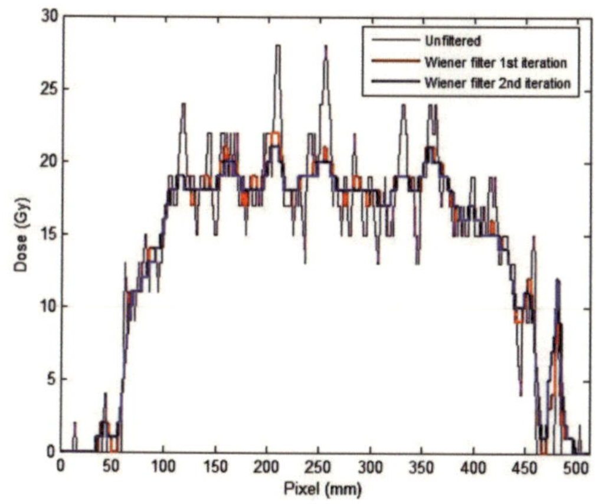

(a)

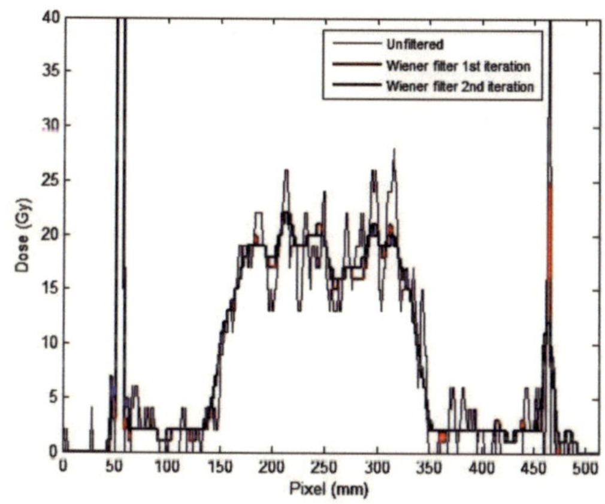

(b)

Fig. 4. (a) Horizontal Dose profiles for unfiltered image and Wiener filtered image 1st and 2nd iteration using 11 x 11 mask. (b) Vertical Dose profiles for unfiltered image and Wiener filtered image 1st and 2nd iteration using 11 x 11 mask.

In general, a tradeoff exists when increasing mask size. A larger mask generally improves image SNR, and hence dose resolution, but simultaneously degrades spatial dose information. In order to illustrate a typical spatial resolution degradation observed when applying a Wiener filter with 11x11 mask, we measured the width of the profiles in figure 4a (i.e. no iterations, 1, and 2 iterations of the Wiener filter). Table 1 illustrates that spatial degradation increases when applying the Wiener filters. Hence, a balance must be struck between image filtering and potential spatial degradation.

In Table 2 Wiener filter applied with different mask is shown. In Table 1 it was shown that as the mask size is increased the dose resolution is improved significantly. However the results of Table 2 showed different results. As the mask size increased the SNR value decreased. The SNR analysis of the unfiltered image showed SNR =1.272, and after Wiener filtering with 11x11 mask, SNR = 39.6049. RMSE and PEARSC values increased as the mask size was increased. By theory higher RMSE and PEARSC indicate poor filtering. The results prove that high filtering technique provides smoothness as shown in figure 3.c but at the same time causes image distortion.

Wiener Mask	RMSE	PEARSC	SNR
3 x 3	0.078	0.0019	2.3274
5 x 5	0.0927	0.0039	4.1234
7 x 7	0.1102	0.0057	7.1857
9 x 9	0.1285	0.0067	10.9466
11 x 11	0.1437	0.0088	19.7918
15 x 15	0.1634	0.018	39.6049
Unfiltered			1.272

Table 2. RMSE, PEARSC and SNR between unfiltered and Wiener filtering. Initial noisy data at SNR = 1.272.

3.2 TPMEM filter

In figure 5 TPMEM filter analysis is shown for the image matrix of size 96 x 96 pixels^2, i.e. the white-space around the gel container was cropped in order to decrease computation time. The actual matrix size of the unfiltered CT image used for noise reduction study was 512 x 512 pixels^2. For the a typical calculation time for a matrix of size 300 x 300 pixels^2 is approximately 20 minutes whereas for matrix of size 90 x 90 pixels^2 is around 40 seconds, hence we rescaled the CT image to a matrix of size 96 x 96 pixels^2. Figure 5 a, b, c and d are the unfiltered and TPMEM filtered CT image of the exposed gel for stopping criterion X = 0.5, 1.4 and 2.5 respectively. Increase in image smoothness can be seen as the stopping criterion is increased.

TPMEM Stopping Criterion	RMSE	PEARSC	SNR
X = 0.2	0.0371	0.001	1.4357
X = 0.5	0.0588	0.004	1.6685
X = 1.4	0.101	0.0145	2.2200
X = 1.6	0.1066	0.0173	2.3511
X = 1.8	0.1171	0.0217	2.4079
X = 2.0	0.1204	0.0239	2.6034
X = 2.5	0.1314	0.0353	3.5329
X – 4	0.1363	0.0394	4.0898
Unfiltered			1.2720

Table 3. RMSE, PEARSC and SNR between unfiltered and TPMEM filtering. Initial noisy data at SNR =1.272.

Other than SNR values, RMSE and PEARSC values for different stopping criterion were analyzed and the results are shown in Table 2. In the previous study it was shown that a truly filtered image will contain high SNR, high PEARS (low PEARSC) and low RMSE [Jirasek et al 2006]. Hence, a tradeoff is again noted between image smoothness and fidelity.

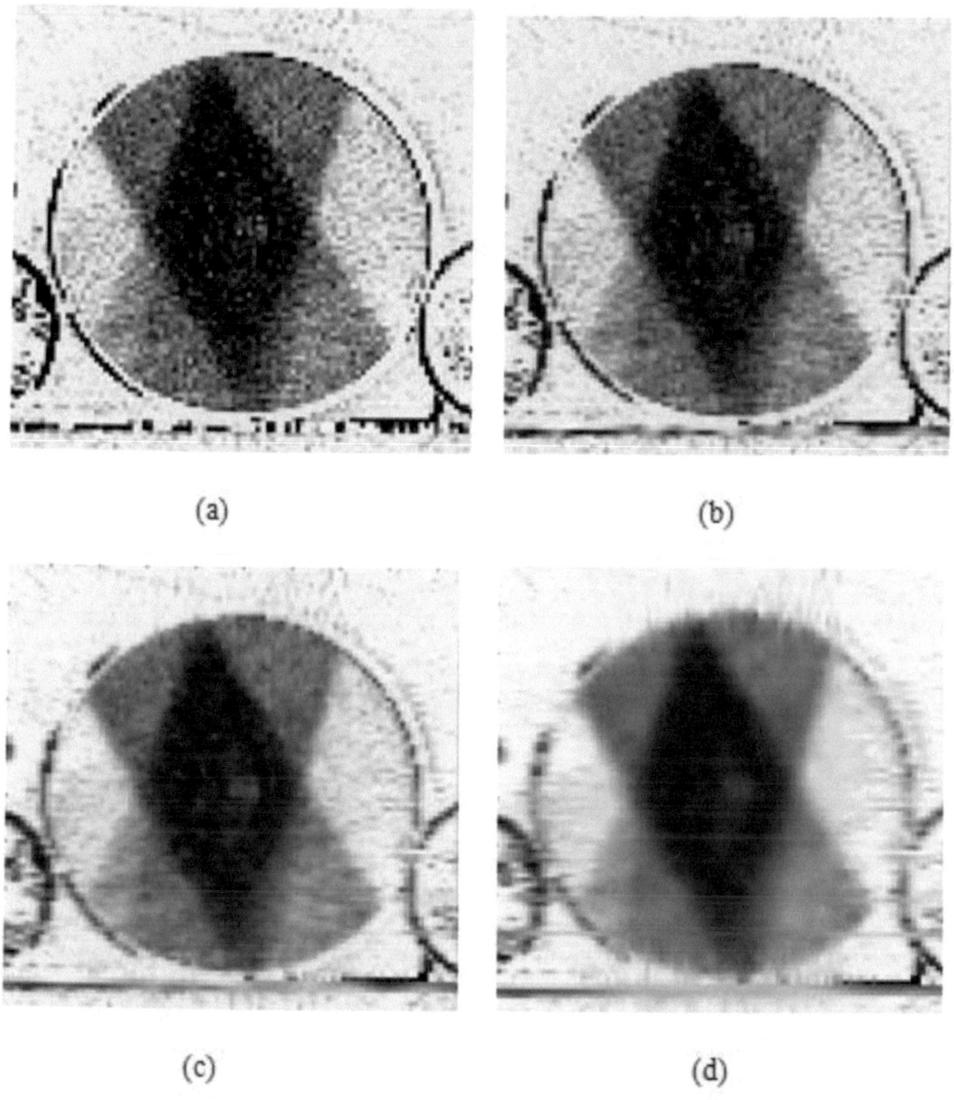

Fig. 5. TPMEM analysis for PAGAT gel image. (a) Unfiltered image. (b) TPMEM filtered image for stopping criterion X = 0.5. (c) TPMEM filtered image for stopping criterion X = 1.4. (d). TPMEM filtered image for stopping criterion X = 2.5.

Table 3 shows the actual SNR for the unfiltered image was SNR = 1.272 and after filtering with TPMEM with stopping criterion X = 0.5, 1.4, 2.5, SNR was observed to be 1.6685, 2.2200, 3.5329 respectively. However as the stopping criterion is increased further spatial degradation is observed as shown in figure 6.

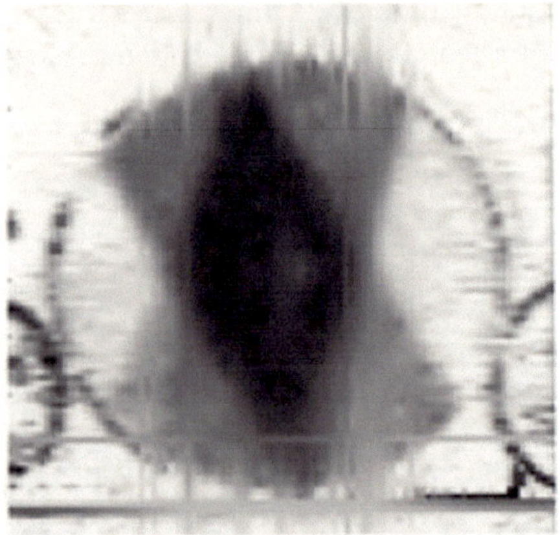

Fig. 6. Spatial degradation of image after filtering with TPMEM stopping criterion X = 4.

3.3 Filter comparison

Figure 7 (a) illustrates profiles taken through unfiltered and TPMEM filtered and Wiener filtered image. The profile comparison was done in the gel image which was rescaled to 96 x 96 pixels^2. The profiles show that Wiener filtered image profile is over smoothened were as the TPMEM filtered image still follows the peak and valley of the unfiltered image.

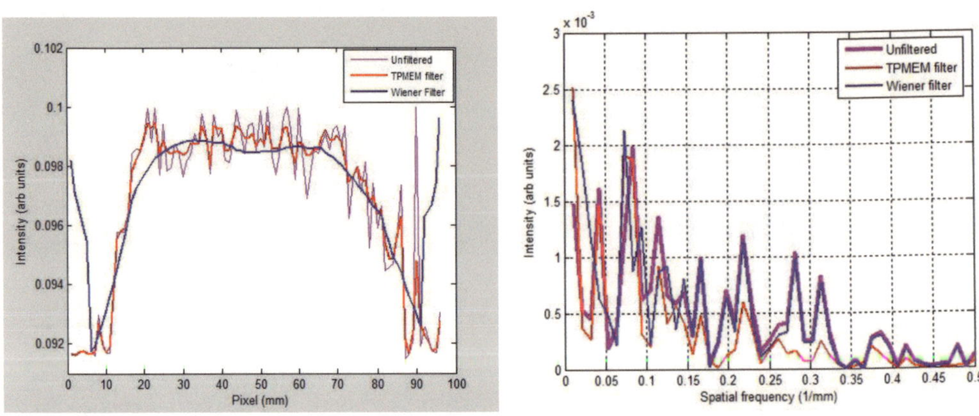

Fig. 7. (a) Profiles extracted through unfiltered, TPMEM filtered and Wiener filtered (11x 11 mask) image. (b) FFT data of the unfiltered and TPMEM filtered and Wiener filtered (11x 11 mask) image are shown in figures

In a previous study the TPMEM was applied to a SRS irradiated gel image [Jirasek et al 2006] which contains high dose gradients whereas in this work filtering was performed on four intersecting 4 x 4 cm^2 6MV fields with lower dose gradients as compared with the SRS study. Fast Fourier transform (FFT) data of the images are shown in figure 7 (b). TPMEM filtering is efficient in removing high frequency noise (> 0.15 mm^{-1}). It was also observed that TPMEM also retains much of the original signal in lower frequency regions (< 0.1 mm^{-1}). FFT results of Wiener filtered did not show any significant alterations in the FFT spectrum of the gel image.

4. Conclusion

In our study we found that dose resolution in CT images can be improved by applying spatial filters for removing noise. Adaptive Mean (Wiener) provided excellent dose resolution. Increasing kernel size and/or the number of filter iterations has a positive effect on noise reduction in filtered images. At the same time increasing kernel size and number of filter iterations become counterproductive, as image distortion (blurring) becomes unacceptably high. A recently developed filtering technique named TPMEM was also tested on the gel image. The TPMEM filter provided good SNR enhancement. Results indicate that CT imaging appears an attractive possibility for polymer gel dosimetry with such techniques available for improving SNR. In general the low filtering technique does not adequately smooth the images. In contrast, the high filtering technique provides smoothness but at the expense of some distortion of the shape. In particular, sharp contour changes are reduced by high filtering. Hence selection of appropriate filter and related parameters which control the fidelity and smoothness of filtered images is required before implementing it for X-ray CT polymer gel dosimetry. At present the TPMEM filters are useful for analysis of images with pixel size less than 300 x 300 pixels whereas the Adaptive mean filter can be applied to images of any size. Both the filters are capable of reducing image noise and improving dose resolution. In future application of filters to 3D image data will be of more importance.

5. Acknowledgements

We wish to gratefully acknowledge the support of AERB, India for this work through research project no N-961.

6. References

Baldock, C., Lepage. M., Back, S A J., Murry, P J., Jayasekara, P M., Porter, D. and Kron, T. (2001) Dose resolution in radiotherapy polymer gel dosimetry: effect of echo spacing MRI pulse sequence. Phys. Med. Biol, 46, 449.

Jirasek, A.,Mathews, Q.,Hilts, M.,Schulze, G., Blades, M. and Turner,R.(2006) Investigation of a 2D two-point maximum entropy regularization method for signal-to-noise ratio enhancement: application to CT polymer dosimetry. Phys. Med. Biol, 51, 2599-2617.

M. Hilts and A.Jirasek. Adaptive mean filtering for noise reduction in CT polymer gel dosimetry. Med. Phys. 35 (1), January 2008.

Hilts, M. and Duzenli, C.(2004) Image noise in x-ray CT polymer gel dosimetry. J Phys Conference Series, 3 252-256.

Hilts, M., Jirasek, A. and Duzenli, C. (2005) Technical considerations for implementation of x-ray CT polymer gel dosimetry, Phy Med Biol, 50, 1727 -1745.

Michelle Hilts., Cheryl Duzenli (2003) Image filtering for improved dose resolution in CT polymer gel dosimetry. Med. Phys. 31 (1), January 2004.

Jirasek. A., Hilts, M., Shaw, C. and Baxter, P. (2006) Investigation of tetrakis hydroxymethyl phosphonium chloride as an antioxidant for use in x-ray computed tomography polyacrylamide gel dosimetry. Physics in Medicine Biology, 51, 1891.

Taylor M L, Franich R D, Johnston P N, Millar R M and Trapp J V. (2007) Systematic variations in polymer gel dosimeter calibration due to container influence and deviations from water equivalence. Phys. Med. Bio, 52, 3991-4005.

R. B. Foist, H. G. Schulze, A. Jirasek, A. Ivanov, and R. F. B. T. Turner. A Matrix-Based Two-Dimensional Regularization Algorithm for Signal-to-Noise Ratio Enhancement of Multidimensional Spec- tral Data. Applied Spectroscopy, 64(11):1209–1219, Oct. 2010.

Part 7

Enhancing Patient Care in RT

19

Information and Support for Patients Throughout the Radiation Therapy Treatment Pathway

Michelle Leech and Mary Coffey
Discipline of Radiation Therapy, School of Medicine, Trinity College, Dublin, Ireland

1. Introduction

Communication is integral to all human interaction. Effective communication is a skill required by all members of the radiotherapy team in order to provide the best care to patients. However, it is an area of practice frequently overlooked amidst all of the technological advances in radiotherapy. This chapter aims to give an overview of the theories and models which underpin communication and then focuses on interactions within the radiotherapy department and how these impact on patient care. There is a particular focus on communication with older patients, children and adolescents-patient groups who can easily be overlooked in the information giving process. The chapter ends with a series of typically presenting patients to the radiotherapy department and invites the reader to consider how they might provide information and support to such patients.

2. Theories of communication

Capella (1) suggests that there are five steps involved in effective communication:

- *Transmission:* Information passing from one individual and assumes that the information will be received by another.
- *Exchange:* During communication, there is an exchange of words, gestures and images, usually termed 'interaction'.
- *Generation of meaning:* Certain words and phrases, as well as other methods of communication have specific meaning in certain cultures or within particular groups of people.
- *Context:* Effective communication must be given within the context of the situation. The location or setting of the communication can impact on this significantly.
- *Discourse.* The message communicated will be within the scope of a recognised format, this is known as the discursive context.

3. Models of communication

A model of communication explains how the communication process actually works in practice; how does the intended message to be communicated actually reach its destination?

Some models of communication include the information theory model, the interactive model, the gatekeeper model and the transactive model. The information theory model of communication was proposed by Shannon (2) in 1948. This model sees a message being created by an information source, a transmitter then carries the message through a carrier or channel, and noise may obscure or confuse the message, which is received either visually or aurally. The message finally reaches its destination when it is processed by the receiver. The interactive model put forward by Weiner (3) in the same year is similar to the information theory model but also includes a feedback loop from the final destination of the initial communicated message to the first information source. The feedback itself then becomes a communicated message and the original information source now becomes the destination of the feedback.

The gatekeeper model was put forward by Katz (4) in 1957 and highlights the role of a 'gatekeeper' or intermediary in the communication process. The proposed message can become distorted or may not reach the intended audience at all. In the radiotherapy setting, the gatekeeper model may become apparent when a third party attends for a radiotherapy consultation with a family member.

The transaction model of communication was described by Hopper (5) in 1992. In this model, two communicators create and consume messages for one another, as during a telephone conversation.

4. Communication in cancer care

Communication is a complex process involving the transfer of information between people, whether intentional or non-intentional, regardless of the model of communication adhered to (6). We can appreciate therefore that effective communication is a core clinical skill in the practice of radiation oncology and health literacy has a central role in cancer patients' ability to discuss their disease and prognosis with their oncologist in a meaningful way. Effective communication has many positive effects on cancer patients' adjustment to their disease and its treatment, whereas poor communication has negative consequences for both health-care professionals and patients (7, 8). Research has demonstrated that effective communication during consultations positively influences patient recovery, pain control, adherence to treatment, satisfaction and psychological functioning (9, 10).

However, it has been reported that patients forget much of the information provided to them (11). Therefore, the timing of information, the amount and content of the information and the method used to provide this information influence how effective the radiation therapy team will be in their communication with patients.

Psychoeducational interventions that are tailored for each patient are more likely to be effective than are interventions that make assumptions about the patient's information needs. Hence content of information should be agreed between the patient, his/her family, as appropriate, and the radiotherapy team. Methods that have been used to deliver information to patients include direct verbal instructions, printed materials, video, group sessions and computer-mediated methods (12)

5. How do we transfer information to patients in the radiotherapy department?

5.1 Verbal information

This is the most frequently used method. The radiation oncologist, radiation therapists and radiotherapy nursing staff meet with the patient regularly during treatment and face to face consultations. For radiation therapists, this is every day over the course of treatment. However, a known problem with verbal information is that about half of the information given is quickly forgotten or perhaps not processed at all (13). A potential advantage however is that it has the potential to lead to an interactive discussion with the patient.

5.2 Print media

Printed materials have the advantage of ensuring uniformity in information given, compared to verbal methods. Solely using printed material can result in gaps in information and the patient-health professional interaction is lost. Literacy can be problematic (14, 15). Therefore, written information should be clear, concise and use simple, easily understood language.

5.3 Video

Reviews have shown that video can be an effective modality for increasing knowledge in particular instances such as knowledge of risks and benefits of treatment options (16). In a study of adult patients receiving radiotherapy (16), the intervention group viewed a video with educational and practical information and was also given a written booklet on radiotherapy. The control group received the booklet only. By the end of treatment the two groups had equal knowledge, although the intervention group did have superior knowledge at the outset of treatment.

5.4 Digital media

Computer-based information allows for high accessibility to information and use of additional tools, such as graphics. However, it should be used with caution, particularly in patient groups who may not be computer literate.

6. Other independent variables that influence information-giving in the radiotherapy department

6.1 Timing of information

Interviews with patients about their information needs have shown that information should be made available to the patient whenever the patient feels a need for it and not when the health professional feels that the patient needs it. It makes sense to provide the information at the time when the patient is actually seeking it and radiotherapy health professionals should be mindful that patients' information needs may change with time, as treatment progresses (17).

6.2 Sociodemographic factors

There is no evidence that education level of patients influences either satisfaction with or recall of information (14) but there is evidence that better-educated patients are more likely to seek information (18). The radiotherapy team should be aware of this and seek to provide relevant information to patients from all sociodemographic groups when it is needed.

7. Communication with the older person

Communication with older patients in the radiotherapy department is often perceived as a challenge. Indeed, older adults diagnosed with cancer are the population considered to be at the highest risk for poor communication with health professionals. Older patients are often less assertive in communicating with health professionals, less likely to ask questions and less inclined to take a controlling role in the decision making process (19).

In a green paper published by the EU in 2005 (6), the then 25 EU countries had a population of 18.2 million people aged 80 and over (4% of the population). This will rise to 24.1 million in 2014 (5.2% of the total population) and by 2020, 70% of all cancers will occur in patients aged 65 and over (20).

Ageing is an individualised process. While 65 is the recognised retirement age in most countries, physiological 'old age' does not begin until 75 years of age. Older people are not homogenous, the needs of those in their 60s varies considerably from those in their 80s. Even within these groups, there are considerable variations on psychological needs, social supports, medical problems and health perspectives (21).

Although treatment for cancer should be based on physiologic rather than chronologic age, there is evidence to suggest that older patients receive less aggressive or appropriate cancer treatment than younger patients (22,23). Often there are misconceptions about the likely survival of the older patient. For example, the life expectancy of a woman aged 50 is 35 years (to 85). Once patients survive to 70 years, their life expectancy is 16 years (to 86) and those aged 80 can expect to survive until 88 (24).

As with all patients, communication with the older patient should be based on that particular patient's needs and personal preference for information. Ascertaining their circumstances and their own personal preferences in relation to their care is critical in the development of a successful therapeutic relationship. In a study of the relationship between health workers and older patients, it was found that being recognised according to their needs and being treated with courtesy and respect was important to older patients (6).

The main changes associated with ageing which may impact on the radiation therapist's communication with the patient and the patient's recall of information include:

- Visual changes such as decrease in visual acuity and contrast sensitivity and increase in glare intolerance.
- Hearing changes, such as presbycusis (decreased hearing of higher frequencies)
- Dementia or neurological damage
- Intergenerational and cultural differences.

Evaluation of such factors is essential when meeting an older patient for the first time. These can be overlooked in general in oncology and frequently compromise the quality of communication (25).

Celik et al (26) report on the attitudes of nursing students towards ageing and older patients. The total sample consisted of 42 all female students. 83.3% of the sample (n=35) stated that they had problems with their older patients. Communication problems due to mental, visual and hearing impairments and chronic diseases were experienced by 38% (n=16) of the participants. A further 42.8% (n=18) had difficulties giving instructions to older patients.

The majority of the participants in this study agreed that nurses caring for older patients need to be knowledgeable about the physical and psychological changes of ageing (26).

7.1 Ageism

Butler and Lewis (27) defined ageism as 'a systematic stereotyping and discrimination against people because they are old'. There has been a history of ageism in cancer treatment such as the under-representation of older patients in cancer trials, lack of attention of management of older patients at conferences and personal biases on the part of practitioners (28). The latter has often lead to a lack of opportunity for older patients to express their own opinions. Shorter interview time and less psychosocial discussion with older patients have been identified as ageist behaviour on the part of health professionals. Patients themselves have been found to be ageist, attributing pathological symptoms to the normal ageing process (21).

7.2 What information does an older patient require?

While one author (29) suggests that older patients may be less interested in knowing their diagnosis than younger patients, others have found that older patients still want information but do not want to be as actively involved in the decision-making process about treatment as younger patients (30).

Studies have shown variability in older patients' desires to actively participate in their cancer treatment (31, 32), while a systematic review revealed that few studies investigated the specific needs of older cancer patients surrounding treatment (33). The review revealed that while older patients prefer to receive information about the most important aspects of their illness and treatment, they are less inclined to look for extremely detailed information.

Posma et al (11) report that when providing information to the older person, it should be given in a structured manner with the most important and relevant information tailored to the patient's personal needs summarised and then repeated. Information should be given in a step like fashion, giving time for the patient to process the information. Language used should be simple and the use of jargon should be avoided. Information should be offered combining different methods (e.g.verbal and written) to improve information retention.

Despite these difficulties in the communication process, it should be noted that older adults may have better psychological resources than younger patients to adapt to their cancer diagnosis (34).

7.3 Role of a family member/friend in communication process

Posma et al (11) found that bringing another person to the consultation was favourable in outcome in helping the older patient remember the information and discuss concerns.

However, Greene et al (35) found that when a relative or friend accompanied the older patient to a consultation, the patient was less likely to ask questions, less assertive and expressive and less likely to be involved in the decision-making process.

Greene and Adelman (36) also concluded in a later paper that an accompanying person can change the dynamic of the communication process either positively or negatively. If the person is present at the patient's request and provides positive support and accurate information to the patient, then their presence is to be welcomed. Butow et al (37) found that over a third of patients preferred to be on their own with their physician when given their diagnosis.

Communication with older patients
• Introduce self and other members of team • Unsolicited use of the patient's first name should be avoided as a matter of respect • Provide verbal information in a structured, step-wise, logical fashion • Do not give an information 'overload'. Summarise the most relevant information for each individual patient • Use simple language and avoid jargon • Repeat the most pertinent points • Provide a summary of information in another format also (e.g. one page information sheet) • Ask the patient if he/she has any questions and answer appropriately

Table 1. Communication with older patients

8. Communication with paediatric patients

When a child is diagnosed with cancer, the initial reaction is to focus on prognosis. Accurate understanding of prognosis is important so that parents and adolescents can make more informed treatment decisions. Parents of children with cancer may be overly optimistic or pessimistic about the outcome of treatment. Parent misconceptions about the likelihood of cure are influenced by many factors including misunderstanding of the information provided to them by healthcare professionals and incorrect information provided to them from other sources, such as the internet (38). Therefore, it is incumbent on all health professionals involved in the cancer care of children to adopt a family centred approach. Family-centred care encompasses the 'professional support of the child and family through a process of involvement, participation and partnership underpinned by empowerment and negotiation' (39).

For children and adolescents with cancer, hospitalisation causes loneliness, losing out on enjoyable aspects of their lives, and concern for their families (40).

A common theme in communication practices with children is in fact the lack of communication expressed by children of all ages. It has been reported that age is not at all a useful guide in determining children's preferences for information (41).

Like adults, not all children want to know everything (42). Some children find information overwhelming and some are content when communication is directed to their parent simply because they fear hearing bad things (43).

Younger children (<10 years) generally interact little during consultations (44). van Dulmen et al (45) found in a Dutch study that during consultations, 35% of children said nothing at all.

Lambert et al (46) found that children's participation in the communication process was based on the children's ability to articulate their desire to engage in the communication process, health professionals' attitudes towards communication with children and the hospital environment itself.

8.1 Preparing a young child for radiotherapy

Younger children are often distressed when faced with a course of radiotherapy. They are alone in unfamiliar surroundings in the dark during set-up, the noise of the linear accelerator can be frightening and they are asked not to move. Children who are being treated in a prone position, for example for craniospinal irradiation in medulloblastoma are increasingly likely to be distressed. The understandable distress of parents who observe their child in these surroundings can also add to the anxiety experienced by the child. Effective communication with the child and parents is critical in minimising trauma to both parents and child.

To prepare the child for the radiotherapy process, the radiation therapy team should have discussions with the child and parents about the steps involved in preparation and treatment. Use of age-appropriate props can assist the child in understanding the process, such as video or picture books. For younger children, an opportunity to bring their favourite soft toy into the treatment room and 'model' the treatment procedure with the soft toy, such as placing stickers with marks on the soft toy, raising the treatment couch, dimming the room lights and switching on the laser alignment system all aids in the child's comprehension of the process. Inclusion of siblings in this process may also assist the family in explaining the radiotherapy process to the entire family.

Allowing the child to play his/her favourite music in the treatment room and facilitating the child's parent to speak to him/her over the intercom system during the treatment are also useful methods to overcome the fear of the noisy treatment room. Similarly, looking at projected images onto the ceiling can distract children being treated in the supine position.

A 'reward' system is used in many radiotherapy departments for children, where the child is given a sticker to place in a booklet after every treatment fraction. Some radiotherapy departments have specialised play therapists who work with the radiotherapy team in treatment preparation. This is especially useful should the child require the construction of an immobilisation device.

A child and his/her parents develop a unique relationship with the radiation therapists over the radiotherapy course as parents are entrusting their child to the radiation therapists every day. It is preferable for the same team of radiation therapists to treat the child every day to allow such a bond to develop and increase the likelihood of effective communication between the team and the family.

Communication with a younger child
• Use age-appropriate props to assist the child in understanding the radiotherapy process • Encourage the child to visit the treatment room prior to treatment and engage in modeling of the treatment using a soft toy • Involve play therapists for communication of difficult procedures such as the construction of a thermoplastic mask for paediatric brain tumours. • Involve other siblings and parents in the communication process • Use music in the treatment room, the intercom system and projected images on the ceiling to relax the child during radiotherapy • Use of a 'reward' system, such as sticker booklets communicates to the child that they have completed their treatment for the day

Table 2. Communication with a younger child.

8.2 Adolescent patients

Effective communication with the adolescent patient can pose a challenge for radiation therapy professionals. Adolescents with cancer require the radiotherapy team to understand intuitively when and how much interaction they want and seek at any given moment in time. Cantrell (40) suggests that knowing when to engage adolescents and when to provide them with time and space is an effective component of care.

In a study of 51 patients with cancer aged between 12-24 years, Dunsmore and Quine (47) found that adolescents with cancer wished to be more informed and involved in treatment decisions. They also found that the ability to listen, having genuine concern, clinical expertise and honesty facilitated healthcare professionals' communication with this population. Negatives reported included that communication was hindered by an impersonal manner, use of technical jargon, haste and the generation gap among healthcare providers. Hedstrom et al (48) interviewed 23 adolescents who described good care as being provided by healthcare professionals who were nice, friendly, supportive and competent. The same author postulates that receiving information assists adolescents with cancer in being active participants in their care and promotes a sense of security and control. Hinds et al (49) identified that humour used by nurses positively affects adolescents' behaviour.

Being authentic in caring is important to adolescent patients and is commented upon by them, well into survivorship. Cantrell (40) reports that survivors knew that nurses cared 'about them' in addition to 'caring for them' when they discussed other topics apart from their illness.

Adolescents have a strong desire for information relating to their illness, but their views on the relative importance of different types of information are not always the same as those of health professionals. Pfefferbaum and Levenson (50) found some significant differences between the groups indicating that health care professionals placed more importance on information about broader, psychological and social issues whereas adolescents placed more emphasis on basic factual information about aetiology, prognosis and late effects. However, there was some evidence of a common core of perceived information needs in the areas of illness severity and long-term effects of treatment.

9. Conflict

Conflict has been defined as a process that occurs within a group, which can take several forms. Some of these include hostility, decreased communication, mistrust, sabotage, verbal abuse and use of coercion (51).

Some level of conflict is inevitable when working in a pressurised, demanding environment such as a radiotherapy department. However, conflict does not have to be viewed as a negative entity; it can challenge our thinking and improve our practice. It can also be an energising experience. However, prolonged conflict in the radiotherapy department can result in a breakdown of communication either between healthcare professionals and patients or within the healthcare team itself. This is understandable when feelings are hurt, morale can become low and tempers flare. This ultimately leads to a negative experience for patients; hence early resolution of conflict through communication is advisable.

9.1 Conflict within the radiotherapy department: Staff

Conflict between healthcare professionals in the radiotherapy department manifests for a number of reasons. Role conflict can occur when the 'traditional' roles of some healthcare workers change due to increased education and hence role development and expansion. An example of this within the radiotherapy department is where radiation therapists have expanded their role from the traditional treatment and psychosocial support of the patient to include some tasks previously undertaken by the radiation oncologist or physicist, such as delineation of organs at risk and target volumes and treatment planning. Such role development for radiation therapists may lead to conflict between disciplines, if the rationale for the development is not clearly explained and the benefits acknowledged. Conversely, such development can motivate other radiotherapy professions to evaluate their own roles and seek to expand and develop them also for the benefit of patient care.

Status and education differences can also be a source of conflict between healthcare professionals. Where one professional believes themselves to be of a higher status than another within a team, conflict is inevitable. Education differences may be problematic where some professionals do not place value on the benefit of further education within the profession. This is due to fear of change as well as the perception that further education is a threat to the current status of the profession. In such circumstances, open communication between departmental managers and healthcare professionals can alleviate any such fears.

Poor communication skills can be detrimental to any working environment. Lack of communication between radiation therapists can ultimately impact negatively on the quality of patient care as teamwork is essential for the successful preparation, implementation and delivery of a course of radiotherapy. Communication of the physical and psychosocial needs and supports of the patient is required to provide optimal care of the patient in the radiotherapy department. A fundamental role of the radiation therapist is critical analysis of the treatment technique and treatment plan and this can lead to conflict when decisions made by others are challenged. Querying decisions of others should be done calmly and politely, without attributing blame. Similarly, queries should be received and answered in the spirit in which they are put forward.

9.2 Managing conflict

There are many styles which healthcare professionals adopt to manage conflict as outlined by Eason and Brown (52).

These include:

Competing: This individual is aggressive and non-compromising and pursues their own goals, disregarding others. In essence, this individual is 'always right'.

Compromising: This individual is equally assertive of their own opinion and accommodating of others.

Avoiding: This individual does not address the conflict situation at all.

Accommodating: This individual is cooperative but is not assertive regarding their own position.

Collaborating: This individual is both assertive and cooperative and strives to find a mutually satisfying endpoint to the decision in question.

Managing conflict in the radiotherapy department is difficult when so many different styles of conflict management are present and inevitably, personalities clash. Departments who always adhere to evidence-based practice can alleviate conflict situations surrounding practice, as seen in case scenario 1. Conflict resolution in other instances relies on professionalism, open communication and a collaborative approach by all involved.

Case Scenario 1
Location: Radiotherapy Department **Staff Involved: Manager of the radiotherapy department, radiation oncologist (Dr. Smyth), radiation therapist (Kate) and physicist (Mark)** **Discussion: The implementation of a new breast technique** **Manager:** Thank you all for attending this meeting. We are here to discuss the suggestion of changing the three field breast technique to a monoisocentric technique. This was brought to me by Kate, a radiation therapist on unit 1. **Kate:** Thank you. I have been concerned for quite some time now about the dosimetric implications of a match-line on skin in our current 3-field breast technique. Off-line verification has indicated that in some patients, the match-line is not stable from one day

Case Scenario 1
to the next. I think we can do better. In the centre where I worked previously, a monoisocentric technique was used to overcome this problem. **Manager:** Mark, what is your opinion? Are Kate's concerns valid in your opinion? **Mark:** Absolutely not. I have been working for 20 years in planning and this has never been an issue until now. Yes, the match-line can tend to be a bit hot but if the radiation therapists got themselves sorted out to use on-line imaging, this would not be a problem. Changing a technique like this has huge implications for planning and we are not prepared to do it just because Kate prefers the technique she used elsewhere before. **Kate:** Mark, this has nothing to do with what I prefer. I am trying to introduce some progress in this department. **(Raises voice)** Just because you clearly are not interested in patient care, don't pretend there isn't a problem with the current technique, I can plan monoisocentric techniques myself if you can't. **Manager:** Let's all take a minute please. Kate and Mark, you clearly have contrasting views on the validity of changing the technique. Dr. Smyth, as the specialist breast radiation oncologist, is there any clinical evidence to suggest that changing the technique is something worth consideration? **Dr. Smyth:** Actually yes. There are articles which illustrate that a monoisocentric technique provides better dose homogeneity and also leads to fewer errors on-set. However, there are drawbacks also associated with this technique, such as the restriction in field length. I am happy to email these references to all here so that we can inform ourselves of the evidence. I understand that Mark has reservations about the workload and it needs to be considered, given the number of patients that are planned in this department each week. I can also see Kate's point of you; the technique as it currently stands is not optimal and it is time that we incorporated best practice in breast radiotherapy. **Manager:** Perhaps, Dr. Smyth you can email those references this afternoon and we can re-convene here again at the same time next Wednesday when we are better informed to have this discussion. Is that reasonable to all? **Mark:** Yes, fine with me. **Kate:** Me too. **Manager:** Thank you all for your input.

9.3 Conflict within the radiotherapy department: Staff and patients

Conflict between patients and healthcare professionals in the radiotherapy department can occur for many reasons. The radiotherapy department can be a difficult environment for patients and exacerbate the stress they are already under. Patients may be feeling unwell, tired from treatment and may be struggling to cope with their diagnosis and its management. Coupled with this, patients may be under financial and personal stress also. They may be faced with lengthy waiting times every day and on occasion, have to cope with the impact of linear accelerator breakdowns on their treatment. Therefore, it is not surprising that patients and healthcare professionals may sometimes come into conflict in the radiotherapy department. A typical instance is given in case scenario 2. The majority of such situations can be diffused through polite and calm dialogue away from the main waiting area.

Case Scenario 2

Location: Radiotherapy Department
Staff and patients' involved: Jill, Radiation Therapist and Head of unit 1; Mr. Murphy, a 45 year old patient receiving post-operative radiotherapy for colorectal cancer and John, a newly-qualified radiation therapist on unit 1 who is working on the unit for the first time today.
Scenario: Jill returns from her lunch break to find Mr. Murphy outside the console area in a heated discussion with John.

Jill: Mr. Murphy, may I help you?
Mr. Murphy: Jill, thank goodness you're back. This young man has left me here waiting for over an hour.
John: Jill, Mr. Murphy had gone to see Dr. Smyth and his appointment card was not in the queue, so I didn't realise he was there
Mr. Murphy: Didn't realise!! **(Raises voice)** You are walking in and out of that room constantly and you didn't realise I was in the waiting area. I am sitting where I always sit, where I have sat every day for the last 3 weeks!
Jill: Let's take this discussion into the patient information room please.

John, would you like to explain to me please what has happened here?
John: Mr. Murphy, I have not treated you before, therefore I was not aware that you always sit in the corner of the waiting area. You know that we always identify patients with their name and date of birth in this department prior to treatment. I did not know that you were at Dr. Smyth's clinic or that you had returned.
Mr. Murphy: Jill, this is unacceptable. I am late now for my business meeting and you know I don't want to tell anyone that I have cancer.
Jill: Mr. Murphy, I can only apologise that you have been inconvenienced to this extent. As you are aware, we are extremely short-staffed in the department today and while this is not an excuse for the oversight, we are not operating as efficiently as we would like. You will be next for treatment and again I can only apologise.
Mr. Murphy: I know you are always run off your feet. Its the thought of facing that business meeting when I am just exhausted....
Jill: Perhaps it's time you considered taking a few days off. You are into the 4th week of treatment at this stage and I can see that the fatigue is starting to take it's toll. I know that work is very important to you and I'm not suggesting you give it up entirely, but perhaps a day off at the beginning and end of the week is something you could consider?
Mr. Murphy: I'll think about it Jill and John I'm sorry for getting so annoyed.
John: I'm so sorry that you were waiting so long; I can assure you that it will not happen again.

10. Masculinity, femininity, body image and sexuality

Men and women have different experiences of cancer, not only because of differing biologies but also from differing expectations about appropriate gendered behaviour (53).

Body changes, such as coping with a stoma in colorectal cancer , gynaecomastia and erectile dysfunction in prostate cancer, breast conservation surgery or mastectomy in breast cancer and facial disfigurement resulting from surgery in head and neck cancers are extremely difficult for patients to cope with. Sensitivity on the part of the radiotherapy team is required when discussing body change and sexuality with patients.

However, masculinity and femininity do not refer solely to sexuality. They are complex phenomena, with sexuality only as one component. Cecil et al (54) reported that financial and employment issues, changed role within the family and community as well as body changes and body image all contribute to an altered sense of masculinity for male cancer patients, over a range of cancer sites. They found that men, in general, did not share their experience of cancer and some had even cut themselves off entirely from their social network. Saini et al (55) studying a group of prostate cancer patients found that depression strongly correlated with poor quality of life, anxiety and sleep disorders. This group also reported that while medics are aware of the metabolic and physical toxicity of androgen deprivation therapy in prostate cancer, little attention is given to its detrimental psychological side effects.

In colorectal cancer, Sharpe et al (56) report that the psychosocial consequences of stoma formation include sexual problems, reduced social functioning, increased level of depression and disturbances to body image. This group found that the stoma itself and its impact on function was not the direct cause of psychosocial problems, but was in fact the extent to which the stoma impacts on body image.

Gilbert et al (57) report that women with gynaecological cancer experience a range of negative feelings relating to their identity including weight gain, loss of femininity, anxiety about sexual attractiveness, concern about their partner's reaction to their illness and changed body and loss of confidence. These can impact negatively on a couple's relationship and can be compounded by a lack of information and support from health professionals about sexuality and sexual well-being (58). Weijmar Schultz et al (59) have argued that if healthcare professionals put an emphasis on the importance of information on sexual function at the time of general information-giving, this brings a sense of 'normality' to the couple's subsequent sexual discussions.

Women with breast cancer experience a range of negative emotions relating to body image and sexual identity. These can include: fear of loss of fertility, feelings of sexual unattractiveness, loss of femininity, depression and anxiety, concern about weight gain or loss, partner's reaction to their changed body appearance (60). Radiotherapy professionals can play a significant role in alleviating concerns surrounding sexuality post treatment, offering information on how to adjust to these changes, the use of sexual enhancement products or differing sexual positions. However, research indicates that further education of healthcare professionals in relation to this is required to address the needs of patients which are currently unmet in this regard (61).

11. Communication in a multicultural society

Effective communication between healthcare professionals and patients in the radiotherapy department can be challenged due to differences between individuals in our now global multicultural society.

11.1 Language barriers

Language barriers are not easily surmounted; it is difficult to establish an easy relationship when the language used by the healthcare professional for the discussion is not the first language of the patient. Interpreters may be used, but confidentiality issues become problematic. Coupled with this, it is not the role of a translator to deliver difficult or sensitive information to a patient. Family members who can speak other languages are useful, but again the health care professional is not quite sure that the patient has understood the information provided, when a third party is included.

11.2 Cultural differences

Differing cultures take differing approaches to healthcare. Generally, in Western society, it is common for patients to have active interactions with healthcare professionals regarding their treatment and illness. Other cultures may take a more passive role and healthcare professionals need to be aware of this and respond appropriately. Approaches to masculinity and femininity may vary from one culture to another and this must be respected by the healthcare professional. An example might be where it is not considered appropriate for a male radiation therapist to treat a female patient. Such cultural differences must be treated with sensitivity and understanding.

11.3 Value differences

Values may differ on how illness is perceived from one culture to another. For many years in Western society, a diagnosis of cancer was perceived as shameful and its treatment hidden from others. However, the role of 'conventional' medicine in Western society is well recognised and accepted; the healthcare professional must be cognisant that there are cultures where this is not necessarily so. Acceptance of death and dying also varies between cultures and when treating palliative patients, the radiotherapy team must be aware of this.

11.4 Religious beliefs

Healthcare professionals must be aware of and respect the religious diversity of all patients. Difficulties can arise when religious beliefs clash with what is believed to be the best treatment choice by the healthcare professional. The importance of informed consent in such an instance cannot be overstated.

12. Task sheet

Consider how you might adapt your communication style to provide information the following patients:

Patient A
Margaret is a 48 year old lady presenting to the radiotherapy department for treatment of an optic nerve glioma. She has lost vision in one eye as a result of her disease.

Patient B
Eric is a 57 year old male presenting with a glioblastoma multiforme to the radiotherapy department. He has significant personality changes and expressive aphasia.
Patient C
John is a 68 year old male who is attending the radiotherapy department for treatment to the prostate. John has been deaf since birth. John has specific communication needs.
Patient D
Adam is a 55 year old male with a T4 N2 tumour of the subglottis. He is attending for post-operative radiotherapy and has a tracheotomy in situ.
Patient E
Michael is a 20 year old male presenting for para-aortic radiotherapy for a testicular tumour. Michael has significant learning difficulties and lives with his parents.

13. Summary

Communication is key to the successful delivery of a course of radiotherapy. Radiotherapy professionals must be mindful of the differing needs of all patients and provide information in a timely, sensitive and supportive fashion, using whatever method is easiest for the patient to understand. Professional communication within the radiotherapy and multi-disciplinary teams is essential in providing the best care for the patient.

14. References

[1] Capella J. Theories of Human Communication. Communication Theory. 2001. 12:165-71.
[2] Shannon C.E and Weaver W. 1949. The Mathematical Theory of Communication. Urbana IL; University of Illinois Press.
[3] Weiner N. 1948. Cybernetics: Or Control and Communication in the Animal and the Machine. Wiley.
[4] Katz E. The Two-step flow of Communication. Public Opinion Quarterly. 1957. 21: 61-78.
[5] Hopper R. Telephone Conversation. 1992. Bloomington, IN: Indiana University Press.
[6] Turk Z and Turk E. Our experience with evaluation of communication among older patients and health workers. Health MED. 2009. 3(3): 195-203.
[7] Fallowfield L and Jenkins V. Effective communication skills are the key to good cancer care. European Journal of Cancer. 1999. 35: 1592-1597.
[8] Williams S.W, Hanson L.C, Boyd C et al. Communication, decision making and cancer: What African Americans want physicians to know. Journal of Palliative Medicine. 2008. 11: 1221-1226.
[9] Stewart M.A. Effective physician-patient communication and health outcomes: A review. Canadian Medical Association Journal. 1995. 152: 1423-1433.

[10] Fallowfield L, Hall A, Maguire G.P et al. Psychological outcomes in women with early breast cancer. British Medical Journal. 1990. 301:1394.
[11] Posma E.R, van Weert J.C.M, Jansen J and Bensiing J.M. Older cancer patients' information and support needs surrounding treatment: An evaluation through the eyes of patients, relatives and professionals. BMC Nursing. 2009. doi: 10.I I 86/1472-6955-8-I.
[12] Bradlyn A.S, Beale I.L and Kato P.M. Psychoeducational interventions with paediatric cancer patients: Part I. Patient information and knowledge.
[13] Ley, P. Towards better doctor-patient communications. 1976. In A. Bennett (Ed). Communications between doctors and patients. 77-96. Oxford: Oxford University Press.
[14] Butow P, Brindle E, McConnell D, Boakes R and Tattersall M. Information booklets about cancer: factors influencing patient satisfaction and utilisation. Patient Education and Counseling. 1998. 33: 129-141.
[15] Mossman J, Boudini M and Slevin M.L. Cancer information: a cost-effective intervention. European Journal of Cancer. 1999. 35:1587-1591.
[16] Krouse H. Video modelling to educate patients. Journal of Advanced Nursing. 2001. 33: 748-757.
[17] Jones J, Nyhof-Young J, Friedman A and Catton P. More than just a pamphlet: development of an innovative computer-based education programme for cancer patients. Patient Education and Counseling. 2001. 44: 271-281.
[18] Cassileth B, Zufkis R, Sutton-Smith K and March V. Information and participation preferences among cancer patients. Annals of Internal Medicine. 1980. 92: 832-836.
[19] Zacharie R, Pedersen C.G, Jensen A.B et al. Association of perceived physician communication style with patient satisfaction, distress, cancer-related self-efficacy and perceived control over the disease. British Journal of Cancer. 2003. 88: 658-665.
[20] Balducci L and Extermann M. Management of Cancer in the Older Person: A Practical Approach. The Oncologist. 2000. 5 (3): 224-237.
[21] Mann S, Sripathy K, Siegler E.L, Davidow A, Lipkin M and Roter D.L. The Medical INterview: Differences between adult and geriatric outpatients. Journal American Geriatrics Society. 2001. 49: 65-71.
[22] Walker L, Kohler R, Heys S and Eremin O. Psychosocial aspects of cancer in the elderly. European Journal of Surgical Oncology. 1998. 24: 375-378.
[23] Turner N, Haward R, Mulley G and Selby P. Cancer in old age- is it adequately investigated and treated? British Journal of Medicine. 1999. 319: 309-312.
[24] Fentiman I.S. Communication with older breast cancer patients. The Breast Journal. 2007. 13 (4): 406-409.
[25] Amalraj S, Starkweather C, Nguyen C and Naeim A. Health literacy, communication and treatment decision-making in older cancer patients. Oncolgy. 2009. 23 (4): 369-375.
[26] Celik S.S, Kapucu S, Tuna Z and Akkus Y. Views and attitudes of nursing students towards ageing and older patients. Australian Journal of Advanced Nursing. 27 (4): 24-30.
[27] Butler R.N and Lewis M.I. Ageing and mental health. Positive Psychosocial approaches. 1973. St. Louis: C.V. Mosley.
[28] Ellis J. Bad old days. Nursing Standard. 2002. 16 (25): 23-24.

[29] Maguire P. Improving communication with cancer patients. European Journal of cancer. 1999. 35: 1415-1422.

[30] van der Molen B Relating information needs to the cancer experience. European Journal of Cancer. 1999. *: 238-244.

[31] Edwards M, Davies M and Edwards A. What are the external influences on information exchange and shared decision-making in healthcare consultations: A meta-analysis of the literature. Patient Education and Counseling. 2009. 75: 37-52.

[32] Say R, Murtagh, M and Thosmson R. Patients' preference for involvement in medical decision making: A narrative review. Patient Education and Counseling. 2006. 60: 102-114.

[33] Jansen J, van Weert J.C.M, van Dulmen A.M, Heeren T.J and Bensing J.M. Patient education about treatment in cancer care. An overview of the literature on older patients' needs. Cancer Nursing. 2007. 30: 251-260.

[34] Brandtstater J and Greve W. The aging self: stabilising and protective processes. Developmental Review. 1994. 14: 52-80.

[35] Greene M.G, Majerovitz S.D, Adelman R.D, and Rizzo C. The effects of the presence of a third person on the physician-older patient medical interview. Journal American Geriatrics Society. 1994. 42: 413-319.

[36] Greene M.G and Adelman R. Physician-older patient communication about cancer. Patient Education and Counseling. 2003. 50: 55-60.

[37] Butow P, Kazemi J, Beeney L, Griffin A, Dunn S and Tattersall M. When the diagnosis is cancer. Cancer. 1996. 77: 2630-2637.

[38] Sung L, Klassen R.J., Dix D, Pritchard S, Yanofsky R, Ethier M.C et al. Parental optimism in poor prognosis paediatric cancers. Psycho-oncology. 2009. 18: 783-788.

[39] Smith L, Coleman V and Bradshaw M. Family-centred Care. 2002. Basingstoke: Palgrave.

[40] Cantrell M.A and Matula C. The meaning of comfort for paediatric patients with cancer. Oncology Nursing Forum. 2009. 36 (6); 303-309.

[41] Young B, Dixon-Woods M, Windridge K.C and Heney D. Managing communication with young people who have a potentially life threatening chronic illness: qualitative study of patients and parents. British Medical Journal. 2003. 326: 305-308.

[42] Boylan P. Children's Voices Project: Feedback from children and young people about teir experience and expectations of healthcare. Commission for Health Improvement. Available at: http://www.chi.gov.uk/_db/_documents/04012717.pdf (Accessed 15 September 2011)

[43] Coyne I, Hayes E, Gallagher P and Regan G. Giving children a voice: Investigation of children's experiences of participation in consultation and decision-making in Irish hospitals. Office of the Minister for Children, Dublin. Available at: http://www.omc.gov.ie/documents/research/Giving_Children_a_Voice.pdf. (Accessed October 4th 2011).

[44] Wissow L.S., Roter D, Bauman L, Crain E, Keresmar C, Weiss K et al. Patient-provider communication during the emergency department care of children with asthma. Medical Care. 1998. 36: 1439-1450.

[45] van Dulmen A.M. Children's contributions to paediatric outpatient encounters. Paediatrics. 1998. 102: 563-568.

[46] Lambert V, Glacken M and McCarron M. 'Visible-ness': the nature of communication for children admitted to a specilaist children's hospital in the Republic of Ireland. Journal of Clinical Nursing. 2008. 17: 3092-3102.

[47] Dunsmore J and Quine S. Information, support and decision-making needs and preferences of adolescents with cancer: Implications for health professionals. Journal of Psychosocial Oncology. 1995. 13 (4): 39-56.

[48] Hedstrom M, Skolin I and von Essen L. Distressing and positive experiences and important aspects of care for adolescents treated for cancer. Adolescent and nurse perceptions. European Journal of Oncology Nursing. 2004. 8 (1): 6-17.

[49] Hinds P.S. Fostering coping by adolescents with cancer. Seminars in Oncology Nursing. 2000. 16 (4): 317-327.

[50] Pfefferbaum B and Levenson P.M. Adolescent cancer patient and physician responses to a questionnaire on patients' concerns. AMerican Journal of Psychiatry. 1982. 139: 348-351.

[51] Cox, K. The Effects of Unit Morale and Interpersonal Relationships on Conflict in the Nursing Unit. Journal of Advanced Nursing. 2001. 35 (1): 17-25

[52] Eason F and Brown S. Conflict Management: Assessing Educational Needs. Journal for Nurses in Staff Development . 1999. 15(3):92-96

[53] Seale C, Ziebland S and Charteris B. Gender, cancer experience and Internet use: a comparative keyboard analysis of interviews and online cancer support groups. Social Science and Medicine. 2006. 62: 2577-2590.

[54] Cecil R, McCaughan E and Parahoo K. 'It's hard to take because I am a man's man': an ethnographic exploration of cancer and masculinity. European Journal of Cancer Care. 2009. 19: 501-509.

[55] Saini A, Berruti A, Cracco C, Sguazzotti E, Porpiglia F, Russo L et al. Psychological distress in men with prostate cancer receiving adjuvant androgen-deprivation therapy. Urologic Oncology. 2011. Article in press. doi: 10.1016/j.urolonc.2011.02.005.

[56] Sharpe L, Patel D and Clarke S. The relationship between body image disturbance and distress in colorectal cancer patients with and without stomas. Journal of Psychosomatic Research. 2011. 70: 395-402.

[57] Gilbert E, Ussher J.M and Perz J. Sexuality after gynaecological cancer: A review of the material, intrapsychic and discursive aspects of treatment on women's sexual well-being. Maturitas. 2011. 70: 42-57.

[58] Stead M.L., Fallowfield L, Selby P and Brown J.M. Psychosexual function and impact of gynaecological cancer. Best Practice and Research Clinical Obstetrics and Gynaecology. 2007. 21 (2): 73-84.

[59] Weijmar Schultz W.C.M., van de Wiel H.B.M., Hahn D.E.E. and Bouma J. Psychosexual functioing after treatment for gynaecological cancer: an integrative model, review of determinant factors and clinical guidelines. International Journal of Gynaecological Cancer. 1992. 2(6): 281-290.

[60] Emilee G, Ussher J.M and Perz J. Sexuality after breast cancer: A review. Maturitas. 2010. 66: 397-407.

[61] Rees C, Bath P and Lloyd-Williams M. The information needs of spouses of women with breast cancer: patients' and spouses' perspectives. Journal of Advanced Nursing. 1998. 28 (6): 1249-1258.